# BACKSCATTERING
# SPECTROMETRY

# BACKSCATTERING
# SPECTROMETRY

### Wei-Kan Chu

IBM
East Fishkill Facility
Hopewell Junction, New York

### James W. Mayer     Marc-A. Nicolet

California Institute of Technology
Pasadena, California

ACADEMIC PRESS   New York  San Francisco  London  1978

A Subsidiary of Harcourt Brace Jovanovich, Publishers

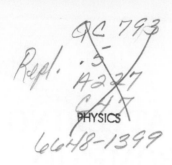

ACADEMIC PRESS, INC.
111 Fifth Avenue, New York, New York 10003

*United Kingdom Edition published by*
ACADEMIC PRESS, INC. (LONDON) LTD.
24/28 Oval Road, London NW1 7DX

Library of Congress Cataloging in Publication Data

Chu, Wei–Kan.
    Backscattering spectrometry.

    Bibliography: p.
    1.  Alpha ray spectrometry.  2.  Backscattering.
I.  Mayer, James W., Date      joint author.
II.  Nicolet, Marc–A., joint author.  III.  Title.
AC793.5.A227C47    539.7'522      78–82418
ISBN 0–12–173850–7

PRINTED IN THE UNITED STATES OF AMERICA

*To the members of the*
*Kaiserlich–Königliche Böhmische Physikalische Gesellschaft,*
*whose field of particle–solid interactions*
*is the basis of backscattering spectrometry*

# Contents

## Chapter 3    Concepts of Backscattering Spectrometry

## Chapter 4    Backscattering Spectrometry of Thin Films

## Chapter 5    Examples of Backscattering Analysis

## Chapter 6    Instrumentation and Experimental Techniques
### by R. A. LANGLEY

# Chapter 7  Influence of Beam Parameters

# Chapter 8  Use of Channeling Techniques

# Chapter 9  Energy-Loss Measurements

# Chapter 10  Bibliography on Applications of Backscattering Spectrometry

# Appendix A  Transformation of the Rutherford Formula from Center of Mass to Laboratory Frame of Reference

# Preface

The conceptual framework on which backscattering spectrometry is based was erected in the years following the discoveries of Rutherford and of Geiger and Marsden (1909–1913). A rapid succession of milestone developments then brought order into the structure of the atom. The nucleus began to attract the attention of increasing numbers in the physics community. Particle accelerators were developed to probe the inner workings of that nucleus. After World War II, the number of accelerators in the 1–3 MeV range increased rapidly. Why, then, did it take about 20 more years before these accelerators came to be used in solving problems outside of the field of nuclear physics? There is probably no single answer to this question. The growth and evolution of interdisciplinary fields of science and technology follow patterns of their own. The rules that govern them and the guidelines one should follow to further such evolutions can perhaps be learned from the study of cases such as that of backscattering spectrometry.

First, one must observe that the nuclear physicists who used these accelerators were fully aware of the analytical power of Rutherford backscattering from the very beginning. For example, it was (and still is) common practice to recognize contaminants of the target by an analysis of backscattered particles. Also, there was a constant trickle of publications over those 20 years to prove that investigators were always conscious of the analytical possibilities that Rutherford backscattering could offer. Throughout the 1960s, applications of the method were proposed by a steadily increasing number of authors. By the end of that decade, backscattering had taken a foothold.

Another development took place independently. In the early 1960s, the channeling of fast particles moving in a crystalline lattice was rediscovered after having been anticipated by W. H. and W. L. Bragg and by J. Stark in the 1910s. The phenomenon attracted attention and brought particle accelerators into the arena of solid-state physics through the other door. By the time backscattering spectrometry was finding acceptance, channeling had already become an integral part of the method.

Clearly, the idea of using Rutherford backscattering had always been alive. The obstacles in the way of its immediate introduction as an analytical tool outside of nuclear physics were elsewhere.

One difficulty was instrumental. At the outset, the only detectors with good energy resolution were the magnetic spectrometers, which are bulky and time-consuming to operate. Around 1960, solid-state detectors became available. These relatively inexpensive devices promised good resolution, good linearity, fast response, and simultaneous analysis over a wide energy range. Their development was correspondingly rapid. At present they constitute the preferred particle detectors in the energy range of interest to backscattering spectrometry.

Another major experimental improvement occurred in the electronic systems for data handling and processing. Speed, accuracy, stability, and generous capacities for data storage and handling became available at reasonable cost. In combination with a solid-state detector, such a system transformed an accelerator into a rapid and efficient analytical instrument.

Planar technology was first introduced to make semiconductor devices in 1960. Because of its inherent advantages, this technology found rapid acceptance, but with it came numerous novel problems in the formation and control of thin layers used for masking and contacting. The fact that backscattering spectrometry was an ideal tool with which to investigate these problems went unnoticed. The *problems existed,* but those equipped to solve them remained unaware of them, and those seeking answers overlooked the tool.

A direct link between planar technology and backscattering spectrometry was finally established with ion implantation. It offered accurate control of the dopants and uniform surface density over a whole wafer, and thus superior yields. The need arose to establish the depth profile of an implanted atom and the amount of disorder produced by the energetic ions. Backscattering spectrometry came as a fairly natural solution to those familiar with ion beams and ion implantation. In early applications an attribute of backscattering spectrometry that had not been fully appreciated became evident, namely, its ability to provide a depth scale to the elements detected. It is this ability more than any other that gives backscattering spectrometry its unique analytical power. The great success of the method in connection with thin films, their structure, composition, and reactions, demonstrates this fact very clearly. Actually, a professional society exists whose purpose is to promote the specific field of particle–solid interaction, of which backscattering spectrometry is a recognized part.

Finally, the pressure to bring MeV accelerators to bear on the problems arising in the semiconductor industry came from the semiconductor industry. Typically, it was not the scientists who had already mastered the tool who sought out the problems, but rather the scientists with the problem who sought out the tool. Without the magnanimous response of those in charge of the accelerators, the interdisciplinary effort would not have unfolded. Where the intellectual curiosity for the solution to a problem at hand overruled the man-made subdivisions of scientific disciplines, the barriers fell and backscattering spectrometry rose to success.

So far, the main beneficiary of the technique has been the semiconductor industry, where thin-film and ion implantation problems abound. In sorts, backscattering spectrometry pays a tribute it owes. It was the semiconductor industry's earlier efforts that had readied the MeV accelerators for this task by providing them with suitable detectors and electronic systems.

# Acknowledgments

This book grew out of a program supported at Caltech by the Office of Naval Research. We are indebted to Larry R. Cooper for providing the encouragement and the continuity of support necessary to develop backscattering spectrometry into an analytical technique. We are also indebted to the staff of the Kellogg Radiation Laboratory at Caltech for access to the 3-MeV van de Graaff accelerator and for generous support of our activities. In particular we express our gratitude to Charles A. Barnes, who gave freely of his time and talent to assist us and fulfilled the role of godfather to backscattering spectrometry. In all of our work, the assistance of Carol Norris, Rob Gorris, and Jeffrey Mallory was invaluable; it is our pleasure to acknowledge their help and to thank them.

In planning and writing this book, we benefited from the assistance of many of our colleagues: Ian V. Mitchell (Chalk River Nuclear Laboratories), who participated in the initial organization of the material (in 1971); Robert A. Langley, who wrote Chapter 6; Jon Mathews, Johnson O. Olowolafe, Tsu-wei Frank Lee, and Woon Tong Nathan Cheung for their contributions to the appendixes; and J. L. 'Ecuyer and R. F. Lever for helpful suggestions. Finally, we thank our many collaborators who were at Caltech, providing inspiration, hard work, and many unforgettable memories of good times.

Chapter

# *1*

# Introduction

## 1.1  INTRODUCTION

To obtain measurable effects, an intense pencil of alpha particles is required. It is further necessary that the path of the alpha particles should be in an evacuated chamber to avoid complications due to the absorption of scattering in air.

This is how Geiger and Marsden (1913)[†] describe the principal conditions that their experiment had to meet. With it they unambiguously confirmed the validity of the new model of an atom proposed by their leader Ernest Rutherford. Figure 1.1 shows a drawing of the simple apparatus that they built to meet these requirements. The year was 1911. The purpose was to test (and prove) a theory.

Figure 1.2 is a sketch of a similar apparatus. It is taken from the final report of the Surveyor Project (Turkevich *et al.*, 1968) and shows the sensor head of

---

[†] References are listed at the end of each chapter. We use the year of publication to identify a reference, followed by *a, b, . . .* , if necessary to avoid ambiguities.

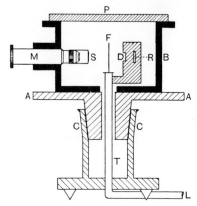

**Fig. 1.1** Drawing of the apparatus used by Geiger and Marsden in 1911–1913 to test and confirm the new model of an atom conceived by Rutherford in 1911. "The apparatus . . . consisted of a strong cylindrical metal box B, which contained the source of alpha particles R, the scattering foil F, and a microscope M to which the zinc-sulphide screen S was rigidly attached. The box was fastened down to a graduated circular platform A, which could be rotated by means of a conical airtight joint C. By rotating the platform the box and microscope moved with it, whilst the scattering foil and radiation source remained in position, being attached to the tube T, which was fastened to the stand L. The box B was closed by the ground-glass plate P, and could be exhausted through the tube T." [from Geiger and Marsden (1913).]

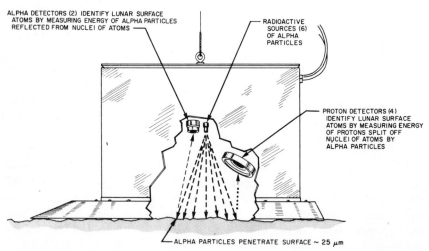

**Fig. 1.2** Diagrammatic view of the internal configuration of the alpha-scattering sensor head deployed on the surface of the moon for the first analysis of the lunar soil, executed as part of the scientific mission of Surveyor V after its soft landing on September 9, 1967. [from Turkevich *et al.* (1968).] This experiment was the first widely publicized application to a problem of nonnuclear interest of the concept of Rutherford scattering introduced some 50 years earlier.

the "alpha-scattering experiment" which was part of the scientific payload of Surveyor V. The year was 1967. The purpose of the alpha-scattering experiment was to analyze the composition of the lunar soil. This experiment probably constitutes the first widely publicized practical application of the ideas of Rutherford, Geiger, and Marsden to a problem of nonnuclear interest.

In the rest of this introduction, we paint an overall picture of the analytical technique of backscattering spectrometry as it exists today. We do not dwell on the details, but rather present the idea of the method; what it can and what it cannot accomplish. The purpose of this chapter is to give a general picture of backscattering spectrometry, a few basic concepts, and some "rules of thumb" to guide in interpreting or reading spectra. Details are given in the following chapters as outlined in Section 1.7. However, the contents of this chapter are intended to convey an impression of the relative strengths and weaknesses of backscattering spectrometry in the framework of materials analysis.

## 1.2   CONCEPT OF A BACKSCATTERING EXPERIMENT AND ITS LAYOUT

Both in its concept and in its elementary execution, Rutherford scattering is quite a simple experiment. A beam of monoenergetic and collimated alpha particles ($^4$He nuclei) impinges perpendicularly on a target. When the sample that constitutes the target is thin, as in the experiment of Geiger and Marsden, almost all of the incident particles reappear at the far side of the target with some slightly reduced energy and only slightly altered direction; that is, the beam is transmitted through the thin target with only very little loss of particles. The situation is sketched in Fig. 1.3. The few alpha particles that are lost undergo large changes in energy and direction, changes due to close encounters of the incident particles with the nucleus of a single target atom. If the sample is thick, only the particles scattered backward by angles of more than 90° from the incident direction can be detected. This is the situation that prevailed in the Surveyor V experiment (Fig. 1.2). It is also that which is adopted in the analytical technique discussed in this book, hence the name *backscattering spectrometry*.[†]

The typical experimental system used today for routine backscattering analyses is considerably more elaborate than the setups shown in Figs. 1.1 and 1.2. Figure 1.4 gives a schematic outline of the major components of

---

[†] An alternative name is *Rutherford backscattering spectrometry*. However, since the scattering cross section can deviate from that given by the Rutherford formula, we use the more general term *backscattering spectrometry*.

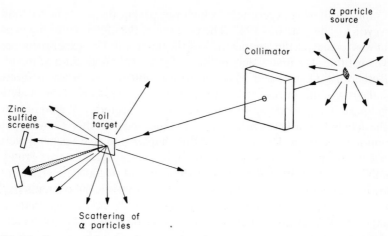

**Fig. 1.3** Conceptual layout of a scattering experiment. In the experiment of Geiger and Marsden (Fig. 1.1) the source was a thin-walled glass tube filled with radon and enclosed in a lead box, shown shaded in the diagram. The collimator consisted of a simple diaphragm. In the experiment of Surveyor V, the six sources were of $^{242}$Cm which emits alpha particles of 6.1 MeV. A short tubular extension of the stainless steel capsule that contained the curium acted as the collimator. The collimator opening was covered with a thin film of aluminum oxide plus polyvinylstyrene, totaling about 1000 Å in thickness, to prevent contamination of the lunar soil or the apparatus by radioactive material.

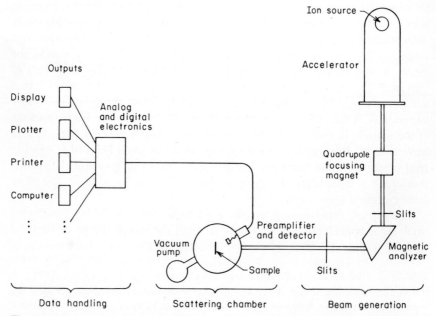

**Fig. 1.4** Schematic diagram of a typical backscattering spectrometry system in use today.

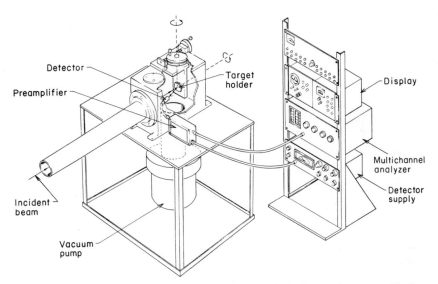

**Fig. 1.5**  Layout of the target chamber and electronics of a backscattering system. The ions impinge on the target in the vacuum chamber. Backscattered particles are analyzed by the detector, and the detector signal is magnified and reshaped in the preamplifier. The electronic equipment in the rack provides power to the detector and preamplifier and stores the data generated by the detector in the form of the backscattering spectra.

such a system. Charged particles are generated in an ion source. Their energy is then raised to several megaelectron volts by an accelerator, usually a van de Graaff (or a similar kind). The high-energy beam then passes through a series of devices which collimate or focus the beam and filter it for a selected type of particle and energy. This equipment replaces the simple source-and-diaphragm arrangement of Figs. 1.1 and 1.2. The immense advantage of this system over the natural source-and-diaphragm apparatus is that the beam parameters can now be varied over a wide range. In particular, higher particle fluxes can be obtained as compared to natural sources; this drastically shortens the measurement time. The beam then enters the scattering chamber and impinges on the sample to be analyzed (Fig. 1.5). Some of the back-scattered particles impinge on the detector, where they generate an electrical signal. This signal is amplified and processed with fast analog and digital electronics. The final stage of the data usually has the form of a (digitized) spectrum, hence the name backscattering *spectrometry*.

   In spite of the sophistication in the beam-generating parts and the data collection end of a backscattering spectrometry system, the chamber in which the backscattering experiment is performed remains simple (Fig. 1.6). Apart from the box and the sample themselves, it has only three elements: the beam, the detector, and the vacuum pump. The requirements on the

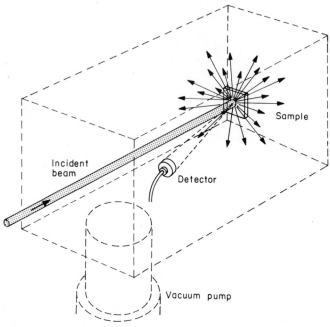

**Fig. 1.6** Even in a sophisticated backscattering spectrometry system, the scattering chamber where the analysis/experiment is actually performed remains simple. Apart from the box forming the chamber and the sample, there are only three other elements: the beam, the detector, and the vacuum pump.

vacuum are quite modest by today's standards: $10^{-5}$ Torr is expedient, and $10^{-6}$ Torr is quite adequate. Such vacua allow simple handling procedures and rapid turn-around times for unloading and reloading samples. A well-functioning backscattering spectrometry system can analyze many samples a day. As a research tool, one system is able to satisfy the demands of a number of people and projects at a time. As a tool for routine surveys, a system can easily be automated for both the execution of the experiment and the reduction of the data.

### 1.3 BASIC PHYSICAL PROCESSES

The translation of individual signals in a backscattering spectrum to depth distributions of atomic concentrations in a sample rests on simple physical principles. Imagine a single self-supporting layer with two elements $M$ and $m$ in equal amounts, $10^{15}$ atoms/cm² each, as shown in Fig. 1.7. Imagine further that a flux of $^4$He particles of 1-MeV energy impinges on this layer. Those few $^4$He particles that do undergo close encounters will be deflected because of the enormous Coulombic force they encounter. If the energy of

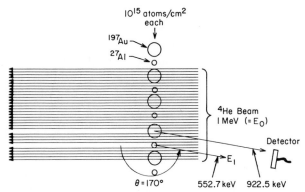

**Fig. 1.7** The kinematic factor $K_M$ gives the ratio of the energy after $(E_1)$ to that before $(E_0)$ an elastic collision of the projectile (here $^4$He) with an atom of mass M (197 amu for Au, 27 amu for Al). The heavier mass reflects the incoming particle more completely, energetically, than the lighter mass, as is the case with billiard balls. Two examples are shown and actual values are given.

the incident $^4$He ion is not too high, nuclear reactions can be ruled out during the collision process as well. The collision then must be an elastic one. The phenomenon is similar to the collision of two hard spheres and can be solved exactly. The kinematic factor $K$ is the ratio of the energy of the projectile after to that before the collision. It is listed in Tables II and III for $^1$H and $^4$He as projectiles.[†] As an example, assume that the two elements are Au and Al, whose atomic masses are 197 and 27 amu, respectively (see Table I). For a scattering angle of 170°, we find from Table III that $K_{Au} = 0.9225$ and $K_{Al} = 0.5527$. A 1-MeV $^4$He particle therefore, has an energy of 922.5 keV after a collision with Au, and an energy of 552.7 keV after a collision with Al.

The probability that a collision will result in a detected particle is given by the differential scattering cross section $d\sigma/d\Omega$, which is tabulated for all elements with $^4$He as a projectile in Table X. For Au, $d\sigma/d\Omega$ is $32.81 \times 10^{-24}$ cm$^2$/sr for each atom; for Al, $d\sigma/d\Omega$ is $0.8512 \times 10^{-24}$ cm$^2$/sr. To find the average scattering cross section $\sigma$ over the field of view of the detector, we must multiply this differential scattering cross section with the solid angle of detection $\Omega$, which we shall assume to be $10^{-3}$ sr (a typical order of magnitude for real systems). Adding up the scattering cross section of all atoms in the layer ($10^{15}$ atoms/cm$^2$ each), we find for Au, $3.3 \times 10^{-11}$ and for Al, $8.5 \times 10^{-13}$. These dimensionless numbers give the probability that a $^4$He projectile will undergo a close encounter with Au or Al in the layer and end up in the detector. Assume that the integrated current of 1-MeV $^4$He$^+$ ions during the exposure of the layer was 1 $\mu$C (which is a typical

[†] Tables I–XI are given in Appendix F.

number used to obtain a backscattering spectrum); the total number of $^4$He
ions that fell on the sample was then $6.2 \times 10^{12}$. The probable number of
events counted after scattering from Au atoms is therefore about 200, and
the number from Al is about five. Note that the charge state of the particles
in the ion beam, whether $^4$He$^+$ or $^4$He$^{++}$ (alpha particles), relates integrated
current to number of incident particles but does not influence scattering
or energy loss cross sections.

Now imagine that the sample is a self-supporting Au film 1000 Å thick
and that the analysis beam consists of 2.0-MeV $^4$He ions (see Fig. 1.8). A
scattering event at the front surface of the film is detected at an energy $K_M E_0$;
the same event at the rear surface is registered at a lower energy. The energy
difference $\Delta E = 133$ keV is nearly ten times the energy resolution of standard
particle detection systems, and hence it is straightforward to determine
whether particles were scattered at the front or rear surfaces of the film.
Scattering events that take place somewhere between front and rear surfaces
are recorded at some intermediate energies. Since the beam is unattenuated,
the scattering probability at any depth is proportional to the number of
atoms of a particular kind present there. This is the way a concentration
profile of a given element is translated into a signal of corresponding height
and decreasing energy in the backscattering spectrum.

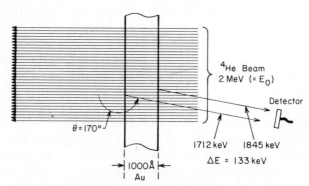

**Fig. 1.8** A swift particle that passes through a dense medium loses some of its energy. As a
consequence, a particle scattered back at the rear surface of a film has less energy when it is
detected than a particle scattered at the front surface. Actual values are given for a 1000-Å-thick
Au film.

The fact that the signal of $^4$He particles scattered from the Au film has a
finite energy width reflects the energy loss of the particles along their inward
and outward paths. Such energy losses can be calculated from the stopping
cross section $\varepsilon$, which is given for $^4$He ions in Table VI in units of electron
volts per $10^{15}$ atoms per square centimeter; for 2.0-MeV $^4$He ions, the value
of $\varepsilon^{Au} = 115.5$. To determine the energy lost along the inward track to the rear

$6 \times 5.9 \times 10^{22} \times 10^{-5} \frac{t}{cm}$

$\frac{N}{115} \times 5.9 \times 10^{22} \times 10^{-5} \frac{t}{cm}$

surface of a film of thickness $t$ and atomic density $N$, one takes the product of $\varepsilon$ and $Nt$, where $Nt$ represents the number of atoms per square centimeter in the film. For the Au film (where $N_{Au} = 5.9 \times 10^{22}$ atoms/cm$^3$ as given in Table I) the value of $Nt = 5.9 \times 10^{17}$ atoms/cm$^2$ and a particle would lose 68.1 keV along the inward path.

The detected energy difference between particles scattered from the front and back surfaces of the film is given by the product of $Nt$ and $[\varepsilon]$, where values of $[\varepsilon]$, the stopping cross section factor, are listed in Table VIII for scattering angles of 170°. In Table VIII, the units of $[\varepsilon_0]$ are electron volts per ($10^{15}$ atoms per square centimeter) and the value for $[\varepsilon_0]^{Au} = 226.2$. For a film with $5.9 \times 10^{17}$ atoms/cm$^2$, the energy width $\Delta E = 133.4$ keV. This energy width could also have been found directly from the values of the energy loss factor $[S]$ given in Table IX for $^4$He in units of electron volts per angstrom. However, the use of an energy-to-depth conversion with units of electron volts per angstrom overlooks the fact that backscattering spectrometry reflects the number of atoms per square centimeter traversed by a particle rather than the physical depth in centimeters. The conversion between the two is direct if the atomic density of the sample is known.

If the energy loss that the particle suffers as it traverses the sample were independent of energy, the relationship between the depth of the collision and the energy of a detected particle would be linear. As a matter of fact, the success of backscattering spectrometry in the analysis of thin films is partly attributable to the small relative change in the energy of the beam as it traverses the film. The energy dependence of the stopping cross section can then be replaced by two fixed values, one along the inward path and one along the backward path across the film. For very thick films where this approximation fails, the analysis of a spectrum is not as simple. However, a large part of this book discusses suitable approximations.

The fact that the projectile loses energy as it penetrates into the sample has another consequence. Scattering cross sections depend on the energy of the impinging projectile as (energy)$^{-2}$. Deeper down in the sample, where the energy of the projectile decreases, the scattering probability increases. The signal of an element which is uniformly distributed in depth is therefore not flat-topped, but rises toward lower energies. This, too, complicates the quantitative analysis of a spectrum.

## 1.4 EXAMPLES AND APPLICATIONS

Applications date back to some of the early nuclear investigations with accelerators, when it was common practice to recognize contaminants of the target by an analysis of backscattered particles (Tollestrup *et al.*, 1949). The earliest applications to problems of nonnuclear interest were the analyses of smog (Rubin and Rasmussen, 1950) and of the bore surfaces of gun barrels

(Rubin, 1954). Other contributions were those of Rubin *et al.* (1957) and Mazari *et al.* (1959), who detected trace elements on thick and on thin targets, respectively, and of Sippel (1959), who measured the diffusion of Au into Cu. In 1960, S. K. Allison suggested the method for the remote analysis of surface composition. Following his suggestion, Turkevich (1961) proved in preliminary investigations that the method was feasible, and Patterson *et al.* (1965) laid the groundwork that culminated in the compositional analysis of the moon's soil by Surveyor V in 1967 (Turkevich *et al.*, 1968).

In this section, however, we present three more recent examples to give a feeling for backscattering spectrometry. The first deals with the detection of contaminants on the surface of Si, while the second shows the depth distribution of a dopant in Si, and the third shows examples of thin film analysis. Other examples are given in Chapter 5.

### 1.4.1  Surface Impurities

As a first example, we present in Fig. 1.9 a schematic energy spectrum of $^4$He backscattered from a Si target with Cu, Ag, and Au on the surface, each in the amount of about $10^{15}$ atoms/cm$^2$. This is of the order of one monolayer of surface coverage. The spectrum was taken with a $^4$He beam of incident energy $E_0 = 2.8$ MeV. The lower abscissa gives the energy scale of the backscattering spectrum. The upper abscissa gives the mass M associated with the positions $K_M E_0$ for the three impurity elements and for Si. Note that the mass-to-energy conversion established via $K_M$ is unique, but nonlinear. Au is the only element in this example that has only one stable isotope (see Table I) and produces only one signal in the spectrum. The two signals of the Ag

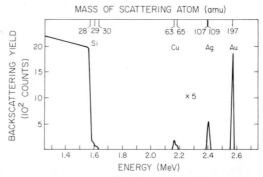

**Fig. 1.9**  Schematic energy spectrum of $^4$He backscattered from a Si substrate with about $10^{15}$ atoms/cm$^2$ each of Cu, Ag, and Au (equivalent to approximately one monolayer of coverage). Projectile: $^4$He$^+$ of 2.8-MeV incident energy; scattering angle of detected particles: 170°; solid angle of detection: 4 msr; total dose (integrated current of incident beam): 10 $\mu$C; energy per channel: 5 keV; resolution: 12.5 keV (FWHM). The ordinate for the signals of Cu, Ag, and Au is magnified five times.

isotopes cannot be distinguished, because the energy resolution of the detection system is too coarse. The signals of the two Cu isotopes are just barely resolved.

The area under each impurity peak is proportional to the number of impurity atoms per unit area and the scattering cross section of the element. Since the surface coverage is about equal for all three impurities, the size of the signals reflects the change in cross section. We are thus able to determine the exact ratio of atoms per square centimeter between these impurities by dividing the area of the signals through the respective scattering cross sections and obtain quantitative results without using standards of calibration. The signals of the two Cu isotopes indicate directly their relative abundance. The Si part of the spectrum is characteristic of a thick sample. Here it is the height of each step at the appropriate energy edge $K_M E_0$ that is proportional to the isotopic abundance (92.2, 4.7, and 3.1% from Table I).

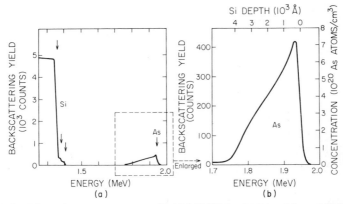

**Fig. 1.10** Schematic energy spectrum of 2.4-MeV $^4$He backscattered [part (a)] from a Si substrate doped with As. The As signal is magnified in a separate plot [part (b)], where the axis of energy (bottom) is converted to depth below the surface (top) and the axis of yield (left) is converted to atomic volume concentration (right). The spectrum was measured with the same system parameters as those given in Fig. 1.9, except for the incident energy ($E_0 = 2.4$ MeV) and the dose (20 $\mu$C).

### 1.4.2  Impurity Distribution in Depth

As a second example, Fig. 1.10 shows a schematic energy spectrum of $^4$He backscattered from a Si sample implanted with As and then heat-treated to diffuse As deeper into the sample. The conversions of the backscattering yield of As to an As concentration as well as the energy axis to one giving the depth of As in Si are given in the enlarged part of the figure. Both conversion scales are linear with only minor corrections.

The concentration scale for the As signal conveys an idea of the sensitivity of backscattering spectrometry in detecting impurities. Compared to other methods—for example, neutron activation analysis or secondary ion mass spectroscopy—backscattering spectrometry is not very sensitive. However, backscattering spectrometry is capable of quantitative measurements without recourse to standards. It can also furnish depth profiles without layer removal by ion sputtering or chemical stripping, which is generally required with other profiling methods.

### 1.4.3   Thickness Measurements

The measurement of film thicknesses is an obvious way of making use of backscattering spectrometry. Figure 1.11 shows schematic spectra of $^4$He backscattered from Ta films of various thicknesses. Several spectra are plotted on the same axes to illustrate the relation between the energy shift and the film thickness: they are nearly proportional. The accuracy of the thickness measurement is directly determined by the accuracy of the energy loss values used for the analysis. Here we have used the values listed in Table IX. As stated previously, the area under each signal is proportional to the total number of Ta atoms in the film. Consequently, one can obtain the film thickness from the area of a signal as well.

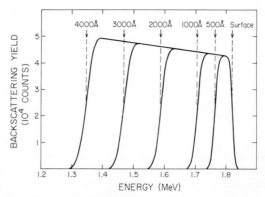

**Fig. 1.11**   Display of five backscattering spectra combined to show how the width of the signal from a thin film reflects the thickness of the films. The incident energy of the $^4$He ions is 2.0 MeV and the five targets were Ta films deposited on $SiO_2$ substrates (the substrate signals are not shown in the spectra).

### 1.5   STRENGTHS AND WEAKNESSES OF
### BACKSCATTERING SPECTROMETRY

The strength of backscattering spectrometry (BS) resides in the speed of the technique, its ability to perceive depth distributions of atomic species below the surface, and the quantitative nature of the results. Furthermore, with single-crystal targets, the effect of channeling also allows the investigation of the crystalline perfection of the sample.

The speed of the data collection is possible in part because the modest requirements on the vacuum permit fast sample changing. The modest vacuum is admissible only because BS measures the bulk of the sample not its surface. Since the typical depth resolution of 100 to 200 Å precludes a study of the first few monolayers, vacua of $10^{-9}$ or $10^{-10}$ Torr, which true surface techniques demand, are unnecessary for BS.

The great increase in sensitivity for heavy elements is an asset for the detection of these elements, but a severe limitation for the detection of light elements. Carbon, nitrogen, and oxygen are ubiquitous elements and therefore of great significance in the near-surface regions of a solid; yet BS is nearly blind to trace quantities of them. This disadvantage is often overcome in studies of thin films by depositing the film on a low atomic mass substrate such as carbon. This approach allows ready identification of signals from oxygen contaminants, for example. Another weakness is the lack of specificity in the signal. After a scattering event, all backscattered particles are alike, save for their energy. Two elements of similar mass cannot be distinguished when they appear together in a sample. This lack of specificity of the signal can be resolved by other analytical tools, such as Auger electron spectroscopy. Finally, one must realize the stringent requirements on lateral uniformity that a sample must meet before the full capability of BS can be utilized. A typical ion beam diameter used for backscattering is 1 mm². If the range of depth analyses is 2000 Å, the width of the beam spot is a factor of $5 \times 10^3$ larger than the thickness of the layers. Scratches, cavities, dust particles, and any other surface nonuniformities can drastically modify the spectrum, if present in sufficient amounts, even if they are of a submicron size. The lateral uniformity of a sample must therefore be assured on the surface as well as in depth.

The most convenient way to establish this uniformity is scanning electron microscopy (SEM), which has excellent lateral resolution and thus constitutes the normal complementary tool for BS. Unfortunately, SEM provides surface topography, without vision below, and with little elemental specificity. X-ray attachments can provide the missing elemental specificity. In this respect, an electron microprobe is an even superior counterpart to BS because it combines good elemental specificity and good lateral resolution. The drawback of the electron microprobe is that the x-ray signal reflects the average composition over depths quite large compared to the depth resolution of a backscattering spectrum.

Another limitation of BS is that chemical information is totally absent. X-ray diffraction of various sorts, in particular the Read camera, has been found most useful for the determination of crystallographic parameters. Usually the combination of atomic composition ratios furnished by BS and the knowledge of diffraction patterns give convincing evidence of the actual nature of the compound present.

Auger electron spectrometry (AES) and secondary ion mass spectroscopy (SIMS) are two other techniques that complement BS well. Both have elemental specificity; but their main drawback is their reliance on ion sputtering for depth profiling. Ion sputtering can modify the sample under investigation and lead to erroneous conclusions (e.g., laterally dissimilar erosion rates or preferential sputtering). The consequences can be particularly severe in AES, where the signal emanates from the uppermost layer of the sputtered area. AES, on the other hand, can be quantified by comparison with reference samples. In SIMS, quantification is still more difficult because the fraction of the ionized (and hence detected) atoms sputtered from the substrate depends on the chemical surrounding of that atom in the sample and on the sputtering gas. In sensitivity, however, SIMS far surpasses most other analytical techniques.

One of the advantages of BS is that it provides depth distributions without the requirements for destruction of the sample by layer removal as in the case of sputter sectioning used with AES or SIMS. However, BS will introduce damage. Whether BS should be considered destructive or not depends on the object analyzed and also on the questions asked. A shallow *pn* junction, for instance, is rapidly destroyed by small doses of irradiation if one looks at the reverse current, but remains essentially unaltered if one considers the doping profile. As a rule, metallurgical structures are quite insensitive to the irradiation doses used in BS.

It is clear that for a full characterization of a sample every possible tool must be brought to bear because each tool has limitations. Only a combination of techniques—fewer if those applied are well adapted to the problem or wisely selected and more of them otherwise—can permit hard conclusions. BS occupies a select place among these tools, in spite of having been a latecomer in the scene, because it is fast, ideally suited for large surveys or routine applications, and quantitative.

### 1.6   HOW TO READ A BACKSCATTERING SPECTRUM

One of the advantages of backscattering spectrometry is that the spectrum can be interpreted rather easily. In this section, we show how the form of a backscattering spectrum provides insight into the composition of a sample. Which physical process is actually responsible for the various characteristics of a spectrum does not concern us at this point. We shall actually proceed backward and assume that the composition of the sample is known, and show by what basic rules this information is translated into a backscattering spectrum. In a practical case, of course, the process is reversed.

Consider a thin film composed of a uniform mixture of two elements, as in the case of a binary compound or two fully miscible solids. To reduce the example to its simplest form, we shall ignore the substrate. For backscattering

ing spectrometry, the masses of the two elements and their atomic numbers are highly significant. We shall therefore characterize the two elements by their masses M and m, rather than by their chemical symbols. To start with, let us assume that the two elements are present in the film in the same proportion; i.e., the atomic concentrations of both elements are the same. This state of affairs is represented graphically in Figs. 1.12a and b. The profile of

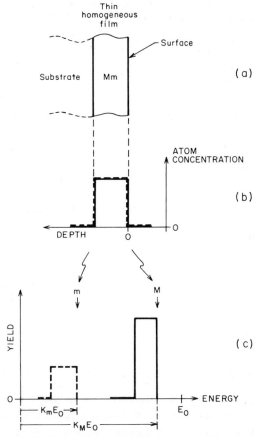

**Fig. 1.12** (a) Translation of concentration profiles to signals in a backscattering spectrum, demonstrated for the example of a thin homogeneous film of a binary compound with elements of a heavy M and a light m atomic mass. (b) The atomic concentrations are the same for both elements. (c) In the backscattering spectrum, the two profiles reappear as two separate signals. The light mass gives a signal at low energies with a low yield. The heavy mass produces a signal at high energies of a high yield. The high-energy edge of each signal (arrows marked m and M) is pegged on the energy axis of the spectrum to the value given by the kinematic factor $K$, where $E_0$ is the energy of the incident particles. The yield ratio of the two signal heights is given (approximately) by the ratio of the scattering cross section $\sigma_M/\sigma_m$ of the two elements, which is proportional to $(Z/z)^2$, the square of the ratio of the atomic numbers $Z$ and $z$ of the heavy and light elements.

atomic concentration versus depth given in Fig. 1.12b translates into the corresponding backscattering spectrum of Fig. 1.12c as follows:

1.   The rectangular profile of the element with the *heavy* mass (M, say) reappears in the backscattering spectrum as a rectangular signal located on the energy axis at *high* energies; the profile of the element with the *light* mass (m, say) gets a place in the backscattering spectrum at *low* energies. The rule for the translation of the abscissas thus is *heavy masses go to high energies; light masses to low energies.*

2.   Atomic concentrations of the same value in Fig. 1.12b are plotted at different levels on the yield axis of the backscattering spectrum in Fig. 1.12c. If the atomic number of the element is high, the yield is high too, and if the atomic number of the element is low, the yield is low. The rule for the translation of the ordinates reads: *High atomic numbers give high yield; low atomic numbers give low yields.* In effect, these two rules amount to saying that each element has its own coordinate system in the backscattering spectrum.

The discussion so far is qualitative. The power of backscattering spectrometry now resides in the fact that the two translations just described can be formulated in quantitative terms. For the $x$ axis of the backscattering spectrum, for instance, there is the so-called *kinematic factor K*, which states where, exactly, the signal of an element of any given mass has its high-energy edge. (The high-energy edge, or "leading" edge, of the signals of the elements of mass M and m in Fig. 1.12 are indicated by arrows marked M and m.) The location of the high-energy edges are indicated by the length of the arrows labeled $K_m E_0$ below the energy axis of the spectrum of Fig. 1.12c.

In very similar fashion, the scattering cross section $\sigma$ gives the scaling factor for the yield axis of different elements. The relative concentration ratio of two elements transforms into relative yields by a ratio given essentially by the cross section ratio of the elements or by $(Z/z)^2$. Some corrections must be applied. These are usually small (less than 10%), but the fact that they do exist has much to do with the reason this book is written. For example, the thicknesses that the two atomic species M and m occupy in the film (Figs. 1.12a and b) are the same; the widths that the signals of these two elements occupy on the energy scale of the backscattering spectrum are not. The range of depth in the film is thus translated into an energy interval on the energy axis of the spectrum, but that interval is not quite the same for each signal. If both intervals were identical, the scaling factor for the two yields would be correctly given by the ratio of the scattering cross sections of the two elements. Generally the intervals differ but not by much (about 10% or less); hence the correction on the yields.

To summarize, the translation of the concentration profiles of the two elements in the film (Fig. 1.12b) into the two signals of the backscattering spectrum of Fig. 1.12c may be viewed in the following way (see Fig. 1.13): There is a coordinate system for each mass in the target, plotting the atomic concentration of that mass as a function of increasing depth below the surface of the sample on which the analyzing beam impinges. Each profile

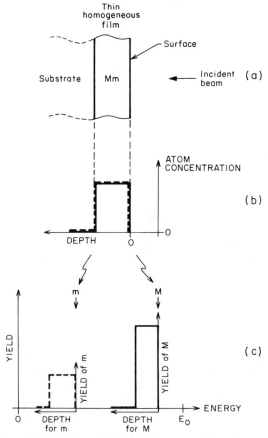

**Fig. 1.13**   (a) Translation of concentration profiles to signals in a backscattering spectrum, demonstrated for the example of a thin homogeneous film of a binary compound with elements of a heavy $M$ (solid line) and a light $m$ (dashed line) atomic mass. (b) The atomic concentration profiles with depth are the same for both elements. (c) In the backscattering spectrum, the two profiles reappear as two separate signals. The position of the coordinate systems for the two signals, and the scaling factor for their ordinates are as described in Fig. 1.12. However, the conversion of the abscissas from depth to energy is generally not the same for the two signals, and the conversion is not exactly linear either. Usually, the nonlinearities are insignificant and the difference in the two scales is not more than 10%.

is reproduced independently of the other in the backscattering spectrum and generates the signal of that mass. The final backscattering spectrum is a linear superposition of these signals. When a concentration profile varies with depth, the height of a signal will vary accordingly. This means that a backscattering spectrum actually constitutes an image of the distribution with depth of the various elements in the sample. Each type of atom of a particular mass is displayed individually. The signal of each has an accurately defined position on the energy scale, which corresponds to the sample surface as a reference point.

If the sample of Figs. 1.12 and 1.13 is thick, the signals of the two masses M and m will extend down to zero energies. The spectrum then has the

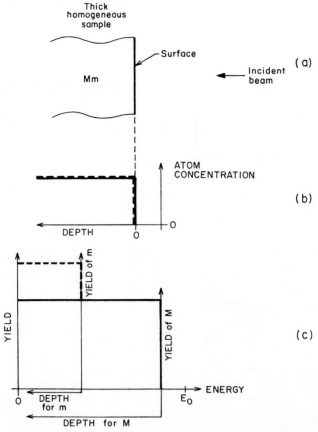

**Fig. 1.14** (a) The signals of a thick sample extend all the way to zero energy. (b) Actual spectra never reach down to the origin of the energy axis, because near zero energy, the yield disappears in a large background of noise. (c) The construction of the spectrum follows exactly the same procedures outlined in Figs. 1.12 and 1.13.

steplike appearance shown in Fig. 1.14. Real spectra never extend to zero energy, because noise in the detection system dominates at these low energies and generates a huge background. Thick-target yields are also never flat-topped as shown here; the reason is the energy dependence of the scattering cross section.

In problems of analysis the situation is reversed. A backscattering spectrum is measured, and the elemental makeup of the sample with depth has to be determined. We shall treat more examples in Chapter 5 to illustrate some of the major characteristics of backscattering analysis. These examples will also demonstrate that backscattering spectrometry is, in essence, mass-sensitive depth microscopy capable of furnishing quantitative information on the sample under investigation.

## 1.7   BOOK OUTLINE

Starting with Chapter 2 we shall repeat the three basic concepts and their mathematical relations to the projectile and to the target parameters in detail. In addition to kinematics, scattering cross sections, and energy loss, we shall discuss energy straggling, which sets the ultimate limit on depth resolution.

Chapter 3 describes how the three basic concepts are combined to produce a backscattering spectrum. This concerns the relation of energy to depth. Also covered in the chapter is how the height of an energy spectrum is related to scattering cross section and energy loss. The emphasis in this chapter is on bulk samples.

Chapter 4 gives backscattering analyses of thin films of various degrees of complications: elemental films, multilayered elemental films, compound films, and layered compound films. Depth and composition analyses at various sophistication levels are given. Different approximations and their justification are also given.

Many examples of backscattering analysis are given in Chapter 5. Formulas developed in Chapters 3 and 4 are applied to real problems. Many examples were chosen to illustrate the capability and limitation of back-scattering. Some of the approximations given in the previous two chapters are also used and compared to give the reader a feeling about the adequacy of the approximations. Since many of the examples have been taken from routine experiments, readers can use them as typical spectra to check their system and their analysis.

Chapter 6 describes the experimental setup. If you do not have a nuclear physics laboratory close by and want to set up a backscattering laboratory, this chapter gives the basic requirements for hardware and electronics. The chapter is also useful in understanding the data-taking system: solid-state detector, preamp, amplifiers, multichannel analyzer, and so on.

Chapter 7 describes the influence of beam parameters. In all the discussions so far we emphasize megaelectron volt $^4$He beams incident perpendicularly on the sample. In this chapter we discuss other alternatives. We shall present mass and depth resolutions and their relationships to the mass and energy of the projectiles. Different geometries for scattering and various problems are also discussed.

Chapter 8 is concerned with backscattering applications when combined with channeling effects. We start with the procedure used to align a crystal and then proceed to half-angle and minimum-yield calculations. The channeling applications dealing with lattice disorder, amorphous layers, and polycrystalline film are discussed. Lattice location and flux peaking of impurities in a crystal are also described.

In the body of the book, we assume that energy loss values are known in the analysis of a backscattering problem. In Chapter 9 we reverse the procedure and use the knowledge of the sample (composition and thickness) to determine stopping cross section values from backscattering measurements. Methods, formulas, and a few examples are given.

Chapter 10 gives a list of references on the applications of backscattering spectrometry. The cut-off date on the citations is August 1976. The references are listed according to various topics; surfaces, bulk, oxide and nitride layers, deposited and grown layers, thin film reactions, and ion implantation in metals and in semiconductors are the main section topics. Subdivision of the references by topic as well as listing the title of each paper provides a useful bibliography for a literature research.

In Appendix F, we provide tables of kinematic factors, scattering cross sections, and various forms of energy loss and energy loss factors. Analyses of examples given in the book are generated by using these tables.

### REFERENCES

Geiger, H., and Marsden, E. (1913). *Phil. Mag.* **25**, 606.

Mazari, M., Velazquez, L., and Alba, F. (1959). *Revista Mexicana de Fisica* **VIII**, 1.

Patterson, J. H., Turkevich, A. L., and Franzgrote, E. J. (1965). *J. Geophys. Res.* **70**, 1311.

Rubin, S., Stanford Res. Inst. Tech. Rep., No. I (1954).

Rubin, S., and Rasmussen, V. K. (1950). *Phys. Rev.* **78**, 83.

Rubin, S., Passell, T. O., and Bailey, E. (1957). *Anal. Chem.* **29**, 736.

Rutherford, E. (1911). *Phil. Mag.* **21**, 669.

Sippel, R. F. (1959). *Phys. Rev.* **115**, 1441.

Tollestrup, A. V., Fowler, W. A., and Lauritsen, C. C. (1949). *Phys. Rev.* **76**, 428.

Turkevich, A. (1961). *Science* **134**, 672.

Turkevich, A. L. *et al.* (1968). *in* Surveyor Project Final Report, Part II. Scientific Results, Nat. Aeronaut. and Space Administration Tech. Rep. 32–1265, pp. 303–387. Jet Propulsion Lab., California Inst. of Technol., Pasadena, California.

Chapter

# 2

# Basic Physical Concepts

## 2.1 INTRODUCTION

Only four basic physical concepts enter into backscattering spectrometry. Each one is at the origin of a particular capability or limitation of backscattering spectrometry and corresponds to a specific physical phenomenon. They are

1. Energy transfer from a projectile to a target nucleus in an elastic two-body collision. This process leads to the concept of the *kinematic factor* and to the capability of mass perception.

2. Likelihood of occurrence of such a two-body collision. This leads to the concept of *scattering cross section* and to the capability of quantitative analysis of atomic composition.

3. Average energy loss of an atom moving through a dense medium. This process leads to the concept of *stopping cross section* and to the capability of depth perception.

4. Statistical fluctuations in the energy loss of an atom moving through a dense medium. This process leads to the concept of *energy straggling* and to a limitation in the ultimate mass and depth resolution of backscattering spectrometry.

In this chapter an introductory treatment of these subjects is provided. Key formulas are given, and functional relationships are examined. The

21

discussion goes as far as the understanding of backscattering spectrometry demands. How these processes actually enter into a backscattering experiment and how they can affect a backscattering spectrum are examined in Chapter 3. When the target is a single crystal, or nearly so, the processes treated in this chapter are combined in a particular fashion, which results in the phenomenon of *channeling*. This effect is discussed in Chapter 8.

## 2.2   KINEMATIC FACTOR $K$

When a particle of mass $M_1$, moving with constant velocity, collides elastically with a stationary particle of mass $M_2$, energy will be transferred from the moving to the stationary particle. In backscattering analysis, mass $M_1$ is that of the projectile atom in the analyzing beam and mass $M_2$ is that of an atom in the target examined. The assumption that the interaction between the two atoms is properly described by a simple elastic collision of two isolated particles rests on two conditions:

(1)   The projectile energy $E_0$ must be much larger than the binding energy of the atoms in the target. Chemical bonds are of the order of 10 eV, so that $E_0$ should be very much larger than that.

(2)   Nuclear reactions and resonances must be absent. This imposes an upper limit to the projectile energy. Nuclear processes depend on the specific choice of projectile and target atoms, so that the upper limit of $E_0$ varies with circumstances. With a $H^+$ beam, nuclear effects can appear even below 1 MeV; with $He^+$, they begin to appear at 2 to 3 MeV.

The simple elastic collision of two masses $M_1$ and $M_2$ can be solved fully by applying the principles of conservation of energy and momentum. Let $\mathbf{v}_0$, $v_0$, and $E_0 = \frac{1}{2}M_1 v_0{}^2$ be the velocity, its value, and the energy of a projectile atom of mass $M_1$ before the collision, while the target atom of mass $M_2$ is at rest. After the collision, let $\mathbf{v}_1$ and $\mathbf{v}_2$ be the velocities and $E_1 = \frac{1}{2}M_1 v_1{}^2$ and $E_2 = \frac{1}{2}M_2 v_2{}^2$ be the energies of projectile and target atoms, respectively. The notation and the geometry of this scattering problem are given in Fig. 2.1, where the scattering angle $\theta$ and the recoil angle $\phi$ are

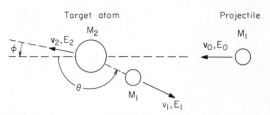

**Fig. 2.1**   Schematic representation of an elastic collision between a projectile of mass $M_1$, velocity $\mathbf{v}_0$, and energy $E_0$ and a target mass $M_2$ which is initially at rest. After the collision, the projectile and the target mass have velocities and energies $\mathbf{v}_1$, $E_1$ and $\mathbf{v}_2$, $E_2$, respectively. The angles $\theta$ and $\phi$ are positive as shown. All quantities refer to a laboratory frame of reference.

defined as positive numbers with the arrows as shown. All quantities refer to a laboratory system of coordinates.

Conservation of energy and conservation of momentum parallel and perpendicular to the direction of incidence are expressed by the equations

$$\tfrac{1}{2}M_1 v_0^2 = \tfrac{1}{2}M_1 v_1^2 + \tfrac{1}{2}M_2 v_2^2, \tag{2.1}$$

$$M_1 v_0 = M_1 v_1 \cos\theta + M_2 v_2 \cos\phi, \tag{2.2}$$

$$0 = M_1 v_1 \sin\theta - M_2 v_2 \sin\phi. \tag{2.3}$$

Eliminating $\phi$ first and then $v_2$, one finds

$$v_1/v_0 = [\pm(M_2^2 - M_1^2 \sin^2\theta)^{1/2} + M_1 \cos\theta]/(M_2 + M_1). \tag{2.4}$$

For $M_1 \leq M_2$ the plus sign holds. We now define the ratio of the projectile energy after the elastic collision to that before the collision as the *kinematic factor K*,

$$K \equiv E_1/E_0. \tag{2.5}$$

From Eq. (2.4) one obtains

$$K_{M_2} = \left[\frac{(M_2^2 - M_1^2 \sin^2\theta)^{1/2} + M_1 \cos\theta}{M_2 + M_1}\right]^2 \tag{2.6a}$$

$$= \left\{\frac{[1 - (M_1/M_2)^2 \sin^2\theta]^{1/2} + (M_1/M_2)\cos\theta}{1 + (M_1/M_2)}\right\}^2, \tag{2.6b}$$

where, following frequent practice, a subscript has been added to $K$ to indicate the target mass $M_2$ for which the factor applies. Another custom uses the chemical symbol of the target atom as the subscript for $K$ (e.g., $K_{Si}$ instead of $K_{28}$). This procedure is less accurate, because elements can have isotopes, and isotopes have slightly different $K$ values. In the center-of-mass system of reference, Eq. (2.6) can be simplified to (Marion and Young, 1968)

$$K = 1 - [2M_1 M_2/(M_1 + M_2)^2](1 - \cos\theta_c), \tag{2.7}$$

where $\theta_c$ is the scattering angle in the center-of-mass coordinates.

The kinematic factor depends only on the ratio of the projectile to the target masses and on the scattering angle $\theta$. The mass ratio $M_1/M_2$ will be abbreviated by $x$. A plot of $K$ versus $M_2/M_1 = x^{-1}$ and $\theta$ as given by Eq. (2.6) is shown in Fig. 2.2. One sees that for any combination of projectile and target mass, i.e., for any value of $x$, $K$ always has its lowest value at 180°. The value there is

$$K(\theta = 180°) = [(M_2 - M_1)/(M_2 + M_1)]^2 = [(1 - x)/(1 + x)]^2. \tag{2.8}$$

At $\theta = 90°$, $K$ is

$$K(\theta = 90°) = (M_2 - M_1)/(M_2 + M_1) = (1 - x)/(1 + x), \tag{2.9}$$

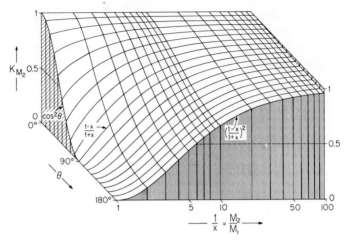

**Fig. 2.2**  The kinematic factor $K$ of Eq. (2.6b) plotted as a function of the scattering angle $\theta$ and the mass ratio $x^{-1} = M_2/M_1$.

that is, the value of the kinematic factor at $\theta = 180°$ is the square of its value at $\theta = 90°$. When the projectile and the target mass are equal ($x = 1$), $K$ is zero for angles larger than $90°$ and increases as $\cos^2 \theta$ when $\theta$ falls below $90°$. This says that a projectile colliding with a stationary atom equal to its own mass cannot be scattered backward, but only forward. This is true also for $M_1 > M_2$ ($x > 1$).

In backscattering spectrometry, angles near $180°$ are of special interest. To describe the behavior of $K$ there, it is convenient to introduce the difference $\delta$ between $\theta$ and $180°$, expressed in units of radians of arc as

$$\delta = \pi - \theta, \tag{2.10}$$

so that $\delta$ measures the deviation of $\theta$ from $\pi$ in units of arc. The kinematic factor then is approximated very well by the first term of an expansion in $\delta$:

$$K \simeq \left(\frac{M_2 - M_1}{M_2 + M_1}\right)^2 \left(1 + \frac{M_1}{M_2} \delta^2\right) = \left(\frac{1 - x}{1 + x}\right)^2 (1 + x\delta^2). \tag{2.11}$$

This equation describes the increase of $K$ along the front edge of Fig. 2.2 for small decreases of $\theta$ from $180°$. The approximation overestimates $K$ by a relative amount which is less than $\delta^4 x(1 - x)^{-2}$. As $\theta$ departs from $180°$, $K$ increases only quadradically with $\delta$. This increase is proportional to the mass ratio $x = M_1/M_2$. When this ratio is small, the factor $[(1 - x)/(1 + x)]^2$ can be approximated by $1 - 4x$, so that in the right corner of Fig. 2.2 the kinematic factor is approximately described by

$$K \simeq 1 - 4x + \delta^2 x. \tag{2.12}$$

This is a convenient formula to estimate $K$ in the region of $\theta$ and $x$ values which are most relevant to backscattering spectrometry. Values of $K$ and $\delta^2$ are given in Tables II–V in Appendix F.

Equations (2.5) and (2.6) contain the essence of how backscattering spectrometry acquires its ability to sense the mass of an atom. Imagine that the primary energy $E_0$ of the projectile atom and its mass $M_1$ are known. Assume that the energy $E_1$ after the elastic scattering event is measured at a known angle $\theta$. Then the mass $M_2$ of the target atom that prompted the scattering is the only unknown quantity in Eq. (2.6). The value of $M_2$ can thus be determined by measuring the energy $E_1$ after the collision if $E_0$, $M_1$, and $\theta$ are known. In effect, the technique amounts to mass spectrometry "by reflection." The method is based on the same laws that govern simple billiard ball physics.

In practice, when a target contains two types of atoms that differ in their masses by a small amount $\Delta M_2$, it is important that this difference produce as large a change $\Delta E_1$ as possible in the measured energy $E_1$ of the projectile after the collision. As Fig. 2.2 shows, a change of $\Delta M_2$ (for fixed $M_1$) gives the largest change of $K$ when $\theta = 180°$ for all but the smallest values of $M_2$. Thus $\theta = 180°$ is the preferred location for the detector. To place a normal detector exactly at $\theta = 180°$ is not possible because the detector would obstruct the path of the incident particles. The detector is thus normally positioned at some steep backward angle, such as 170°. It is this particular experimental arrangement that has given the method its name of *backscattering* spectrometry. With annular detectors, scattering angles very near 180° can be reached; these special solid-state detectors have a hole along the center axis through which the primary beam passes before impinging on the target.

In quantitative terms, $\Delta E_1$ and $\Delta M_2$ are related to each other by

$$\Delta E_1 = E_0 (dK/dM_2) \Delta M_2. \tag{2.13}$$

In the vicinity of $\theta = 180°$, i.e., $\theta = \pi - \delta$, $K$ is very closely approximated by Eq. (2.11), so that

$$\frac{\Delta E_1}{E_0} = \frac{1 - x}{(1 + x)^3} [4(1 + x\delta^2) - \delta^2(1 - x^2)] x \frac{\Delta M_2}{M_2}. \tag{2.14}$$

For $M_2 \gg M_1$, which is most often the case, this reduces further to

$$\Delta E_1 = E_0 (4 - \delta^2)(M_1/M_2{}^2) \Delta M_2. \tag{2.15}$$

Every practical detection system has a finite resolution. If $\Delta E_1$ falls below this limit, the distinction between two masses is lost. To obtain good mass resolution, it is therefore desirable that the coefficient of $\Delta M_2$ be as large as possible. To accomplish this, one can

   (i)   Increase the primary energy $E_0$;
   (ii)  Use a projectile of large mass $M_1$ (Note, however, that $M_2$ masses smaller than $M_1$ will not produce any backscattering signal.);
   (iii) Measure at scattering angles approximately $180°$ (small $\delta$).

We also notice that mass resolution is inherently better for light target atoms than for heavy ones, the effect going as $M_2^{-2}$.

### 2.3  SCATTERING CROSS SECTION $\sigma$

The preceding section established the connection between the energy $E_0$ of the incident particle of mass $M_1$ and the energy $K_{M_2}E_0$ that this particle possesses at any angle $\theta$ after an elastic collision with an initially stationary mass $M_2$. How frequently such a collision actually occurs and ultimately results in a scattering event at a certain angle $\theta$ remains open.

The differential scattering cross section $d\sigma/d\Omega$ is the concept introduced to answer this. Its definition is derived from a simple conceptual experiment. A narrow beam of fast particles impinges on a thin uniform target that is wider than the beam. At an angle $\theta$ from the direction of incidence, let an ideal detector count each particle scattered in the differential solid angle $d\Omega$ (see Fig. 2.3). If $Q$ is the total number of particles that have hit the target and $dQ$ is the number of particles recorded by the detector, then the *differential scattering cross section $d\sigma/d\Omega$* is defined as

$$d\sigma/d\Omega = (1/Nt)[(dQ/d\Omega)/Q], \tag{2.16}$$

where $N$ is the volume density of atoms in the target and $t$ is its thickness. Thus $Nt$ is the number of target atoms per unit area (areal density). The

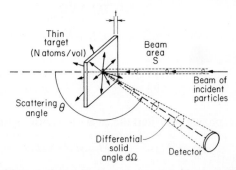

**Fig. 2.3** Simplified layout of a scattering experiment to demonstrate the concept of the differential scattering cross section. Only primary particles that are scattered within the solid angle $d\Omega$ spanned by the detector are counted.

definition implies that the solid angle $d\Omega$ is so small that the scattering angle $\theta$ is well defined. The definition also assumes that the thickness $t$ is minimal and that, therefore, the energy loss of the particles in the target is so small that the energy of the particles is virtually the same at any depth in the target. Finally, the total number of incident particles $Q$ must be so large that the ratio $dQ/Q$ has a well-determined value.

The differential scattering cross section $d\sigma/d\Omega$ has the dimension of an area ("cross section") whose meaning is based on a geometrical interpretation of the probability that the scattering will result in a signal at the detector. One imagines that each nucleus of an atom presents an area $d\sigma/d\Omega$ to the beam of incident particles. It is also assumed that this area is quite small and that the atoms within the target are randomly distributed in such a way that the differential cross sections $d\sigma/d\Omega$ of the nuclei do not overlap. Let $S$ be the surface area of the target illuminated uniformly by the beam. Then the total number of atoms eligible for a scattering collision in the target is $SNt$.[†] The ratio of the total cross-sectional area of all eligible atoms $SNt\,d\sigma/d\Omega$ to the area $S$ actually exposed is then interpreted as the probability that the scattering event will be recorded by the detector; that is, this ratio is set equal to $(1/d\Omega)\,dQ/Q$. Equation (2.16) then follows. The multiplication with $(d\Omega)^{-1}$ is introduced because doubling the solid angle $d\Omega$ would obviously double the number of counts $dQ$. By dividing $dQ$ with $d\Omega$, this geometrical contribution to the number of counts $dQ$ is eliminated. The cross section defined in this way thus becomes a value per unit of solid angle; hence the name *differential scattering cross section*, and therefore the notation $d\sigma/d\Omega$. Other equally valid interpretations of the meaning of a differential scattering cross section can be found in various textbooks (Leighton, 1959; Goldstein, 1959).

When one inquires as to the number of scattering events falling within a finite solid angle $\Omega$ rather than a differential solid angle $d\Omega$, the probability of a successful event is described by the *integral scattering cross section* $\Sigma$:

$$\Sigma = \int_{\Omega} (d\sigma/d\Omega)\,d\Omega. \tag{2.17}$$

Its geometrical interpretation is analogous to that of the differential scattering cross section. In backscattering spectrometry, the solid angle $\Omega$ of a typical detector system with a surface-barrier detector is fairly small ($10^{-2}$ sr or less)

---

[†] One can also conceive of situations where the picture of randomly distributed cross sections over an area $S$ and a uniform illumination of this area $S$ by the incident particles breaks down. When the target is single-crystalline, the cross sections are clustered along sets of lines in space. If the incident particles move in a direction parallel to such lines, and if the flux of these particles is concentrated in the voids ("channels") surrounding these lines, the probability of a scattering collision is obviously reduced. This is the situation commonly referred to as "channeling" (see Chapter 8).

and the scattering angle $\theta$ is well defined. It is then convenient[†] to introduce the *average differential scattering cross section $\sigma$*:

$$\sigma \equiv (1/\Omega) \int_\Omega (d\sigma/d\Omega)\, d\Omega. \qquad (2.18)$$

For very small detector angles $\Omega$, $\sigma \to d\sigma/d\Omega$. The average differential scattering cross section is the value ordinarily used in backscattering spectrometry. It is usually called *scattering cross section* in the literature. We follow this convention.

For the experimental condition given in Fig. 2.4, in which a uniform beam impinges at normal incidence on a uniform target that is larger than the area of the beam, the *total number of detected particles $A$* can be written from Eqs. (2.16) and (2.18) as

$$A = \sigma\Omega \cdot Q \cdot Nt. \qquad (2.19)$$

$$\binom{\text{number of}}{\text{detected particles}} = \sigma\Omega \cdot \binom{\text{total number of}}{\text{incident particles}} \cdot \binom{\text{number of target}}{\text{atoms per unit area}}.$$

This equation shows that when $\sigma$ and $\Omega$ are known and the numbers of incident and detected particles are counted, the number of atoms per unit area in the target, $Nt$, can be determined. The ability of backscattering spectrometry to provide quantitative information on the number of atoms present per unit area of a sample stems from Eq. (2.19) and the fact that the average scattering cross section $\sigma$ of the elements is known quite accurately.

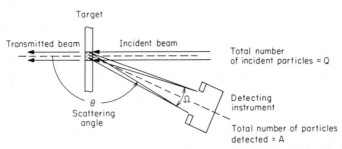

**Fig. 2.4** Schematic layout of a backscattering experiment, showing a thin target, the transmitted portion of the beam, and the fraction of the backscattered beam that is intercepted and counted by the detector.

---

[†] In nuclear physics, the symbol $\sigma$ is used to refer to the integral ("total") scattering cross section, called $\Sigma$ in Eq. (2.17). The use of $\sigma$ for the right-hand side of Eq. (2.18) is inconsistent with this older tradition, which would have required a symbol such as $\langle d\sigma/d\Omega \rangle$ instead. On the other hand, the newer (inconsistent) convention of Eq. (2.18) simplifies the writing of many equations to $\sigma\Omega$ rather than the clumsy $\langle d\sigma/d\Omega \rangle \Omega$.

To calculate the differential cross section for an elastic collision, the principles of conservation of energy and momentum must be complemented by a specific model for the force that acts during the collision between the projectile and the target masses. In most cases, this force is very well described by the Coulomb repulsion of the two nuclei as long as the distance of closest approach is large compared with nuclear dimensions, but small compared with the Bohr radius $a_0 = \hbar/m_e e = 0.53$ Å. When these assumptions are made, the differential scattering cross section is given by Rutherford's formula (Rutherford, 1911; Goldstein, 1959; Leighton, 1959):

$$(d\sigma/d\Omega)_c = \left[ \frac{Z_1 Z_2 e^2}{4E_c \sin^2(\theta_c/2)} \right]^2, \tag{2.20}$$

where the subscript c indicates that the values are given with respect to the center-of-mass coordinates. Here $Z_1$ is the atomic number of the projectile atom with mass $M_1$, $Z_2$ is the atomic number of the target atom with mass $M_2$, $e$ is the electronic charge ($e = 4.80 \times 10^{-10}$ statC),[†] and $E$ is the *energy of the projectile immediately before scattering*. This formula is valid also for values in the laboratory frame of reference, but only when $M_1 \ll M_2$. For the general case, the transformation of this formula from the center-of-mass to the laboratory frame of reference yields (Darwin, 1914)

$$\frac{d\sigma}{d\Omega} = \left( \frac{Z_1 Z_2 e^2}{4E} \right)^2 \frac{4}{\sin^4 \theta} \frac{\{[1 - ((M_1/M_2)\sin\theta)^2]^{1/2} + \cos\theta\}^2}{[1 - ((M_1/M_2)\sin\theta)^2]^{1/2}}. \tag{2.22}$$

A detailed execution of this transformation is given in Appendix A. The order of magnitude of this differential scattering cross section is predominantly given by the first factor $(Z_1 Z_2 e^2/4E)^2$. As an example, consider 1-MeV He ($Z_1 = 2$) impinging on Ni ($Z_2 = 28$); then $(Z_1 Z_2/4)^2 = 196$. In electrostatic cgs units, the electronic charge has the value $e = 4.80286 \times 10^{-10}$ statC and the unit of potential is the statV $= 299.79$ V. For 1 MeV, the value of $(e^2/E)^2$ is therefore $(e/10^6\,\text{V})^2 = (4.80286 \times 10^{-10} \times 299.79/10^6)^2(\text{statC})^2/(\text{statV})^2 = 2.0731 \times 10^{-26}\,(\text{statC/statV})^2$. The ratio statC/statV has the value of the

---

[†] It is customary in the nuclear physics literature to use cgs units. To avoid confusion and to help in identifying the system of units adopted for an equation, we shall use $e$ throughout when electrostatic units are assumed and $q$ throughout when mks units are used. To translate an equation from one set of units to another, one substitutes

$$e^2 \rightleftarrows q^2/4\pi\varepsilon_0 \tag{2.21}$$

where $e = 4.80286 \times 10^{-10}$ statC, where $q = 1.60206 \times 10^{-19}$ C, and $\varepsilon_0 = 8.85434 \times 10^{-12}$ Asec/V m. A convenient constant to remember in connection with Eq. (2.21) is that $e^2 = 1.4398 \times 10^{-13}$ MeV cm $\simeq 1.44 \times 10^{-13}$ MeV cm. This permits quick estimates of $d\sigma/d\Omega$ when $E$ is given in mega electron volts, as usual.

unit length 1 cm, so that $(e^2/E)^2 = 2.0731 \times 10^{-26}$ cm$^2$ = 0.020731 b. Note that the conversion $e^2 = 1.4398 \times 10^{-13}$ MeV cm yields this result directly. The product $(Z_1 Z_2/4)^2 \cdot (e^2/E)^2$ thus is $196 \times 0.020731$ b = 4.06328 b for a unit steradian. Performed in mks units, the same calculation starts from the formula $(Z_1 Z_2 q^2/4\pi\varepsilon_0 4E)^2$, where the electronic charge has the value $q = 1.60206 \times 10^{-19}$ A sec  and  $\varepsilon_0 = 8.85434 \times 10^{-12}$ A sec/V m. The  ratio $(q^2/4\pi\varepsilon_0 E)^2$ for $E = 10^6$ qV then becomes $(1.60206 \times 10^{-19}/4\pi \times 8.85434 \times 10^{-12} \times 10^6)^2$(A sec)$^2$/(A sec/V m)$^2$ = $2.0731 \times 10^{-30}$ m$^2$, which is again 0.020731 b. (1 b = 1 barn = $10^{-24}$ cm$^2$)

If we disregard the factor $(Z_1 Z_2 e^2/4E)^2$, the Rutherford differential scattering cross section depends only on the ratio $M_1/M_2$ of the projectile and target masses and on the scattering angle $\theta$. A plot of $d\sigma/d\Omega$ versus $M_2/M_1 = x^{-1}$ and $\theta$ as given by Eq. (2.22) is shown in Fig. 2.5. For any combination of projectile and target mass, $d\sigma/d\Omega$ always has its lowest value at 180°. Expressed in units of $(Z_1 Z_2 e^2/4E)^2$, this minimum value is $[1 - (M_1/M_2)^2]^2 = (1 - x^2)^2$. In the vicinity of 180°, i.e., along the front edge of Fig. 2.5, where $\theta = \pi - \delta$, the Rutherford differential cross section increases quadratically with $\delta$:

$$(d\sigma/d\Omega)/(Z_1 Z_2 e^2/4E)^2 = (1 - x^2)^2 + \tfrac{1}{2}b\,\delta^2, \qquad (2.23)$$

where $b = 1 - 3x^4 + 2x^6$. The formula shows that near 180°, the scattering cross section does not change much with the scattering angle. This fact enables one to use the average acceptance angle of the particle detector and still obtain an accurate value for the calculated cross section near 180° [see

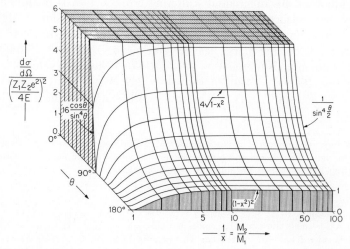

**Fig. 2.5**  The dependence of the Rutherford differential scattering cross section given by Eq. (2.22) as a function of the scattering angle $\theta$ and the mass ratio $x^{-1} = M_2/M_1$.

Eq. (2.18)]. For $M_1 \ll M_2$, i.e., in the lower right corner of Fig. 2.5, the angular dependence of the right-hand side of Eq. (2.22) can be expanded in the power series (Marion and Young, 1968)

$$\frac{d\sigma}{d\Omega} \simeq \left(\frac{Z_1 Z_2 e^2}{4E}\right)^2 \left[\sin^{-4}\frac{\theta}{2} - 2\left(\frac{M_1}{M_2}\right)^2 + \cdots\right],\qquad (2.24)$$

where the first omitted term is of the order of $(M_1/M_2)^4$. The last expression reveals the significant functional dependences of the Rutherford differential scattering cross sections:

(i)   $d\sigma/d\Omega$ is proportional to $Z_1{}^2$. The backscattering yield obtained from a given target atom with a He beam ($Z_1 = 2$) is four times as large as with a proton beam ($Z_1 = 1$) but only a ninth of that produced by a carbon beam ($Z_1 = 6$).

(ii)   $d\sigma/d\Omega$ is proportional to $Z_2{}^2$. For any given projectile, heavy atoms are very much more efficient scatterers than light atoms. Therefore, backscattering spectrometry is much more sensitive to heavy elements than to light ones.

(iii)   $d\sigma/d\Omega$ is inversely proportional to the square of the projectile energy ($\propto E^{-2}$). The yield of scattered particles rises rapidly with decreasing bombarding energy.

(iv)   $d\sigma/d\Omega$ is axially symmetrical with respect to the axis of the incident beam; i.e., $d\sigma/d\Omega$ is a function of $\theta$ only.

(v)   $d\sigma/d\Omega$ is approximately inversely proportional to the fourth power of $\sin(\theta/2)$ when $M_1 \ll M_2$. This dependence gives rapidly increasing yields as the scattering angle $\theta$ is reduced.

Values of $d\sigma/d\Omega$ for various elements $Z_2$ and energies are tabulated in Table X. For He in the MeV energy range, Rutherford differential scattering cross sections are typically within an order of magnitude or two of barns ($1\text{ b} = 10^{-24}\text{ cm}^2$) per unit steradian. A monolayer of a solid typically contains about $10^{15}$ atoms/cm$^2$. A 1-MeV He particle will thus typically traverse very many monolayers before being scattered out of its path by a nuclear collision.

Deviations of the differential scattering cross section from the Rutherford formula do exist.

For $\theta \to 0$, the Rutherford cross section tends to infinity, which of course violates the initial assumption that the cross sections of the target nuclei should be so small that they do not overlap. Small scattering angles correspond to large fly-by distances between the projectile and the target nuclei, that is, distances greater than the radius of the innermost electron shell of the target atom. At these distances the electrostatic interaction does not take place between bare nuclei as Rutherford's formula assumes ($d\sigma/d\Omega \simeq Z_1 Z_2 e^2$).

A similar situation exists when a low-energy projectile collides with a heavy atom. In such instances, one must use scattering cross sections derived from a potential which includes electron screening. Examples are the Born potential (Everhart *et al.*, 1955), the Born–Mayer potential (Abrahamson, 1969; Robinson, 1974) or the Firsov potential (Firsov, 1959). The validity of the Rutherford scattering approximation has been tested by calculation using different potentials (Everhart *et al.*, 1955) and by measurements with 100-keV $^1H^+$ and $^4He^+$ on Au (Van Wijngaarden *et al.*, 1970). Barely detectable departure from the Rutherford differential cross section was obtained in the latter case.

For sufficiently high energies $E$, the distance of closest approach between the projectile and the target nuclei reduces to the dimensions of nuclear sizes. The short-range nuclear forces then begin to influence the scattering process, and deviations from the Rutherford scattering cross sections appear. When the scattering process is inelastic, the energy of the scattered particle differs from $KE_0$ as well. In other cases, the scattering process is elastic still, but the differential scattering cross section departs from the Rutherford value, sometimes by a large factor. In either case, the value of the differential scattering cross section is strongly dependent on energy, on the scattering angle, and on the particular combination of projectile and target nuclei.

Apparent deviations from the Rutherford differential cross section can occur with electrostatic and magnetic analyzers. These analyzers are often desirable at low energies because of their good resolution and precision. In contrast to solid-state detectors, however, they detect particles of only one charge state at a time. The charge of the projectile atom after backscattering and escape from the target is a strong function of the escape velocity of the projectile (Marion and Young, 1968). Adjustments are therefore required to correct the observed particle counts for the undetected fraction of the scattered particles at any given energy for a given target.

## 2.4 ENERGY LOSS AND STOPPING CROSS SECTION

### 2.4.1 Energy Loss $dE/dx$

An energetic particle that impinges on a target will penetrate into it. This is so because the large-angle Rutherford scattering collision discussed in the previous section is highly unlikely. The fate of an impinging particle is overwhelmingly determined by the processes that control the penetration into the target, rather than by the large-angle scattering collisions. Back-scattering spectrometry is an analytical method to secondary process; the first-order process is the implantation of the beam particles into the target.

The concepts used to describe how a swift particle penetrates into matter arise from energetic considerations. As the particle pushes its way through

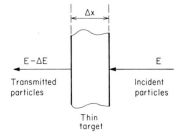

**Fig. 2.6** Schematic of a transmission experiment to measure the $\Delta E/\Delta x$ loss of a swift particle in a dense medium.

the target, it slows down and its kinetic energy $E = \frac{1}{2}M_1 v^2$ decreases. The amount of energy $\Delta E$ lost per distance $\Delta x$ traversed depends on the identity of the projectile, on the density and composition of the target, and on the velocity itself. The simplest experiment that can be conceived to determine this energy loss is to take a very thin target of thickness $\Delta x$ and of known composition. A beam of monoenergetic particles is directed at this target (see Fig. 2.6). The energy difference $\Delta E$ of the particles before and after transmission through the target is measured. The *energy loss* per unit length, also called sometimes the *specific energy loss*, and frequently abbreviated *dE/dx loss*, at the energy $E$ of the incident beam is then defined as

$$\lim_{\Delta x \to 0} \Delta E/\Delta x \equiv \frac{dE}{dx}(E) \tag{2.25}$$

for that particular particle and energy in that medium. Note that this expression gives an energy loss that is a positive quantity.

Since the early days of nuclear physics, measurements of the energy loss per unit length have been performed for many projectile atoms, for a multitude of compounds, for most elements, and over a very wide range of energies. A list of available compilations of experimental energy loss information is given in Appendix D. For backscattering spectrometry, it is the energy loss of ⁴He in the elements at energies between 0.5 and 3 MeV that is of chief concern, because beams of ⁴He in that energy range are most frequently used. Typical $dE/dx$ values for ⁴He of that energy range lie between 10 and 100 eV/Å. Additional information on the subject is provided in Section 2.4.2.

For the present we shall assume that $dE/dx$ is known at any energy, and we wish to establish the energy $E$ of the projectile at any depth $x$ below the surface of a thick sample into which the particle penetrates with an initial energy $E_0$. Generally, $dE/dx$ is a function of energy and has the form sketched in Fig. 2.7a. The energy $E$ at any depth $x$ below the surface is then given by

$$E(x) = E_0 - \int_0^x (dE/dx)\, dx. \tag{2.26}$$

As the functional parentheses $(E)$ in Eq. (2.25) point out, $dE/dx$ is defined and normally given as a function of $E$, not of $x$. The preceding integral thus

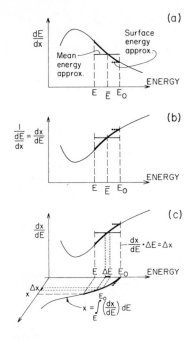

**Fig. 2.7** (a) Typical dependence of $dE/dx$ as a function of the kinetic energy $E$ of the projectile. To obtain the depth of penetration $x$ at which the particle energy has been reduced from $E_0$ to $E < E_0$, one takes the reciprocal of $dE/dx$, as shown in (b), and integrates this function from $E$ to $E_0$, as represented in (c). In the surface energy approximation, $dE/dx$ is replaced by its value at $E_0$ (heavy dashed line). In the mean energy approximation, the constant value of $dE/dx$ is chosen at the mean energy $\bar{E} = \frac{1}{2}(E + E_0)$.

cannot be evaluated without the knowledge of the energy as a function of $x$, $E(x)$. But $E(x)$ is the unknown in the equation. The difficulty is resolved by regarding $x$ as a function of $E$, rather than $E$ as a function of $x$; then

$$dx = \frac{dx}{dE}(E) \cdot dE, \qquad (2.27)$$

so that

$$x = \int_E^{E_0} (dx/dE)\, dE = \int_E^{E_0} (dE/dx)^{-1}\, dE. \qquad (2.28)$$

To find $x(E)$, one thus integrates over the function $(dE/dx)^{-1}$. The situation is sketched graphically in Figs. 2.7b and c. Note that the upper limit $E_0$ is fixed and the lower limit $E$ varies; hence $x$ increases as $E$ decreases.

It is frequently convenient to replace the actual $dE/dx$ function by an approximation. The simplest procedure is to replace $dE/dx$ by its value at the energy $E_0$ of the incident particle, as indicated by the dashed line in Fig. 2.7. Either Eq. (2.26) or Eq. (2.28) can then be used to determine $x(E)$:

$$E = E_0 - \frac{dE}{dx}\bigg|_{E_0} x \qquad \text{or} \qquad x = (E_0 - E)\left(\frac{dE}{dx}\right)^{-1}\bigg|_{E_0}. \qquad (2.29)$$

This method provides good estimates only in the uppermost or surface region of the target, and is thus called the *surface energy approximation.*

Another approximation replaces $dE/dx$ by its value at the energy $\bar{E} = \frac{1}{2}(E + E_0)$. One then obtains, from Eq. (2.26) or Eq. (2.28),

$$E = E_0 - \frac{dE}{dx}\bigg|_{\bar{E}} x \qquad \text{or} \qquad x = (E_0 - E)\left(\frac{dE}{dx}\right)^{-1}\bigg|_{\bar{E}}, \qquad (2.30)$$

so that $x$ again increases linearly with $(E_0 - E)$. This procedure is called the *mean energy approximation* and is sketched in Fig. 2.7 as well. The mean energy approximation provides good estimates at intermediate depths of penetration. Figure 2.7c shows how the two approximations are related to the exact solution given by Eq. (2.28).

The accuracy of the linear approximation can obviously be increased by selecting the specific value for $dE/dx$ that reproduces the magnitude of $x$ when this specific value is substituted for the integrand in Eq. (2.28). The $dE/dx$ curve takes on this specific value at some suitably selected energy $\bar{E}$ intermediate to $E$ and $E_0$. As an example, Warters (1953) assumes that the functional dependence of $dE/dx$ can be approximated by

$$dE/dx = CE^{-a(E)}, \qquad (2.31)$$

where $C$ is a constant and the exponent $a(E)$ varies only slowly, so that it may be set to a fixed value for any given energy interval $\Delta E = E_0 - E$. According to Eqs. (2.28) and (2.30), the specific value $\bar{E}$ to choose is that which will satisfy the condition

$$\Delta x = \int_{E_0 - \Delta E}^{E_0} dE/CE^{-a} = \Delta E/C\bar{E}^{-a}. \qquad (2.32)$$

The integration over $dE$ yields, as the condition that $\bar{E}$ must meet,

$$(a + 1)^{-1}E_0^{a+1}\{1 - [1 - (\Delta E/E_0)]^{a+1}\} = \Delta E\,\bar{E}^a. \qquad (2.33)$$

Expanding the left-hand side to second orders of $\Delta E/E_0$, dividing by $E_0^a$, and extracting the root gives

$$[1 - \tfrac{1}{2}a(\Delta E/E_0)]^{1/a} = \bar{E}/E_0 \qquad (2.34)$$

or

$$\bar{E}/E_0 = 1 - \tfrac{1}{2}(\Delta E/E_0) + \cdots . \qquad (2.35)$$

To the extent that Eq. (2.31) approximates $dE/dx$ adequately and as long as $\Delta E \ll E_0$, the best choice of $\bar{E}$ is thus midway between $E_0$ and $E_0 - \Delta E$. This is the same value specified in the mean energy approximation.

### 2.4.2  Stopping Cross Section ε

The energy loss $dE/dx$ accounts for the energy a fast particle expends as
it passes through the electron cloud of the atoms that lie along its path or
as it suffers numerous small-angle collisions with nuclei lying along its route.
The value of $dE/dx$ can be viewed as an average over all possible energy-
dissipative processes activated by the projectile on its way past a target atom.
It is natural, then, to interpret $dE/dx$ as the result of independent contributions
of every atom exposed to the beam. This number is $SN \, \Delta x$ if $\Delta x$ is the thickness
of the target, $S$ is the target area illuminated by the beam, and $N$ the atom
density in the target. The projection of all these atoms on the area $S$ produces
a surface density of atoms $SN \, \Delta x/S = N \, \Delta x$. This quantity increases linearly
with $\Delta x$, as does the energy loss $\Delta E = (dE/dx) \, \Delta x$. We therefore set $\Delta E$
proportional to $N \, \Delta x$ and define the proportionality factor as the *stopping
cross section* ε:

$$\varepsilon \equiv (1/N)(dE/dx). \qquad (2.36)$$

The conventional unit for ε is electron volts · square centimeters per atom
usually abbreviated eV cm$^2$.

The distinction between $dE/dx$ and ε is most evident when one considers
two targets made up of the same number of atoms per unit area. Assume
that in one case the atoms are closely packed and form a high volume density.
In the other case they are loosely assembled in a spongelike structure of low
volume density. The energy $\Delta E$ transferred to the target by a fast particle
must be the same in both cases as long as the energy loss is an atomic property,
that is, independent of the packing density of the atoms. A larger value of
$dE/dx$ will be assigned to the denser target, however, because that energy
$\Delta E$ is deposited over the shorter distance $\Delta x$. But $\Delta E/N \, \Delta x$ has the same value
in both instances since the difference in the densities is caused by the different
values of $\Delta x$ in the two cases; in other words, $N \propto 1/\Delta x$, so that $N \, \Delta x = $ const.
Hence $\Delta E/N \, \Delta x = \varepsilon$ is constant in the two cases. The subject is discussed
also in Section 3.9.

Another definition which is used predominantly in the nuclear physics
literature sets

$$\varepsilon^* \equiv (1/\rho)(dE/dx), \qquad (2.37)$$

where $\rho$ is the mass density (grams per cubic centimeter) of the target and
$\varepsilon^*$ is usually given in units of kiloelectron volts · square centimeters per gram.
The symbol $\varepsilon^*$ is introduced here to distinguish between the two definitions
of Eqs. (2.36) and (2.37), but the literature does not make that differentiation.
Which definition applies in a particular case can always be established from
dimensional considerations. The two quantities can be converted into each

other by the relationship

$$\rho = N(M/N_0), \tag{2.38}$$

so that $\varepsilon^* = \varepsilon N_0/M$. Here $M$ is the atomic weight (grams per mole) of the element and $N_0 = 6.025 \times 10^{23}$ atoms/mole is Avogadro's number.

The advantage of using the stopping cross section $\varepsilon$ rather than the $dE/dx$ is evident when one compares the energy loss of neighboring elements in the periodic table. Table 2.1 lists data for Na and Al for 2-MeV $^4$He. The ratios of the atomic numbers $Z_2$ and of the atomic masses $M_2$ are within 4% of the $\varepsilon$ ratio, but the $dE/dx$ ratio is larger by more than a factor of two. It is mainly the difference in the atomic density of Na and Al that is responsible for this difference. Atomic densities vary over almost an order of magnitude. Interpolations from one element to another are thus much more reliably performed on $\varepsilon$ than on $dE/dx$ when direct information is unavailable.

**TABLE 2.1**

Comparison of Energy Loss per Unit Length $dE/dx$ and
Stopping Cross Section $\varepsilon$ for 2.0-MeV $^4$He in Na and Al

|  | $Z_2$ | $M_2$ | $N$ (atoms/cm$^3$) | $\varepsilon$ (eV cm$^2$) | $dE/dx$ (eV/Å) |
|---|---|---|---|---|---|
| Na | 11 | 22.99 | $2.65 \times 10^{22}$ | $39.6 \times 10^{-15}$ | 10.5 |
| Al | 13 | 26.98 | $6.02 \times 10^{22}$ | $44.3 \times 10^{-15}$ | 26.6 |
| $\dfrac{\text{Al}}{\text{Na}}$ ratio | 1.18 | 1.17 | 2.27 | 1.12 | 2.53 |

For backscattering spectrometry, interest in stopping cross-section values centers predominantly on $^4$He because this is the most frequently used ion for the analyzing beam. Ziegler and Chu (1974) have surveyed the literature and tabulated semiempirical tables of stopping cross sections for $^4$He in all elements and from 0.4 to 4.0 MeV. Their tables are reproduced as Table VI of Appendix F. A graphical display of the values from 0.4 to 2.0 MeV is shown in Fig. 2.8. As can be seen, the stopping cross section of all elements vary with energy in much the same way. The curves have a broad maximum somewhere near 1 MeV. For constant energy, $\varepsilon$ tends to increase with $Z_2$, but there are strong variations superimposed on this trend. In their fine structure these variations are irregular, but their overall features are closely correlated with the electronic configuration of the element. This is particularly pronounced at 400 keV, where the three transition metal groups show up as regions of reduced $\varepsilon$ values.

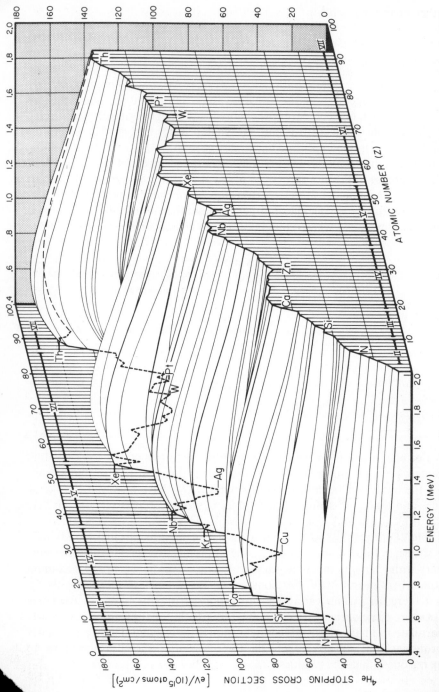

**Fig. 2.8** Stopping cross sections for ⁴He in all elements from 0.4 to 2.0 MeV. The plot gives the semiempirical values of the tables of Ziegler and Chu (1974), here reproduced as Table VI in Appendix F.

Many sections of this figure have been obtained by interpolation to fill in nonexistent data, and some of the data available may be revised in the future. Details in Fig. 2.8 will then change; but it is clear that as a whole the dependence of the stopping cross section on energy and target element in the range of interest to backscattering spectrometry is complicated. This is the reason why theoretical calculations of stopping cross sections turn out to be difficult to do accurately. The following subsection therefore presents in detail only the simplest classical picture of electronic energy loss. The approach offers some physical insight but no quantitative accuracy.

### 2.4.3 Physical Models

The theory of the fast particle interaction in dense media began with the work of Bohr (1913) and is still an active field of investigation. Much is now known, particularly for amorphous materials. For the light projectile atoms and the energy range of interest to backscattering spectrometry, the two dominant processes of energy loss are the interactions of the moving ion with the bound or free electrons in the target, and the interactions of the moving ion with the screened or unscreened nuclei of the target atoms. One can thus set

$$\varepsilon = \varepsilon_e + \varepsilon_n. \tag{2.39}$$

Figure 2.9 shows schematically how these two contributions depend on the projectile energy. Nuclear stopping originates from the multitude of small-angle scattering collisions of the projectile with the atomic nuclei of the target. Electronic stopping comes from the "frictional resistance" that the projectile encounters on its pass through the electron clouds surrounding each target atom.

**Fig. 2.9** Typical dependences of electronic $\varepsilon_e$ and nuclear $\varepsilon_n$ contributions to the stopping cross section $\varepsilon$ as a function of the incident particle energy $E$. The Bethe–Bloch equation [Eq. (2.46)] is a good approximation only at high energies beyond the maximum in the stopping cross section.

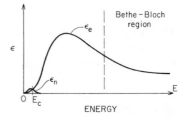

In very simplified terms, both interactions may be viewed as taking place between two isolated particles that interact electrostatically. Assume that the direction and speed of the incident particle are perturbed only slightly by the interaction. If the projectile has a mass $M_1$, a charge $Z_1 e$, and a velocity $v_1$, and if the target particle of mass $M_2$ and charge $Z_2 e$ is initially stationary, then the momentum transferred to the mass $M_2$ in a direction perpendicular to the path of the projectile is

$$P_\perp = 2(Z_1 Z_2 e^2 / b v_1) \tag{2.40}$$

for this simplified model. Here, the *impact parameter* $b$ is the distance of closest approach between the two particles if the mass $M_2$ were held fixed in place while the projectile flew past it along a straight trajectory. The energy transferred to the stationary particle thus is

$$E_\perp = P_\perp^2 / 2M_2 \tag{2.41}$$
$$= (2/M_2)(Z_1 Z_2 e^2 / b v_1)^2. \tag{2.42}$$

The energy lost by the projectile is very closely equal to $E_\perp$ when the perturbation is small, as presently assumed. It is thus evident that electrons with their light mass ($M_2 = m_e$) absorb much more energy per encounter than the nuclei do.

From this value of $E_\perp$ one can readily obtain the electronic energy loss $\Delta E$ incurred by the projectile over a length $\Delta x$ of the target. Statistically, the probability of an encounter with the impact parameter between $b$ and $b + db$ is $2\pi b\, db$ per unit area, since the electron may lie anywhere on a circle of radius $2\pi b$ around the particle track. The number of electrons per unit area over the length $\Delta x$ of the track is $N Z_2 \Delta x$. The average number $d\,\Delta n(E_\perp)$ of encounters that will generate a quantum $E_\perp$ of energy loss is therefore

$$d\,\Delta n(E_\perp) = N Z_2\, \Delta x \cdot 2\pi b\, db. \tag{2.43}$$

Together, these losses contribute the average differential amount $d\,\Delta E$ to the total energy loss $\Delta E$ across $\Delta x$; hence,

$$d\,\Delta E = N Z_2\, \Delta x\, [2(Z_1 e^2)^2 / m_e v_1^2] 2\pi\, (db/b). \tag{2.44}$$

If the impact parameter can range from $b_{min}$ to $b_{max}$, and $\Delta x$ tends to the limit $dx$, one finds after integration:

$$(dE/dx)|_e = N Z_2 [4\pi (Z_1 e^2)^2 / m_e v_1^2] \ln(b_{max}/b_{min}). \tag{2.45}$$

This result closely matches the Bethe–Bloch formula (2.46).

This simple picture of scattering in a cloud of free electrons neglects the fact that electrons are bound to atomic nuclei. Even in a metal, most electrons are bound to atoms. The ionization energy required to separate the electron from the atom has to be accounted for, and the scattering process becomes an inelastic one. The correct calculation of the average energy transferred to an electron is thus a problem for which we must consider every possible energetic state of an electron in the target and which depends additionally on the average population of each of these states. Also, the problem has to be treated quantum mechanically.

A number of approximations have been developed over the years to perform this averaging. They provide very useful analytical expressions for

$dE/dx|_e$. A well-known result is that the electronic stopping can be cast in the general form

$$(dE/dx)|_e = NZ_2[4\pi(Z_1e^2)^2/m_ev_1{}^2]L, \tag{2.46}$$

where $L$ is called the *stopping number*. According to quantum-mechanical calculations of Bethe (1930), its value is given by

$$L = \ln(2m_ev_1{}^2/I), \tag{2.47}$$

where the energy $I$ is an average over the various excitations and ionizations of the electrons in a target atom. Exact calculations of this *mean excitation potential* are difficult to perform, and $I$ is usually regarded as an empirical parameter. Bloch (1933) also made a quantum mechanical analysis and showed that $I$ is approximately proportional to $Z_2$; that is, $I = KZ_2$, where $K$ is an empirical parameter known as *Bloch's constant* and is of the order of approximately 10 eV. Equation (2.46) is commonly referred to as the *Bethe–Bloch formula* for the specific energy loss. The formula describes the experimental energy loss well only at energies beyond the maximum of the $dE/dx$ curve (see Fig. 2.9). Equations (2.46) and (2.47) state that for any elemental target the electronic component of $dE/dx$ has the generic form

$$dE/dx|_e = NZ_2(Z_1e^2)^2f(v_1{}^2) \tag{2.48}$$
$$= NZ_2(Z_1e^2)^2f(E/M_1), \tag{2.49}$$

where $f(E/M_1)$ is a function that depends only on the target element, not on the type of projectile, and also describes the energy dependence of $dE/dx|_e$. Equation (2.49) states that $dE/dx$ is proportional to the atomic density $N$ (as discussed in connection with Table 2.1). The equation also states that in any given element the electronic energy loss of $^4$He ($M_1 = 4$, $Z_1 = 2$) at an energy $E$ is four times larger than the energy loss of protons at an energy $E/4$. Neither statement is exactly correct, but both are very useful rules.

Electronic stopping depends on the electronic states in the target so that, in principle, the gaseous, liquid, and solid phases of the same element must have different stopping cross sections. The nature of the chemical binding in a target affects the electronic states and should thus also affect electronic stopping. Such effects, although they have been reported (Matteson *et al.*, 1976) are weak. They are ignored in the theoretical treatments previously discussed. These effects are expected to be significant mainly at low projectile energies and for light targets, where the number of core electrons are few. One theoretical model of $dE/dx|_e$ actually assumes that the valence electrons may be treated as a Fermi gas with a plasma frequency $\omega_p = (4\pi n_e e^2/m_e)^{1/2}$, where $n_e$ is the density of the electron gas (Lindhard *et al.*, 1964). The analysis, performed in terms of a complex dielectric constant, again leads to Eq. (2.46) for high energies, where $NZ_2$ stands for $n_e$ and $L$ now has the value $\ln(2m_ev_1{}^2/h\omega_p)$. Recent calculations based on wavefunctions of a Hartree–Fock–Slater model have proven fruitful in explaining the systematic variations of $dE/dx|_e$ with $Z_2$ for a fixed projectile (Rousseau et al., 1971; Chu and Powers, 1972), as shown for $^4$He in Fig. 2.8. The rather remarkable decrease in the stopping cross section shown in Fig. 2.8 from Ca to Cu, Nb to Ag, and past Xe is due to the fact that when d-shell electrons are added in the sequence of transition elements, the electron density near the atom increases enough to reduce the average electron density seen by an energetic particle traversing the material.

As long as the particle moves through matter so fast that the velocity $v_1$ is large compared with the speed $Z_1 v_0$ of its electrons in their innermost orbit, where $v_0 = e^2/\hbar = 2.2 \times 10^8$ cm/sec, the particle is effectively stripped of electrons and moves as an ion through the medium. At these velocities the simple model of charge $Z_1 e$ interacting elastically (or inelastically) with free (or bound) electrons in the target applies [Eq. (2.49)]. As the particle slows down, however, the probability that an electron is captured by the moving ion increases (Bohr, 1940, 1941; Northcliffe, 1960) and the effective charge of the projectile decreases. Also, the most tightly bound electrons of the target atoms play a gradually declining role in the stopping process. As a result, $dE/dx|_e$ increases less rapidly with falling energy $E$, and eventually turns around and actually decreases. The maximum of the stopping curve lies in the general vicinity of the "*Thomas–Fermi*" *velocity* $Z_1^{2/3} v_0$ and usually somewhat above it. This velocity is a convenient reference point when comparing the electronic energy loss of different projectiles.[†]

At these low energies, the Bethe–Bloch formula [Eq. (2.49)] breaks down. The reduction of the number of electrons contributing to the energy loss gives very large corrections. Also, the neutralization probability of the projectile becomes large. In this low energy range, the electronic energy loss becomes proportional to the velocity of the projectile. Lindhard et al., (1963, abbreviated as LSS in the literature), and Firsov (1959) gave theoretical descriptions for this energy range. The LSS expression is based on elastic scattering of free target electrons in the static field of a screened point charge which describes the projectile. Firsov's expression is based on a simple geometric model of momentum exchange between the projectile and the target atom during the interpenetration of the electron clouds surrounding the two colliding atoms. Both theories adequately describe the general behavior of the stopping power with regard to the energy dependence and the magnitude.

---

[†] The velocity $v_0 = e^2/\hbar = 2.2 \times 10^8$ cm/sec imparted to one nucleon corresponds to 25 keV of energy. The Thomas–Fermi velocity $Z_1^{2/3} e^2/\hbar$ thus corresponds to $Z_1^{4/3} \times 25$ keV per nucleon of the projectile. This amounts to 25 keV for $^1$H and 250 keV for $^4$He. Maxima of electronic stopping for $^4$He occur more typically at 0.6 to 1.0 MeV (see Fig. 2.8).

At very low velocities, an additional energy loss process occurs. Energy can be transferred from the nucleus of the projectile to that of a target atom by electrostatic interaction between the screened charges of the two nuclei. This *nuclear energy loss*, as it is usually called, may be viewed as an elastic interaction between two free particles, except for the very last collisions, where the chemical binding energy ($\sim 10$ eV) must be considered. As suggested by Bohr (1948) and later developed by Lindhard et al. (1963), the nuclear energy loss becomes another major component of energy loss at low energies, especially for heavy projectile atoms. To a good approximation, nuclear and electronic energy loss are roughly independent of each other, as is stated by Eq. (2.39).

With regard to megaelectron volt backscattering spectrometry, the situation is that, for $^1$H and $^4$He as projectiles, nuclear stopping is negligible everywhere except at the very lowest energies, that is, at the very end of the track of the projectile in the material.

In summary, it is fair to say that accurate numerical predictions of stopping cross sections from theory are difficult, at best, because of the large number of possible interactions that can conceivably take place. Atomic collisions are violent disturbances of atoms, and one would expect that effects due to chemical bonding and shell structure should normally be of minor importance. It has indeed turned out that approximate results come out rather easily, but accurate calculations are exceedingly difficult to obtain. The most trustworthy values of $\varepsilon$ are therefore semiempirical compilations that combine theoretically evaluated dependences with the most reliable experimental data, such as the recent table of Ziegler and Chu reproduced in Table VI of Appendix F.

A number of reviews and reports on the subject of energy loss of charged particles in matter have been written over the years. The reader is referred to these and their references for further information on the subject (Bohr, 1948; Fano, 1963; Lindhard et al., 1963; Lindhard, 1969; Northcliffe, 1963; Datz et al., 1967; Sauter and Bloom, 1972; Schiøtt, 1973) and to the bibliography of published tables given in Appendix D.

Special effects occur in $dE/dx$ when the beam is channeled in a single crystal target. The subject is treated in recent reviews (Gibbons, 1968; Mayer et al., 1970; Dearnaley et al., 1973; Gemmell, 1974).

## 2.5   LINEAR ADDITIVITY OF
### STOPPING CROSS SECTIONS (BRAGG'S RULE)

The preceding section on energy loss is restricted to elemental targets. The present section deals with energy loss in compound targets.

To a simple approximation, the process by which a particle loses energy when it moves swiftly through a medium consists of a random sequence of independent encounters between two particles: the moving projectile and an electron attached to an atom in the case of electronic energy loss, or the moving projectile and an atomic core in the case of nuclear energy loss. To the extent that this picture is correct, the situation presented by a target that contains more than one element differs only with respect to the type of atoms the projectile encounters. The energy lost to the electrons or to the atomic core in each encounter should be the same at a given projectile velocity, regardless of the further surrounding of the target atoms, since the interaction is considered to take place with only one atom at a time. This is, in essence, the idea contained in the *principle of additivity of stopping cross sections*, according to which the energy loss in a medium composed of various atomic species is the sum of the losses in the constituent elements, weighted proportionately to their abundance in the compound. The principle was postulated first by Bragg and Kleeman (1905) for the special case of molecules. Their postulate is now known as *Bragg's rule*. It states that the stopping cross section $\varepsilon^{A_m B_n}$ of a molecule $A_m B_n$ or a mixture with an equivalent composition of $A_m B_n$ is given by[†]

$$\varepsilon^{A_m B_n} = m\varepsilon^A + n\varepsilon^B, \qquad (2.50)$$

where $\varepsilon^A$ and $\varepsilon^B$ are the stopping cross sections of the atomic constitutents A and B. Let the volume density of the molecular units $A_m B_n$ in a compound be $N^{A_m B_n}$; then the specific energy loss of the material is

$$dE^{A_m B_n}/dx = N^{A_m B_n}\varepsilon^{A_m B_n}. \qquad (2.51)$$

This formula, completely analogous to Eq. (2.36) for an element, states that the energy $dE$ dissipated over the distance $dx$ is proportional to the number of molecular units $A_m B_n$ traversed over this distance, the proportionality constant being $\varepsilon^{A_m B_n}$. Often, to simplify notation, the clumsy form $A_m B_n$ as a superscript or subscript is abbreviated AB, e.g., $\varepsilon^{AB}$ for $\varepsilon^{A_m B_n}$, or $N^{AB}$ for for $N^{A_m B_n}$; the symbol AB then refers to a molecular unit of the compound composed of atoms of A and B.

For high-velocity protons ($v \gg v_0$), the rule is valid within about 1% (Fano, 1963; Burlin, 1968). For $^4$He in the 1–2-MeV range, good agreement has been reported in metallic alloys and compounds (Feng *et al.*, 1973;

---

[†] We prefer superscripts to denote the stopping medium. Subscripts then always indicate the identity of the partner in a collision (as in $K_{Si}$, $[\varepsilon]_{Si}^{SiO_2}$). Since this convention is not followed consistently in the literature, some care is indicated when formulas of different sources are compared.

Baglin and Ziegler, 1974). There are indications that violations can occur in gaseous organic compounds (Lodhi and Powers, 1974) and in oxides, nitrides, or other compounds in which one element is a gas in elemental form (Ziegler *et al.*, 1975). Generally, the departures are 10% or less.

## 2.6   ENERGY STRAGGLING

An energetic particle that moves through a medium loses energy via many individual encounters. Such a quantized process is subject to statistical fluctuations. As a result, identical energetic particles, which all have the same initial velocity, do not have exactly the same energy after passing through a thickness $\Delta x$ of a homogeneous medium. The energy loss $\Delta E$ is subject to fluctuations. The phenomenon, sketched in Fig. 2.10, is called *energy straggling*. Energy straggling places a finite limit for the precision with which energy losses, and hence depths can be resolved by backscattering spectrometry. The ability to identify masses is also impaired, except for atoms located at the surface of the target. The reason is that the beam energy $E$ before a collision with a specific mass $M_2$ at some depth within the target is no more monoenergetic, even if it was so initially, so that the ratio $E_1/E_0$, and hence the identification of $M_2$, become uncertain as well. For these reasons, it is important to have quantitative information on the magnitude of energy straggling for any given combination of energy, target material, target thickness, and projectile.

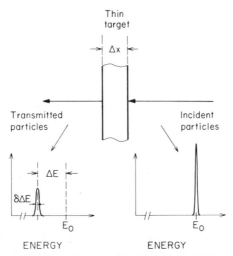

**Fig. 2.10**   A monoenergetic beam of energy $E_0$ loses energy $\Delta E$ in traversing a thin film of thickness $\Delta x$. Simultaneously, energy straggling broadens the energy profile.

Light particles such as $^1$H or $^4$He in the megaelectron volt range lose energy primarily by encounters with the electrons in the target, as discussed in Section 2.4. One would thus expect that the dominant contribution to energy straggling is the consequence of these electronic interactions too. This is indeed the case. One can therefore calculate the main contribution to energy straggling with the help of the same classical model employed in Section 2.4 to describe the process of electronic energy loss. It is shown there [Eq. (2.43)] that the average number $d\Delta n(E_\perp)$ of encounters that generate an energy loss $E_\perp$ over the distance $\Delta x$ is $NZ_2 \Delta x \cdot 2\pi b \, db$, where $b$ is the impact parameter for such an encounter. The actual number of encounters will fluctuate statistically about this average value $d\Delta n(E_\perp)$. If one assumes that the actual numbers of these encounters have a Poisson distribution, the standard deviation of $d\Delta n(E_\perp)$ is $[d\Delta n(E_\perp)]^{1/2}$. In turn, the deviation of these numbers from their average value causes deviations from the average differential value $d\Delta E$ that these encounters contribute to $\Delta E$. Let the deviations from the average contribution $d\Delta E$ be called $d\delta\Delta E$. Their standard deviation will be $E_\perp [d\Delta n(E_\perp)]^{1/2}$. The variance of encounters with an impact parameter between $b$ and $b + db$ is therefore

$$d\langle(\delta\Delta E)^2\rangle = E_\perp^2 NZ_2 \Delta x 2\pi b \, db. \qquad (2.52)$$

Encounters with other impact parameters produce similar fluctuations. As long as these fluctuations are independent, their corresponding variances add up incoherently, and the overall variance $\langle(\delta\Delta E)^2\rangle$ of $\delta\Delta E$ will be given by

$$\langle(\delta\Delta E)^2\rangle = NZ_2 \, \Delta x \, 2\pi \int_{b_{\max}}^{b_{\min}} E_\perp^2 b \, db. \qquad (2.53)$$

For an impact parameter $b$, the energy loss $E_\perp$ has the value $E_\perp = (2/m_e)(Z_1 e^2/bv_1)^2$ [see Eq. (2.42)]. The integral thus yields

$$\langle(\delta\Delta E)^2\rangle = NZ_2 \, \Delta x \, 2\pi \frac{(Z_1 e^2)^2}{m_e v_1^2} (E_{\max} - E_{\min}), \qquad (2.54)$$

where $E_{\max}$ and $E_{\min}$ are the energy losses corresponding to encounters with minimum and maximum impact parameters $b_{\min}$ and $b_{\max}$, respectively. The largest possible energy transfer in a collision between the ion of mass $M_1$ and an electron of mass $m_e \ll M_1$ is $2m_e v_1^2$, so that if $E_{\min} \ll E_{\max}$, then

$$\langle(\delta\Delta E)^2\rangle = NZ_2 \, 4\pi(Z_1 e^2)^2 \, \Delta x. \qquad (2.55)$$

This result was first derived by Bohr (1915) with the help of the same simple classical model discussed here. It is usually referred to as the *Bohr value* $\Omega_B^2$ of *energy straggling*.[†] For a layer of thickness $t$, Bohr straggling thus

---

[†] The common notation in the literature is $\Omega_B$. We use $\underline{\Omega}_B$ to distinguish between the standard deviation of an energy distribution $\underline{\Omega}$ and a solid angle of detection $\Omega$.

has a variance

$$\underset{\sim}{\Omega}_B{}^2 = 4\pi(Z_1 e^2)^2 N Z_2 t. \tag{2.56}$$

We introduce the abbreviation

$$s^2 = 4\pi(Z_1 e^2)^2 N Z_2 \tag{2.57}$$

with which the Bohr value of energy straggling has the simple form

$$\underset{\sim}{\Omega}_B{}^2 = s^2 t. \tag{2.58}$$

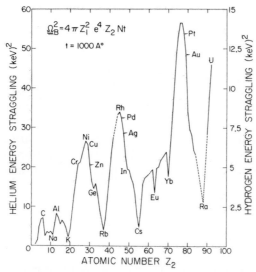

**Fig. 2.11**   The value of the variance $\Omega_B{}^2 = 4\pi Z_1{}^2 e^4 Z_2 N t$ for $t = 1000$ Å for energy straggling according to the classical model of Bohr for electronic energy loss versus the atomic number of the target atom. The pronounced structure reflects the difference in the atomic density of the elements.

Bohr's theory predicts that energy straggling does not depend on the energy of the projectile and that the rms value of the energy variation increases with the square root of the electron density per unit area $N Z_2 t$ in the target. A plot of $\Omega_B{}^2$ as a function of $Z_2$ is given in Fig. 2.11. The proportionality with the number of electrons per atom $Z_2$ accounts for the general increase of $\Omega_B{}^2$ with $Z_2$, but the pronounced structure in the plot is caused by the differences in the density $N$ of the elements. This variation is removed by considering $\Omega_B{}^2/Nt$. On finds that this quantity is numerically equal to $Z_2$ within 4% when expressed in units of $10^{-12}$ (eV cm)$^2$. This fact can be remembered for quick estimates of energy straggling.

**TABLE 2.2**

Experimentally Observed Values of the Standard Deviation $\Omega_{exp}$ of
Energy Straggling Compared to the Energy Loss $\Delta E$ of $^4$He Traversing Films of
Al, Ni, or Au at an Energy $\bar{E}$ of 1.0 and 2.0 MeV[a]

| | Thickness $\Delta x$ traversed | | $\Delta E$ (keV) | $\Omega_{exp}$ (keV) | $\Omega_{exp}/\Delta E$ (%) | $(\bar{E}/\Delta E)^{1/2}$ (%) | $\bar{E}$ (MeV) |
|----|----|----|----|----|----|----|----|
| | ($\mu$gm/cm$^2$) | (Å) | | | | | |
| Al | 120 | 4300 | 125 | 7.0 | 5.6 | 4 | 2.0 |
| Ni | 180 | 2000 | 125 | 5.6 | 4.5 | 4 | 2.0 |
| Au | 370 | 1900 | 125 | 5.1 | 4.1 | 4 | 2.0 |
| Al | 60 | 5900 | 200 | 7.0 | 3.5 | 2.2 | 1.0 |
| Ni | 260 | 2900 | 200 | 5.6 | 2.8 | 2.2 | 1.0 |
| Au | 520 | 2700 | 200 | 5.1 | 2.6 | 2.2 | 1.0 |

[a] The film thicknesses are chosen to produce the same $\Delta E$ in all three elements. Experimental values are derived from Harris and Nicolet (1975a).

Another useful relationship can be obtained by comparing the variations in $\Delta E$ given by the value of $\Omega_B$ with $\Delta E$ itself. For an estimate, one uses the Bethe–Bloch formula [Eq. (2.46)] for $dE/dx$ and substitutes some average value $\overline{v_1^2}$ for the velocity along the track, say, $\bar{E} = \frac{1}{2}M_1\overline{v_1^2}$, and compares this with the value of $\Omega_B$; the result is

$$\frac{\Omega_B}{\Delta E} = \left( \frac{\bar{E}}{\Delta E} \frac{2}{L} \frac{m_e}{M_1} \right)^{1/2}. \tag{2.59}$$

For $^4$He, the ratio $(m_e/M_1)^{1/2}$ is about $10^{-2}$. In this case, neglecting the factor $2/L$, one thus finds

$$\Omega_B/\Delta E \simeq (\bar{E}/\Delta E)^{1/2} \times 10^{-2}, \tag{2.60}$$

so that $\Omega_B$ itself is approximately 1% of the geometrical mean of $\Delta E$ and $\bar{E}$. Helium ions of 2 MeV undergoing an energy loss of 125 keV thus have a standard deviation of energy straggling that is about $(2.0/0.125)^{1/2}\% = 4\%$. Table 2.2 shows the experimentally observed values of $(\Omega_{exp}/\Delta E)^{1/2}$ for this and another example, with Al, Ni, and Au as targets. As can be seen, this ratio is indeed quite constant; it agrees in the order of magnitude predicted by the formula above, although the actual value differs from the estimate. The formula is thus a good rule of thumb, but does not yield quantitatively trustworthy numbers.

Bohr's model assumes that an individual energy transfer takes place between a free stationary electron and a fully ionized projectile of charge $Z_1e$. These assumptions are fulfilled only in the Bethe–Bloch region (see Fig. 2.9).

At energies in the vicinity of the maximum of the $dE/dx$ curve and below, the assumption of a fully ionized projectile is no longer valid. The fact that electrons are bound to atoms and are not free and stationary, as assumed, also becomes increasingly important as the projectile energy decreases. To account for this, Lindhard and Scharff (1953) extended Bohr's theory and derived a correction factor for low- and medium-energy projectiles. They obtained

$$\Omega^2 = \Omega_B^2 \tfrac{1}{2} L(\chi) \qquad \text{for} \quad \chi \leq 3,$$
$$\Omega^2 = \Omega_B^2 \qquad \text{for} \quad \chi \geq 3,$$

(2.61)

where $\chi$, a reduced energy variable, is

$$\chi = v^2/Z_2 v_0^2.$$

(2.62)

Here $v$ is the velocity of the projectile, $v_0 \equiv e^2/\hbar = 2.2 \times 10^8$ cm/sec, and $L(\chi)$ is the stopping number, which appears in the Bethe–Bloch formula, Eq. (2.46). Bonderup and Hvelplund (1971) have improved Lindhard and Scharff's expression by using a more refined description than had been used previously for the atomic charge distribution and for the process of energy straggling. They compare their calculations with experimental results of energy straggling for $^1$H and $^4$He in various gases (Bonderup and Hvelplund 1971; Hvelplund, 1971) and conclude that the Lindhard–Scharff formulation gives a fair account of the observed overall energy dependence of straggling. They also observe that when one plots $\Omega_{\text{exp}}^2/Nt$ against the projectile energy for various gases, the curve exhibits oscillations versus $Z_2$ similar to those observed in Fig. 2.8 for the stopping cross sections. These oscillations have been explained both for $dE/dx$ (Chu and Powers, 1972) and for energy straggling (Chu, 1976) by using atomic charge distributions of the Hartree–Foch–Slater type and incorporating them into the theory of Lindhard and Winther (Lindhard and Winther, 1964) for $dE/dx$, and the theory of Bonderup and Hvelplund (1971) for energy straggling. Where the measurements of energy straggling are sufficiently reliable, an acceptable agreement with these calculations is obtained.

In the energy range 1–2 MeV, which is of primary interest to backscattering spectrometry, almost all of the available experimental data on energy straggling pertain to $^1$H in gases. The advent of backscattering spectrometry as an analytical tool has generated renewed interest in experimental information on straggling in this energy range, particularly for $^4$He in solids. Presently, the only data available are for Al, Ni, Pt, and Au (Harris and Nicolet, 1975a,b). The results show only a weak energy dependence which is in qualitative agreement with the theories of Lindhard and Scharff, of Bonderup and Hvelplund, and of Chu. Numerically, Bohr's value $\Omega_B$ is within 40% of the data. Until more experimental data are available, the standard deviation $\Omega_B$ thus is the most appropriate value to use in estimating energy straggling in solids in the 1 to 2 MeV range.

Bohr's theory of energy straggling not only gives the standard deviation $\Omega_B$ of a beam which has traversed a medium, but also predicts that the distribution is Gaussian. This is a consequence of the assumption that the number of collisions is large and follows a Poisson distribution. The result is clearly approximate, as a Gaussian has a finite amplitude at any energy, but the transmitted beam surely cannot contain particles of energy larger than $E_0$. An accurate description of energy straggling must therefore necessarily lead to a distribution function that is not symmetrical with respect to the mean. This is born out by theoretical studies of energy straggling in beams passing through very thin absorbers (Landau, 1944; Vavilov, 1957; Tschalär, 1968; Kolata, 1968; and others; for a recent contribution, with references, see Bichsel and Saxon, 1975; Deconninck and Fouilhe, 1976), and by recent transmission measurements of protons through Si. In the energy range of 1 to 2 MeV for $^1$H and $^4$He, the effect is below the resolution of conventional solid-state detection systems. For the purposes of backscattering spectrometry, the Gaussian distribution thus describes energy straggling satisfactorily (also see Appendix B).

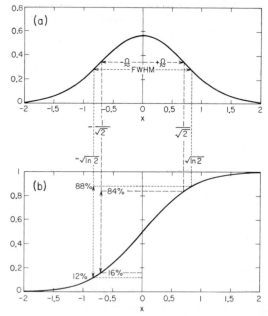

**Fig. 2.12**  Plot of (a) the Gaussian distribution $(2\pi\Omega^2)^{-1/2}\exp[-(x^2/2\Omega^2)]$ with $\Omega = 1/\sqrt{2}$, and (b) the corresponding error function integral $\operatorname{erf}(x) = (2\pi\Omega^2)^{-1/2}\int_{-\infty}^{x}\exp[-(x^2/2\Omega^2)]\,dx$ with $\Omega = 1/\sqrt{2}$.

Backscattering spectra most often display the integral of the Gaussian distribution, the error function

$$\text{erf}(x) = (2\pi\Omega^2)^{-1/2} \int_{-\infty}^{x} \exp[-(x^2/2\Omega^2)]\, dx \qquad (2.63)$$

rather than the Gaussian distribution

$$(2\pi\Omega^2)^{-1/2} \exp[-(x^2/2\Omega^2)]. \qquad (2.64)$$

The relation between the two is graphically shown in Fig. 2.12a and b for $\Omega^2 = \frac{1}{2}$. As can be seen, the full width at half maximum (FWHM) of a Gaussian corresponds to the 12 to 88% range of the error function and the $\pm \Omega$ points in the Gaussian correspond to the 16 to 84% points. The FWHM is wider than $\Omega$ by a factor of $2(2\ln 2)^{1/2} = 2.355$.

## 2.7  LINEAR ADDITIVITY OF ENERGY STRAGGLING

Experimental data on energy straggling below 2 MeV for $^1$H and $^4$He in solid elemental targets are few. For solid compound targets, no experimental data exist at all. The need for information is obvious. Until such results become available, statements on energy straggling in solid compounds must necessarily be conjectural.

The most obvious suggestion as to how energy straggling behaves in a compound or a mixture $A_m C_n$ proceeds as follows (Chu, 1976). Let $N_A$ and $N_C$ be the volume densities of the individual elements A and C, and let $N^{A_m C_n}$ be the volume density of compositional units $A_m C_n$ in the mixture or compound. Assume that for a thickness $t$, the energy straggling in elements A and C individually is [Eq. (2.56)]:

$$(\Omega_B^A)^2 = 4\pi(Z_1 e^2)^2 N_A Z_A t, \qquad (2.65)$$

$$(\Omega_B^C)^2 = 4\pi(Z_1 e^2)^2 N_C Z_C t. \qquad (2.66)$$

This means that $(\Omega_B^A)^2/N_A Z_A t = (\Omega_B^C)^2/N_C Z_C t = 4\pi(Z_1 e^2)^2$ is independent of the target, the ratio being simply the square of the energy variance per electron in a unit area of the target with thickness $t$. An extension of Bohr's model to a compound target then predicts that this quantity should apply independently of the composition of the target, or

$$\frac{(\Omega_B^{A_m C_n})^2}{\left(\begin{array}{l}\text{number of electrons per unit} \\ \text{area of the target of thickness } t\end{array}\right)} = 4\pi(Z_1 e^2)^2, \qquad (2.67)$$

and therefore

$$(\Omega_B^{A_m C_n})^2 = 4\pi(Z_1 e^2)^2 N^{A_m C_n}(m Z_A + n Z_C) t. \qquad (2.68)$$

The last three factors give the number of electrons per unit area in the target. This equation can also be written as

$$\frac{(\Omega_B^{A_m C_n})^2}{N^{A_m C_n} t} = m \frac{(\Omega_B^A)^2}{N_A t} + n \frac{(\Omega_B^C)^2}{N_C t}, \tag{2.69}$$

which clearly bears out the assumption of additivity. Until measurements are made, these equations must be considered as hypothetical and should be used as guidelines only. Their validity has yet to be tested.

## REFERENCES

Abrahamson, A. A. (1969). *Phys. Rev.* **178**, 76.
Baglin, J. E. E., and Ziegler, J. F. (1974). *J. Appl. Phys.* **45**, 1413.
Bethe, H. A. (1930). *Ann. Phys.* **5**, 325.
Bichsel, H., and Saxon, R. P. (1975). *Phys. Rev. A* **11**, 1286.
Bloch, F. (1933). *Ann. Phys.* **16**, 285; *Z. Phys.* **81**, 363.
Bohr, N. (1913). *Phil. Mag.* **25**, 10.
Bohr, N. (1915). *Phil. Mag.* **30**, 581.
Bohr, N. (1940). *Phys. Rev.* **58**, 654.
Bohr, N. (1941). *Phys. Rev.* **59**, 270.
Bohr, N. (1948). *Mat. Fys. Medd. Dan. Vid. Selsk.* **18**, No. 8.
Bonderup, E., and Hvelplund, P. (1971). *Phys. Rev. A* **4**, 562.
Bourland, P. D., and Powers, D. (1971). *Phys. Rev. B* **3**, 3635.
Bourland, P. D., Chu, W. K., and Powers, D. (1971). *Phys. Rev. B* **3**, 3625.
Bragg, W. H., and Kleeman, R. (1905). *Phil. Mag.* **10**, S318.
Burlin, T. E. (1968). *in* "Radiation Dosimetry" (F. H. Attix and W. C. Roesch, eds.), Vol. I, p. 331. Academic Press, New York.
Chu, W. K. (1976). *Phys. Rev. A* **13**, 2057.
Chu, W. K., and Powers, D. (1972). *Phys. Lett.* **38A**, 267.
Darwin, C. G. (1914). *Phil. Mag.* **28**, 499.
Datz, S., Erginsoy, C., Leibfried, G., and Lutz, H. O. (1967). *Ann. Rev. Nucl. Sci.* **17**, 129.
Dearnaley, G., Freeman, J. H., Nelson, R. S., and Stephen, J. (1973). "Ion Implantation." North-Holland Publ., Amsterdam.
Deconninck, G., and Fouilhe, Y. (1976). *In* "Ion Beam Surface Layer Analysis" (O. Meyer, G. Linker, and F. Käppeler, eds.), p. 87. Plenum Press, New York.
Everhart, E., Stone, G., and Carbone, R. J. (1955). *Phys. Rev.* **99**, 1287.
Fano, U. (1963). *Ann. Rev. Nucl. Sci.* **13**, 1.
Feng, J. S.-Y., Chu, W. K., and Nicolet, M-A. (1973). *Thin Solid Films* **19**, 227.
Feng, J. S.-Y., Chu, W. K., and Nicolet, M-A. (1974). *Phys. Rev. B* **10**, 3781.
Firsov, O. B. (1959). *Zh. Eksp. Teor. Fiz.* **36**, 1517.
Gemmell, D. S. (1974). *Rev. Mod. Phys.* **46**, 129.
Gibbons, J. F. (1968). *Proc. IEEE* **56**, 295.
Goldstein, H. (1959). "Classical Mechanics." Addison-Wesley, Reading, Massachusetts.
Harris, J. M., and Nicolet, M-A. (1975a). *Phys. Rev. B* **11**, 1013.
Harris, J. M., and Nicolet, M-A. (1975b). *J. Vac. Sci. Technol.* **12**. 439.
Hvelplund, P. (1971). *Mat. Fys. Medd. Dan. Vid. Selsk.* **38**, 4.
Kolata, J. J. (1968). *Phys. Rev.* **176**, 484.
Landau, L. (1944). *J. Phys.* **8**, 201.

Leighton, R. B. (1959). "Principles of Modern Physics." McGraw-Hill, New York.

Lindhard, J. (1969). *Proc. Roy. Soc. London* **A311**, 11.

Lindhard, J., and Scharff, M. (1953). *Mat. Fys. Medd. Dan. Vid. Selsk.* **27**, No. 15.

Lindhard, J., Scharff, M., and Schiøtt, H. E. (1963). *Mat. Fys. Medd. Dan. Vid. Selsk.* **33**, No. 14.

Lindhard, J., and Winther, A. (1964). *Mat. Fys. Medd. Dan. Vid. Selsk.* **34**, No. 4.

Lodhi, A. S., and Powers, D. (1974). *Phys. Rev. A* **10**, 2131.

Marion, J. B., and Young, F. C. (1968). "Nuclear Reaction Analysis, Graphs and Tables." Wiley, New York.

Matteson, S., Chau, E. K. L., and Powers, D. (1977). *Phys. Rev.* **A15**, 856.

Mayer, J. W., Eriksson, L., and Davies, J. A. (1970). "Ion Implantation in Semiconductors." Academic Press, New York.

Northcliffe, L. C. (1960). *Phys. Rev.* **120**, 1744.

Northcliffe, L. C. (1963). *Ann. Rev. Nucl. Sci.* **13**, 67.

Powers, D., Chu, W. K., Robinson, R. J., and Lodhi, A. S. (1972). *Phys. Rev. A* **6**, 1425.

Powers, D., Lodhi, A. S., Lin, W. K., and Cox, H. L. (1973). *Thin Solid Films* **19**, 205.

Robinson, M. T., (1974). Oak Ridge Rep. ONRL-4556.

Rousseau, C. C., Chu, W. K., and Powers, D. (1971). *Phys. Rev. A* **4**, 1066.

Rutherford, E., (1911). *Phil. Mag.* **21**, 669.

Sauter, G. D., and Bloom, S. C. (1972). *Phys. Rev. B* **6**, 669.

Schiøtt, H. E. (1973). Interaction of Energetic Charged Particles with Solids, p. 6. Brookhaven Nat. Lab. Rep. No. BNL-50336.

Thompson, D. A., and Mackintosh, W. D. (1971). *J. Appl. Phys.* **42**, 3969.

Tschalär, C. (1968). *Nucl. Instrum. Methods* **61**, 141.

Van Wijngaarden, A., Brimmer, E. J., and Baylis, W. E. (1970). *Can. J. Phys.* **48**, 1835.

Vavilov, P. V. (1957). *JETP* **5**, 749.

Warters, W. D. (1953). Ph. D. Thesis, Caltech (unpublished).

Ziegler, J. F., and Chu, W. K. (1973). IBM Res. Rep. RC 4288.

Ziegler, J. F., and Chu, W. K. (1974). *At. Data Nuc. Data Tables* **13**, 463.

Ziegler, J. F., Chu, W. K., and Feng, J. S.-Y. (1975). *Appl. Phys. Lett.* **27**, 387.

Chapter

# 3

# Concepts of
# Backscattering Spectrometry

The purpose of this and the following chapter is to describe in principle how a backscattering spectrum is generated and how it is interpreted in terms of the basic concepts introduced in Chapter 2. The concern here is with general notions. In Chapter 4 these concepts are applied to thin films and layered structures. Detailed examples are presented in Chapter 5.

## 3.1  INTRODUCTION

The components of a backscattering system are shown in Fig. 3.1. The source generates a beam of collimated and monoenergetic particles of energy $E_0$. A typical case is a current of 10 to 100 nA of 2.0-MeV He$^+$ ions in a

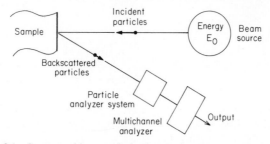

**Fig. 3.1**  Conceptual layout of a backscattering spectrometry system.

54

1-mm$^2$ area. These particles impinge on the sample (or target) which is the object to be analyzed. Almost all of the incident particles come to rest within the sample. A very few (much less than one in $10^4$) are scattered back out of the sample. Of these, a small fraction is incident on the area defined by the aperture of an analyzing system. The output of that system is an analog signal. This signal is processed by a multichannel analyzer, which subdivides its magnitude into a series of equal increments. Each increment is numbered and referred to as a *channel*. Modern multichannel analyzers contain thousands of channels. An event whose magnitude falls within a particular channel is registered there as a *count*. At the termination of the experiment, each channel has registered a certain number of counts. The output of the multichannel analyzer is thus a series of counts contained in the various channels.

A segment of such a series from channels 132–136 is shown in Fig. 3.2. We shall refer to the counts contained in channel $i$ as $H_i$. This digital information can be recorded in various ways. The graphical display is advantageous for quick interpretation. Digital outputs are used for numerical

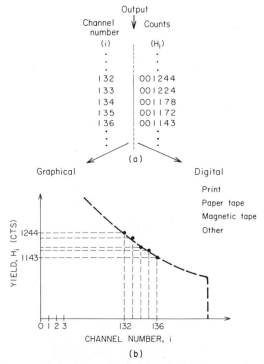

**Fig. 3.2** Basic content of a backscattering spectrum and some methods of recording. (a) The ordinal number (left) identifies each channel, which contains a certain number of counts. (b) Various ways of recording a spectrum.

analysis. Computer facilities with graphical display terminals can combine both. Such a series of counts versus channel number constitutes a *backscattering spectrum*. In the graphical display, the ordinate is frequently labeled *yield* or *backscattering yield*.

The analog signal generated by the analyzer contains quantitative information on one particular parameter of the detected particle. As shown in Fig. 3.3, there are a number of parameters—energy, momentum, etc.—that can be used to characterize the backscattered particles. For example, magnetic spectrometers measure momentum. The backscattering spectrum obtained with such an analyzer is a *backscattering momentum spectrum*. A semiconductor particle detector produces an analog signal proportional to the energy of the backscattered particle. Correspondingly, a spectrum obtained with such a detector is a *backscattering energy spectrum*.

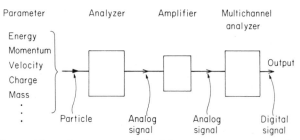

**Fig. 3.3** The particle analyzer system of Fig. 3.1 may measure any one of several distinct parameters that characterize a backscattered particle. This analyzer generates an analog signal. The multichannel analyzer measures that signal and registers the value as a count in the appropriate channel.

The particular analyzing system assumed in the rest of this book consists of an energy-sensitive analyzer followed by amplifiers and a multichannel analyzer, as shown in Fig. 3.3. This analyzing system is the most commonly used in backscattering spectrometry, but there are other methods to obtain a spectrum. For instance, the multichannel analyzer can be replaced by a single channel whose position is changed sequentially so as to scan the range of the parameter measured. Regardless of their inner working, the common feature of all such systems is an output consisting of a set of counts corresponding to a sequence of channels.

In general, whatever the analyzer, there should exist a one-to-one correspondence between the channel number and the magnitude of the particle parameter to be measured by the analyzer. The most desirable property of this relationship is that it be exactly linear and stable in time. Additionally, for convenience, one likes fast acquisition of data and detectors of small physical size. The semiconductor surface-barrier detector combined with a charge-sensitive preamplifier meets these criteria best among current options. It is therefore used almost universally in backscattering spectrometry. Consequently, in this book we shall be concerned almost exclusively with back-scattering energy spectra. The descriptive term, "backscattering energy spectrum," will thus often be shortened to *backscattering spectrum* or *spectrum*. In those rare cases where energy is not displayed, one should explicitly identify the parameter measured.

The relation between the energy of a backscattered and detected particle and the channel number in which that particle is counted is a characteristic of the system and must be determined experimentally (as described in Section 5.2). Figure 3.4 shows this relation schematically. The abscissa gives the channel number $i$. The ordinate gives the energy $E_1$ of a detected particle, where $E_{1,i}$ is the energy of particles that produce counts in channel $i$. We shall assume a linear relationship, as indicated in the figure. The slope of the line will be denoted by $\mathscr{E}$, *the energy interval corresponding to one channel*. The offset of the line is always adjustable by changing the gain settings of the electronics in the analyzer system. This allows one to display a selected part of the energy spectrum over the full range of the multichannel analyzer. (Typical numbers for $\mathscr{E}$ are about 4 keV with megaelectron volts He ions, and the offset is some hundreds of kiloelectronvolts.) As defined previously, $\mathscr{E}$ is the slope of a straight line, and hence constant. In the rest of the book we shall assume that this holds in a given experimental situation. When $\mathscr{E}$ is a function of energy, i.e., when the relation between channel number and particle energy is not a linear one as shown in Fig. 3.4, equations describing an energy spectrum must be modified.

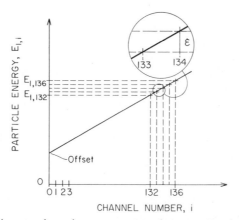

**Fig. 3.4**   Ideally, in an analyzer that senses energy, the energy $E_1$ of a detected particle is related exactly linearly to the channel number that identifies the channel in which the event is registered as one count. The slope of the line is characterized by the energy interval $\mathscr{E}$ of one channel.

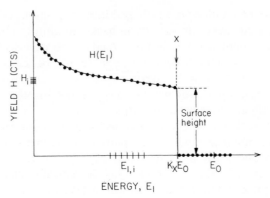

**Fig. 3.5**   The conversion of channel number to energy $E_1$ (shown in Fig. 3.4) transforms the abscissa of a backscattering spectrum (as shown in Fig. 3.2) from channel number to energy $E_1$, where $E_0$ gives the energy of the incident particles and $K_X E_0$ is called the *edge* of element X in the spectrum.

   With the relation of channel number to energy established, one can convert the abscissa of a backscattering spectrum from channel numbers to particle energy $E_1$, as shown in Fig. 3.5. This plot is a typical form of a backscattering energy spectrum. Sometimes spectra are plotted in terms of channel numbers only. In such a case, one should specify $\mathscr{E}$ and the energy offset; otherwise the information provided is not complete.

   One frequently interprets such a spectrum in terms of a continuous function $H$ of the continuous variable $E_1$.[†] The expression $H(E_1)$ then stands for the counts $H_i$ in channel $i$ which corresponds to the energy $E_{1,i}$. The terms $H_i$ and $H$ are both referred to as the *height of the spectrum*. The terms *yield* and *backscattering yield* are sometimes used with the same meaning.

   In Section 2.2 it is shown that the energy of particles scattered from an atom at rest cannot have energies above $KE_0$, where $E_0$ is the energy of the incident particle. For particles backscattered from a monoisotopic elemental sample, the spectrum has a step at an energy $E_1 = KE_0$ corresponding to scattering from surface atoms; this step is referred to as the *edge* of the element and is frequently indicated with an arrow or a line, as in Fig. 3.5. In the vicinity of $KE_0$, the height of the spectrum is frequently called the *surface height*.

   If there is more than one element in the sample, the spectrum contains counts generated by particles scattered from the different elements. The counts generated from a given element are called the *signal* of this element in the spectrum.

---

   [†] Certain analyzing methods actually generate a permanent continuous record whose ordinate gives an analog signal of $H$ (e.g., photographic records). A digital output can then be formed by subsequent digitalization and multichannel analysis.

The purpose of backscattering spectrometry is to extract quantitative information on the elemental composition of the sample. Since the edges are well defined, one can usually readily identify some of the elements present in the outermost layers of the sample. Since the primary particles penetrate into the sample virtually unattenuated, scattering occurs from atoms located below the surface as well. The energy immediately before the scattering is less than $E_0$ because energy is lost along the incident path. After scattering, the particles escaping the sample lose energy along the outward path. Consequently, the energy of the detected particles depends on the depth at which scattering occurred. The backscattering yield at that energy depends on the number of atoms present at that depth. The problem in backscattering analysis, therefore, consists of properly interpreting the measured back-scattering spectrum in terms of distributions of atoms in depth below the surface. This, then, is the topic to which we shall address ourselves in the rest of this chapter.

We assume, of course that the sample is laterally uniform. When that assumption cannot be made, the analysis of the spectrum becomes vastly more difficult.

## 3.2 DEPTH SCALE FOR AN ELEMENTAL SAMPLE

This section describes how one relates the energy $E_1$ of the detected particle to the depth $x$ at which the backscattering event occurs in a mono-isotopic elemental sample. In Fig. 3.6 the energy of the incident particles is $E_0$, the *energy immediately before scattering at a depth* $x$ *is* $E$, and the energy of the particle emerging from the surface is $E_1$. The incident beam is smaller than the target. The incident particle, the exiting particle and the normal of the sample are all contained in one plane, so that the scattering angle in the laboratory frame of reference is given by $\theta = 180° - \theta_1 - \theta_2$, where $\theta_1$ and $\theta_2$ are the angles between the sample normal and the direction of the incident beam and of the scattered particle, respectively. Note that

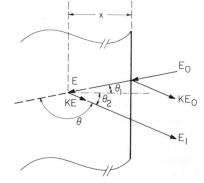

**Fig. 3.6** Symbols used in the description of backscattering events in a sample (or target) consisting of a monoisotopic element. The angles $\theta_1$ and $\theta_2$ are positive regardless of the side on which they lie with respect to the normal of the sample. The incident beam, the direction of detection, and the sample normal are coplanar.

both $\theta_1$ and $\theta_2$ are defined as positive numbers whether they are located on one or the other of the sample normal. (Other geometrical arrangements are described in Section 7.5.) According to Section 2.4, we can relate the energy $E$ to the length $x/\cos\theta_1$ of the incident path by

$$x/\cos\theta_1 = -\int_{E_o}^{E} dE/(dE/dx). \tag{3.1}$$

where the negative sign arises because $E$ is smaller than $E_0$ and $dE/dx$ is taken as a positive quantity. Similarly, the path length $x/\cos\theta_2$ of the outward path is related to $KE$ and $E_0$ by

$$x/\cos\theta_2 = -\int_{KE}^{E_1} dE/(dE/dx). \tag{3.2}$$

A graphical interpretation of these two equations is given in Fig. 3.7. Part (a) shows $dE/dx$ as a function of energy as a light line. The heavy segments give the $dE/dx$ values for the inward path from $E_0$ to $E$ and for the outward

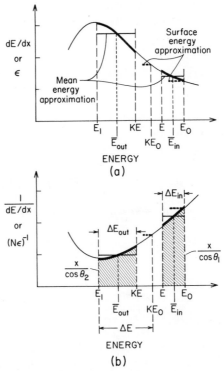

**Fig. 3.7**  Graphical representation of the energy loss of particles along their inward and outward paths (heavy line) through a sample consisting of a monoisotopic element. The light line is the functional form of $dE/dx$ versus $E$ in (a) and of $(dE/dx)^{-1}$ versus $E$ in (b). Since $dE/dx = N\varepsilon$, the plot in (a) applies to $\varepsilon$ versus $E$ as well.

path from $KE$ to $E_1$. The difference $E_0 - E$ is the *energy loss along the inward path* $\Delta E_{in}$; similarly, $KE - E_1$ is the *energy loss along the outward path* $\Delta E_{out}$. According to Eqs. (3.1) and (3.2), it is the reciprocal of $dE/dx$ that must be integrated over these two segments. This reciprocal curve is shown in part (b) of Fig. 3.7, with the heavy segments again indicating values for the inward and outward paths. By Eqs. (3.1) and (3.2), the two shaded areas give the path lengths $x/\cos\theta_1$ for the inward path and $x/\cos\theta_2$ for the outward path. If $\theta_1 = \theta_2$, these two areas are exactly equal.

To relate the energy $E_1$ of the detected particle to the depth $x$ at which the backscattering event occurs, it is necessary to find the value of the shaded areas. The problem is that the energy $E$ before scattering is not an experimentally accessible quantity, but $E_0$ and $E_1$ are. One thus desires to find $x$ in terms of $E_0$ and $E_1$. There are three ways of doing this:

1.  Use tabulated values of $dE/dx$ and execute the integrations numerically to find corresponding sets of $E$ and $x$, and subsequently $KE$ and $E_1$. This approach, generally carried out with computers, is described in Section 3.4.

2.  Assume that $dE/dx$ is constant over each path. Equations (3.1) and (3.2) can then be integrated and $E$ can be eliminated. This is discussed in the following section.

3.  Assume some functional dependence for $dE/dx$. Matching pairs of $E$ and $x$ and of $x$ and $E_1$ can then be obtained analytically.

### 3.2.1  Energy Loss Factor [$S$]
### and Stopping Cross Section Factor [$\varepsilon$]

If one assumes a constant value for $dE/dx$ along the inward and outward paths, the two integrals in Eqs. (3.1) and (3.2) reduce to

$$E = E_0 - \frac{x}{\cos\theta_1}\left.\frac{dE}{dx}\right|_{in} \tag{3.3}$$

and

$$E_1 = KE - \frac{x}{\cos\theta_2}\left.\frac{dE}{dx}\right|_{out}, \tag{3.4}$$

where the subscripts "in" and "out" refer to the (constant) values of $dE/dx$ along the inward and outward paths (Fig. 3.7). By eliminating $E$ from these two equations, we have

$$KE_0 - E_1 = \left[\frac{K}{\cos\theta_1}\left.\frac{dE}{dx}\right|_{in} + \frac{1}{\cos\theta_2}\left.\frac{dE}{dx}\right|_{out}\right]x. \tag{3.5}$$

The energy $KE_0$ is the edge of the backscattering spectrum (Fig. 3.5) and corresponds to the energy of particles scattered from atoms at the surface of the target. The energy $E_1$ is the measured value of a particle scattered from an atom at depth $x$. If one introduces the symbol $\Delta E$ for the *energy difference between $E_1$ and $KE_0$* (Fig. 3.7), i.e.,

$$\Delta E = KE_0 - E_1, \tag{3.6}$$

then one can write

$$\Delta E = [S]x, \tag{3.7}$$

where

$$[S] \equiv \left[ \frac{K}{\cos\theta_1} \frac{dE}{dx}\bigg|_{in} + \frac{1}{\cos\theta_2} \frac{dE}{dx}\bigg|_{out} \right] \tag{3.8}$$

is called the *energy loss factor* or *S factor*. An equivalent set of equations can be given in terms of stopping cross sections rather than $dE/dx$:

$$\Delta E = [\varepsilon]Nx, \tag{3.9}$$

where

$$[\varepsilon] = \left[ \frac{K}{\cos\theta_1} \varepsilon_{in} + \frac{1}{\cos\theta_2} \varepsilon_{out} \right] \tag{3.10}$$

is called the *stopping cross section factor* or *ε factor*.

The assumption of constant values for $dE/dx$ or $\varepsilon$ along each track thus leads to a linear relationship between the energy $\Delta E$ below the edge $KE_0$ and the depth at which scattering occurs. One can therefore assign a linear depth scale to the energy axis, as indicated in Fig. 3.8.

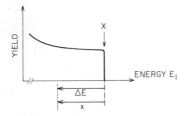

Fig. 3.8  When one assumes the energy loss to be constant along the inward and outward paths, then the energy $\Delta E$ can be linearly related to the depth $x$ through $\Delta E = [S]x$ as indicated in the abscissa of the backscattering spectrum.

This result is derived under the assumption that $dE/dx$ or $\varepsilon$ is constant along the inward and outward path. Since this is an approximation, the resulting depth scale also applies only approximately. However, it is also clear from inspection of Fig. 3.7 that for any given $E_0$ and $E_1$ a pair of unique values of $(dE/dx)_{in}$ and $(dE/dx)_{out}$ exist for which this linear scale gives one exact value of depth at which scattering occurs. These two particular values of $dE/dx$ are those for which the product of $(dE/dx)^{-1}$ and the energy intervals $\Delta E_{in}$ and $\Delta E_{out}$ exactly coincide with the values of the corresponding integrals (shaded areas in Fig. 3.7). In Section 3.3 an iterative procedure is described by which these particular values of $dE/dx$ can be sought. By applying this procedure point by point, an accurate relation between the backscattering depth and $E_1$ can be constructed. In the next subsection, useful approximation methods of finding values of $(dE/dx)_{in}$ and $(dE/dx)_{out}$ or $\varepsilon_{in}$ and $\varepsilon_{out}$ are discussed. We also use $\varepsilon(E_{in})$ and $\varepsilon(E_{out})$ to indicate the energy at which $\varepsilon$ is evaluated.

Figure 3.9 describes graphically the connection between the energy loss factor $[S]$ and the actual depth $x$ at which backscattering occurs for a given energy loss $\Delta E$. The exact relationship between $\Delta E$ and $x$ derived from Eqs. (3.1) and (3.2) is generally not linear. The energy loss factor provides a linear approximation [Eq. (3.7)] which is exact at one point.

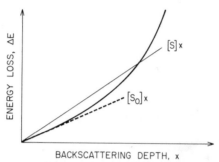

**Fig. 3.9** The solid curve shows the general relation between the energy loss $\Delta E$ and the depth $x$ at which backscattering occurs. The linear relation $\Delta E = [S]x$ is exact at one depth. The symbol $[S_0]$ refers to the surface energy approximation discussed in Section 3.2.2, with the dashed line representing $[S_0]x$. The incident energy is $E_0$.

### 3.2.2 Approximations to $[S]$ and $[\varepsilon]$

**a. Surface Energy Approximation.** For regions near the surface, the thickness $x$ is small and the relative change of energy along the incident path is small also. Therefore $(dE/dx)_{in}$ is evaluated at $E_0$. Similarly, $(dE/dx)_{out}$ is taken at $KE_0$ (see heavy dotted lines in Fig. 3.7). In this *surface energy approximation*, one thus sets

$$[S_0] = \left[ \frac{K}{\cos\theta_1} \frac{dE}{dx}\bigg|_{E_0} + \frac{1}{\cos\theta_2} \frac{dE}{dx}\bigg|_{KE_0} \right] \qquad (3.11)$$

or

$$[\varepsilon_0] = \left[ \frac{K}{\cos\theta_1} \varepsilon(E_0) + \frac{1}{\cos\theta_2} \varepsilon(KE_0) \right] \qquad (3.12)$$

where the stopping cross sections $\varepsilon(E_0)$ and $\varepsilon(KE_0)$ are evaluated at energies $E_0$ and $KE_0$, respectively. This particular approximation is used so frequently that the symbols $[S_0]$ and $[\varepsilon_0]$ are introduced to refer to it. The connection between $[S_0]$ and the exact $\Delta E$ versus $x$ dependence is shown in Fig. 3.9.

**b.   Mean Energy Approximation.**   When the path length becomes appreciable, the surface approximation degrades (Fig. 3.9). As can be seen from Fig. 3.7, a better approximation can be obtained by selecting a constant value of $dE/dx$ or $\varepsilon$ at an *energy $\bar{E}$ intermediate to that which the particle has at the end points of each track.* We define

$$[\bar{S}] \equiv \left[ \frac{K}{\cos \theta_1} \frac{dE}{dx}\bigg|_{\bar{E}_{in}} + \frac{1}{\cos \theta_2} \frac{dE}{dx}\bigg|_{\bar{E}_{out}} \right] \tag{3.13}$$

or

$$[\bar{\varepsilon}] \equiv \left[ \frac{K}{\cos \theta_1} \varepsilon(\bar{E}_{in}) + \frac{1}{\cos \theta_2} \varepsilon(\bar{E}_{out}) \right]. \tag{3.14}$$

In the *mean energy approximation*, one assumes that

$$\bar{E}_{in} = \tfrac{1}{2}(E + E_0) \tag{3.15}$$

and

$$\bar{E}_{out} = \tfrac{1}{2}(E_1 + KE). \tag{3.16}$$

The value of $E$ in the preceding equations is unknown, but can be estimated in various ways. General methods are described in Section 3.3.

For quick estimates one can assume that the energy difference $\Delta E = KE_0 - E_1$ is known and that this loss is subdivided symmetrically between the incident path and the outward path, so that $E$ is approximately $E_0 - \tfrac{1}{2}\Delta E$. The values $\bar{E}_{in}$ and $\bar{E}_{out}$ are then given by

$$\bar{E}_{in} \simeq E_0 - \tfrac{1}{4}\Delta E \tag{3.17}$$

and

$$\bar{E}_{out} \simeq E_1 + \tfrac{1}{4}\Delta E. \tag{3.18}$$

When these values are used to complete the definitions of $[\bar{S}]$ or $[\bar{\varepsilon}]$, the method is called the *symmetrical mean energy approximation*. This approximation, which is particularly good when $K \simeq 1$ and $\theta_1 \simeq \theta_2$, has the advantage of simplicity. It serves well as a quick estimate of the probable error of the surface approximation.

### 3.3   ENERGY $E$ BEFORE SCATTERING

In the previous section the energy $E$ immediately before scattering at the depth $x$ is needed for the mean energy approximation. This energy $E$ is needed not only for depth calculations, but also to evaluate the scattering cross section $\sigma(E)$ in depth profile applications. In that latter case, fairly accurate estimates are required because scattering cross sections vary inversely with the square of the energy $E$. Cruder approximations to $E$ suffice for the evaluation of the depth at which scattering occurs, since $dE/dx$ is not a strong function of energy. In this section we enumerate methods for finding $E$ that have been used in the analysis of backscattering spectra.

### 3.3.1 Energy Loss Ratio Method

A simple but very useful procedure to obtain $E$ as a function of $E_1$ and $E_0$ has been described by Lever (1976). One assumes that the ratio $\alpha$ of the energy lost along the outward track $\Delta E_{out}$ to that lost along the inward track $\Delta E_{in}$ is independent of depth, i.e.,

$$\alpha = \Delta E_{out}/\Delta E_{in} = \text{const} \tag{3.19}$$

(see Fig. 3.7). The energy losses $\Delta E_{out}$ and $\Delta E_{in}$ are $\Delta E_{out} = KE - E_1$ and $\Delta E_{in} = E_0 - E$. The ratio $\alpha$ then is $\alpha = (KE - E_1)/(E_0 - E)$, which gives

$$E = (E_1 + \alpha E_0)/(K + \alpha). \tag{3.20}$$

An approximate value for $\alpha$ can be determined from the surface energy approximation, which assumes that Eqs. (3.1) and (3.2) can be written as $(E_0 - E)/\varepsilon(E_0)N = x/\cos\theta_1$ and $(KE_0 - E_1)/\varepsilon(KE_0)N = x/\cos\theta_2$, respectively, so that

$$\alpha \simeq [\varepsilon(KE_0)/\varepsilon(E_0)]\beta, \tag{3.21}$$

where

$$\beta \equiv \cos\theta_1/\cos\theta_2. \tag{3.22}$$

This value of $\alpha$ can readily be computed from tabulated stopping cross sections and substituted into Eq. (3.20) to find $E$. This method is most accurate for the analysis of thin-film spectra where the surface approximation holds. It is also useful for thicker films where the surface approximation is poor, because the ratio $\alpha$ of the energy losses changes less rapidly than $\varepsilon$.

### 3.3.2 Iterative Method

This method starts with the surface energy approximation in which $\bar{E}_{in} \simeq E_0$ and $\bar{E}_{out} \simeq KE_0$ and sets $[\bar{S}] = [S_0]$ or $[\bar{\varepsilon}] = [\varepsilon_0]$ to obtain a zeroth-order depth $x$ at which scattering occurs by using Eq. (3.7) and a given value of $\Delta E = KE_0 - E_1$. Then, one calculates a zeroth-order $E$ using $dE/dx$ or $\varepsilon$ evaluated at $E_0$ [Eq. (3.3)]. With this value of $E$, a new and improved estimate of $\bar{E}_{in}$ and $\bar{E}_{out}$ is obtained with Eqs. (3.15) and (3.16). These improved values of $\bar{E}_{in}$ and $\bar{E}_{out}$ define a first-order $[\bar{S}]$ or $[\bar{\varepsilon}]$ [Eqs. (3.13) and (3.14)]. The process can now be iterated to find still better estimates of $x$, $E$, and $[\bar{S}]$ or $[\bar{\varepsilon}]$. The method converges rapidly and an accurate depth scale can be established.

### 3.3.3   Analytical Methods

To obtain analytical formulas for $E$, the functional dependence of $dE/dx$ or $(dE/dx)^{-1}$ must be known analytically. Two methods have been described in the literature.

**a.   Taylor Expansion of $\varepsilon$.**   Since $E_0$ and $E_1$ are the experimentally accessible energies, it is natural to expand $\varepsilon$ around those two points. These expansions can be used to find the values of $\varepsilon$ at the mean energies $\bar{E}_{in}$ and $\bar{E}_{out}$. It is assumed that these energies are given by the mean energy approximation, i.e., $\bar{E}_{in} = \frac{1}{2}(E + E_0)$ and $\bar{E}_{out} = \frac{1}{2}(E_1 + KE)$. By eliminating $x$ from Eqs. (3.3) and (3.4) one obtains

$$\frac{(E_0 - E)}{(KE - E_1)} = \frac{(dE/dx)|_{E_{in}}}{(dE/dx)|_{E_{out}}} \beta^{-1} = \frac{\varepsilon(\bar{E}_{in})}{\varepsilon(\bar{E}_{out})} \beta^{-1}. \tag{3.23}$$

Note that the energy loss ratio method of Section 3.3.1 is based on the same relationship [Eq. (3.21)] except that $\bar{E}_{in}$ and $\bar{E}_{out}$ are evaluated at the surface values $E_0$ and $KE_0$.

The Taylor expansion of $\varepsilon$ below $E_0$ gives

$$\varepsilon(\bar{E}_{in}) = \varepsilon(E_0) - \tfrac{1}{2}(E_0 - E)\varepsilon'(E_0) + \cdots, \tag{3.24}$$

where $\varepsilon'(E_0)$ is the derivative of $\varepsilon$ with respect to energy taken at $E_0$. The expansion of $\varepsilon$ above $E_1$ gives, similarly,

$$\varepsilon(\bar{E}_{out}) = \varepsilon(E_1) + \tfrac{1}{2}(KE - E_1)\varepsilon'(E_1) + \cdots. \tag{3.25}$$

By substituting these expansions for the ratio $\varepsilon(\bar{E}_{in})/\varepsilon(\bar{E}_{out})$, one obtains a quadratic expression for $E$:

$$aE^2 + bE + c = 0, \tag{3.26}$$

where

$$a = \tfrac{1}{2}K[\varepsilon'(E_0)\beta^{-1} + \varepsilon'(E_1)], \tag{3.27}$$

$$b = [K\varepsilon(E_0)\beta^{-1} + \varepsilon(E_1)] - \tfrac{1}{2}(KE_0 + E_1)[\varepsilon'(E_1) + \varepsilon'(E_0)\beta^{-1}], \tag{3.28}$$

and

$$c = \tfrac{1}{2}E_0E_1\left[\varepsilon'(E_0)\beta^{-1} + \varepsilon'(E_1)\right] - E_0\varepsilon(E_1) - E_1\varepsilon(E_0)\beta^{-1}. \tag{3.29}$$

The coefficients $a$, $b$, and $c$ are expressed in terms of the known quantities $E_0$, $E_1$, $K$, $\beta$, and $\varepsilon$ and $\varepsilon'$ at $E_0$ and $E_1$. Values of $\varepsilon$ and $\varepsilon'$ are given in Tables VI and VII. Therefore, $a$, $b$, and $c$ can be written in terms of $E_0$ and $E_1$, and $E$ can be solved from Eq. (3.26). The method was introduced by Chu and Ziegler (1975).

**b.  Power Law Assumption for $dE/dx$.**  Another way to obtain the relation between the energy $E$ before scattering and the depth $x$ at which scattering takes place is to assume a functional dependence for $\varepsilon$ or $dE/dx$ such that the integrals in Eqs. (3.1) and (3.2) can be solved analytically. One approach (Behrisch and Scherzer, 1973) assumes that $dE/dx$ can be approximated over an energy region by a power law in $E$, as $dE/dx = A_v E^v$. They assume that the exponent $v$ of the power dependence is a constant with values equal to $\frac{1}{2}$, 0, or $-1$ depending on the energy region where $dE/dx$ is evaluated. When this power law expression for $dE/dx$ is substituted into Eqs. (3.1) and (3.2) and the variable $x$ is eliminated, one obtains

$$E = \left( \frac{E_1^{1-v} + \beta E_0^{1-v}}{K^{1-v} + \beta} \right)^{1/(1-v)}. \tag{3.30}$$

### 3.4  NUMERICAL METHODS TO FIND THE ENERGY $E$ BEFORE SCATTERING

Numerical methods proceed from Eqs. (3.1) and (3.2) with tabulated values for $dE/dx$. With the first equation, (3.1), one computes a table of $x$ values versus $E$ values for the incident path. With the second equation, (3.2), one computes $E_1$ for each pair of values of $x$ and $E$. This establishes a set of corresponding values of $x$, $E$, and $E_1$ for a given $E_0$.

In practice, there are two different ways to do the numerical calculation. One approach is to divide the depth into many slabs of equal width $\Delta x$, as shown in Fig. 3.10. The calculation starts from the surface layer. The thickness $\Delta x$ is made thin enough so that $dE/dx$ is practically constant over the width $\Delta x$. The energies at the two boundaries of the $(n + 1)$th slab can be related to each other by the recursion relation

$$_{(n+1)}E = {}_n E - \left. \frac{dE}{dx} \right|_{nE} \left( \frac{x}{\cos \theta_1} \right). \tag{3.31}$$

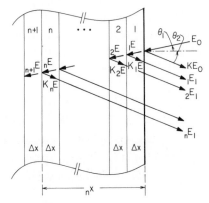

**Fig. 3.10**  Concept and symbols used in the numerical method of calculating the energy $E$ before scattering at depth $x$ and the corresponding detected energy $E_1$ at the detector.

In this way one obtains the energy of the incident particles before scattering at each slab boundary. Upon scattering, the energy of the particle is reduced by the kinematic factor $K$. Along the outgoing path, the energy lost in each slab is equal to the product of $dE/dx$ evaluated at the local energy and the effective path length $x/\cos \theta_2$. The emerging particles will have energies $_1E_1, \, _2E_1, \, \ldots, \, _nE_1$, and $_{n+1}E_1$, etc., where $_nE_1$ is the energy of a particle emerging after a collision in the $n$th slab; therefore,

$$_1E_1 = K_1 E - \frac{\Delta x}{\cos \theta_2} \frac{dE}{dx}\bigg|_{K_1 E} \qquad (3.32)$$

The energy $_2E_1$ of an emerging particle scattered after traversing inward and outward through two slabs is

$$_2E_1 = \left( K_2 E - \frac{\Delta x}{\cos \theta_2} \frac{dE}{dx}\bigg|_{K_2 E} \right) - \frac{\Delta x}{\cos \theta_2} \frac{dE}{dx}\bigg|_{\genfrac{}{}{0pt}{}{\text{energy at 1,2}}{\text{interface}}}. \qquad (3.33)$$

The energy at the 1,2 interface at which the last term must be evaluated is identical to that given in the parentheses preceding that last term. Iterating this procedure, one can write

$$_nE_1 = \left( \left( \left( \left( K_n E - \frac{\Delta x}{\cos \theta_2} \frac{dE}{dx}\bigg|_{K_n E} \right) - \frac{\Delta x}{\cos \theta_2} \frac{dE}{dx}\bigg|_{(E)} \right) - \frac{\Delta x}{\cos \theta_2} \frac{dE}{dx}\bigg|_{((E))} \right) \right.$$
$$\left. - \frac{\Delta x}{\cos \theta_2} \frac{dE}{dx}\bigg|_{(((E)))} \right) - \cdots, \qquad (3.34)$$

where each $dE/dx$ is evaluated at a local energy which is given in the parentheses preceding the term and from which it is subtracted.

The other approach is to divide the sample into thin slabs of differing thicknesses chosen such that particles scattered from the two boundaries of all slabs have a fixed energy difference $\mathscr{E}$ at the detector. This procedure has the advantage that it reproduces the subdivision of the energy $E_1$ into equal increments, as a multichannel analyzer really does.

It is also convenient to peform numerical calculations when $dE/dx$ can be expressed as a function of $E$ analytically. This is usually done by fitting a polynomial to $dE/dx$ or $\varepsilon$. For the purpose of numerical calculations such fits are presented in Table VII.

### 3.5  HEIGHT OF AN ENERGY SPECTRUM
### FOR AN ELEMENTAL SAMPLE

In the previous sections we have discussed the relation between the energy of the detected backscattered particle (abscissa of an energy spectrum) and the depth within the target where the backscattering events occurred. In the next few sections we develop the relation between the height of the energy spectrum (ordinate of an energy spectrum) and the number of scattering centers per unit area within the sample where backscattering occurs. In the remaining sections of this chapter, only stopping cross sections $\varepsilon$ and stopping cross section factors $[\varepsilon]$ will be used. A conversion to $dE/dx$ or $[S]$ can always be made [Eqs. (3.7) and (3.9)].

According to the preceding sections, the energy axis of a backscattering spectrum and the depth below the surface of a sample are uniquely related to each other by a functional dependence such as that shown in Fig. 3.9. Each energy width $\mathscr{E}$ of a channel $i$ in the multichannel analyzer is thus imaged within the sample by a slab $i$ of thickness $\tau_i$ from which all the backscattering events recorded in channel $i$ emanate. The number of counts $H_i$ in channel $i$ is thus determined by two factors: the thickness $\tau_i$ of the slab and the number of scattering centers (atoms) in that slab. The basic problem then is to relate the number of counts $H_i$ to the number of scattering centers per unit area $N\tau_i$ in the slab of thickness $\tau_i$ at depth $x_i$ which corresponds to the energy width $\mathscr{E}$ and the position $E_{1,i}$ of channel $i$ in the energy spectrum, as indicated in Fig. 3.11.

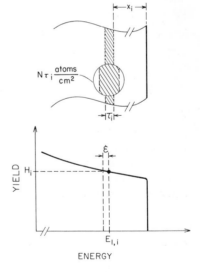

**Fig. 3.11** Schematic showing the correspondence between (a) slab $i$ at depth $x_i$ in a monoisotopic sample and (b) channel $i$ at energy $E_{1,i}$. The width $\mathscr{E}$ of every channel is the same, but the width $\tau_i$ of the slabs is not.

Assume for the time being that the width $\tau_i$ is known. (The method for determining this width is explained later in this section.) It then follows from Eq. (2.19) that for a beam of normal incidence the total number of particles detected in channel $i$ is $H_i = \sigma(E_i)\Omega Q N\tau_i$, where $\sigma(E_i)$ is the differential cross section evaluated at energy $E_i$ and averaged over the finite solid angle $\Omega$ spanned by the detector, $Q$ is the total number of particles incident on the sample, and $N$ is the atomic density of the sample element. This result is correct only for a normal incidence of the beam, because then $N\tau_i$ correctly gives the number of target atoms in a unit of area perpendicular to the beam. For other angles of incidence, i.e., for $\theta_1 > 0$, the trajectory of the beam across the slab $i$ has a length $\tau_i/\cos\theta$, not $\tau_i$. The number of atoms per unit area as seen by the beam is therefore increased by $1/\cos\theta_1$, so that for this general case

$$H_i = \sigma(E_i)\Omega QN\tau_i/\cos\theta_1. \tag{3.35}$$

It will be seen that the value and the position of $\tau_i$ also change as the beam is tilted from a normal to a slanted incidence, e.g., (3.35) holds for the value of $\tau_i$ applicable to the particular geometrical arrangement under consideration.

The shape and the height of the backscattering energy spectrum were first treated in the early 1950s (Wenzel, 1952). Several different versions of the analytical form of the backscattering yield exist and are well documented (Wenzel and Whaling, 1952; Van Wijngaarden et al., 1970; Powers, 1961). Although the notation differs, all the approaches are conceptually the same. Approximations have been applied in some cases to simplify the problem and the mathematics. We will start with the simplified case and progress to the general form.

### 3.5.1 Spectrum Height for Scattering from the Top Surface Layer

Consider the backscattering spectrum obtained from a thick sample and focus attention on the backscattering events that take place either at the surface of the sample or near the surface region. For this region the analysis is simplified because the energy before scattering can be taken as $E_0$ and is therefore known. Figure 3.12 gives a schematic of the backscattering processes in this surface region, and the resulting spectrum. The notation adopted for the near-surface region is $H_0$ and $\tau_0$ in contrast to $H_i$ and $\tau_i$ for regions within the sample. For the surface region Eq. (3.35) then becomes

$$H = \sigma(E_0)\Omega QN\tau_0/\cos\theta_1 \equiv H_0. \tag{3.36}$$

The subscript is often dropped from $H$ because of the widespread use of this symbol in this particular context. The thickness $\tau_0$ is defined by the energy width $\mathscr{E}$ of a channel. Particles scattered from atoms within $\tau_0$ will have energies between $KE_0$ and $KE_0 - \mathscr{E}$. From Eqs. (3.9) and (3.12) the depth scale at the surface is given by

$$\mathscr{E} = [\varepsilon_0]N\tau_0. \tag{3.37}$$

The corresponding expression in terms of the energy loss factor is $\mathscr{E} = [S_0]\tau_0$. As stated previously, we shall retain only the formulation in terms of $\varepsilon$ and

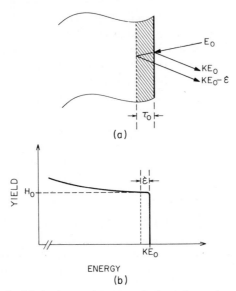

**Fig. 3.12** Schematic of the backscattering process in the surface region of a sample consisting of a monoisotopic element (a) and the resulting spectrum (b).

$[\varepsilon]$ in the remainder of this chapter. Substituting Eq. (3.37) in (3.36) to eliminate $N\tau_0$ yields

$$H_0 = \sigma(E_0)\Omega Q \mathscr{E}/[\varepsilon_0]\cos\theta_1. \qquad (3.38)$$

This equation states that the height of the energy spectrum at the surface is directly proportional to

(i)  $Q$, the total number of incident projectiles bombarding the sample;

(ii)  $\sigma(E_0)$, the average differential scattering cross section between the projectile and the sample evaluated at the incident energy $E_0$;

(iii)  $\Omega$, the solid angle spanned by the detector aperture;

(iv)  $\mathscr{E}$, the energy width of a channel, which is determined by the electronic setting of the detecting system; and

(v)  $([\varepsilon_0]\cos\theta_1)^{-1}$, the inverse of the stopping cross section factor evaluated at the surface for a given scattering geometry multiplied by the cosine of the angle of incidence of the beam against the sample normal.

The direct proportionality of $H_0$ to $Q$, $\sigma$, $\Omega$, and $\mathscr{E}$ is physically evident. The inverse proportionality of $H_0$ to $[\varepsilon_0]\cos\theta_1$ can be understood by considering the energy that particles lose on their inward and outward paths through the surface layer. Consider first the case of normal incidence. If the stopping cross section is high, then so is the stopping cross section factor $[\varepsilon_0]$. A fixed energy is then dissipated by the moving particle over fewer atomic layers than if $[\varepsilon_0]$ were small. This means that the larger $[\varepsilon_0]$ is, the smaller will be the number of scattering processes for the fixed energy interval $\mathscr{E}$. For example, compare two target materials A and B, where A has a larger stopping cross section factor than B. For the same energy loss, the projectile will have fewer encounters with A atoms than B. Thus there will be fewer backscattering events that produce counts within a given channel for target A than for target B (neglecting differences in $\sigma$).

Consider next the case of fixed stopping cross section but varying angle of incidence $\theta_1$. Changing $\theta_1$ has a twofold effect: the thickness of the slab corresponding to the single channel of the multichannel analyzer undergoes a change expressed by the factor $[\varepsilon_0]^{-1}$, and the number of atoms per unit of an area perpendicular to the beam undergoes a change expressed by the factor $(\cos\theta_1)^{-1}$. These two effects tend to cancel because one of the two terms of $[\varepsilon_0]$ goes as $(\cos\theta_1)^{-1}$ [Eq. (3.11)]. This is the reason for considering the product $[\varepsilon_0]\cos\theta_1$ rather than the individual terms when discussing the dependence on the angle of incidence $\theta_1$ of the beam. In general, signal heights depend on the product $([\varepsilon]\cos\theta_1)^{-1}$, whereas depth-to-energy-loss conversions depend on $[\varepsilon]$ only [compare, e.g., Eqs. (3.7) or (3.9) and (3.44)]. Because signal heights depend on the product $[\varepsilon]\cos\theta_1$, some authors introduce the *effective stopping cross section factor* $\varepsilon_{\text{eff}} \equiv [\varepsilon]\cos\theta_1$, which is the natural parameter to introduce when the interest focuses on the height of a spectrum.

Observe that the height $H_0$ does not depend on the atom density $N$ of the sample. This is a general property of backscattering yields. The matter is discussed in Section 3.9.

### 3.5.2   Spectrum Height for Scattering at a Depth

The essence of depth profiling is to relate a spectrum height $H_i$ to a slab of material with thickness $\tau_i$ and number of atoms per unit area $N\tau_i$ at depth $x_i$. From Eq. (3.35) the height is

$$H_i = \sigma(E_i)\Omega QN\tau_i/\cos\theta_1. \qquad (3.39)$$

The cross section $\sigma$ is evaluated here at the energy $E_i$ of the projectile immediately before scattering at depth $x_i$ (see Section 3.3). The amount of material $N\tau_i$ is defined by the energy width $\mathscr{E}$ such that the particles backscattered from the slab will emerge from the sample with energies between $E_{1,i}$ and $E_{1,i} - \mathscr{E}$. It would be wrong to conclude therefore that the energy width $\mathscr{E}_i'$ of these particles immediately after scattering is also $\mathscr{E}$. The reason is that particles with slightly different energies after scattering at $x_i$ undergo slightly different energy losses on their outward path, so that $\mathscr{E}_i' \neq \mathscr{E}$. To be precise, the energy lost along the outgoing path reduces $KE_i$ to $E_{1,i}$ while $(KE_i - \mathscr{E}_i')$ is reduced to $(E_{1,i} - \mathscr{E})$. This is sketched in Fig. 3.13a.

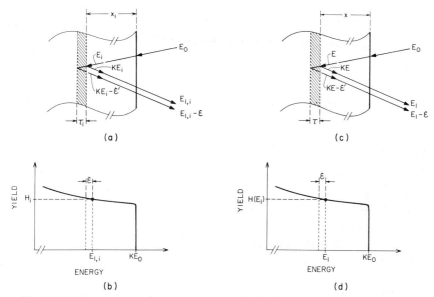

**Fig. 3.13**  Schematics of and nomenclature for (a) the backscattering process at depth $x_i$ within a monoisotopic sample in the language of discrete functions, and (b) the resulting spectrum. (c) and (d) give the corresponding schematics and nomenclature in the language of continuous functions.

Before developing further the subject of the spectrum height for scattering below the surface, it is appropriate at this point to introduce a more efficient notation. The subscript $i$ in the preceding equations indicates that the quantities considered refer to a specific slab $i$ and its corresponding channel $i$ at energy $E_{1,i}$ in the multichannel analyzer. With this subscript, the equations are cumbersome to read and to write. When it is understood that the quantities discussed here are really discrete, the subscript $i$ need not be retained, and Eq. (3.39) can be written as

$$H(E_1) = \sigma(E)\Omega QN\tau/\cos\theta_1. \qquad (3.40)$$

The cross section $\sigma$ is now a function of a continuous variable $E$, the energy of the particle immediately before scattering at any depth $x$ within the sample. Similarly, $H$ is a function of the continuous variable $E_1$, the energy of a detected particle, and $\tau$ is the thickness of a slab (at any depth $x$) that produces particles detected in the energy interval $\mathscr{E}$, the energy width of a channel in the multichannel analyzer. As was previously explained, this energy interval differs from the interval $\mathscr{E}'$ that these same particles span immediately after scattering from a slab of thickness $\tau$ at depth $x$. These definitions and the new notation are explained in Figs. 3.13c and d.

We now return to the derivation of the spectrum height $H(E_1)$ for scattering at a depth $x$ below the surface. The surface energy approximation cannot be used, since the energy $E$ before the collision may differ noticeably from $E_0$. Consequently, the thickness $\tau$ of a slab at depth $x$ may differ from that of $\tau_0$ at the surface. We shall therefore solve the problem by first calculating the thickness $\tau$ of a slab in terms of the energy interval $\mathscr{E}'$. Then we shall express the energy interval $\mathscr{E}'$ in terms of the interval $\mathscr{E}$ located at an energy $E_1$ in the energy spectrum.

To find the relation between $\tau$ and $\mathscr{E}'$, note that the particles scattered at opposite interfaces of the slab at depth $x$ can be viewed as a backscattering process at a surface covered by a layer of thickness $x$. The particles incident on this surface have an energy $E$, and the energy difference corresponding to scattering at the opposite interfaces of the slab there is $\mathscr{E}'$. Exactly the same condition would prevail at the actual surface of the sample if the incident energy $E_0$ were reduced to $E$ and the energy width per channel were set to $\mathscr{E}'$ rather than $\mathscr{E}$ at the multichannel analyzer. It therefore follows from Eq. (3.37) that

$$\mathscr{E}' = [\varepsilon(E)]N\tau. \tag{3.41}$$

The stopping cross section factor $[\varepsilon(E)]$ which appears in this equation is defined in analogy to Eq. (3.12) as

$$[\varepsilon(E)] \equiv \frac{K}{\cos\theta_1}\varepsilon(E) + \frac{1}{\cos\theta_2}\varepsilon(KE) \tag{3.42}$$

and there exists a corresponding energy loss factor

$$[S(E)] \equiv \frac{K}{\cos\theta_1}\frac{dE}{dx}\bigg|_E + \frac{1}{\cos\theta_2}\frac{dE}{dx}\bigg|_{KE} \tag{3.43}$$

defined in analogy to Eq. (3.11). The interpretation of this energy loss factor in terms of Fig. 3.9 is as follows: $[S(E)]$ is the slope (dashed curve) at the origin of the energy loss versus depth curve, which is measured for particles of incident energy $E$, rather than $E_0$. In other words, $[S(E)]$ gives the depth scale of a spectrum in the surface energy approximation when the incident energy of the particles is $E$. With $\tau$ expressed in terms of $\mathscr{E}'$, the height $H(E_1)$ of the spectrum becomes

$$H(E_1) = \sigma(E)\Omega Q(\mathscr{E}'/[\varepsilon(E)]\cos\theta_1). \tag{3.44}$$

This expression for $H$ is incomplete in that $\mathscr{E}'$ is not an experimentally accessible quantity, while $\mathscr{E}$ is. The second step is thus to express $\mathscr{E}'$ in terms of the energy interval $\mathscr{E}$ at a position $E_1$ on the energy scale of a spectrum. The answer is obtained by considering the energy loss of backscattered particles along their outward path. Consider two particles whose energies

**Fig. 3.14** Graphical interpretation of Eq. (3.46). The light line gives $\varepsilon^{-1}$ versus energy. The heavy segments indicate the $\varepsilon^{-1}$ and energy values for two particles along their inward and outward tracks. One particle loses slightly more energy than the other. A difference $\mathscr{E}'$ in the particle energy immediately after scattering produces an energy difference $\mathscr{E}$ when the particles emerge from the sample. The two shaded areas for the outward paths must be equal.

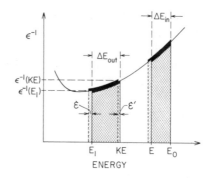

immediately after backscattering at depth $x$ differ by $\mathscr{E}'$. The energy loss along the outward path is given by Eq. (3.2) or

$$Nx/\cos\theta_2 = -\int_{KE}^{E_1} dE/\varepsilon. \tag{3.45}$$

Since the slab $\tau$ is very thin compared to the depth of scattering, one can assume that scattering from the same depth with different energies approximates closely the real situation where particles scatter from the front and rear surfaces of the slab. The outward path thus is essentially the same for both particles and the right-hand side has the same value in both cases; hence

$$\int_{KE}^{E_1} dE/\varepsilon = \int_{KE-\mathscr{E}'}^{E_1-\mathscr{E}} dE/\varepsilon \tag{3.46}$$

must hold. By assumption of the model, $\mathscr{E}'$ and $\mathscr{E}$ are small compared to $KE$ and $E_1$ and can be treated as differentials, so that $\mathscr{E}/\varepsilon(E_1) = \mathscr{E}'/\varepsilon(KE)$, or

$$\mathscr{E}'/\mathscr{E} = \varepsilon(KE)/\varepsilon(E_1). \tag{3.47}$$

The graphical interpretation of this result is sketched in Fig. 3.14. Because the two particles traverse the same layer on their outward path, it is the area under the $\varepsilon^{-1}$ curve that must be conserved. Equation (3.47) then follows at once.

If there is little difference between $\varepsilon(KE)$ and $\varepsilon(E_1)$, a linear interpolation between these two values provides a reasonable approximation to $\varepsilon$. Then $\varepsilon(KE) \simeq \varepsilon(E_1) + \Delta E_{\text{out}}\varepsilon'(\bar{E}_{\text{out}})$, where $\varepsilon'(\bar{E}_{\text{out}})$ is the derivative of $\varepsilon$ with respect to energy, evaluated at some intermediate energy $\bar{E}_{\text{out}}$ along the outward path. When this expression for $\varepsilon(KE)$ is used for the ratio $\varepsilon(KE)/\varepsilon(E_1)$, one obtains (Feng *et al.*, 1973)

$$\frac{\mathscr{E}'}{\mathscr{E}} = 1 + \frac{\Delta E_{\text{out}}}{\varepsilon(E_1)} \varepsilon'(\bar{E}_{\text{out}}) \tag{3.48}$$

$$\simeq 1 + \frac{Nx}{\cos\theta_2} \varepsilon'(\bar{E}_{\text{out}}). \tag{3.49}$$

This expression shows that the difference between $\mathscr{E}'$ and $\mathscr{E}$ increases with the length $x/\cos\theta_2$ of the outward path. It also shows that $\mathscr{E}' \simeq \mathscr{E}$ for particles scattered from the surface region of the sample because $x$ is small there.

With Eq. (3.47), the yield $H(E_1)$ from a slab located at depth $x$ given by Eq. (3.44) becomes

$$H(E_1) = \sigma(E)\Omega Q \, \frac{\mathscr{E}}{[\varepsilon(E)]\cos\theta_1} \frac{\varepsilon(KE)}{\varepsilon(E_1)}. \tag{3.50}$$

In the discrete notation of Fig. 3.13a and b, this formula takes the form

$$H_i = \sigma(E_i)\Omega Q \, \frac{\mathscr{E}}{[\varepsilon(E_i)]\cos\theta_1} \frac{\varepsilon(KE_i)}{\varepsilon(E_{1,i})}. \tag{3.51}$$

The physical interpretation of this result is as follows: As the incident beam penetrates the sample the energy of the projectiles decreases. As a consequence the scattering cross section $\sigma(E)$ increases. This effect tends to increase the yield $H(E_1)$ with decreasing energy $E_1$ of the detected particles. On the other hand, the stopping cross section $\varepsilon$ also varies with $E$. In general this dependence is not as strong as that for $\sigma(E)$, but $\varepsilon$ can either increase or decrease with decreasing values of $E$ (Fig. 3.7). Consequently, the effect of the change in $\varepsilon$ on the backscattering yield may either enhance or counteract the effect of the change in $\sigma$. Specifically, when $\varepsilon$ increases with decreasing energy the effect is to decrease the yield as expressed by the inverse proportionality to $[\varepsilon(E)]$. The contribution from the change in the ratio $\varepsilon(KE)/\varepsilon(E_1)$ is of lesser importance. [The application of Eq. (3.50) will be discussed in Section 5.5.2.]

Alternative derivations of the thick-target yield have been given for uniform targets (Wenzel, 1952) and for nonuniform targets (Wenzel and Whaling, 1952; Powers and Whaling, 1962). Recent work on the thick-target yield has emphasized specific aspects such as the influence of energy straggling (Van Wijngaarden et al., 1970; Brice, 1973), the influence of scattering geometry (Jack, 1973), the dependence on energy loss (Behrisch and Scherzer, 1973; Siritonin et al., 1971, 1972), and analytical formulations (Chu and Ziegler, 1975).

### 3.6  DEPTH SCALE FOR A HOMOGENEOUS SOLID CONTAINING MORE THAN ONE ELEMENT (COMPOUND SAMPLE)

In this section, we shall discuss the backscattering spectrum of a sample composed of a homogeneous mixture of several elements. For simplicity we denote the material as a *compound sample* although it could be either a mixture or a chemical compound. This case differs from that of the monoisotopic elemental sample considered thus far in this chapter in two significant ways. First, as the probing particles penetrate the film, they lose energy as the result of interactions with more than one element. Consequently, the stopping cross section depends on the composition of the sample. Second, when the probing particles with energy $E$ are scattered at a specific depth within the sample, the value of the kinematic factor $K$ and the scattering cross section $\sigma$ will depend on the particular mass (atomic number) of the atom they strike. Since the stopping cross section varies with energy, the energy that the particles lose along identical outward tracks also depends on the atom struck in the scattering collision. For a compound sample, the yield of the backscattering spectrum and the energy-to-depth conversion thus depend on the element struck in the collision. All counts generated by back-scattering from a given element constitute the signal of this element in the spectrum.

In the rest of this section and in Section 3.7 we shall consider the particular case of a sample composed of two monoisotopic elements A and B. The extension to the general case of a multielemental compound sample is straightforward. We also assume that the sample is homogeneous, i.e., of uniform composition both in lateral dimensions and in depth.

### 3.6.1   Stopping Cross Section Factor $[\varepsilon]$

To relate the energy $E_1$ of the detected particle to the depth $x$ at which the backscattering event occurs, we shall follow the formalism described in Section 3.2 for the depth scale of an elemental target. We use a subscript to indicate the atom struck, so that $E_{1A}$ and $E_{1B}$ denote the energies of detected particles scattered from atoms A and B, respectively. Superscripts are used to denote the stopping medium, so that $\varepsilon^{A_m B_n}$ is the stopping cross section of a material containing elements A and B in the atomic ratio $m/n$. For a compound, $m$ and $n$ are integers; for a solid solution, for example, they need not be. In the spirit of Section 2.5 we shall give preference to the abbreviated notation $\varepsilon^{AB}$ for $\varepsilon^{A_m B_n}$, even if $m$ and $n$ are not unity. From Chapter 2, the stopping cross section $\varepsilon^{A_m B_n}$ of the sample is given by $m\varepsilon^A + n\varepsilon^B$, assuming that Bragg's rule for the linear additivity of stopping cross sections holds true. Examples of the application of that rule are given in Section 5.4.

For the scattering geometry shown in Fig. 3.15, a particle penetrating the sample to a depth $x$ undergoes an energy loss $\Delta E_{in}$ along the inward path given by

$$\Delta E_{in} = (N^{AB}x/\cos\theta_1)\varepsilon_{in}^{AB}, \tag{3.52}$$

where $N^{AB}$ is the number of molecules $A_m B_n$ per unit volume. The energy loss $\Delta E_{out}$ along the outgoing path depends on the collision partner. There-fore, the energy difference $\Delta E$ between particles scattered at the front surface

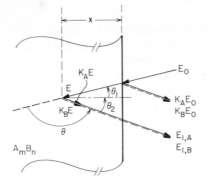

**Fig. 3.15** Symbols used in the description of backscattering events in a compound sample composed of a homogeneous mixture of two monoisotopic elements A and B.

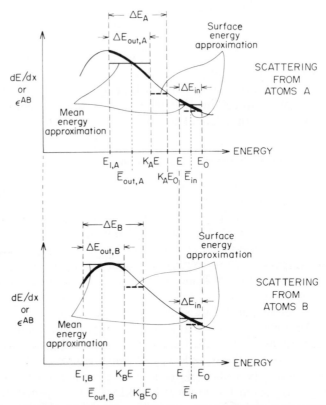

**Fig. 3.16** Graphical representations of the energy loss of particles along their inward and outward paths (heavy lines) through a sample composed of a homogeneous mixture of two monoisotopic elements A and B. The light line is the functional form of $dE/dx$ versus $E$. Since $dE/dx = N^{AB}\varepsilon^{AB}$, the same plots apply to $\varepsilon^{AB}$ versus $E$ as well. Particles scattered at the two elements cover different energy ranges along their outward paths. The top of the figure applies for scattering by the heavy atom A; the bottom of the figure is for scattering by the atom B which is lighter than A. (Compare this with the corresponding parts of Fig. 3.7 for a monoisotopic sample.)

and at a depth $x$ can have two values, $\Delta E_A$ or $\Delta E_B$, depending on whether the particles scatter from atom A or atom B. The situation is represented graphically in Fig. 3.16 in a way corresponding to Fig. 3.7a. Notice the two different energy regions covered by particles scattered by atoms A and B. Thus there are now two depth scales, one attached to each signal, as shown for a single element in Fig. 3.8. These scales are in general different, but not by more than 10% in most cases for megaelectron volts of $^4$He.

In analogy with the result of Section 3.2 for an elemental sample, we thus have

$$\Delta E_A = [\varepsilon]_A^{AB} N^{AB} x \tag{3.53}$$

and

$$\Delta E_B = [\varepsilon]_B^{AB} N^{AB} x, \tag{3.54}$$

where

$$[\varepsilon]_A^{AB} = \frac{K_A}{\cos\theta_1} \varepsilon_{in}^{AB} + \frac{1}{\cos\theta_2} \varepsilon_{out,A}^{AB}, \tag{3.55}$$

$$[\varepsilon]_B^{AB} = \frac{K_B}{\cos\theta_2} \varepsilon_{in}^{AB} + \frac{1}{\cos\theta_2} \varepsilon_{out,B}^{AB}. \tag{3.56}$$

These *generalized stopping cross section factors* contain the special case of an elemental sample composed of elements A or B only as $[\varepsilon]_A^A$ or $[\varepsilon]_B^B$ [see Eq. (3.10)]. As shown in Section 3.2 for elemental stopping cross section factors, approximations can be used to evaluate the stopping cross sections on the inward path $\varepsilon_{in}^{AB}$ and on the outward paths $\varepsilon_{out,A}^{AB}$ and $\varepsilon_{out,B}^{AB}$ for particles scattered from atoms A or B. The discussion given there applies to the present case of a compound sample as well. The next section repeats this treatment in brief.

### 3.6.2 Approximations to [$\varepsilon$]

For regions near the surface, the thickness $x$ is small and the relative changes of energy along the incident and outward path are small also. Therefore, in analogy to Eq. (3.12) one gets

$$[\varepsilon_0]_A^{AB} = \frac{K_A}{\cos\theta_1} \varepsilon^{AB}(E_0) + \frac{1}{\cos\theta_2} \varepsilon^{AB}(K_A E_0) \tag{3.57}$$

and

$$[\varepsilon_0]_B^{AB} = \frac{K_B}{\cos\theta_1} \varepsilon^{AB}(E_0) + \frac{1}{\cos\theta_2} \varepsilon^{AB}(K_B E_0), \tag{3.58}$$

where the symbols $[\varepsilon_0]_A^{AB}$ and $[\varepsilon_0]_B^{AB}$ are used to denote the surface energy approximation to the stopping cross section factor for particles scattered from atoms A and B, respectively.

Similarly, one defines $[\bar{\varepsilon}]_A^{AB}$ and $[\bar{\varepsilon}]_B^{AB}$ as the mean energy approximation. For the inward path, $\bar{E}_{in} = \frac{1}{2}(E + E_0)$, as given in Eq. (3.15). However, the intermediate energy $\bar{E}_{out}$ along the outward path is different for particles scattered from atoms A and B and must be specified for each case. Following Eq. (3.16),

$$\bar{E}_{out,A} = \tfrac{1}{2}(E_{1,A} + K_A E) \tag{3.59}$$

and

$$\bar{E}_{out,B} = \tfrac{1}{2}(E_{1,B} + K_B E), \tag{3.60}$$

where $E_{1,A}$ and $E_{1,B}$ refer to the detected energy of particles scattered at a depth $x$ from atoms A and B, respectively. The locations of $\bar{E}_{out,A}$ and $\bar{E}_{out,B}$ for the mean energy approximation are shown in Fig. 3.16 also.

The value of $E$ can be found from the methods described in Section 3.3 or estimated from the symmetrical mean energy approximation, in which case the values of $\bar{E}_{in}$ and $\bar{E}_{out}$ for the signals from A and B are then given by

$$\bar{E}_{in,A} = E_0 - \tfrac{1}{4}\Delta E_A, \tag{3.61}$$
$$\bar{E}_{in,B} = E_0 - \tfrac{1}{4}\Delta E_B, \tag{3.62}$$
$$\bar{E}_{out,A} = E_{1,A} + \tfrac{1}{4}\Delta E_A, \tag{3.63}$$
$$\bar{E}_{out,B} = E_{1,B} + \tfrac{1}{4}\Delta E_B, \tag{3.64}$$

in analogy with Eqs. (3.17) and (3.18). Note that in this case different values of $E$ and $\bar{E}_{in}$ are used for the different collision partners.

### 3.7  HEIGHT OF AN ENERGY SPECTRUM FOR A HOMOGENEOUS SOLID CONTAINING MORE THAN ONE ELEMENT (COMPOUND SAMPLE)

In the preceding section we established the connection between the energy of a detected backscattered particle and the depth within the homogeneous compound sample where scattering occurs. In this section we shall discuss the height of the backscattering spectrum of such a compound sample. Again we shall consider in detail the case of a mixture of two monoisotopic elements A and B. The extension to a multielemental compound sample is straightforward.

The backscattering spectrum of such a compound sample is sketched in Fig. 3.17b. This energy spectrum consists of a superposition of the two signals

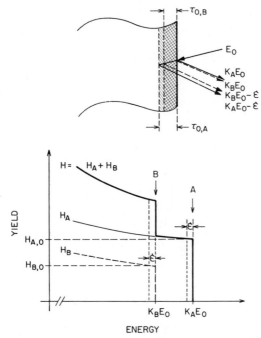

**Fig. 3.17** Schematic of the backscattering process in the surface region of a sample composed of a homogeneous mixture of two monoisotopic elements A and B (a), and the resulting spectrum (b).

generated by the elements A and B in the sample. The edge of each signal is defined by the kinematic factor $K$ of these two elements. For the example shown, $K_A > K_B$; that is, $A$ is the heavier of the two atomic species. If $H_A(E_1)$ and $H_B(E_1)$ are the heights of the individual signals generated by particles detected with energy $E_1$ after scattering from elements A and B, the height of the total spectrum $H$ at that energy is given generally by

$$H(E_1) = H_A(E_1) + H_B(E_1). \tag{3.65}$$

We shall develop the shape of this total spectrum by first considering scattering from the top surface region.

### 3.7.1 Spectrum Height for Scattering from the Top Surface Layer

For backscattering processes near the sample surface, the energy before scattering can be taken as $E_0$. The expression for scattering from elements

A and B can then be taken directly from that for an elemental target [Eq. (3.36)] to give

$$H_{A,0} = \sigma_A(E_0)\Omega Q N_A^{AB}(\tau_{A,0}/\cos\theta_1) \qquad (3.66)$$

and

$$H_{B,0} = \sigma_B(E_0)\Omega Q N_B^{AB}(\tau_{B,0}/\cos\theta_1), \qquad (3.67)$$

where $N_A^{AB}$ and $N_B^{AB}$ are the number of atoms A and B per unit volume. The thicknesses $\tau_{A,0}$ and $\tau_{B,0}$ are chosen such that particles scattered within these slabs will have energies between $K_A E_0$ and $K_A E_0 - \mathscr{E}$ or $K_B E_0$ and $K_B E_0 - \mathscr{E}$. There are two such surface slabs now, because the energy lost along the outward path for particles scattered by atom A differs from that for particles scattered by atom B. This is shown schematically in the diagram of Fig. 3.17a. These two widths thus satisfy the conditions

$$\mathscr{E} = [\varepsilon_0]_A^{AB} N^{AB} \tau_{A,0} \qquad (3.68)$$

and

$$\mathscr{E} = [\varepsilon_0]_B^{AB} N^{AB} \tau_{B,0}, \qquad (3.69)$$

where $N^{AB}$ is the number of molecular units $A_m B_n$ per unit volume. Since $N_A^{AB} = mN^{AB}$ and $N_B^{AB} = nN^{AB}$, the surface heights can be written as

$$H_{A,0} = \sigma_A(E_0)\Omega Q m(\mathscr{E}/[\varepsilon_0]_A^{AB}\cos\theta_1) \qquad (3.70)$$

and

$$H_{B,0} = \sigma_B(E_0)\Omega Q n(\mathscr{E}/[\varepsilon_0]_B^{AB}\cos\theta_1). \qquad (3.71)$$

The ratio of these heights is

$$\frac{H_{A,0}}{H_{B,0}} = \frac{\sigma_A(E_0)}{\sigma_B(E_0)}\frac{m}{n}\frac{[\varepsilon_0]_B^{AB}}{[\varepsilon_0]_A^{AB}}. \qquad (3.72)$$

To determine the ratio $m/n$ from a backscattering spectrum, the ratio $[\varepsilon_0]_B^{AB}/[\varepsilon_0]_A^{AB}$ can be taken as unity in a zeroth-order approximation. This ratio actually approaches unity within 10% in most cases for He ion energies of 1 to 2 MeV; thus

$$m/n \simeq [H_{A,0}/\sigma_A(E_0)]/[H_{B,0}/\sigma_B(E_0)]. \qquad (3.73)$$

From this zeroth-order approximation, one can then obtain a better estimate of the ratio $[\varepsilon_0]_B^{AB}/[\varepsilon_0]_A^{AB}$ and hence a first-order approximation to $m/n$. Typically, this first iteration is sufficient to give a value of the ratio $m/n$ within the errors of the experimental data.

### 3.7.2 Spectrum Height for Scattering at a Depth

The calculation of the spectrum height $H(E_1)$ for particles detected at energy $E_1$ is complicated by the fact that the signals generated by scattering from atoms A and atoms B have different depth scales. That is, particles escaping the sample with the same detected energy $E_1$ are scattered from atoms A at a depth $x_A$, whereas those scattered from atoms B come from a depth $x_B \neq x_A$ (see Fig. 3.18a). Thus the energies $E_A$ and $E_B$ of the particles immediately before scattering will differ. In analogy to Eq. (3.44), the height of each signal can be written as

$$H_A(E_1) = \sigma(E_A)\Omega Q m(\mathscr{E}_A'/[\varepsilon(E_A)]_A^{AB} \cos \theta_1) \qquad (3.74)$$

and

$$H_B(E_1) = \sigma(E_B)\Omega Q n(\mathscr{E}_B'/[\varepsilon(E_B)]_B^{AB} \cos \theta_1), \qquad (3.75)$$

where $\mathscr{E}_A'$ and $\mathscr{E}_B'$ are the energy intervals spanned by particles immediately after scattering within the slabs of thickness $\tau_A$ and $\tau_B$ at depth $x_A$ and $x_B$. One

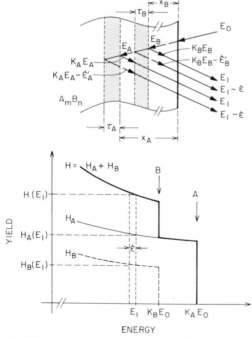

**Fig. 3.18** Schematic of the backscattering process at some depth within a sample composed of a homogeneous mixture of two monoisotopic elements A and B (a), and the resulting spectrum (b).

can relate the energies $\mathscr{E}_A'$ and $\mathscr{E}_B'$ to $\mathscr{E}$ by the procedure developed in connection with Eq. (3.47). The result is

$$H_A(E_1) = \sigma_A(E_A)\Omega Qm \frac{\mathscr{E}}{[\varepsilon(E_A)]_A^{AB}\cos\theta_1} \frac{\varepsilon^{AB}(K_A E_A)}{\varepsilon^{AB}(E_1)}, \qquad (3.76)$$

$$H_B(E_1) = \sigma_B(E_B)\Omega Qn \frac{\mathscr{E}}{[\varepsilon(E_B)]_B^{AB}\cos\theta_1} \frac{\varepsilon^{AB}(K_B E_B)}{\varepsilon^{AB}(E_1)}, \qquad (3.77)$$

and corresponds to Eq. (3.50) for the elemental case. As is true there, the last factor with the ratio in stopping cross sections is of lesser importance. The main changes as compared with the surface heights $H_{A,0}$ and $H_{B,0}$ come from variations in the cross sections $\sigma_A$ and $\sigma_B$ and the stopping cross section factors $[\varepsilon]_A^{AB}$ and $[\varepsilon]_B^{AB}$ with energy.

### 3.8  HIGH-ENERGY EDGE OF AN ENERGY SPECTRUM FOR AN ELEMENTAL SAMPLE WITH SEVERAL ISOTOPES

In the preceding discussions we treated the sample as composed of monoisotopic elements. In general, an element has several stable isotopes of the same atomic number, but different atomic mass. The kinematic factor differs for each isotope. As a consequence, a sample of such an element has a backscattering spectrum with steps in the high-energy edge as shown in Fig. 3.19. The formalism required to develop the spectrum height of such a

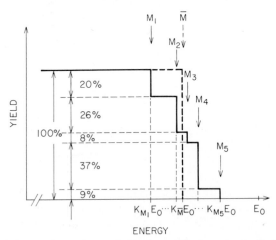

**Fig. 3.19**  Each isotope in an elemental sample contributes a step to the high-energy edge commensurate with its natural abundance. These isotopic steps are often so close to each other that the high-energy edge of an isotopic mixture can be replaced by a single step at some average location.

sample is like that for a compound sample. The equations are simplified by the fact that the stopping cross sections $\varepsilon$ and scattering cross sections $\sigma$ are practically the same for each isotope. However, since the kinematic factor $K$ is different for each isotope, the stopping cross section factor $[\varepsilon]$ will differ for each isotope also. Assume that the mass $M_i$ is present in a fractional abundance $m_i$. Then the kinematic factor $K_{M_i}$ will specify a stopping cross section factor, which we can denote as $[\varepsilon]_{M_i}$. The ratio of the spectrum heights at the surface for any two isotopes, such as $M_3$ and $M_4$, is then

$$\frac{H_{M_3,0}}{H_{M_4,0}} = \frac{m_3}{m_4} \frac{[\varepsilon_0]_{M_4}}{[\varepsilon_0]_{M_3}} \tag{3.78}$$

by Eq. (3.72). For backscattering of $^4$He in the megaelectron volt energy range, the ratio of isotopic $[\varepsilon]$'s is very close to unity and the ratio of the spectrum heights at the surface equals the ratio of the fractional abundances, as indicated in Fig. 3.19 for a sample containing five isotopes, such as Ge.

With $^4$He ions in the megaelectron volt range and conventional solid-state detection systems, the isotopic steps in the high-energy edge are difficult to resolve when the element is of medium or heavy mass. The spectrum is then often interpreted as a single step of an average mass $\bar{M} = \sum_i m_i M_i$ at the position $K_{\bar{M}}E_0$ in the energy scale. This procedure is actually incorrect, because the kinematic factor is not a linear function of $M$. Strictly taken, the mean of the isotopic steps is located at $\bar{K}E_0 = (\sum_i m_i K_{M_i})E_0$ and $\bar{K} \neq K_{\bar{M}}$ in general. The difference is insignificant for target masses much larger than the projectile mass, and usually $K$ is used for $\bar{K}$ or $K_{\bar{M}}$. We follow this usage in this book as well. The table of Ziegler (1973) gives $K_{\bar{M}}$, not $\bar{K}$.

## 3.9 ENERGY LOSS AND YIELD RESPOND TO ATOMS PER UNIT AREA

Up to this point in our development of the subject we have derived general formulas for converting energy to depth and for calculating the height of a backscattering spectrum. The purpose of this section is to emphasize the facts that (i) depth has a specific meaning in backscattering spectrometry which is not that of distance, as commonly associated with the word, but, rather refers explicitly to atoms per unit area, and (ii) the height of a back-scattering spectrum does not depend on the atomic volume density of the target.

The fact that the energy loss that particles incur when they penetrate through a sample does not depend on the atomic density can be seen with the help of a conceptual experiment shown in Fig. 3.20, where a beam of

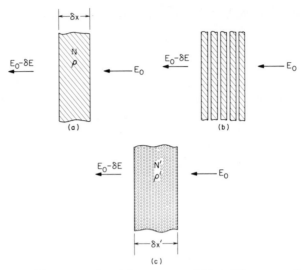

**Fig. 3.20**  Three different samples of the same material but different overall density all of which generate the same backscattering spectrum.

particles is shown incident perpendicularly with energy $E_0$ on three different samples. In the first case (Fig. 3.20a), the sample is a thin film of atomic density $N$, specific gravity $\rho$, and thickness $\delta x$. After transmission through the target, the particle energy is $E_0 - \delta E$. Imagine next that the physical thickness of this sample is increased to a value $\delta x'$ by slicing it into thin slabs and spacing them. Clearly, no energy is lost as the beam crosses these spaces, since no matter is present there. Hence $\delta E$ is unchanged. Imagine now that this procedure is carried to the limit so that the additional volume is distributed microscopically and uniformly throughout the sample. Again, $\delta E$ is unchanged, but the atomic density has been reduced to a value $N' < N$. Similarly, the specific gravity is now $\rho' < \rho$. This shows that the energy loss depends only on the amount of material traversed regardless of the physical thickness.

The number of atoms traversed is expressed by $N \, \delta x = N' \, \delta x'$. The energy loss $\delta E$ is given by $\delta E = (dE/dx)dx = (dE/dx)' \, \delta x'$. This shows that $dE/dx$ depends on the atomic density of the target. On the other hand, the energy loss can also be written $\delta E = \varepsilon N \, \delta x = (\varepsilon)' N' \, \delta x'$. Since $N \, \delta x = N' \, \delta x'$, then $\varepsilon = (\varepsilon)'$. The formal description of energy loss as

$$\delta E = \varepsilon N \, \delta x \qquad\qquad (3.79)$$

has the advantage of expressing the energy loss in terms of the two physically relevant quantities: $\varepsilon$, the specific energy loss per atom, and $N \, \delta x$, the number of atoms per unit area. For this same reason we prefer the formulation

$$\Delta E = [\varepsilon] N \, \Delta x \qquad (3.80)$$

rather than $\Delta E = [S] \Delta x$ for energy loss in backscattering. Whenever a measurement made by backscattering spectrometry expresses depth in units of length, the knowledge of the density has been assumed. The word "depth" used in connection with backscattering therefore indicates a distance only when the density is known; otherwise "depth" stands as an abbreviation for the number of atoms per unit area $N \, \Delta x$ over the distance $\Delta x$ traversed.

The three samples shown in Fig. 3.20 all have the same number of atoms per unit area. This is stated by the equality $N \, \delta x = N' \, \delta x'$. It also must follow, then, that the total numbers of counts generated by these samples in backscattering measurements are the same. Since the energy widths of the backscattering signals from these samples are the same, the spectra of all three are indistinguishable. Therefore the height (counts per channel) of a backscattering signal is independent of the atomic density of the sample. For example, backscattering measurements on a sample of evaporated silicon that has an atomic density less than that of bulk silicon will give spectra identical to those obtained from bulk silicon. One should note, however, that density changes generated by additional atoms of a different species do change the spectrum, as discussed further in Section 5.3. Such modifications are not of the type described by Fig. 3.20, because the additional volume contains energy-absorbing atoms, not voids.

### 3.10   NUMERICAL METHODS TO COMPUTE BACKSCATTERING SPECTRA

Many laboratories engaged in backscattering analysis have developed computer programs to calculate backscattering spectra. Most of these programs are tailored to meet the specific needs of the respective laboratories.

One program is available in documented form.[†] It is written in Fortran and can accommodate samples consisting of up to 10 distinct layers with up to 10 elements. The program considers only beams of normal incidence and does not incorporate energy straggling. Bragg's rule of additivity of stopping cross sections is assumed to be valid, and the composition and thickness of each layer in the sample are constant.

---

[†] The program is available from Rome Air Development Center, Air Force Systems Command, Griffiss Air Force Base, New York, as Report RADC-TR-76-182 (June 1976), entitled "Computer Program to Synthesize Backscattering Spectra for Samples Composed of Successive Layers of Uniform Thickness and Composition," by P. Børgesen, J. M. Harris, and B. M. U. Scherzer.

**REFERENCES**

Behrisch, R., and Scherzer, B. M. U. (1973). *Thin Solid Films* **19**, 247.

Brice, D. K. (1973). *Thin Solid Films* **19**, 121.

Chu, W. K., and Ziegler, J. F. (1975). *J. Appl. Phys.* **46**, 2768.

Feng, J. S.-Y., Chu, W. K., Nicolet, M.-A., and Mayer, J. W. (1973). *Thin Solid Films* **19**, 195.

Jack, H. E. Jr. (1973). *Thin Solid Films* **19**, 267.

Lever, R. F. (1976) *in* "Ion Beam Surface Layer Analysis" (O. Meyer, G. Linker, and F. Käppeler, eds.), Vol. 1, p. 111. Plenum Press New York.

Powers, D. (1961). Ph.D. Thesis, California Inst. of Technol.

Powers, D., and Whaling, W. (1962). *Phys. Rev.* **126**, 61.

Siritonin, E. I., Tulinov, A. F., Fiderkevich, A., and Shyskin, K. S. (1971). *Vestnik MGU (Ser. Fiz. Astr.)* **12**, 541; see also (1972). *Radiat. Effects* **15**, 149.

Van Wijngaarden, A., Brimmer, E. J., and Baylis, W. E. (1970). *Can. J. Phys.* **48**, 1835.

Wenzel, W. A. (1952). Ph.D. Thesis, California Inst. of Technol.

Wenzel, W. A., and Whaling, W. (1952). *Phys. Rev.* **87**, 449.

Ziegler, J. F. (1973). *Thin Solid Films* **19**, 289.

# *4*

# Backscattering Spectrometry of Thin Films

## 4.1 INTRODUCTION

One of the main applications of backscattering spectrometry is the analysis of thin films and layered structures. This chapter is therefore devoted specifically to the discussion of backscattering spectra obtained from such samples. The outstanding feature of such a spectrum is that both the front surface *and* the interfaces below are identifiable in the backscattering signals. This is in contrast to spectra of thick samples, where only the surface is clearly recognizable as the high-energy edge of the signals in the spectrum. Since both the outer and the inner interfaces of the sample can be identified in a backscattering signal, two independent numbers can be extracted from it: (i) the energy width $\Delta E$ between two edges of the signal corresponding to adjacent interfaces and (ii) the total number of counts $A$ contained in all channels of the signal between these edges. These two independent quantities each specify the number of atoms per unit area contained in the film. When the film contains more than one element in the form of a uniform mixture, the backscattering spectrum contains a signal for each element and the atomic composition of the film can be derived as well.

The energy loss of the projectile within the thin film is usually small relative to the incident energy $E_0$. The surface energy approximation discussed in Section 3.2.2 then offers a most convenient way to evaluate the spectra.

Because thin-film spectra do image sharp interfaces that are below the front surface of the sample, energy straggling is clearly visible in these spectra. This energy straggling limits the depth resolution. The subject is discussed at the end of the chapter.

### 4.2 ENERGY SPECTRUM OF A THIN ELEMENTAL FILM

In this section we shall discuss the backscattering spectrum of a thin film containing $Nt$ atoms of a single element per unit area. We assume that the film is self-supporting or that it is deposited on a substrate of atomic mass lighter than the element in the film. The signal from the substrate can then be ignored for the purpose of this discussion, and the spectrum appears as sketched in Fig. 4.1. The two quantities of interest are the energy width $\Delta E$ of the signal and $A$, the total number of counts added over all channels in the signal. Both quantities are directly related to the number of atoms per unit area contained in the film.

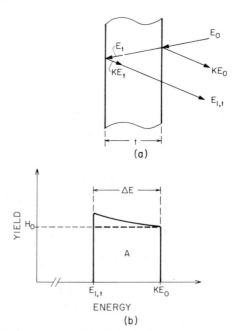

Fig. 4.1 (a) Schematic representation of the backscattering process in a self-supporting monoisotopic thin-film sample and (b) the resulting backscattering signal. The figure also shows the meaning of the symbols used in the text.

### 4.2.1  Energy Width $\Delta E$ between High- and Low-Energy Edges of the Signal

As shown in Fig. 4.1, particles backscattered from atoms at the front surface of the film generate counts at an energy $KE_0$; those backscattered from atoms at the rear produce counts at an energy $E_{1,t}$. According to Section 3.2 the energy difference $\Delta E = KE_0 - E_{1,t}$ is related to the number of atoms per unit area in the film by

$$\Delta E = [\bar{\varepsilon}]Nt. \tag{4.1}$$

When the film is very thin, the surface approximation for $[\bar{\varepsilon}]$ is adequate and Eq. (3.12) can be used in lieu of Eq. (3.14). From the measured value $\Delta E$ and the knowledge of $[\bar{\varepsilon}]$, one can thus determine the number of atoms per unit area $Nt$ in the film.

### 4.2.2  Total Number of Counts in the Signal

The other simple quantity that one can extract from the backscattering spectrum of a thin film is the *total number of counts*

$$A \equiv \sum_i H_i \tag{4.2}$$

summed over all channels $i$ of the signal. If $H_i$ is expressed in terms of $E_{1,i}$, as in Eq. (3.51), the summation becomes unwieldy. The reason is that scattering cross sections are given as a function of the energy $E_i$ immediately before the scattering. This energy is most readily arrived at by computing the energy the particle loses along its incident path, which is $N(x_i/\cos\theta_1)\varepsilon(\bar{E}_{in})$, and subtracting it from the incident energy $E_0$:

$$E_i = E_0 - (Nx_i/\cos\theta_1)\varepsilon(\bar{E}_{in}). \tag{4.3}$$

To compute $A$, it is therefore more convenient to go back to Eq. (3.35), which gives the backscattering yield $H_i$ in terms of the scattering cross section at the energy $E_i$ and the number of atoms per unit area in the $i$th slab as $H_i = \sigma(E_i)\Omega QN\tau_i/\cos\theta_1$. The total number of counts then becomes

$$A = \sum_i \sigma(E_i)\Omega QN\tau_i/\cos\theta_1. \tag{4.4}$$

In the limit of the continuous variables $E$ and $x$, $\tau_i \to dx$. For a film of thickness $t$, the last two expressions then take the form

$$E = E_0 - (Nx/\cos\theta_1)\varepsilon(\bar{E}_{in}) \tag{4.5}$$

and

$$A = (\Omega QN/\cos\theta_1)\int_0^t \sigma(E)\,dx. \tag{4.6}$$

With these two equations, the total number of counts in the backscattering signal of a thin elemental film can be determined when the relation between $E$ and $x$ is known. Several simple cases can be treated in closed form and are useful.

**a. Total Number of Counts in the Surface Energy Approximation.** The simplest case is that of a film so thin that the energy lost by an incident particle on traversing it is negligible compared to its initial energy $E_0$; that is, $E \simeq E_0$ throughout the film. The scattering cross section then has essentially the same value $\sigma(E_0)$ everywhere in the film, and the stopping cross section $\varepsilon$ is practically constant also and equals $\varepsilon(E_0)$. Under these conditions the backscattering signal of the film has the form of a step of constant height as shown by the dashed line in Fig. 4.1. The integral in Eq. (4.6) has the value $\sigma(E_0)t$, and the total number of counts becomes

$$A_0 = \sigma(E_0)\Omega QNt/\cos\theta_1 \qquad (4.7)$$

or

$$Nt = [A_0/\sigma(E_0)\Omega Q]\cos\theta_1. \qquad (4.8)$$

The subscript on $A$ recalls the fact that this equation is valid only when $E \simeq E_0$; that is, in the surface energy approximation.

Formula (4.8) provides the number of atoms per unit area $Nt$ without knowledge of the stopping cross section in the film. However, the solid angle of detection $\Omega$ and the total number $Q$ of particles incident on the target must be known; this requirement presupposes an absolutely calibrated system.

To illustrate the reason why the stopping cross section is not needed in this measurement, it is instructive to rederive $A_0$ in another way. In the surface energy approximation, the backscattering yield $H_i$ is $\sigma(E_0)\Omega QN\tau_0/\cos\theta_1$ for all channels. The total number of counts $A_0$ is thus equal to $nH_i$, where $n$ is the number of channels in the backscattering signal. But $n = \Delta E/\mathscr{E}$, so that

$$A_0 = H_0 n = \sigma(E_0)\Omega QN(\tau_0/\cos\theta_1)\,(\Delta E/\mathscr{E}). \qquad (4.9)$$

Now, generally, $\Delta E = [\bar{\varepsilon}]Nt$ [Eq. (4.1)] and $\mathscr{E} = [\varepsilon(E)][\varepsilon(E_1)/\varepsilon(KE)]N\tau_i$ [Eqs. (3.41) and (3.47)], so that

$$\frac{\Delta E}{\mathscr{E}} = \frac{[\bar{\varepsilon}]}{[\varepsilon(E)]}\frac{\varepsilon(KE)}{\varepsilon(E_1)}\frac{t}{\tau_i}. \qquad (4.10)$$

In the surface energy approximation, $\tau_i = \tau_0$, $[\bar{\varepsilon}] \simeq [\varepsilon(E)] \simeq [\varepsilon(E_0)]$, and $E_1 \simeq KE_0$, so that $n = \Delta E/\mathscr{E} \simeq t/\tau_0$ and $A_0 = \sigma(E_0)\Omega QNt/\cos\theta_1$ follows again. The crucial fact is the equality of $\Delta E/\mathscr{E}$ and $t/\tau_i$. A spread of the total number of counts $A_0$ over the $n$ channels in the energy width $\Delta E$ clearly does not change that total as long as each slab $\tau_i$ corresponding to one channel contains the same fractional number of scattering centers and each scattering center has the same scattering cross section.

**b. Total Number of Counts for a More General Case.** Generally, the scattering cross sections vary inversely with the square of the energy: $\sigma(E) = \sigma(E_0)(E_0/E)^2$ [Eq. (2.22)]. For films which are thick enough that the decrease of the energy $E$ before scattering becomes significant, the cross section increases as a function of depth. This causes an increase in the signal height toward decreasing energies, as sketched in Fig. 4.1 (solid line). To compute the total number of counts $A$, this energy dependence of the scattering cross section must be accounted for. By substituting the $E^{-2}$ energy dependence of the cross section into the integral for $A$, expressing $E$ in terms of $x$, and assuming that $\varepsilon(\bar{E}_{in})$ is a constant, the integration yields

$$A = \sigma(E_0)\Omega Q N(t/\cos\theta_1)\{1 - [Nt\varepsilon(\bar{E}_{in})/E_0]\}^{-1} \qquad (4.11)$$

or

$$Nt = [A/\sigma(E_0)\Omega Q]\cos\theta_1 \{1 + [\varepsilon(\bar{E}_{in})A/\sigma(E_0)\Omega Q E_0 \cos\theta_1]\}^{-1} \quad (4.12)$$

The first part of the right-hand side of Eq. (4.12) is formally identical with Eq. (4.8), except for $A$ replacing $A_0$. With the abbreviation

$$(Nt)_0 = A\cos\theta_1/\sigma(E_0)\Omega Q \qquad (4.13)$$

one can rewrite the expression for $Nt$ in the form

$$Nt = (Nt)_0 \{1 + (Nt)_0[\varepsilon(\bar{E}_{in})/E_0 \cos\theta_1]\}^{-1} \qquad (4.14)$$

For thin films, $\bar{E}_{in}$ can be estimated quite adequately by the symmetrical mean value $E_0 - \frac{1}{4}\Delta E$, or even simply by $E_0$.

The factor $(Nt)_0\varepsilon(\bar{E}_{in})/\cos\theta_1$ is a zeroth-order estimate of the energy loss $\Delta E_{in}$ along the incident track. The second term of the denominator thus is of the order of magnitude of $\Delta E_{in}/E_0$. For sufficiently thin films, the correction factor can thus be replaced by $1 - (\Delta E_{in}/E_0)$. Whether the correction is significant or not thus depends on the magnitude of $\Delta E_{in}/E_0$. It is useful, therefore, to estimate $\Delta E_{in}$ when evaluating thin-film spectra. One way is to assume that the energy loss $\Delta E$ is subdivided equally along the incoming and outgoing tracks so that $\Delta E_{in} = \frac{1}{2}\Delta E$. Other methods are discussed in Section 3.3.

The noteworthy feature of Eqs. (4.12) and (4.14) for $Nt$ is that the stopping cross section $\varepsilon$ enters only as a correction. A determination of the number of atoms per unit area in the film based on the total number of counts $A$ is therefore largely independent of the knowledge of the stopping cross section for the element in the film. As in the preceding case, however, the solid angle of detection $\Omega$ and the total number $Q$ of particles incident on the target must be known; hence an absolutely calibrated system is required.

The preceding discussion establishes that there are two independent ways to determine the number of atoms per unit area in a thin film. One method [Eq. (4.1)] relies on the measurement of the energy width $\Delta E$. In this case, the stopping cross section factor $[\varepsilon]$, and therefore $\varepsilon$, must be known, but the solid angle of detection $\Omega$ and the total number $Q$ of incident particles do not enter in the evaluation of $Nt$. The other method makes use of the total number of counts $A$ [Eq. (4.12) or (4.14)]. Here $\Omega$ and $Q$ must be known, but the stopping cross section $\varepsilon$ enters only as a correction. Stopping cross sections are rarely known to better than 5% accuracy. The method based on the measurement of total counts in the signal is therefore potentially more accurate, provided that the calibration of the system is better than the uncertainty in $\varepsilon$. The prudent course of action is to apply both methods.

### 4.3  ENERGY SPECTRUM OF MULTILAYERED ELEMENTAL FILMS

In this section we shall discuss the backscattering spectra of structures consisting of a sequence of elemental thin films. The analysis can be separated into two parts. The first part deals with the top layer. The backscattering spectrum of this layer is not affected by the layers underneath (except when the spectra overlap as described below). Therefore the analysis of this layer follows the treatment given in the preceding section, where the elemental film is discussed. The second part of the analysis deals with the other elemental films beneath the top surface layer. For those remaining films, the surface layer can be viewed as an absorber, which lowers the energy of the incoming projectile and also reduces the energy of particles escaping to the detector.

A qualitative discussion of this problem is illustrated in Fig. 4.2. The thin sample shown there consists of an elemental film A on top of an elemental film B based on an elemental substrate S. For backscattering spectrometry, the most favorable situation is when element A is heavier than B, and B is heavier than S, i.e., $K_S < K_B < K_A$. The backscattering energy spectrum for this particular case is sketched in Fig. 4.2a. The signal of A reaches to the edge $K_A E_0$ of $A$, but the signals of B and of S are shifted to energies below their respective edges $K_B E_0$ and $K_S E_0$ because of the energy loss in the outer layer A. To illustrate this point, Fig. 4.2b gives a spectrum of a sample without layer B. The signal of A is unchanged, because the spectrum of a surface layer is not influenced by the underlying material. The signal of S is shifted toward higher energies. This is so because both A and B act as

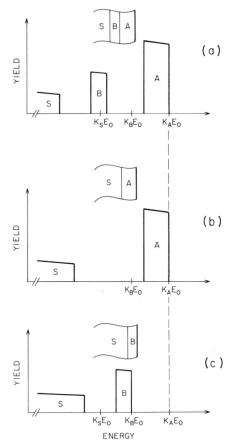

**Fig. 4.2**   (a) Schematic representation of the backscattering spectrum of a bilayered film on a substrate S. The monoisotopic element A is the heaviest, B is intermediate, and S is the lightest. (b) Spectrum for a sample without the intermediate layer B. (c) Spectrum for a sample without the top layer A.

energy absorbers for S; the removal of B reduces the absorbing layer to the thickness of A only. If A rather than B is removed, the signals of S and B change their positions from those given in Fig. 4.2a to those shown in Fig. 4.2c. The high-energy edge of the signal of B now appears at $K_B E_0$ because B is at the surface. To a first approximation, the signals of B and S are both shifted toward higher energies by an amount that roughly equals the width of the absorber signal A. We discuss the effect of an absorber layer rigorously in subsequent parts of this section.

When the element of the surface layer A is lighter in mass than element B below, complications occur sometimes. Figure 4.3 illustrates this case

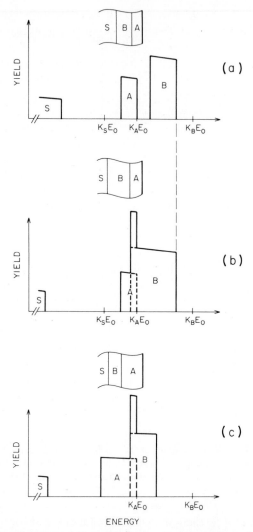

**Fig. 4.3**    (a) Schematic representation of the backscattering spectrum of a bilayered film on a substrate S. The monoisotopic element B is the heaviest, A is intermediate, and S is the lightest. (b) Spectrum for a sample in which the layer B is thicker than in (a). (c) Spectrum for a sample in which the top layer A is thicker than in (a).

$(K_S < K_A < K_B)$. In Fig. 4.3a, both film A and film B are thin. Layer A acts as absorber to B, and the signal of B is therefore shifted below the edge $K_B E_0$ of B. The energy separation between the signals of A and B is considerably narrower than in Fig. 4.2a; in fact, the gap between these two signals may disappear entirely. This comes about because the high-energy edge of signal A is fixed in this position at $K_A E_0$ since A is on the surface. The high-energy edge of signal B is also fixed in position, since the shift of this signal below $K_B E_0$ is given by the thickness of the absorber A. If we now let the layer B grow thicker while the thickness of A is fixed, the width of the signal B must widen, and this will reduce the gap between A and B. There comes a point at which the low-energy edge of signal B merges into the high-energy edge of signal A. Beyond this point, the signals of A and B will overlap (Fig. 4.3b). Figure 4.3c shows that an overlap of signal A and B could also occur for a thicker surface layer A with the thickness of B remaining constant.

The shifting of signals from lower-lying layers with a change in absorber thickness has a useful application. Imagine that a backscattering spectrum such as that shown in Fig. 4.3a has been measured with the conventional geometrical arrangement in which the beam impinges perpendicularly to the sample surface. There is a simple way to establish experimentally which one of the two signals corresponds to the top layer. A second spectrum is taken at slanted incidence; that is, the sample is tilted so that the surface normal moves away from the direction of the incident beam. In such a position, the trajectory of the incident beam across any of the films in the sample is lengthened by the secant of the tilting angle. In effect, the absorber layer now appears correspondingly thicker, and the signal from the under-lying elemental film is shifted toward lower energies. The signal of the absorber layer widens correspondingly, but the high-energy edge of that signal remains fixed on the energy scale. The signal whose high-energy edge does not change position upon tilting the sample thus identifies the top layer.

The conditions under which this overlapping of signals occurs depend on the incident energy of the projectile, the films' thicknesses, and the kinematic factor of their element. When signals overlap, the spectrum has to be separated into individual signals before the analysis can proceed. The separation can be done graphically on a plot of the spectrum or numerically from the data. When the steps in the spectrum are sharp and well defined, individual signals can be identified easily. However, if the detector resolution is poor and/or energy straggling is large, interfaces become difficult to identify in the spectrum and the individual signals are then correspondingly hard to define.

For ease of discussion, we shall assume in the following that the signals of the individual layers do not overlap and that the sample contains two layers, A on top and B underneath. In Fig. 4.4 we repeat the energy spectrum

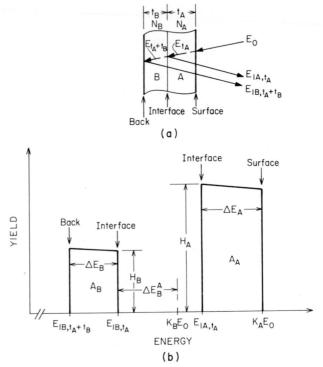

**Fig. 4.4**  (a) Schematic representation of the backscattering process in a self-supporting bilayered film composed of a layer of monoisotopic heavy element A on top of a layer of monoisotopic light element B and (b) the resulting backscattering spectrum. The figure also shows the meaning of the symbols used in the text.

of such a two-layered sample, assuming that the heavier element constitutes the top layer. The spectrum ignores the substrate. Notations are introduced to define the sample and the spectrum: $t_A$ and $t_B$ are the thicknesses of the two layers with atom density $N_A$ and $N_B$; $\Delta E_A$ and $\Delta E_B$ are the widths of the two signals; $A_A$ and $A_B$ are the total number of counts in each signal; and $H_A$ and $H_B$ are the heights of the two signals measured at the edges that correspond to the interface between the two films.

The signal of the top layer A is independent of layer B. The analysis of that layer thus proceeds as described in Section 4.2. The analysis of layer B has several aspects. First, the absorber A causes the high-energy edge of signal B to undergo a shift $\Delta E_B^A$ from the position $K_B E_0$ to lower energies. That shift is larger the thicker the layer A is. Then, there is the conversion of depth in layer B to energy in the shifted signal of B. Finally, there is the height of signal B and the total number of counts in that signal. All of them are influenced by the absorber A. We discuss this in the succeeding subsections.

### 4.3.1   Energy Shift Due to Elemental Absorber

In Fig. 4.4 the *leading edge*, or high-energy edge, of signal B is labeled $E_{1B,t_A}$. The subscript "1" in this notation indicates that this is a detected energy; the subscript B identifies the scattering atom; and $t_A$ gives the depth at which scattering occurred. The projectile energy before scattering at that depth is $E_{t_A}$, and is given by

$$E_{t_A} = E_0 - (N_A t_A/\cos\theta_1)\varepsilon^A(\bar{E}_{in}). \tag{4.15}$$

At that depth, let the projectile scatter back from an atom B. The escaping particle thus has the energy

$$E_{1B,t_A} = K_B E_{t_A} - (N_A t_A/\cos\theta_2)\varepsilon^A(\bar{E}_{out}). \tag{4.16}$$

The unknowns $\bar{E}_{in}$ and $\bar{E}_{out}$ which appear in the argument of the stopping cross section $\varepsilon^A$ are energies intermediate to those which the particle has at the end point of each track. They could be evaluated by the surface energy or the mean energy approximations. The reason for retaining the general formulation in terms of $\bar{E}_{in}$ and $\bar{E}_{out}$ is that the energy shift

$$\Delta E_B^A \equiv K_B E_0 - E_{1B,t_A} \tag{4.17}$$

then has the form

$$\Delta E_B^A = K_B(N_A t_A/\cos\theta_1)\varepsilon^A(\bar{E}_{in}) + (N_A t_A/\cos\theta_2)\varepsilon^A(\bar{E}_{out}) \tag{4.18}$$

which is simply

$$= N_A t_A [\bar{\varepsilon}]_B^A, \tag{4.19}$$

according to Section 3.6. In the corresponding approximation the width $\Delta E_A$ of the absorber is given by $N_A t_A [\bar{\varepsilon}]_A^A$. The two widths are thus generally not equal, but are of a related magnitude.

### 4.3.2   Depth Scale of Underlying Film

In discussing the depth-to-energy conversion of the underlying layer B, it is conceptually useful to imagine that the absorber layer A has been physically separated from the layer B, as shown in Fig. 4.5a. We imagine further that the energy of the incident and backscattered particles can be measured there and label them as indicated. Figure 4.5b shows schematically the stopping cross sections as a function of energy for the two layers (light lines). The heavy lines identify the ranges of these curves covered by a particle scattered from an atom B at the interface, and at a depth $t_A + x_B$.

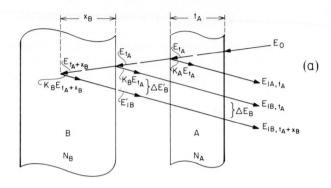

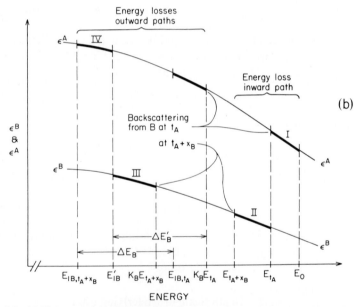

**Fig. 4.5**  (a) Blown-up view of the bilayered sample of Fig. 4.4 showing details of the back-scattering process and the meaning of the symbols used in the text, and (b) the cross section versus energy curves $\varepsilon^A(E)$ and $\varepsilon^B(E)$ of the two layers (light lines) with heavy segments to indicate the ranges covered by incoming and backscattered particles. The figure also shows the meaning of the symbols used in the text.

For the incident particles at the interface, the effect of the absorber is to reduce the incident energy from $E_0$ to $E_{t_A}$ (ignoring energy straggling). The relationship between $\Delta E_B'$ and $x_B$ is thus exactly that which would be observed in a backscattering experiment with particles of energy $E_{t_A}$ incident on B; hence

$$\Delta E_B' = [\varepsilon(\bar{E}^B)]_B^B N_B x_B, \tag{4.20}$$

where the subscript B has been added to the averaging bar to express the fact that the energies $\bar{E}_{in}$ and $\bar{E}_{out}$ at which the stopping cross section factor is evaluated must be taken along the inward and outward paths within layer B (regions II and III of Fig. 4.5b). For example, in the surface energy approximation—which here actually becomes an *interface energy approximation*—the stopping cross section factor would have the value $[K_B \varepsilon(E_{t_A})/\cos\theta_1] + [\varepsilon(K_B E_{t_A})/\cos\theta_2]$.

As the backscattered particles cross the absorber layer A on their outward path, additional energy is lost. This loss is not the same for a particle backscattered from an atom B at the interface as for one scattered deeper within B, because the two particles have different energies when they traverse A, and $\varepsilon^A$ differs for the two (see Fig. 4.5b). For scattering at the interface, the energy lost is $K_B E_{t_A} - E_{1B,t_A}$; for scattering within B, the loss is $E_{1B}' - E_{1B,t_A+x_B}$. In the interface energy approximation, the difference between these two energy losses depends on the difference of $\varepsilon^A$ at the energies $K_B E_{t_A}$ and $E_{1B}'$ of the particles before escaping through layer A on their way out. Since neither of these two energies is experimentally accessible, another approximation is to evaluate $\varepsilon^A$ at the detected energies $E_{1B,t_A}$ and $E_{1B,t_A+x_B}$, thus setting $K_B E_{t_A} - E_{1B,t_A} \simeq \varepsilon^A(E_{1B,t_A})N_A t_A/\cos\theta_2$ and $E_{1B}' - E_{1B,t_A+x_B} \simeq \varepsilon^A(E_{1B,t_A+x_B}) \cdot N_A t_A/\cos\theta_2$. The measured energy difference between two particles scattered from atoms B at the interface and at depth $x_B$ within the layer B then is

$$\Delta E_B = \Delta E_B' + \frac{N_A t_A}{\cos\theta_2}\left[\varepsilon^A(E_{1B,t_A+x_B}) - \varepsilon^A(E_{1B,t_A})\right] \tag{4.21}$$

or, with the previous equation,

$$\Delta E_B = N_B x_B[\varepsilon(\bar{E}^B)]_B^B + \frac{N_A t_A}{\cos\theta_2}\left[\varepsilon^A(E_{1B,t_A+x_B}) - \varepsilon^A(E_{1B,t_A})\right]. \tag{4.22}$$

The second term on the right-hand side of (4.22) expresses explicitly the effect of the absorber layer on $\Delta E_B$ due to energy losses in the outward paths. The effect of the absorber due to energy losses in the inward path is implicitly contained in the energy $\bar{E}^B$ at which the stopping cross section factor in B is evaluated. The second term depends on the difference of stopping cross sections $\varepsilon^A$, and is thus typically a correction. One way to see this is to expand

$\varepsilon^A(E)$ in a Taylor series, setting $E_{1B,t_A+x_B} = E_{1B,t_A} - \Delta E_B$. The difference in the $\varepsilon^A$s then becomes proportional to $\Delta E_B$, and one finds

$$\Delta E_B = \frac{[\varepsilon(\bar{E}^B)]_B^B}{1 + (N_A t_A/\cos\theta_2)(d\varepsilon^A/dE)_{E_{1B,t_A}}} N_B x_B. \qquad (4.23)$$

To this approximation, the effect of the absorber is thus twofold: (1) The stopping cross section factor $[\varepsilon]_B^B$ of the layer B has to be evaluated at energies (for the inward and outward tracks within the layer B) that are reduced because of the energy-absorbing effect of layer A on the incident particles; and (2) The scale is altered by a corrective multiplier that depends on the derivative of $\varepsilon^A(E)$ because of the energy-absorbing effect of layer A on the backscattered particles. As the absorber layer vanishes, the energy-to-depth conversion factor becomes $[\varepsilon(\bar{E})]_B^B$, as expected.

The treatment given here is similar in concept to that given in Section 3.5.2, where the connection between $\mathcal{E}'$ and $\mathcal{E}$ is derived for an elemental sample. The main difference is that in the present case $\Delta E_B'$ and $\Delta E_B$ are much larger than $\mathcal{E}'$ and $\mathcal{E}$, so that the differential treatment given for $\mathcal{E}'$ and $\mathcal{E}$ is not usually valid [compare Eqs. (3.49) and (4.2)].

### 4.3.3   Signal Heights at the Interface

Signal heights are related to the channel width $\mathcal{E}$. When expressed as an energy of a channel, $\mathcal{E}$ is always very small compared with the particle energy itself. For the present problem, it will thus be assumed that $\mathcal{E}$ and $\mathcal{E}'$, the energy width of a channel as projected back into the sample, are differential quantities. The treatment then becomes essentially that given in Section 3.5.2 for an elemental sample. The reader is referred to that section for a detailed comparison of the corresponding equations.

Let $H_A$ be the height of signal A at energy $E_{1A,t_A}$ as shown in Fig. 4.4. These events correspond to backscattering collisions from atoms A within a slab of thickness $\tau_A$ from the interface. The height of the signal thus is

$$H_A = \sigma_A(E_{t_A})\Omega Q N_A(\tau_A/\cos\theta_1). \qquad (4.24)$$

Similarly, one has for the slab $\tau_B$ within the interface

$$H_B = \sigma_B(E_{t_A})\Omega Q N_B(\tau_B/\cos\theta_1). \qquad (4.25)$$

Now $\tau_A$ and $\tau_B$ are defined such that particles backscattered within these slabs will all be registered in one channel when detected *outside* of the sample. *Inside* of the sample, at the depth $t_A$, the energy difference spanned by these two groups of particles is not $\mathcal{E}$, but $\mathcal{E}_A'$ and $\mathcal{E}_B'$, respectively. The connection between $\mathcal{E}_A'$ and $\tau_A$ is

$$\mathcal{E}_A' = [\varepsilon(E_{t_A})]_A^A N_A \tau_A \qquad (4.26)$$

and similarly

$$\mathscr{E}_B{}' = [\varepsilon(E_{t_A})]_B^B N_B \tau_B. \tag{4.27}$$

To determine how $\mathscr{E}_A{}'$ and $\mathscr{E}_B{}'$ relate to $\mathscr{E}$, one observes that all particles must escape through the same layer of absorber A along the same outward path. This constraint leads to

$$\mathscr{E}_A{}'/\mathscr{E} = \varepsilon^A(K_A E_{t_A})/\varepsilon^A(E_{1A,t_A}) \tag{4.28}$$

and

$$\mathscr{E}_B{}'/\mathscr{E} = \varepsilon^A(K_B E_{t_A})/\varepsilon^A(E_{1B,t_A}). \tag{4.29}$$

Expressing $N\tau$ in terms of $\mathscr{E}'$ and $\mathscr{E}'$ in terms of $\mathscr{E}$ leads to

$$H_A = \sigma_A(E_{t_A})\Omega Q \frac{\mathscr{E}}{[\varepsilon(E_{t_A})]_A^A \cos\theta_1} \frac{\varepsilon^A(K_A E_{t_A})}{\varepsilon^A(E_{1A,t_A})}, \tag{4.30}$$

$$H_B = \sigma_B(E_{t_A})\Omega Q \frac{\mathscr{E}}{[\varepsilon(E_{t_A})]_B^B \cos\theta_1} \frac{\varepsilon^A(K_B E_{t_A})}{\varepsilon^A(E_{1B,t_A})}. \tag{4.31}$$

The last factor in these expressions is the ratio of $\varepsilon^A$ taken at the energy of the backscattered particles as they penetrate the absorber after the collision at the interface, and as they leave the absorber and escape the sample. That ratio is typically close to unity and thus amounts to a correction. The ratio of the two heights is independent of $Q\Omega\mathscr{E}$;

$$\frac{H_A}{H_B} = \frac{\sigma_A(E_{t_A})}{\sigma_B(E_{t_A})} \frac{[\varepsilon(E_{t_A})]_B^B}{[\varepsilon(E_{t_A})]_A^A} \frac{\varepsilon^A(K_A E_{t_A})}{\varepsilon^A(E_{1A,t_A})} \frac{\varepsilon^A(E_{1B,t_A})}{\varepsilon^A(K_B E_{t_A})}. \tag{4.32}$$

Since both cross sections are evluated at the same energy $E_{t_A}$, their ratio is simply $(Z_A/Z_B)^2$. The last two correction factors tend to cancel each other so that, to first approximation,

$$\frac{H_A}{H_B} = \left(\frac{Z_A}{Z_B}\right)^2 \frac{[\varepsilon(E_{t_A})]_B^B}{[\varepsilon(E_{t_A})]_A^A}. \tag{4.33}$$

This ratio is that of the signal heights of A and of B if they were measured individually at the high-energy edge of a backscattering spectrum taken with an incident beam of energy $E_{t_A}$. The main effect of the absorber on the ratio of the signal heights for scattering at the interface is to reduce the incident beam energy from $E_0$ to $E_{t_A}$ in the stopping cross section factor for A and B.

### 4.3.4 Signal Height of Underlying Film at a Depth

The approach followed in this section is again basically that of Section 3.5.2. At any depth $x_B$ within layer B, i.e., at a depth of $t_A + x_B$ below the surface of the sample, a slab of thickness $\tau_B$ produces all those backscattering events which will be detected in a single channel of energy width $\mathscr{E}$. Immediately after the backscattering collision, however, these particles span an energy width $\mathscr{E}_B'$ that differs from $\mathscr{E}$. The slab thickness $\tau_B$ and $\mathscr{E}_B'$ are related to each other by the stopping cross section factor evaluated at the local projectile energy before scattering, which is $E_{t_A+x_B}$. The number of counts in the channel corresponding to the slab at depth $t_A + x_B$ thus is

$$H_B(E_{1B,t_A+x_B}) = \sigma_B(E_{t_A+x_B})\Omega Q(\mathscr{E}_B'/[\varepsilon(E_{t_A+x_B})]_B^B \cos\theta_1). \qquad (4.34)$$

The argument that leads to the relationship between $\mathscr{E}_B'$ and $\mathscr{E}$ has to be applied twice now, since the outward track passes through at thickness $x_B$ of B as well as through the full thickness $t_A$ of the absorber A. One finds

$$\mathscr{E}_B' = [\varepsilon^B(K_B E_{t_A+x_B})/\varepsilon^B(E_{1B}')][\varepsilon^A(E_{1B}')/\varepsilon^A(E_{1B,t_A+x_B})]\mathscr{E}, \qquad (4.35)$$

where $E_{1B}'$ is the energy of the backscattered particle at the point at which it crosses the interface between B and A (see Fig. 4.5a). For $x_B = 0$, this energy is $K_B E_{t_A}$ and the last equation reduce to Eq. (4.29) at the interface.

The last two formulas together give the number of counts per channel $H_B$ for the signal of B. The second formula describes the change in the energy width spanned by particles collected in one channel from the point of collision at $t_A + x_B$ to the detector. Usually, that change is of the order of unity, so that the first equation giving $H_B$ as a function of $\mathscr{E}_B'$ contains the main terms.

As these equations and those of the preceding subsection show, the key to the analysis of signal heights is to keep track of the projectile energies before scattering and after scattering at various points along the outward track (compare Fig. 4.5a and the energies in the preceding equations). The strongest effect of the absorber is on the value of the scattering cross section, because of the reduction of the energy along the inward pass across the usually less. The energy loss along the outward path across the absorber affects the energy width of particles collected in one channel. That contribution is usually the least significant one.

### 4.3.5   Total Number of Counts in Signal of Underlying Film

The assumption made in all calculations of backscattering spectra is that the attenuation of a beam penetrating into a sample affects only the energy of the particles, not their number. This is an excellent assumption because the probability of a large-angle scattering process is so low. With the help of this notion, it is simple to see how an absorber layer modifies the total number $A_B$ of counts in the signal of the underlying layer B.

$A_B$ would be the same if the backscattered particles had not traversed the absorber, because none of the particles headed for the detector are lost in the absorber (although their energy is changed). As far as the incident particles are concerned, the effect of the absorber A is to reduce their energy from $E_0$ to $E_{t_A}$, disregarding energy straggling. The total number of counts in the signal of B in the presence of the absorber is thus equal to the total number of counts observed without the absorber, but with an incident particle energy $E_{t_A}$ instead of $E_0$. With that modification, all the results derived in Section 4.2 for an elemental film without an absorber are valid also in the presence of an absorber film. The energy $E_{t_A}$ itself is computed from

$$E_{t_A} = E_0 - (N_A t_A/\cos\theta_1)\varepsilon^A(\bar{E}_{in}), \tag{4.36}$$

where $\bar{E}_{in}$ is an energy taken somewhere along the incident path across A. In the surface energy approximation, $\bar{E}_{in} = E_0$.

In conclusion, we stress that layer A acts on the incident particles and on the particles scattered back from layer B as an absorber of energy, not as an attenuator of the particle flux. What medium actually causes the energy loss does not matter. The results derived here and in the previous subsection for the total number of counts in the signal of B and in the previous subsection for the number of counts in a channel thus apply also for an absorber of multielemental composition. (Note that in these two subsections, A never appears as a subscript, only as a superscript.) The subject of a compound absorber is taken up again in Section 4.5.

### 4.4 ENERGY SPECTRUM OF A HOMOGENEOUS THIN FILM CONTAINING MORE THAN ONE ELEMENT (COMPOUND FILM)

In this section, we discuss the backscattering spectrum of a thin film composed of a homogeneous mixture of several elements. This case differs from that of a simple elemental film in two significant ways. First, as the probing particles penetrate the film, they lose energy as the result of interactions with more than one element. Consequently, the stopping cross section depends on the composition of the film and is therefore initially an unknown quantity. Second, when the probing particles are scattered at a specific point within the film, their remaining kinetic energy will depend on the mass of the particular atom they struck. This mass is different for the various elements in the film. The stopping cross section varies with energy, so that the energy the scattered particles lose along an identical outward track also depends on the atom struck in the scattering collision.

Assume, for simplicity, that the film contains two elements A and B in the atomic ratio $A_m B_n$. The results can be extended afterward to films with more than two elements. For a compound, $m$ and $n$ are integers, whereas for a solid solution, for example, they need not be. The backscattering spectrum of such a homogeneous film with two elements is sketched in Fig. 4.6b. Two signals

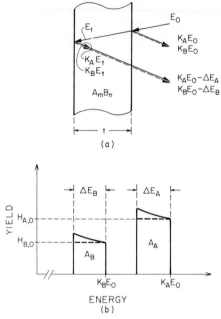

**Fig. 4.6** (a) Schematic representation of the backscattering process in a self-supporting compound thin-film sample, and (b) the resulting backscattering spectrum. The figure also shows the meaning of the symbols used in the text.

are observed, corresponding to scattering from the heavy atom A and the light atom B. As in the preceding section, the signal from the substrate is ignored.

### 4.4.1   Energy Width $\Delta E$ between High- and Low-Energy Edges of the Signals

The energy widths $\Delta E_A$ and $\Delta E_B$ generally differ by as much as 10% in spite of the fact that there is only one film thickness $t$, because, generally, $[\varepsilon]_A^{AB} \neq [\varepsilon]_B^{AB}$. This difference means that the depth-to-energy conversions are not the same for the two signals. Hence, the number of molecules per unit area (or molecular units per unit area in the case of a mixture) can be found in two different ways:

$$N^{AB}t = \Delta E_A / [\varepsilon]_A^{AB} \tag{4.37}$$

or

$$N^{AB}t = \frac{\Delta E_B}{[\varepsilon]_B^{AB}}. \tag{4.38}$$

### 4.4.2 Total Number of Counts in the Signals

As was pointed out initially, the composition of the film is generally unknown. This means that $[\varepsilon]_A^{AB}$ and $[\varepsilon]_B^{AB}$ cannot be computed, even when the elements A and B are known, because $m$ and $n$ are not. Equations (4.37) and (4.38) then are of no help in determining the number of molecules (or molecular units) per unit area in the film. As was shown in Section 4.2.2, however, the total number of counts in a signal is related to the number per unit area of the atoms in the target that generate the signal. The ratio $A_A/A_B$ of the total number of counts $A_A$ and $A_B$ in the spectrum of Fig. 4.6b is therefore related to the ratio $m/n$. The connection between the total numbers of counts $A_A$ or $A_B$ in signals A or B and the number $N_A^{AB}t$ and $N_B^{AB}t$ of atoms per unit area in the film thus constitutes the point of main interest in the backscattering spectrum of Fig. 4.6b. This relationship is investigated next.

The analysis follows along the lines of Section 4.2.2. The energy immediately before the collision is

$$E = E_0 - (N^{AB}x/\cos\theta_1)\varepsilon^{AB}(\bar{E}_{in}) \tag{4.39}$$

and the total number of counts $A_A$ in the signal of A is

$$A_A = (\Omega Q m N^{AB}/\cos\theta_1) \int_0^t \sigma_A(E)\,dx. \tag{4.40}$$

The factor $m$ appears in this equation since each compositional unit of the volume density $N^{AB}$ contains $m$ atoms of A. Formally, the last two relationships flow out of Eqs. (4.5) and (4.6) by the substitutions

$$N \to N^{AB}, \qquad \varepsilon(\bar{E}_{in}) \to \varepsilon^{AB}(\bar{E}_{in}), \qquad \sigma(E) \to \sigma_A(E), \qquad A \to A_A/m. \tag{4.41}$$

Their interpretation is straightforward, except perhaps for the last substitution $A \to A_A/m$. The division by $m$ comes about because $N^{AB}$ is the volume concentration of molecular units. When $N$ in Eqs. (4.5) and (4.6) is reinterpreted as the concentration $N^{AB}$ of such molecular units, as this substitution does, then the total number of counts $A$ in Eq. (4.6) becomes that which would be generated with *one* scattering center per molecular unit. But each molecular unit contains $m$ atoms of type A. The total number of counts $A_A$ actually measured is thus $m$ times larger than that. It is thus $A_A/m$ that corresponds to the total counts $A$ in the right-hand side of Eq. (4.6). An analogous result holds for B:

$$A_B = (\Omega Q n N^{AB}/\cos\theta_1) \int_0^t \sigma_B(E)\,dx. \tag{4.42}$$

With the set of substitutions just listed, the results derived in Section 4.2.2 for a monoelemental film can be translated immediately for the present case. Alternatively, the derivations outlined there may be repeated, starting with Eqs. (4.39) and (4.40).

**a.  Total Number of Counts in the Surface Energy Approximation.**  In this case, the two signals form essentially rectangular steps (dashed lines in Fig. 4.6). Let $A_{\mathrm{A},0}$ and $A_{\mathrm{B},0}$ denote the total number of counts in those signals. One finds, in an analogy to Eq. (4.8), that

$$mN^{\mathrm{AB}}t = (A_{\mathrm{A},0}/\sigma_{\mathrm{A}}(E_0)\Omega Q)\cos\theta_1 \qquad (4.43)$$

and

$$nN^{\mathrm{AB}}t = (A_{\mathrm{B},0}/\sigma_{\mathrm{B}}(E_0)\Omega Q)\cos\theta_1. \qquad (4.44)$$

The ratio of the atomic densities $N_{\mathrm{A}}^{\mathrm{AB}} = mN^{\mathrm{AB}}$ and $N_{\mathrm{B}}^{\mathrm{AB}} = nN^{\mathrm{AB}}$ is given by

$$N_{\mathrm{A}}^{\mathrm{AB}}/N_{\mathrm{B}}^{\mathrm{AB}} = m/n = [A_{\mathrm{A},0}/\sigma_{\mathrm{A}}(E_0)]/[A_{\mathrm{B},0}/\sigma_{\mathrm{B}}(E_0)]. \qquad (4.45)$$

Since all quantities on the right-hand side are experimentally accessible, the ratio $m/n$ can be obtained directly from the thin-film spectrum. The ratio of the concentrations of atoms A and B follows without the knowledge of the solid angle of detection $\Omega$ and the total number $Q$ of incident particles. The explanation of this result resides simply in the fact that both types of atoms are exposed to the beam simultaneously, and the backscattered particles are detected at the same time with the same detector system. In the ratio, these common factors cancel. If $\Omega$ and $Q$ are known, as is the case for an absolutely calibrated system, then the individual atomic concentrations per unit area can be derived as well from

$$N_{\mathrm{A}}^{\mathrm{AB}}t = (A_{\mathrm{A},0}/\sigma_{\mathrm{A}}(E_0)\Omega Q)\cos\theta_1 \qquad (4.46)$$

or

$$N_{\mathrm{B}}^{\mathrm{AB}}t = (A_{\mathrm{B},0}/\sigma_{\mathrm{B}}(E_0)\Omega Q)\cos\theta_1. \qquad (4.47)$$

**b.  Total Number of Counts for a More General Case.**  When the signals of A and B are not rectangular (solid lines in Fig. 4.6), an inverse quadratic energy dependence of the cross sections $\sigma_{\mathrm{A}}(E)$ and $\sigma_{\mathrm{B}}(E)$ provides an improved approximation for the analysis. With the substitutions discussed above and with $N_{\mathrm{A}}^{\mathrm{AB}} = mN^{\mathrm{AB}}$, $N_{\mathrm{B}}^{\mathrm{AB}} = nN^{\mathrm{AB}}$, one then finds from Eq. (4.12) that

$$N_{\mathrm{A}}^{\mathrm{AB}}t = \frac{A_{\mathrm{A}}}{\sigma_{\mathrm{A}}(E_0)\Omega Q}\cos\theta_1\{1 + [(\varepsilon^{\mathrm{AB}}(\bar{E}_{\mathrm{in}})A_{\mathrm{A}}/m)/\sigma_{\mathrm{A}}(E_0)\Omega Q E_0\cos\theta_1]\}^{-1}$$

$$(4.48)$$

or

$$N_{\mathrm{B}}^{\mathrm{AB}}t = \frac{A_{\mathrm{B}}}{\sigma_{\mathrm{B}}(E_0)\Omega Q}\cos\theta_1\{1 + [(\varepsilon^{\mathrm{AB}}(\bar{E}_{\mathrm{in}})A_{\mathrm{B}}/n)/\sigma_{\mathrm{B}}(E_0)\Omega Q E_0\cos\theta_1]\}^{-1}.$$

$$(4.49)$$

The ratio $N_A^{AB}t/N_B^{AB}t$ is the atomic ratio $m/n$, but when this ratio is formed with these two equations, $m$ and $n$ are seen to appear on the right-hand side as well. Now $\varepsilon^{AB} = m\varepsilon^A + n\varepsilon^B$, so that the right-hand sides are functions of the ratio $m/n$ only and an explicit solution for $m/n$ exists. One finds

$$m/n = [A_A/\sigma_A(E_0)]/[A_B/\sigma_B(E_0)], \tag{4.50}$$

a relation which can also be obtained directly from the ratio of Eqs. (4.40) and (4.42) since the energy dependence is the same for both cross sections. This result states that even when the signals of A and B in a backscattering spectrum of a thin film are not flat-topped, the atomic ratio of the two constituting elements is given to an advanced approximation by the ratio of the total number of counts in each signal, properly normalized to a common cross section. Neither the stopping cross section of the medium nor the total number of incident particles and the solid angle of detection have to be known. The identity of the elements A and B can be deduced from the position of the high-energy edges of each signal via the kinematic factor $K_M$. The scattering cross sections then follow from tables. Hence, the atomic ratio of the two atoms that constitute a uniform film can be deduced from a backscattering spectrum such as Fig. 4.6 alone.

Once the atomic ratio and the identity of each of the two elements, A and B, are known, the stopping cross sections $\varepsilon^{AB}/m = \varepsilon^A + (n/m)\varepsilon^B$ and $\varepsilon^{AB}/n = (m/n)\varepsilon^A + \varepsilon^B$, which appear in expressions (4.48) and (4.49), can be computed. The number of atoms per unit area $N_A^{AB}t$ or $N_B^{AB}t$ for elements A and B can then be calculated from the total number of counts in each signal if the total number of incident particles $Q$ and the solid angle of detection $\Omega$ are known. If they are unknown, the number of molecular units per unit area $N^{AB}t$ that contain $m$ atoms of A and $n$ atoms of B can be deduced from the energy widths $\Delta E_A$ and $\Delta E_B$ and Eqs. (4.37) and (4.38).

As for the monatomic film, there are thus two ways to determine the amount of substance contained in the compound film. One approach is based on the total number of counts in a signal and yields the number of atoms A or B per unit area in the film, but the total number of incident particles and the solid angle of detection must be known. When these are not known, the other approach will yield the number of molecular units $mA + nB$ per unit area $N^{AB}t$ from the measurement of $\Delta E_A$ or $\Delta E_B$, and the values of $[\varepsilon]_A^{AB}$ or $[\varepsilon]_B^{AB}$.

The backscattering spectrum of a uniform thin film containing more than two atomic species is interpreted in a similar fashion. A molecular unit then consists of $mA + nB + \cdots = \sum_i n_i A_i$ atoms in all, and the volume density of these units is $N^{mol}$. The stopping cross section is $\varepsilon^{mol} = m\varepsilon^A + n\varepsilon^B + \cdots = \sum_i n_i \varepsilon^{A_i}$, assuming that Bragg's rule holds. With these changes, all the relationships derived in this section remain valid for the general case, A and B standing for any pair of atom types in the target.

### 4.4.3   Ratio of Signal Heights in
### Surface Energy Approximation

The ratio $m/n$ of the relative concentrations $m$ and $n$ of the elements A and B in a homogeneous bielemental film can also be obtained from the heights $H_{A,0}$ and $H_{B,0}$ of the signals of A and B (see Fig. 4.6). Scattering from the top surface layer of such a sample is the same as for a bulk sample. The treatment given in Section 3.7.1 thus applies equally well to the present case of a thin compound film.

There are therefore two different ways of finding the ratio $m/n$. One approach (Section 4.4.2) uses the total number of counts in the signals of A and of B, and the stopping cross section factors need not be known. The other approach uses the height of the signals at their high-energy edge (as shown in Section 3.7.1). In this case, the ratio $[\varepsilon_0]_A^{AB}/[\varepsilon_0]_B^{AB}$ of the stopping cross section factors enters into the result. That ratio is usually close to unity, and can be attained by iteration.

From the knowledge of $m/n$, the relative compositions $m$ and $n$ follow with the condition that $m + n = 1$. It is not possible, however, to state that therefore the sample is composed of the chemical compound $A_m B_n$, even if $m$ and $n$ are fractions of small integers. Backscattering spectra only provide information on relative atomic composition. How these atoms are combined, i.e., the chemical constitution of the sample, must be deduced from other experiments, such as chemical analyses or x-ray diffraction.

### 4.4.4   Overlap of Individual Signals

As the compound film gets thicker, the width $\Delta E$ for each element in the film increases ($\Delta E_A$ and $\Delta E_B$ in Fig. 4.6). The high-energy edge of every signal is fixed in its position on the energy axis ($K_A E_0$ and $K_B E_0$ in Fig. 4.6) because, in a homogeneous sample, all elements are present right to the surface of the sample. As the width of a signal increases, it will eventually overlap with the signal of a lighter element. Specifically, for the case of Fig. 4.6, the signal of A will overlap with the signal of B if $\Delta E_A \geq (K_A - K_B)E_0$. Now $\Delta E_A$ increases with the thickness of the film as $\Delta E_A = N^{AB} t [\varepsilon]_A^{AB}$. The thickness at which overlap between the signals of A and B will occur thus is

$$N^{AB}t = (K_A - K_B)E_0/[\varepsilon]_A^{AB}. \tag{4.51}$$

Whether two signals overlap depends on the difference of their kinematic factors and on the incident energy $E_0$. The stopping cross section factor $[\varepsilon]_A^{AB}$ typically decreases a little with increasing $E_0$ so that slightly overlapping signals can be separated by increasing the incident energy of the beam. As the thickness of the film gets large, all signals overlap and the spectrum becomes that of a thick compound sample. In the case of two elements, the spectrum of Fig. 4.6 will change to that of Fig. 3.17. The thickness $t$ in the preceding expression (4.51) then gives the depth of a thick sample which can be probed by the signal of A without an interference from the signal of B. Beyond this depth, the signal of A must be obtained by first subtracting the signal of B from the total spectrum, which reduces the accuracy of the information derived from the spectrum.

## 4.5   ENERGY SPECTRUM OF MULTILAYERED FILMS CONTAINING MORE THAN ONE ELEMENT (LAYERED COMPOUND FILMS)

This section combines the subject of multilayered elemental films (Section 4.3) with that of compound film (Section 4.4). There are many different cases of such multilayered compound films: a compound film on an elemental film, an elemental film on a compound film, a compound film on another compound film, films of an element common to the compound film or of an element foreign to the compound film, the heaviest element being either in the top layer or beneath, and so on. The formulas describing the spectrum of all these examples will be different in each case, but the concepts from which they are derived are the same in every case. We shall therefore give a detailed treatment of one particular case only to demonstrate the procedure by way of that example.

In general, the spectrum of a multilayered sample with compound films is complex because the signals of the various elements in the layers overlap. To identify certain features of the spectrum with signals generated from a specific element in a specific layer of the target is usually a nontrivial task. This job is greatly facilitated if several spectra of the same sample are taken under different experimental conditions (sample normal and tilted with respect to the incident beam, several incident energies, etc.). We are not concerned here with these questions of proper interpretation of a spectrum. Rather, we assume that the subdivision of the spectrum in its individual signals has been accomplished already. It is then convenient for the discussion to assume that the signals of the various elements do not overlap, as will be done in the succeeding sections although this condition is generally an exception rather than the rule.

The top layer of a multilayered sample can be analyzed as described in Section 4.2 if the layer is elemental, or as in Section 4.4 if it is a compound film, since the underlying layers do not affect the spectrum of the top layer, other than possibly an overlap of signals. Once the top layer is analyzed, the energy spectrum of the next underlying films can be treated as shown in Section 4.3. There the top layer is assumed to be elemental, but, as was pointed out in the end of that section, the treatment is the same for a compound absorber. After the second layer has been analyzed, the first two layers together are considered as one absorber to the third layer. The analysis of that layer is undertaken next, and the process is iterated as necessary until all layers have been analyzed. Although this process is logically simple, its execution and the formulas rapidly become long and cumbersome. A comparison with numerically computed spectra may be easier.

In this section, we demonstrate this process in detail for the particular case of a sample composed of a thin film of element A on top of a compound film of composition $A_m B_n$. This situation is encountered when a thin film of A reacts at the interface with the substrate B and starts to form a compound $A_m B_n$ there. A schematic diagram of the backscattering energy spectrum of such a sample is shown in Fig. 4.7. For simplicity, A is assumed to be the heavier element of the two, and the total amount of A is taken to be small enough so that no signals overlap. The figure also gives the notation, which is fairly cumbersome but necessary. In the superscripts, the compound $A_m B_n$ is abridged to AB. Recall that superscripts identify the medium and that

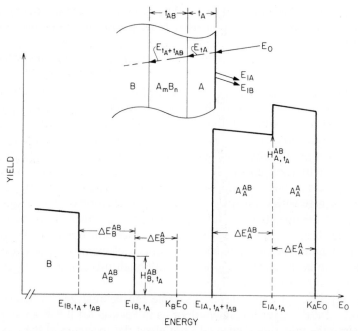

**Fig. 4.7**  (a) Schematic representation of the backscattering process in a sample composed of a thin surface film of a monoisotopic heavy element A, a substrate of a monoisotopic light element B, and a compound layer $A_m B_n$ between, and (b) the resulting backscattering spectrum. The figure also shows the meaning of the symbols used in the text.

subscripts give the element to which the quantity refers. The succeeding subsections discuss the main quantities of interest in this spectrum.

### 4.5.1   Energy Shift Due to Elemental Absorber

In Section 4.3.1 we showed that when an elemental absorber A is placed in front of an elemental film B, the backscattering signal of that film is shifted to lower energies. What is different here is that the underlying layer is made of a compound $A_m B_n$ rather than of an element B only. There are, correspondingly, two shifts to be considered here, namely $\Delta E_B^A$ and $\Delta E_A^A$. For $\Delta E_B^A$, the situation is the same as in Section 4.3.1 and one has

$$\Delta E_B^A, = N_A t_A [\varepsilon]_B^A. \tag{4.52}$$

When the scattering at the interface between the absorber layer A and the compound layer $A_m B_n$ is from an atom of type A, rather than B, one has

$$\Delta E_A^A = N_A t_A [\varepsilon]_A^A. \tag{4.53}$$

The stopping cross section factors in the last two equations differ only by the energy of the outgoing path and are usually within about 10% of each other.

### 4.5.2   Depth Scale of an Underlying Compound Film

A quantity of interest for the compound film is $N^{AB} t_{AB}$, the number of molecular units per unit area in the thickness $t_{AB}$ of the film. Expressions relating this quantity to the signal width of the underlying layer are given in Section 4.3.2. There the underlying layer is elemental, so that there is only one signal width $\Delta E_B$, and the expressions contain the elemental stopping cross section $\varepsilon^B$ and the elemental stopping cross section factor $[\varepsilon]_B^B$. In the present case, the underlying layer contains two elements, A and B, so that the layer generates two signals of widths $\Delta E_A^{AB}$ and $\Delta E_B^{AB}$, and the expressions relating $N^{AB} t_{AB}$ to these widths will depend on the stopping cross section $\varepsilon^{AB}$ and the stopping cross section factors $[\varepsilon]_A^{AB}$ and $[\varepsilon]_B^{AB}$ of the compound. For both signals, the derivation and the formulas are analogous to those given in Section 4.3.2. For example, Eq. (4.22) becomes

$$N^{AB} t_{AB} = \frac{1}{[\varepsilon(\bar{E}^{AB})]_A^{AB}} \left\{ \Delta E_A^{AB} + \frac{N_A t_A}{\cos \theta_2} \left[ \varepsilon^A(E_{1A,t_A}) - \varepsilon^A(E_{1A,t_A + t_{AB}}) \right] \right\} \tag{4.54}$$

or

$$= \frac{1}{[\varepsilon(\bar{E}^{AB})]_B^{AB}} \left\{ \Delta E_B^{AB} + \frac{N_A t_A}{\cos \theta_2} \left[ \varepsilon^A(E_{1B,t_A}) - \varepsilon^A(E_{1B,t_A + t_{AB}}) \right] \right\}. \tag{4.55}$$

The superscript AB is attached to the averaging bar of $E$ as a reminder that the energies in the argument of the stopping cross section factor have to be taken along the inward and outward paths within the compound layer AB. The first of these equations expresses $N^{AB}t_{AB}$ in terms of the signal from atom A in the compound layer, and the second formula expresses it in terms of the signal from atom B. If $\varepsilon^A$ is a weak function of energy or $N_A t_A$ is small enough, expressions (4.54) and (4.55) reduce to those derived in Section 4.4.1 for the analysis of a compound layer at the sample surface [cf. Eqs. (4.37) and (4.38)].

### 4.5.3   Signal Heights of Underlying Compound Film

The ratio $m/n$ can be obtained from the heights of two signals taken at energies corresponding to the same depth. In a thin-film spectrum, energies corresponding to the same depth are readily identified by the edges of signals associated with the same interface. The present example contains two interfaces, one at $x = t_A$ and the other at $x = t_A + t_B$. Signals of A and of B in the compound layer coming from either of these two interfaces can be used to determine $m/n$. Let us choose the interface at $x = t_A$, for example. This choice will obviously lead to the simpler equations, because for scattering from that interface the only absorber to consider is the elemental layer A. The signal heights associated with that interface are labeled $H^{AB}_{A,t_A}$ and $H^{AB}_{B,t_A}$ in Fig. 4.7.

The problem now is analogous to that treated in Section 4.3.3. One difference is that now both signals emanate from the underlying compound layer. It is therefore $[\varepsilon(E_{t_A})]^{AB}_{A\,or\,B}$ that will appear in connection with the energy widths $\mathscr{E}'_{A\,or\,B}$. The other difference is that each molecular unit contains $m$ atoms of A and $n$ of B, so that a multiplicative factor is introduced—of $m$ for $\sigma_A$ and $n$ for $\sigma_B$. One then finds, from Eqs. (4.30) and (4.31), that

$$H^{AB}_{A,t_A} = m\sigma_A(E_{t_A})\Omega Q \frac{\mathscr{E}}{[\varepsilon(E_{t_A})]^{AB}_A \cos \theta_1} \frac{\varepsilon^A(K_A E_{t_A})}{\varepsilon^A(E_{1A,t_A})}, \qquad (4.56)$$

$$H^{AB}_{B,t_A} = n\sigma_B(E_{t_A})\Omega Q \frac{\mathscr{E}}{[\varepsilon(E_{t_A})]^{AB}_B \cos \theta_1} \frac{\varepsilon^A(K_B E_{t_A})}{\varepsilon^A(E_{1B,t_A})}, \qquad (4.57)$$

as can be verified by following the derivation given in Section 4.3.3; hence

$$\frac{m}{n} = \frac{H^{AB}_{A,t_A}}{H^{AB}_{B,t_A}} \frac{\sigma_B(E_{t_A})}{\sigma_A(E_{t_A})} \frac{[\varepsilon(E_{t_A})]^{AB}_A}{[\varepsilon(E_{t_A})]^{AB}_B} \frac{\varepsilon^A(E_{1A,t_A})}{\varepsilon^A(K_A E_{t_A})} \frac{\varepsilon^A(K_B E_{t_A})}{\varepsilon^A(E_{1B,t_A})}. \qquad (4.58)$$

This equation is analogous to Eq. (4.32). The last two terms contain the effect of the absorber. These two factors can usually be ignored, since their product is usually within a few percent of unity. The rest of the equation gives the height ratio for signals generated by backscattering from the surface of a compound $A_m B_n$ with particles of incident energy $E_{t_A}$ [Eq. (3.72)].

If the height ratio had been formed for the two signals of A and B associated with the interface at $t_A + t_{AB}$, the derivation would follow the same pattern, but the final formula would be complicated by the effect of yet another absorber layer in the form of the compound layer itself. The signal height for such a case is derived in Section 4.3.4 [Eqs. (4.34) and (4.35)]. One can readily see that the signal heights at the interface $t_A + t_{AB}$ are given by equations such as those given previously, but with the following modifications: (i) A local energy $E_{t_A + t_{AB}}$ must be used for the cross sections $\sigma_A$ and $\sigma_B$ and for both $[\varepsilon]_A^{AB}$ and $[\varepsilon]_B^{AB}$; (ii) The energies at which the ratio of the stopping cross sections $\varepsilon^A$ must be taken are those before ($E'_{1A}$ and $E'_{1B}$) and after ($E_{1A, t_A + t_{AB}}$ and $E_{1B, t_A + t_{AB}}$), the outward path of the projectile through the absorber layer A; (iii) An additional ratio of stopping cross sections $\varepsilon^{AB}$ will appear, taken at energies immediately after the backscattering collision at the interface ($K_A E_{t_A + t_{AB}}$ and $K_B E_{t_A + t_{AB}}$) and after the outward traverse through the compound layer ($E'_{1A}$ and $E'_{1B}$). The effect of these two correction factors for the two absorbers on the heights of the signals will be cumulative, but the correction on the ratio of $H_{A, t_A + t_{AB}}^{AB}$ and $H_{B, t_A + t_{AB}}^{AB}$ will be much less because the factors tend to cancel each other.

When the correction factors for the absorber are ignored in the ratio of the heights of the signals of A and B coming from the same depth $x$, the ratio of $m/n$ takes the simple form

$$\frac{m}{n} \simeq \frac{H_{A,x}^{AB}}{H_{B,x}^{AB}} \left( \frac{Z_B}{Z_A} \right)^2 \frac{[\varepsilon(E_x)]_A^{AB}}{[\varepsilon(E_x)]_B^{AB}}, \tag{4.59}$$

which constitutes a close and useful approximation analogous to Eq. (4.33).

### 4.5.4  Total Number of Counts in Signals of Underlying Compound Film

The total number of counts in the signals of A and B from the compound layer beneath the absorber is best derived by combining the arguments given in Section 4.4.2 for the total number of counts in the signals of a compound layer at the surface, and in Section 4.3.5 for the influence of an absorber on the total number of counts in a signal below an absorber. There it is argued that, as far as the total number of counts in a signal is concerned, an absorber layer only attenuates the energy of the incoming particles from $E_0$ to $E_{t_A}$. Along the outgoing path, the effect of the absorber is to lower the particle energy still further, but the number of particles detected is unchanged by that process. In Section 4.4.2 it is shown that because the energy dependence of the Rutherford cross section is the same for all elements, the ratio $m/n$ for a compound layer at the surface is given by the ratio of $A/\sigma(E_0)$ of each signal [Eq. (4.50)]. Combined, these two conclusions state that in the present case

$$m/n = [A_A^{AB}/\sigma_A(E_{t_A})]/[A_B^{AB}/\sigma_B(E_{t_A})], \tag{4.60}$$

which is equal to

$$m/n = (A_A^{AB}/A_B^{AB})/(Z_B/Z_A)^2. \tag{4.61}$$

The ratio of the total number of counts in two signals originating from a compound film is independent of the presence of an absorber.

### 4.6  INFLUENCE OF ENERGY STRAGGLING AND SYSTEM RESOLUTION

So far in this chapter, spectra have been treated as having sharp, steplike discontinuities corresponding to backscattering events at the surface, or at interfaces below the surface, as sketched in Fig. 4.8a for a thin elemental

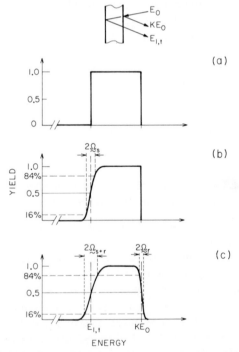

**Fig. 4.8**  (a) Ideal energy spectrum of a thin film, neglecting energy straggling and system resolution. (b) Spectrum of (a) as modified by the energy straggling $\Omega_s$. (c) Spectrum of (b) as modified by the system resolution $\Omega_r$.

film. This idealized situation does not correspond to reality. There are two quite unrelated reasons.

One reason is energy straggling. This effect smoothes out the steps of backscattering signals emanating from interfaces beneath the sample surface. Backscattering from the sample surface itself does not involve penetration of the sample, so that the backscattering signal from the surface of the sample remains unaltered by energy straggling. A schematic backscattering spectrum of a thin elemental film including energy straggling is shown in Fig. 4.8b.

The other reason is purely instrumental. Every parameter of an experimental system is subject to statistical fluctuations of some kind. In backscattering spectrometry, the fluctuations that most directly affect a spectrum are variations in the energy $E_0$ of the incident particles, and noise in the detector and in the signal processing chain. To describe the effect of these fluctuations on a backscattering spectrum accurately, it would be necessary to know the cause and the statistical nature of each of these fluctuations. Fortunately, this amount of elaboration is seldom required. In a typical case, the fluctuations from all major experimental causes can be lumped together as if generated by one single source. This source is almost always very closely Gaussian and can be characterized by the standard deviation $\Omega_r$. This quantity, or some linear multiple of it, is commonly referred to as the *system resolution*.

The system resolution introduces variations in the measured energy of all particles. As a consequence, the high-energy edge of a signal is now smoothed out as well, as sketched in Fig. 4.8c, and edges at lower energies are broadened further because energy straggling and system resolution both contribute to the energy fluctuations. In this section we discuss these effects in some detail.

### 4.6.1  Influence of Energy Straggling on a Thin-Film Spectrum

Energy straggling is present in the spectrum of any sample, thick or thin, but the effect is most apparent in thin elemental films and layered structures. This discussion thus deals with energy straggling in thin films in particular.

In Section 2.6, we mention that not many accurate measurements of energy straggling exist. Because of this lack of a data base, Bohr's theory of energy straggling is frequently used as a guideline for energy straggling evaluation. The theory also has the advantage of being simple. Its predictions are known to be in error, perhaps by as much as a factor of two or more in the worst cases (e.g., for $^4$He at incident energies below 1 MeV and samples of heavy atoms). Fortunately, energy straggling is a second-order effect and rarely needs to be known with accuracy.

According to this theory, monoenergetic particles of incident energy $E_0$, which penetrate a distance $x$ into a target, have a Gaussian of energy distribution with a variance $\Omega_B^2 = 4\pi(Z_1e^2)^2Z_2Nx$ centered on an average energy $E_0 - \varepsilon Nx$ (see Section 2.6). To apply this result to the backscattering signal of a thin film of thickness $t$, we consider the particles that are backscattered at the rear interface of the film. They travel a total length

$$l = (t/\cos\theta_1) + (t/\cos\theta_2) \qquad (4.62)$$

through the film. Since these particles contribute most of the counts in the vicinity of the low-energy edge of the signal at $E_{1,t}$, one would expect that this edge has a width that is determined by the variance

$$\Omega_s^2 = 4\pi(Z_1e^2)^2Z_2Nl. \qquad (4.63)$$

Indeed, if those particles that are scattered back in the film shortly before they reach the rear interface also had this same variance $\Omega_s^2$, the problem would be one of a simple convolution of the low-energy step of the ideal signal with a Gaussian of variance $\Omega_s^2$. In that case, the functional form of the broadened signal is an error function

$$g(E_1) = \tfrac{1}{2} + \tfrac{1}{2}\operatorname{erf}((E_1 - E_{1,t})/\sqrt{2\Omega_s}). \qquad (4.64)$$

In fact, the problem is more complicated, for two reasons. First, particles that travel a slightly shorter distance than $l$ have a slightly smaller variance than $\Omega_s^2$. Second, the particles lose some of their energy in the backscattering collision. As a consequence, the energy variations originating along the incident path weigh less than those originating along the outward path. These questions are treated in more detail in Appendix B. It is found that the error function $g(E_1)$ given in Eq. (4.64) approximates the real situation very closely in all practical cases. It is also shown there that because of the energy loss in the backscattering collision, the variance is generally given by

$$\Omega_s^2 = K^2\Omega_{in}^2 + \Omega_{out}^2, \qquad (4.65)$$

where $\Omega_{in}^2$ and $\Omega_{out}^2$ are the variances associated with the inward and outward path, respectively, i.e.,

$$\Omega_{in}^2 = 4\pi(Z_1e^2)^2Z_2Nt/\cos\theta_1 \qquad \text{and} \qquad \Omega_{out}^2 = 4\pi(Z_1e^2)Z_2Nt/\cos\theta_2.$$

Figure 4.9 shows the effect of energy straggling on backscattering spectra of gold films of various thicknesses. The spectra have been computed with the program of Ziegler et al. (1976), using Bohr's theory to account for energy straggling. The corresponding ideal spectra, free of energy straggling, are shown as well. The error function solution previously discussed follows the computed low-energy edges of the spectra very closely, as indicated by the two points given for each curve at 16 and 84% of the ideal step height. These points lie on the error function $g(E_1)$ at $\pm\Omega_s$ on either side of the position $E_{1,t}$ of the ideal step.

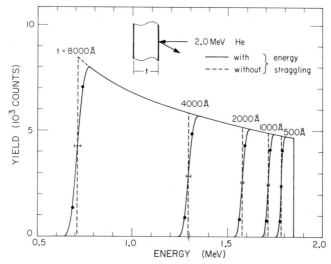

**Fig. 4.9**   Backscattering spectra of 2.0 MeV He computed for Au films of 500 to 8000 Å thickness. The dashed lines are ideal spectra. The solid lines include energy straggling as given by Bohr's theory, i.e., $\Omega_B{}^2 = s^2 t$, where $s^2 = 4\pi(Z_1 e^2)^2 N z_2$. The two points at 16 and 84% of the ideal step height lie on the erf-approximation at $\pm \Omega_s$ and show that this approximation is very close to the computed solution. The error bars at the half-height points give the width $\pm\sqrt{2\bar{E}\,\Delta E} \times 10^{-2}$ estimated from Eq. (4.67).

The rule of thumb developed in Section 2.6 to find the magnitude of $\Omega_B$ can be applied to thin-film spectra as well. To this end, $\Omega_{in}^2$ and $\Omega_{out}^2$ in the last displayed expression are estimated from the energy loss $\Delta E_{in}$ and $\Delta E_{out}$ according to $\Omega_B/\Delta E = 10^{-2}(\bar{E}/\Delta E)^{1/2}$ (Eq. 2.60), where the factor $10^{-2}$ is valid for $^4$He projectiles. In general, $\Delta E_{in}$ and $\Delta E_{out}$ are different, but, in the spirit of an estimate, one can apply the symmetrical mean approximation and assume that they are both equal. It then follows that

$$\Omega_s/\Delta E = (\bar{E}/\Delta E)^{1/2}\sqrt{2} \times 10^{-2}[(K^2 + 1)^{1/2}/(K + 1)], \qquad (4.66)$$

$$\simeq (\bar{E}/\Delta E)^{1/2}\sqrt{2} \times 10^{-2}, \qquad (4.67)$$

where $\Delta E$ now is the width of the backscattering signal of the thin film. The factor $\sqrt{2}$ accounts for the twofold traverse of the particle through the film. The spread of $\pm \Omega_s$, estimated by this formula for the various thicknesses of gold films shown in Fig. 4.9, is also given there as error bars at the half-height point of the low-energy edge. As can be seen, the estimate is in very fair agreement with the actual width of the signal edge. Equation (4.67) is a handy formula for quick estimates of the magnitude of energy straggling at the low-energy edge of a backscattering signal.

The essential result of this section is summarized in Figs. 4.8a and b. In the ideal case, that is, when the statistical fluctuations in the energy loss mechanism of the projectiles in the target are neglected, and when the finite system resolution is ignored, a backscattering spectrum of a thin film has sharp, steplike front and rear edges, as shown in Fig. 4.8a. Energy straggling produces a smooth transition in the low-energy edge of the signal, as indicated in Fig. 4.8b. The width of this transition is characterized by the standard deviation $\Omega_S$ of the energy straggling. This value increases with the thickness of the film, that is, with increasing $E_t$. Backscattering from the surface of the film is not affected by the statistical fluctuations of $dE/dx$ losses since penetration into the target has not occurred. The high-energy edge of the backscattering signal thus remains sharp.

### 4.6.2  Influence of System Resolution on a Thin-Film Spectrum

It is very convenient to represent formally all random fluctuations of experimental origin by a single source whose energy distribution is Gaussian and whose standard deviation defines $\Omega_r$. In fact, however, $\Omega_r$ may be the result of a number of independent causes. How they combine to a resultant total of $\Omega_r$ must be investigated in each particular case.

As an example, assume that the incident beam is not monoenergetic, but has an energy profile with standard deviation $\Omega_{beam}$. Furthermore, say that the detection system can be characterized by a standard deviation $\Omega_{det}$. Let both these distributions be closely Gaussian. If energy straggling along the incident path has a variance $\Omega_{in}^2$ at some depth $x$, the total variance in the particle energy at that depth is $\Omega_{beam}^2 + \Omega_{in}^2$. After a backscattering collision, this variance is reduced to $K^2(\Omega_{beam}^2 + \Omega_{in}^2)$. Along the outward path, energy straggling adds the contribution $\Omega_{out}^2$ to the variance, and after detection it has increased once more by $\Omega_{det}^2$. The total variance thus is

$$\Omega_{s+r}^2 = K^2(\Omega_{beam}^2 + \Omega_{in}^2) + \Omega_{out}^2 + \Omega_{det}^2 \tag{4.68}$$

or

$$\Omega_{s+r}^2 = (K^2\Omega_{in}^2 + \Omega_{out}^2) + (K^2\Omega_{beam}^2 + \Omega_{det}^2). \tag{4.69}$$

The first term is the contribution $\Omega_s^2$ of energy straggling [Eq. (4.65)]. This term increases with increasing depth at which scattering occurs. The second term is independent of the scattering depth and represents the system resolution,

$$\Omega_r^2 = K^2\Omega_{beam}^2 + \Omega_{det}^2, \tag{4.70}$$

so that

$$\Omega_{s+r}^2 = \Omega_s^2 + \Omega_r^2. \tag{4.71}$$

Even in the absence of energy straggling ($\Omega_s = 0$), the system resolution causes a finite spread in a backscattering signal. This is seen most readily on the high-energy edge of a backscattering signal. Particles scattered back from the surface of a target do not undergo energy straggling, since they do not penetrate into the target. For these particles, $\Omega_s$ is thus zero. The recorded signal will nevertheless have a finite spread because of the finite system resolution. Figure 4.8c shows this schematically. Conversely, the high-energy edge of a monoisotopic target can be used to determine the magnitude of $\Omega_r$ by the procedure presented in Fig. 2.12.

This last observation, though derived on a specific example, is generally valid. Energy straggling is absent in the high-energy edge of a backscattering signal. All other contributions to noise are contained and measured there in the same way they actually enter into the recorded data. The standard deviation of the high-energy edge measured on the signal of a monoisotopic target thus provides the value of $\Omega_r$ that must be entered in Eq. (4.71) to give the correct $\Omega_{s+r}$, regardless of the actual origin of the fluctuations. It is not necessary to know the physical origin of the system resolution to measure and account for it.

The system resolution also broadens the low-energy edge of a thin-film spectrum. Since the standard deviations of the energy and the system resolution add up quadratically, the resulting total deviation $\Omega_{s+r}$ is usually determined predominantly by one of two processes. For thick films, the width of the low-energy edge of a signal is dominated by energy straggling. For very thin films, the system resolution prevails.

The influence of the system resolution on the signal of very thin films deserves particular attention. Figure 4.10 shows a sequence of computed

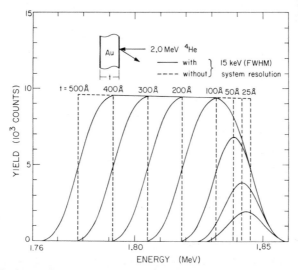

**Fig. 4.10**   Backscattering spectra of 2.0 MeV $^4$He computed for thin Au films of 25 to 500 Å. The dashed lines are ideal spectra. The solid lines include a system resolution of 15-keV FWHM ($\Omega_r = 6.4$ keV). As the width of the ideal spectrum approaches $\Omega_r$ and becomes much less than that, the actual spectrum approaches more and more a Gaussian with standard deviation $\Omega_r$.

backscattering spectra for 2-MeV $^4$He and Au films ranging from 25 to 500 Å. A system resolution of 15-keV FWHM is assumed. For the thickest film, the height of the ideal signal (dashed line) can be extracted directly from the backscattering spectrum. The width of the ideal signal is also readily found, because its steps intersect the actual signal at the 50% height points of each edge. As the film gets thinner, however, the maximum of the actual spectrum never reaches the height of the ideal signal; nor do the positions of the high- and low-energy steps of the ideal signal and its width agree with the 50% height points of the actual signal. It then becomes a nontrivial exercise to extract the correct position and width of the ideal signal from the actual one. The solution to this problem is based on the fact that system resolution alone produces a signal that is the convolution of the ideal step with the Gaussian function of the system resolution. The question is treated further in Appendix C.

### REFERENCES

Ziegler, J. F., Lever, R. F. and Hirvonen, J. K. (1976). *In* Ion Beam Surface Layer Analysis" (O. Meyer, G. Linker, and F. Käppeler, eds.), Vol. 1, p. 163. Plenum Press, New York.

Chapter

# 5

# Examples of Backscattering Analysis

### 5.1 INTRODUCTION

In this chapter we apply the formulas developed in Chapters 3 and 4 to real examples, to illustrate the analytical approach and the magnitude of such quantities as spectrum heights, typical energy losses, and measurable amounts of impurities in the samples. The examples were chosen to illustrate analytical methods rather than to describe applications; therefore, many of them are academic rather than practical. Often, two or more different approximations are applied to the same example. Comparison of the results will help the reader decide whether to make a crude approximation to obtain a quick answer or to make a detailed analysis.

For most of the examples a 2-MeV $^4$He ion beam has been used, with scattering through $\theta = 170°$ ($\theta_2 = 10°$). In all cases, the beam is normal to the sample ($\theta_1 = 0$). The scattered particles were analyzed with a solid-state detector located about 10 cm from the target and with a solid angle $\Omega$ between 3 and 4 msr. The detector resolution is between 15 and 20 keV, and the multichannel analyzer is set up with a channel width $\mathscr{E}$ between 3 and 5 keV. The spectra were obtained under routine experimental conditions; no special effort was made to optimize those conditions.

In these examples, we use stopping cross section values given in Table VI. Although these values may be found to be in error as more refined measurements are made, they serve as a basis for demonstrating different approaches to numerical calculations.

## 5.2   SURFACE IMPURITY ON AN ELEMENTAL BULK TARGET

### 5.2.1   System Calibration

Backscattering can be used to detect surface impurities on a light-element substrate. For example, a carbon substrate is often used as a control sample to check the quality of a vacuum-deposited layer. Any surface impurity with an atomic mass greater than that of carbon will be visible in a backscattering spectrum. These samples can also be used to determine the channel width $\mathscr{E}$ of the multichannel analyzer.

Figure 5.1 shows a spectrum for 2-MeV $^4$He ions backscattered at an angle $\theta$ of 170° from a carbon target on which oxygen, silicon, and gold are present as surface impurities. The carbon substrate returns a thick target signal whose leading edge is at channel number 98 (located at half-height) and three peaks for the impurities at channel numbers 141, 222, and 366 (at the midpoints of the full widths at half the peak heights). The right-hand scale gives kinematic values for scattering, with $K = 0.2526$ for C, 0.3625 for O, 0.5657 for Si, and 0.9225 for Au (kinematic values are given in Table V). The slope of the dashed line gives a value of 5.00 keV for the channel width,

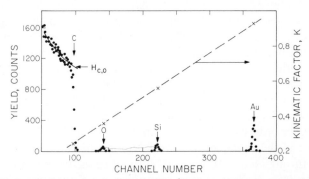

**Fig. 5.1**   Backscattering spectrum for 2.0-MeV $^4$He ions incident on a carbon substrate containing a surface layer of O, Si, and Au. Scattering geometry is set up for normal incidence and $\theta = 170°$, $\Omega = 4.11$ msr, and beam current of 15 nA for $Q = 10$ $\mu$C $= 6.25 \times 10^{13}$ ions. The crosses on the dashed line show the position of the impurity peaks and carbon edge versus the kinematic factor $K$.

and the intercept is at 18 keV. Hence the detected energy $E_1$ is obtained from 5.00 keV times the channel number plus 18 keV.

It is necessary to ensure that the contaminants are on the surface rather than buried under it. This can be done by tilting the target; the energy positions of signals from species on the surface will not be affected, but signals from species below the surface will shift to lower energies.

### 5.2.2 Number of Impurities per Square Centimeter

The number $Nt$ of impurities per square centimeter can be calculated directly from the area of the signals $A$ given by the total number of counts integrated over the region of interest. For a given impurity denoted by the subscript i, the area $A_i$ can be expressed from the equations in Chapter 4 [for example, Eq. (4.7)] for normal incidence as

$$A_i = \sigma_i \Omega Q (Nt)_i. \tag{5.1}$$

The value of $\sigma_i$ can be calculated from Eq. (2.22) for an impurity with atomic number $z_2 = z_i$. The scattering cross sections given in Table 5.1 were obtained by using the conversion factors $e^2 = 1.44 \times 10^{13}$ MeV cm and $\csc^4(\theta/2) = 1.0154$ for $\theta = 170°$. The cross sections are also given in Table X for the condition that $E_0 = 1.0$ MeV. The solid angle $\Omega$ is determined from the experimental setup, and the amount of charge $Q$ collected during the measurement is read from a current integrator. For the spectrum in Fig. 5.1, a solid-state detector was used with an active area of 49 mm$^2$, at a distance of 109.2 mm from the target; the solid angle is 4.11 msr. The total charge collected was 10 $\mu$C, $Q = 6.25 \times 10^{13}$ ions. The total number of impurity

**TABLE 5.1**

Analysis of Fig. 5.1[a]

| Element | Mass (amu) | $K_i$ | $K_i E_0$ (keV) | $\sigma_i$ ($\times 10^{-24}$ cm$^2$) | $A_i$ (counts) | $(Nt)_i$ ($10^{15}$ atoms/cm$^2$) Eq. (5.1) | Eq. (5.2) |
|---------|------------|-------|-----------------|----------------------------------------|----------------|----------------------------------------------|-----------|
| C | 12 | 0.2526 | 505 | 0.037 | 1100[b] | | |
| O | 16 | 0.3625 | 725 | 0.074 | 390 | 20.5 | 20.9 |
| Si | 28 | 0.5657 | 1131 | 0.248 | 460 | 7.2 | 7.4 |
| Au | 197 | 0.9225 | 1845 | 8.200 | 1400 | 0.67 | 0.68 |

[a] The experimental conditions are $E_0 = 2.0$ MeV, normal incidence with $\theta = 170°$, $\Omega = 4.11$ msr, $Q = 10 \mu$C $= 6.25 \times 10^{13}$ ions, $\mathscr{E} = 5.0$ keV, and $[\varepsilon_0]_C = 42.4 \times 10^{-15}$ eV cm$^2$. Values for $K_i$ are taken from Table V and for $\sigma_i$ are from Table X corrected by a factor $(1/2)^2$.

[b] $H_{C,0}$ is the surface height of the carbon signal.

atoms per unit area $(Nt)_i$ for a given species on the surface can be calculated by use of Eq. (5.1); values are given in Table 5.1.

The value of $\Omega$ was determined from the solid angle sustained by the area of the detector, under the assumption that there were no "dead" spots on the detector surface. Under prolonged exposure to energetic particles, the detector can degrade and develop dead regions. The charge collection $Q$ is based on suppression of secondary electrons; total suppression is sometimes difficult to achieve.

Another way to determine $(Nt)_i$ is to use the thick-target signal of the substrate as a reference. The yield $H_{C,0}$ for scattering from the carbon surface is given in Eq. (3.38). By combining Eqs. (3.38) and (5.1) for normal incidence we obtain

$$(Nt)_i = (A_i/H_{C,0})(\sigma_C/\sigma_i)(\mathscr{E}/[\varepsilon_0]_C^C),  \qquad (5.2)$$

where the subscripts $C, 0$ refer to the carbon substrate and surface energy approximation. Values of the number of impurities per square centimeter determined in this manner are also given in Table 5.1. The stopping cross section factor $[\varepsilon_0]_C^C$ for 2.0-MeV He ions in a carbon substrate is $42.4 \times 10^{-15}$ eV cm$^2$ if we use the surface energy approximation [Eq. (3.12)]. Values for the stopping cross section factor are given in Table VIII. The height of the carbon signal, $H_{C,0} = 1100$ counts, the carbon signal was found by drawing a line through the scatter of points for the signal from the carbon substrate and extrapolating the line to channel 98 $(E_1 = K_C E_0)$. The product $\Omega Q$ is not required when one uses Eq. (5.2), but is required when one uses Eq. (5.1).

Figure 5.1 and Table 5.1 give us some estimate of the sensitivity of backscattering in the determination of impurities (a monolayer is of the order of $10^{15}$ atoms/cm$^2$). In principle, in a low-noise system—that is, one in which there are no background counts—one can detect infinitesimal amounts of impurities $(Nt)_i$ on the surface simply by increasing $Q$ without limit. In reality, however, the larger the $Q$, the larger the background noise. Any value quoted for sensitivity will have to depend on the experimental conditions and the criteria used to define the sensitivity. In a routine operation such as the present one, the sensitivity of backscattering for 2-MeV $^4$He ions can be estimated on a purely empirical basis by

$$(Nt)_i \approx \left[\frac{Z(\text{substrate})}{Z(\text{impurity})}\right]^2 \times 10^{14} \text{ impurity atoms/cm}^2.  \qquad (5.3)$$

This equation gives an estimate of the minimum amount of surface impurities on a lighter substrate—$Z(\text{impurity}) > Z(\text{substrate})$—that can be detected by 2-MeV $^4$He backscattering. For gold as an impurity on carbon, Eq. (5.3) gives the detectable minimum as $10^{12}$ atoms/cm$^2$ or $\approx 10^{-3}$ monolayers.

A trace amount of impurity on a substrate that has a mass number larger than that of the impurity cannot be detected because the signal from the impurity is buried under that from the thick-target yield. By using channeling to depress the thick-target yield, the ratio of the impurity signal to the background can be improved. This is discussed further in Chapter 8.

### 5.3 ELEMENTAL SAMPLES CONTAINING UNIFORM CONCENTRATIONS OF IMPURITIES

#### 5.3.1 Low-Impurity Concentrations

The amount of surface impurity was expressed as $(Nt)_i$, the number of impurity atoms per unit area. In this section, dealing with uniform concentrations of impurities in a bulk sample, we will use $N_i$ to denote the number of impurity atoms per unit volume. Further, we will use the surface energy approximation since we assume uniform concentrations.

For silicon uniformly doped with arsenic as shown in Fig. 5.2 (Chu *et al.*, 1973a), the height of the arsenic signal is proportional to the concentration

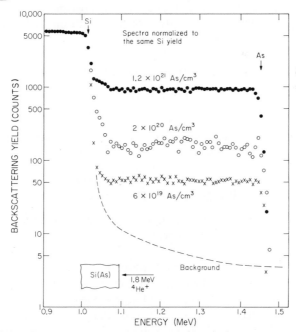

**Fig. 5.2** Composite backscattering spectra for 1.8-MeV $^4$He ions incident on Si samples containing different concentrations of As atoms. The spectra were normalized to the Si signal height. The dashed line shows the background at energies greater than $K_{Si}E_0$ obtained on samples without As. [From Chu *et al.* (1973a).]

of the arsenic in the silicon ($H_{As} \propto N_{As}$). The basic relation for the height of a signal is given in Eq. (3.36) as $H_0 = \sigma(E_0)\Omega Q N \tau_0$ for normal incidence, where $\tau_0$ is the thickness of a slab corresponding to one channel. The signal height for the arsenic atoms $H_{As}$ can be written in the surface energy approximation as

$$H_{As,0} = \sigma_{As}(E_0)\Omega Q (N_{As}/N_{Si})(\mathscr{E}/[\varepsilon_0]^{Si}_{As}). \tag{5.4}$$

This formula follows from Eq. (3.70) by observing that the number $m$ of As atoms per unit of the "compound" As–Si is very closely given by $N_{As}/N_{Si}$. The factor $[\varepsilon_0]^{Si}_{As}$ is defined as the stopping cross section factor (in the surface energy approximation) for arsenic in a silicon matrix [Eq. (3.58)] as

$$[\varepsilon_0]^{Si}_{As} = K_{As}\varepsilon^{Si}(E_0) + (1/\cos\theta_2)\varepsilon^{Si}(K_{As}E_0) \tag{5.5}$$

and $\varepsilon^{Si}$ is the stopping cross section in the stopping medium, here silicon. The small amount of arsenic does not influence $\varepsilon^{Si}$, but does enter $[\varepsilon_0]^{Si}_{As}$ through collision kinematics. The stopping cross section factor is labeled with a subscript to denote the scattering atom and a superscript to denote the stopping medium; thus for the present example it is $[\varepsilon_0]^{Si}_{As}$. When the scattering atom and the stopping medium are the same, we sometimes use only the subscript. Thus for an analysis of a thick silicon target, the notation would be $[\varepsilon_0]_{Si}$, implying that both the scattering atom and the stopping medium are silicon.

We can eliminate the values of $\Omega$ and $Q$ by taking the height ratio:

$$\frac{N_{As}}{N_{Si}} = \frac{H_{As,0}}{H_{Si,0}} \frac{\sigma_{Si}(E_0)}{\sigma_{As}(E_0)} \frac{[\varepsilon_0]^{Si}_{As}}{[\varepsilon_0]^{Si}_{Si}}. \tag{5.6}$$

The $[\varepsilon]$ ratio can be calculated from $\varepsilon$ values given in Table VI for 2-MeV $^4$He ions at $\theta = 170°$ to give

$$\frac{[\varepsilon_0]^{Si}_{As}}{[\varepsilon_0]^{Si}_{Si}} = \frac{95.3 \times 10^{-15} \text{ eV cm}^2}{92.6 \times 10^{-15} \text{ eV cm}^2}.$$

The $[\varepsilon]$ ratio is within 3% of unity in this case. In general, for megaelectron volt He ions, this type of ratio is within 10% of unity for a wide variety of impurities and substrates.

The ratio $H_{As,0}/H_{Si,0}$ can be directly measured from Fig. 5.2, and therefore the concentration ratio can be obtained. For $N_{Si} = 4.98 \times 10^{22}$ atoms/cm$^3$, $\sigma_{As}(1.8 \text{ MeV}) = 1.76 \times 10^{-24}$ cm$^2$, and $\sigma_{Si}(1.8 \text{ MeV}) = 0.306 \times 10^{-24}$ cm$^2$, the value of the concentration given by crosses in Fig. 5.2 was calculated to be $6 \times 10^{19}$/cm$^3$.

The sensitivity of backscattering for measuring bulk samples depends on the problem and the experimental conditions. For typical conditions we have found that the detectable height of an impurity signal is about one-thousandth of the height of the substrate target signal; that is, $H_{imp}/H_{sub} = 10^{-3}$. If we assume that the $[\varepsilon]$ ratio is unity and that the $\sigma$ ratio is approximately equal to the $Z^2$ ratio, Eq. (5.6) then gives a sensitivity limit of

$$N_{imp}/N_{sub} \gtrsim (Z^2_{sub}/Z^2_{imp}) \times 10^{-3}. \qquad (5.7)$$

As an example, the amount of impurity that can be detected in silicon is $\approx 10^{19}$ atoms/cm$^3$ if the impurity is arsenic, or $1.5 \times 10^{18}$ atoms/cm$^3$ if it is gold. If the substrate is CdTe ($Z = 50$ as an average), the amount of gold impurity that can be detected is about $2 \times 10^{19}$ atoms/cm$^3$.

### 5.3.2  High-Impurity Concentrations

If the amount of impurity is too high, the sample can no longer be treated as a pure element as far as the value of the stopping cross section is concerned. As a rough estimate, impurity concentrations above one atomic percent will make a detectable change in the stopping cross section.

For an Al sample alloyed with Cu, we denote the mixture as $Al_{1-x}Cu_x$, where the atomic ratio of Cu to Al is given by $x/(1 - x)$. For simplicity, we abbreviate the nomenclature as AlCu. The stopping cross section is given by

$$\varepsilon^{AlCu} = (1 - x)\varepsilon^{Al} + x\varepsilon^{Cu}, \qquad (5.8)$$

where we assume Bragg's rule of linear additivity. The ratio of the Cu to the Al signals in the spectrum of $Al_{1-x}Cu_x$ (as shown in Fig. 5.3) is given by

$$\frac{H^{AlCu}_{Cu,0}}{H^{AlCu}_{Al,0}} = \frac{x}{1 - x} \frac{\sigma_{Cu}}{\sigma_{Al}} \frac{[\varepsilon]^{AlCu}_{Al}}{[\varepsilon]^{AlCu}_{Cu}}. \qquad (5.9)$$

As in Eq. (5.5), the superscripts denote the stopping medium AlCu, and the subscripts denote the collision partner Al or Cu, and hence the choice of

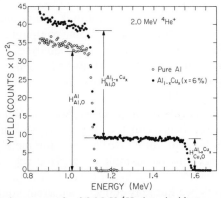

**Fig. 5.3**  Backscattering spectra for 2.0-MeV $^4$He ions incident on an Al sample ($\bigcirc$) and an Al–6% Cu sample ($\bullet$). Scattering geometry is set up for normal incidence with $\theta = 170°$, $\Omega = 4.11$ msr, $\mathscr{E} = 5$ keV, and $Q = 10$ $\mu$C $= 6.25 \times 10^{13}$ ions. [From Howard et al. (1976).]

the kinematic factor used in calculating $[\varepsilon]$. As in the treatment in Section 5.3.1, the $[\varepsilon]$ ratio for particles scattered from different elements in the same medium is close to unity and is not changed appreciably by changes in the atomic concentration ratios. Consequently, one determines zeroth-order values of $x$ and $1 - x$ by first setting the $[\varepsilon]$ ratio equal to unity. The calculation can be improved by using the zeroth-order values of $x$ and $1 - x$ to calculate better $[\varepsilon]$ values, which can then be used in Eq. (5.9) to give new values of $x$ and $1 - x$. Generally, this first iteration does not change the values of $x$ and $1 - x$ by more than a few percent.

One cannot treat the $[\varepsilon]$ ratio as unity when calculating ratios of signal heights for the same element in different matrices. For example, the ratio of the Al heights in AlCu to Al is given by

$$H_{Al,0}^{AlCu}/H_{Al,0}^{Al} = (1 - x)[\varepsilon_0]_{Al}^{Al}/[\varepsilon_0]_{Al}^{AlCu}. \tag{5.10}$$

For the sample given by the spectra in Fig. 5.3, the composition corresponds to values of $x = 0.06$ (Howard *et al.*, 1976). However, the measured height ratio of $H^{AlCu}/H_{Al}$ equals 0.90; this would imply a value of $x = 0.10$ if the $[\varepsilon]$ ratio were unity. That is, one could make an error of nearly a factor of two in assigning composition values if the change in stopping cross section factors were neglected. This procedure was also found to be necessary in evaluating the composition of GaAlAs, in order to compare the height ratio of GaAlAs to GaAs (Mayer *et al.*, 1973).

## 5.4  COMPOSITION OF HOMOGENEOUS SAMPLES CONTAINING MORE THAN ONE ELEMENT

### 5.4.1  Two Elements

Backscattering can be used to analyze a bulk compound or mixture. The method is straightforward, except that for some compounds containing both heavy and light elements the accuracy of the analysis is reduced because the signals from the light elements are always superimposed on the signals from the heavy elements. As in impurity analysis, the concentration is determined from the signal height.

We treat a sample of known composition, $SiO_2$, to illustrate the method of calculating spectrum heights in the surface energy approximation. The $SiO_2$ target is made of fused quartz with a very thin metal layer on the surface for charge integration. The sample is analyzed with an incident beam on ${}^4He^+$ ions at 2.0 MeV, with a beam current of 15 nA and a total dose of 10 $\mu$C, as measured by a Faraday cup with a current integrator. The backscattering spectrum thus obtained is plotted in Fig. 5.4. The signal heights are designated as $H_{Si}^{SiO_2}$ and $H_{O}^{SiO_2}$, with values determined from the spectrum to be 1500 and 980 counts, respectively.

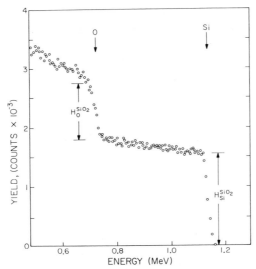

**Fig. 5.4**  Backscattering spectrum for 2.0-MeV $^4$He ions incident on a sample of $SiO_2$. The scattering geometry is set up for normal incidence with $\theta = 170°$, $\Omega = 4.11$ msr, $Q = 10$ $\mu$C = $6.25 \times 10^{13}$ ions, and $\mathscr{E} = 5.4$ keV.

There are two equivalent methods of treating the stopping cross section in the medium by using Bragg's rule of linear additivity: on a molecular basis,

$$\varepsilon^{SiO_2} = \varepsilon^{Si} + 2\varepsilon^{O}, \tag{5.11}$$

where one considers the stopping cross section per molecule of $SiO_2$ or on an atomic basis for a mixture,

$$\varepsilon^{Si_xO_{1-x}} = 0.33\varepsilon^{Si} + 0.66\varepsilon^{O}, \tag{5.12}$$

where $x = 0.33$ and one considers the effective cross section per atom within the molecule. Figure 5.5 shows the stopping cross section for the two methods. For normal incidence the signal height for the oxygen component, for example, is given for the compound on a molecular basis by

$$H_{0,0}^{SiO_2} = \sigma_O(E_0)\Omega Q \frac{N_O^{SiO_2}}{N^{SiO_2}} \frac{\mathscr{E}}{[\varepsilon_0]_O^{SiO_2}}, \tag{5.13}$$

where $(N_O^{SiO_2}/N^{SiO_2}) = 2$ since there are two oxygen atoms per $SiO_2$ molecule. This formula is the same as Eq. (3.70) with $m = 2$ and $\cos \theta_1 = 1$. On an atomic basis, the height of the oxygen signal in the mixture $Si_xO_{1-x}$ is

$$H_{0,0}^{Si_xO_{1-x}} = \sigma_O(E_0)\Omega Q \frac{N_O^{Si_xO_{1-x}}}{N^{Si_xO_{1-x}}} \frac{\mathscr{E}}{[\varepsilon_0]_O^{Si_xO_{1-x}}}, \tag{5.14}$$

where $(N_O^{Si_xO_{1-x}}/N^{Si_xO_{1-x}}) = 0.66$. Because $[\varepsilon_0]_O^{Si_xO_{1-x}}$ is now the effective stopping cross section factor per atom, not per molecule, the concentration of oxygen atoms must be taken with respect to one atom of $Si_xO_{1-x}$, whose concentration is three times larger than that of the $SiO_2$ molecules.

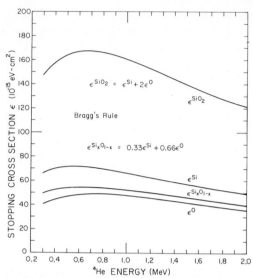

**Fig. 5.5** Stopping cross sections for $^4$He ions in Si, O, and $SiO_2$. The oxide stopping cross section was determined both on a molecular basis $\varepsilon^{SiO_2}$ and on an atomic basis $\varepsilon^{Si_xO_{1-x}}$, on the assumption that Bragg's rule of linear additivity holds.

**TABLE 5.2**

Different Methods of Formulating Expressions for $\varepsilon$, $H$, and $[\varepsilon]$ for $SiO_2$ Using the Surface Energy Approximation[a]

| Molecular basis (one molecule of $SiO_2$) | Atomic basis (0.33 atoms Si + 0.66 atoms O) |
|---|---|
| $N^{SiO_2} = 2.3 \times 10^{22}$ cm$^{-3}$ [b] | $N^{Si_xO_{1-x}} = 6.9 \times 10^{22}$ cm$^{-3}$ |
| $N_O^{SiO_2} = 4.6 \times 10^{22}$ cm$^{-3}$ | $N_O^{Si_xO_{1-x}} = 4.6 \times 10^{22}$ cm$^{-3}$ |
| $\varepsilon^{SiO_2} = \varepsilon^{Si} + 2\varepsilon^O$ | $\varepsilon^{Si_xO_{1-x}} = 0.33\varepsilon^{Si} + 0.66\varepsilon^O$ |
| $[\varepsilon_0]_{Si}^{SiO_2} = 226 \times 10^{-15}$ eV cm$^2$ | $[\varepsilon_0]_{Si}^{Si_xO_{1-x}} = 75.3 \times 10^{-15}$ eV cm$^2$ |
| $[\varepsilon_0]_O^{SiO_2} = 213 \times 10^{-15}$ eV cm$^2$ | $[\varepsilon_0]_O^{Si_xO_{1-x}} = 71.0 \times 10^{-15}$ eV cm$^2$ |
| $H_{Si}^{SiO_2} = \sigma_{Si}\Omega Q \dfrac{\mathscr{E}}{[\varepsilon_0]_{Si}^{SiO_2}} = 1522$ | $H_{Si}^{Si_xO_{1-x}} = 0.33\sigma_{Si}\Omega Q \dfrac{\mathscr{E}}{[\varepsilon_0]_{Si}^{Si_xO_{1-x}}} = 1522$ |
| $H_O^{SiO_2} = 2\sigma_O\Omega Q \dfrac{\mathscr{E}}{[\varepsilon_0]_O^{SiO_2}} = 966$ | $H_O^{Si_xO_{1-x}} = 0.66\sigma_O\Omega Q \dfrac{\mathscr{E}}{[\varepsilon_0]_O^{Si_xO_{1-x}}} = 966$ |

[a] These values were obtained for normal incidence.

[b] From A. S. Grove, "Physics and Technology of Semiconductor Devices." Wiley, New York, 1967, p. 102.

The different formulations are shown in Table 5.2. The values in the table were calculated for 2.0-MeV $^4$He ions in the surface energy approximation for normal incidence with $\theta = 170°$, $Q = 6.25 \times 10^{13}$ particles (10 $\mu$C), $\Omega = 4.11$ msr, and $\mathscr{E} = 5.4$ keV. The cross section values are $\sigma_{Si} = 0.248 \times 10^{-24}$ cm$^2$ and $\sigma_O = 0.742 \times 10^{-25}$ cm$^2$. The tabulated values show that consistent spectrum heights can be obtained if consistent values for $[\varepsilon]$ and $N^{AB}$ are chosen.

### 5.4.2 Multielemental Samples

We demonstrate next the analysis of a bulk sample from measurements of signal heights, as shown in Fig. 5.6 for 2.4-MeV $^4$He ions incident in a nonchanneled direction, from an alkali halide crystal made of $(KCl)_1(KBr)_x$. We solve for the unknown $x$ by measuring the surface heights of the backscattering signals due to chlorine, potassium, and bromine, and assume that there is a K atom associated with each Cl or Br atom. In Fig. 5.6 lines are drawn over the points of the spectrum to form a ladder. The vertical positions of the ladder—that is, the half height at the leading edges—are at 1.53, 1.60, and 1.97 MeV, corresponding to the energies of particles scattered from Cl, K, and Br in the surface layer of the compound.

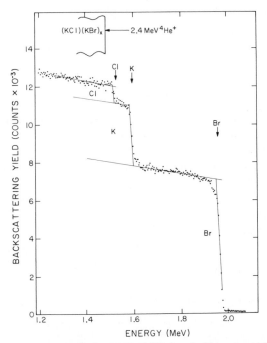

**Fig. 5.6** Backscattering spectrum for 2.4-MeV $^4$He ions incident on an alkali halide crystal of $(KCl)_1(KBr)_x$ for normal incidence with $\theta = 170°$.

The surface height of the signal due to scattering from a given element with a beam of normal incidence is

$$H_{elem} = \sigma_{elem}(E_0)\Omega Q(N_{elem}/N_{comp})(\mathscr{E}/[\varepsilon_0]^{comp}_{elem}). \tag{5.15}$$

Here $[\varepsilon_0]^{comp}_{elem}$ is the stopping cross section factor of the compound material with scattering from a given element evaluated in the surface energy approximation, $N_{elem}$ is the atomic density of the given element, and $N^{comp}$ is the density of the unit of compound considered in evaluating the stopping cross section $\varepsilon^{comp}$ of the compound. In the present example, it is natural to consider one KCl pair and its associated atoms as the unit of the compound.

The density ratios of the elements to the compound $(KCl)_1(KBr)_x$ then are

$$N_{elem}/N^{comp} = \begin{cases} 1 + x & \text{for K} \\ 1 & \text{for Cl} \\ x & \text{for Br} \end{cases} \tag{5.16}$$

and the height ratios are

$$\frac{H_K}{H_{Cl}} = \frac{\sigma_K}{\sigma_{Cl}}(1 + x)\frac{[\varepsilon_0]^{comp}_{Cl}}{[\varepsilon_0]^{comp}_K}, \tag{5.17}$$

$$\frac{H_K}{H_{Br}} = \frac{\sigma_K}{\sigma_{Br}}\frac{(1 + x)}{x}\frac{[\varepsilon_0]^{comp}_{Br}}{[\varepsilon_0]^{comp}_K}, \tag{5.18}$$

$$\frac{H_K + H_{Cl}}{H_{Br}} = \frac{\sigma_K}{\sigma_{Br}}\frac{1 + x}{x}\frac{[\varepsilon_0]^{comp}_{Br}}{[\varepsilon_0]^{comp}_K} + \frac{\sigma_{Cl}}{\sigma_{Br}}\frac{1}{x}\frac{[\varepsilon_0]^{comp}_{Br}}{[\varepsilon_0]^{comp}_{Cl}}. \tag{5.19}$$

If we know the values of $[\varepsilon_0]$, then any of these three equations will give a solution for $x$, because the heights $H$ can be measured directly from the spectrum and the cross sections $\sigma$ can be calculated. We do not know a priori the values of $[\varepsilon_0]$ for a given element, but $[\varepsilon_0]$ ratios for scattering from the different elements in the compound are normally within 10% of unity, regardless of composition.

Approximating the $[\varepsilon_0]$ ratios by unity and using values for $\sigma$ listed in Table X, we obtain

$$\frac{H_K}{H_{Cl}} = 1.255(1 + x), \tag{5.20}$$

$$\frac{H_K}{H_{Br}} = 0.290\frac{1 + x}{x}, \tag{5.21}$$

$$\frac{H_K + H_{Cl}}{H_{Br}} = 0.290\frac{1 + x}{x} + 0.231\frac{1}{x}. \tag{5.22}$$

The left-hand sides of these equations are measured directly from the signal heights in the spectrum $H_{Cl} = 3 \pm 0.5$, $H_K = 10 \pm 0.5$, and $H_{Br} = 20.9 \pm 0.5$. All heights are given in arbitrary units. Substituting the height numbers in Eqs. (5.20)–(5.22), we solve for $x$ and obtain three different values of $x = 1.66$, 1.54, and 1.57, with an average value of 1.59. These three values are zeroth-order approximations, because the $[\varepsilon_0]$ ratio terms in Eqs. (5.17)–(5.19) have been ignored. Now, from a zeroth-order value of $x$, we can calculate values for $[\varepsilon_0]$ and make a first-order calculation for $x$, using the ratio of these $[\varepsilon_0]$ values. In calculating $\varepsilon$, we assume Bragg's rule and we must consider as the unit of the compound one pair of KCl with its associated atoms:

$$\varepsilon^{comp} = \varepsilon^{KCl_1(KBr)_{1.59}} = 2.59\varepsilon^K + \varepsilon^{Cl} + 1.59\varepsilon^{Br}. \tag{5.23}$$

The values of $[\varepsilon_0]^{comp}_{elem}$ are then computed by using Eq. (5.23) with the elemental $\varepsilon$ values given in Table VI: for 2.4-MeV $^4$He,

$$[\varepsilon_0]^{comp}_K = 608.5 \times 10^{15} \quad eV \; cm^2,$$
$$[\varepsilon_0]^{comp}_{Cl} = 607.5 \times 10^{15} \quad eV \; cm^2,$$
$$[\varepsilon_0]^{comp}_{Br} = 614.8 \times 10^{15} \quad eV \; cm^2,$$

and the $[\varepsilon_0]$ ratios become

$$[\varepsilon_0]^{comp}_{Cl}/[\varepsilon_0]^{comp}_K = 0.999,$$
$$[\varepsilon_0]^{comp}_{Br}/[\varepsilon_0]^{comp}_K = 1.010,$$
$$[\varepsilon_0]^{comp}_{Br}/[\varepsilon_0]^{comp}_{Cl} = 1.011.$$

The $[\varepsilon_0]$ ratios are not a sensitive function of $x$. For example, when $x$ changes from 0.1 to 10.0, the terms $[\varepsilon]^{comp}_{Cl}/[\varepsilon]^{comp}_K$ and $[\varepsilon]^{comp}_{Br}/[\varepsilon]^{comp}_{Cl}$ change by 1% or less.

Substituting values of the $[\varepsilon_0]$ and $\sigma$ ratios into Eqs. (5.17)–(5.19) and solving for $x$, we obtain $x = 1.66$, 1.58, and 1.60. The average value, $\bar{x} = 1.61$, differs from the zeroth-order value of 1.59 by 1.3%. This is less than the experimental uncertainty in the height ratios and indicates that a zeroth-order analysis is quite adequate.

A last example for bulk analysis by backscattering is shown in Fig. 5.7. Here the sample is a magnetic bubble material grown on garnet (Nicolet and Chu, 1975). The bubble material is a film 10 $\mu$m thick, which is thicker than the range of the 2.0-MeV $^4$He beam, and thus, in effect, acts like a bulk material. Because the bubble material is an insulator, a thin film of Al was deposited on it before the analysis to provide a return path for beam current to ground.

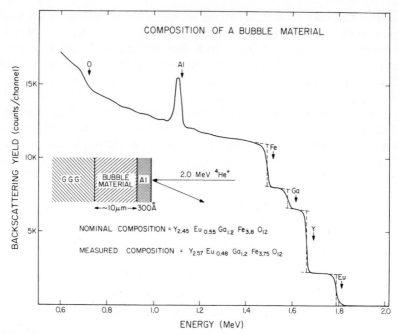

**Fig. 5.7**   Backscattering spectrum of 2.0-MeV $^4$He ions incident on a thick target consisting of a magnetic bubble material with a thin surface of Al. The bubble material was known to have the garnet composition $X_8O_{12}$ with the nominal composition = $Y_{2.45}Eu_{0.55}Ga_{1.2}Fe_{3.8}O_{12}$ and measured composition = $Y_{2.57}Eu_{0.48}Ga_{1.2}Fe_{3.75}O_{12}$. [From Nicolet and Chu (1975).]

The garnet was known to consist of molecular units $X_8O_{12}$. We therefore treat this example on a molecular basis; that is, we consider the molecule $X_8O_{12}$ as the unit of the compound; hence, as in Eq. (5.13), we have

$$H_A^{comp} = (N_A/N)\sigma_A Q\Omega(\mathscr{E}/[\varepsilon]_A^{comp}). \qquad (5.24)$$

Here the subscript A indicates one of the elements in the material, $N_A/N$ is the number of A atoms in a molecular unit, and $[\varepsilon_0]_A^{comp}$ is the stopping cross section of a molecular unit.

On the assumption that the $[\varepsilon]$ ratios are unity, the spectrum height ratios are

$$(H_A/H_{A'}) = (N_A/N_{A'})(\sigma_A/\sigma_{A'}), \qquad (5.25)$$

where A and A′ are any two of the four elements, iron, gallium, yttrium, and europium, contained in the bubble material. For elements with high atomic mass, one can simplify Eq. (5.25) to [see Eq. (2.24)]

$$(H_A/H_{A'}) = (N_A/N_{A'})(Z_A{}^2/Z_{A'}^2). \qquad (5.26)$$

The values of $N_A/N_{A'}$ can be obtained from the signal heights in Fig. 5.7 which are 2820 counts for Fe, 1280 for Ga, 4340 for Y, and 2120 for Eu.

The molecule of the bubble material is known to be $X_8O_{12}$. This gives an additional condition:

$$(N_{Fe}/N) + (N_{Ga}/N) + (N_Y/N) + (N_{Eu}/N) = 8. \qquad (5.27)$$

Substituting the values of $N_A/N_{Fe}$ from Eq. (5.26) into Eq. (5.27), we have

$$(N_{Fe}/N)(1 + 0.320 + 0.685 + 0.128) = 8, \qquad (5.28)$$

which gives $N_{Fe}/N = 3.75$. This value gives $N_{Ga}/N = 1.2$, $N_Y/N = 2.57$, and $N_{Eu}/N = 0.48$, in good agreement with the nominal compositions quoted by the material supplier (see Fig. 5.7).

### 5.5 IMPURITIES DISTRIBUTED IN DEPTH IN AN ELEMENTAL SAMPLE

#### 5.5.1 Ion-Implanted Samples

The first major application of backscattering spectrometry to semiconductor problems was in the investigation of ion implantation processes. Ion implantation has advanced rapidly over the past years, and implantation methods are firmly established in semiconductor technology. Backscattering spectrometry with or without channeling offers independent methods of measuring the implantation dose, the range profile, and the lattice location of the impurities, and of studying damage; therefore, backscattering spectrometry has become a major method for characterizing the implantation process. We shall illustrate the method with a very simple example.

Figure 5.8 shows an energy spectrum (Sigmon et al., 1975) of 2.0-MeV ${}^4$He ions backscattered from a silicon target implanted with ${}^{75}$As at 250 keV to a dose of $1.2 \times 10^{15}$ As/cm$^2$. The silicon signal gives a step with leading edge at 1.13 MeV, and the arsenic signal (plotted on an amplified scale) has a Gaussian distribution with a peak at 1.55 MeV and an FWHM of 60 keV. The peak is shifted by $\Delta E_{As} = 68$ keV below the energy edge $K_{As}E_0 = 1.618$ MeV of the As at the surface. The data from Fig. 5.8 are given in Table 5.3.

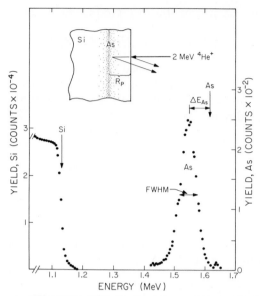

**Fig. 5.8**  Energy spectrum of 2-MeV $^4$He ions backscattered from a silicon crystal implanted with a nominal dose of $1.2 \times 10^{15}$ As ions/cm$^2$ at 250 keV. The vertical arrows indicate the energies of particles scattered from surface atoms of $^{28}$Si and $^{75}$As. [From Sigmon *et al.* (1975).]

**TABLE 5.3**

Data Extracted from Fig. 5.8 with Backscattering Parameters
Based on the Surface Energy Approximation[a]

| Data | Parameters |
|---|---|
| $H_{\text{Si,O}} = 27{,}000$ counts | $[\varepsilon_0]_{\text{Si}}^{\text{Si}} = 92.6 \times 10^{-15}$ eV cm$^2$ |
| $H_{\text{As}}^{\text{Si}} = 250$ counts (at peak) | $[\varepsilon_0]_{\text{As}}^{\text{Si}} = 95.3 \times 10^{-15}$ eV cm$^2$ |
| $A_{\text{As}} = 3350$ counts | $\sigma_{\text{As}} = 1.425 \times 10^{-24}$ cm$^2$ |
| $\Delta E_{\text{As}} = 68$ keV | $\sigma_{\text{Si}} = 0.248 \times 10^{-24}$ cm$^2$ |
| $(\text{FWHM})_{\text{As}} = 60$ keV | $K_{\text{As}} = 0.809$ |
| | $K_{\text{Si}} = 0.566$ |

[a] Values are for $E_0 = 2.0$ MeV, normal incidence with $\theta = 170°$ and $\mathscr{E} = 5.0$ keV.

From Fig. 5.8, we can start a zeroth-order analysis—that is, a surface energy approximation—and calculate the dose, range, and range distribution of arsenic in silicon. That is, we assume the As is so shallow that the surface energy approximation can be used in calculating the stopping cross section and the differential scattering cross section. The implantation dose can be calculated from Eq. (5.2), with the implanted arsenic treated as a surface impurity. The dose of arsenic is then

$$(Nt)_{As} = \frac{A_{As}}{H_{Si}} \frac{\sigma_{Si}(E_0)}{\sigma_{As}(E_0)} \frac{\mathscr{E}}{[\varepsilon_0]_{Si}} = 1.2 \times 10^{15} \quad As/cm^2 \tag{5.29}$$

in agreement with the nominal value of the implanted dose.

The maximum concentration of arsenic in silicon can be estimated from the peak height of the arsenic signal. Using the formula derived for the bulk impurities and data given in Table 5.3, we have

$$\frac{N_{As}}{N_{Si}} = \frac{H_{As}}{H_{Si}} \frac{\sigma_{Si}(E_0)}{\sigma_{As}(E_0)} \frac{[\varepsilon_0]_{As}^{Si}}{[\varepsilon_0]_{Si}^{Si}} = 0.166 \quad at. \% \tag{5.30}$$

or $N_{As} = 8.3 \times 10^{19}$ atoms/cm$^3$ using $N_{Si} = 4.98 \times 10^{22}$ atoms/cm$^3$.

To obtain a concentration profile, we use the stopping cross section factor $[\varepsilon_0]_{As}^{Si}$, which gives an energy-to-depth conversion for scattering from arsenic in a silicon matrix. The peak position of the arsenic is shifted by $\Delta E_{As} = 68$ keV below the surface edge, and

$$N_{Si}R_p = \Delta E/[\varepsilon_0]_{As}^{Si} = 7.14 \times 10^{17} \quad atoms/cm^2, \tag{5.31}$$

where $R_p$ is the projected range of the implanted arsenic. We thus obtain $R_p = 1430$ Å with $N_{Si} = 4.98 \times 10^{22}$ atoms/cm$^3$. The depth scale in angstroms is more convenient than that in atoms per square centimeter, but the latter is an intrinsic unit in depth for backscattering.

When the implant distribution is Gaussian, the depth profile can be described by a projected range $R_p$ and a range straggling $\Delta R_p$, which is the standard deviation of the Gaussian distribution in depth. The standard deviation is related to the FWHM of a Gaussian distribution by

$$FWHM = 2(2 \ln 2)^{1/2} \times (standard\ deviation)$$
$$= 2.355 \times (standard\ deviation), \tag{5.32}$$

as shown in Fig. 2.12. The FWHM of the energy spectrum for arsenic is measured to be 60 keV. This FWHM contains not only the depth distribution of the arsenic, but also the energy resolution of the backscattering system and the energy straggling of the $^4$He ions.

The energy resolution of the backscattering system is quite independent of the detected energy and can be measured from the slope of the silicon step in Fig. 5.8. If we differentiate the step near the silicon surface, we obtain a negative Gaussian (negative because the yield decreases when the energy increases). The FWHM of this negative Gaussian is the energy resolution of the backscattering system. As is discussed at the end of Section 2.6, it too can be obtained easily, without differentiating the spectrum, by simply measuring the energy spread of the step from 12 to 88% of the step height (see Fig. 2.12). In Fig. 5.8, the energy spread of the silicon step from 12 to 88% of the height is 22 keV.

The energy straggling of $^4$He ions in silicon has not been measured, but can be estimated from Bohr's theory (Chapter 2) to be about 3.2 keV in the implanted region. The FWHM of this energy straggling is then $2.355 \times 3.2 = 7.5$ keV.

From the measured FWHM of arsenic, it is necessary to deconvolute the measured FWHM system resolution (22 keV) and the energy straggling (7.5 keV). Since all three distributions are assumed to be Gaussian, the deconvolution process is simply a subtraction in quadrature:

$$\text{FWHM (corrected)} = [(60 \text{ keV})^2 - (22 \text{ keV})^2 - (7.5 \text{ keV})^2]^{1/2}$$
$$= 55.3 \quad \text{keV.} \tag{5.33}$$

This value represents the real spread of arsenic in the energy scale. It can be readily converted into a depth scale by using Eq. (5.31):

$$\Delta R_p = \text{FWHM(corrected)}/2.355N[\varepsilon_0]_{As}^{Si} = 500 \text{ Å.} \tag{5.34}$$

Up to this point our analysis has been based on the surface energy approximation. That is, we have evaluated the stopping cross section $[\varepsilon]_{As}^{Si}$ at a surface energy $E_0 = 2.0$ MeV, using $\varepsilon$ values shown in Fig. 5.9. The curve plotted there is based on the values listed in Table VI. If we use the mean energy approximation described in Section 3.2.2, then

$$[\bar{\varepsilon}]_{As}^{Si} = K_{As}\varepsilon(\bar{E}_{in}) + (1/\cos\theta_2)\varepsilon(\bar{E}_{out}) \tag{5.35}$$

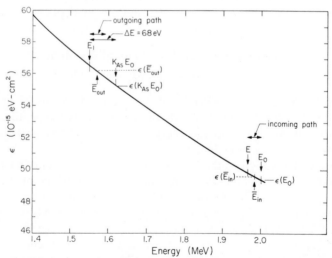

**Fig. 5.9** Stopping cross section of $^4$He ions in Si used to evaluate the As depth profile given in Fig. 5.8. The curve shows the values listed for $\varepsilon^{Si}$ in Table VI.

with $\bar{E}_{in} = \frac{1}{2}(E + E_0)$ and $\bar{E}_{out} = \frac{1}{2}(E_1 + KE)$, where $E$ is the energy immediately before scattering. To obtain the average energies $\bar{E}_{in}$ and $\bar{E}_{out}$ we use the symmetric mean energy approximation:

$$\bar{E}_{in} = E_0 - \tfrac{1}{4}\Delta E = E_0 - \tfrac{1}{2}NR_p\varepsilon(E_0) = 1982 \quad \text{keV} \tag{5.36}$$

and

$$\bar{E}_{out} = E_1 + \tfrac{1}{4}\Delta E = E_1 + \tfrac{1}{2}NR_p\varepsilon(E_1) = 1569 \quad \text{keV.} \tag{5.37}$$

By substituting these energies into Eq. (5.35) for $[\bar{\varepsilon}]_{As}^{Si}$, we have $[\bar{\varepsilon}]_{As}^{Si} = 97.1 \times 10^{-15}$ eV cm$^2$. The new stopping cross section factor is only 2% larger than $[\varepsilon_0]_{As}^{Si} = 95.3 \times 10^{-15}$ eV cm$^2$ calculated from the surface approximation (Table 5.3). Therefore, the new range and range straggling will be 2% lower than the values obtained.

The results of the two analyses, one performed by the surface energy approximation and the other by the mean energy approximation, are summarized in Table 5.4. The differences are small. We conclude that for shallow depths, where $\Delta E$ is small compared to $E$ and $\varepsilon$ changes little over $\Delta E$, the surface energy approximation is adequate for the analysis of depth distributions. For a higher-order calculation there can be a correction for the total dose of implanted As atoms. In Eq. (5.29), both $\sigma_{Si}$ and $\sigma_{As}$ are evaluated at a surface energy $E_0 = 2.0$ MeV. Since the arsenic signal area $A_{As}$ is distributed over an energy interval, we should evaluate $\sigma_{As}$ at the various energies. If a fixed value of $\sigma_{As}$ is to be taken, $\sigma_{As}$ should be evaluated at $E$ rather than at $E_0$, such that

$$E = E_0 - NR_p\varepsilon(\bar{E}_{in}) = 1965 \quad \text{keV.} \tag{5.38}$$

The scattering cross section is related to energy by

$$\sigma_{As}(E)/\sigma_{As}(E_0) = E_0^2/E^2 = 1.018.$$

Therefore, the total implanted dose is 1.8% lower when calculated with energy $E$ than when calculated with the surface energy $E_0$. As an alternative

**TABLE 5.4**

Approximations Used in the Analyses of Fig. 5.8, for $\varepsilon$ Values Shown in Fig. 5.9

| Method | Incoming energy (keV) | Outgoing energy (keV) | $[\varepsilon]_{As}^{Si}$ $(10^{-15}$ eV/cm$^2)$ | $R_p$ (Å) | $\Delta R_p$ (Å) |
|---|---|---|---|---|---|
| Surface energy approximation | $E_0 = 2000$ | $K_{As}E_0 = 1617$ | 95.3 | 1434 | 500 |
| Symmetrical mean energy approximation | $\bar{E}_{in} = 1982$ | $\bar{E}_{out} = 1569$ | 97.1 | 1406 | 490 |

procedure to Eq. (5.38), the energy ratio method, Eq. (3.20), could be used to
determine $E$.

### 5.5.2  Diffusion Profiles

In the discussion of the dose and depth distribution of As implanted in Si,
it was apparent that the surface energy approximation was adequate. When
one evaluates impurity distributions that extend 0.5 to 1 $\mu$m below the
surface, the mean energy approximation should be used.

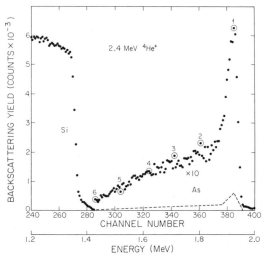

**Fig. 5.10**  Backscattering spectrum of 2.4-MeV $^4$He ions incident on a silicon sample implanted with a nominal dose of $3.4 \times 10^{16}$ As/cm$^2$ and then heat treated to produce a depth
diffusion profile of As. Scattering geometry is set for normal incidence: $\theta = 170°$, $\Omega = 4.11$ msr,
and $Q = 20$ $\mu$C $= 1.25 \times 10^{14}$ ions. The data points for the As signal are scaled by a factor of
ten above the original signal height (dashed line).

As an example of a deeper profile, consider an As-implanted Si wafer after
a drive-in diffusion process step. Figure 5.10 shows the backscattering spectrum (with the As signal magnified in the region between 1.4 to 2.0 MeV) for
2.4-MeV He ions incident at a total dose of 20 $\mu$C ($\theta_1 = 0$, $\theta = 170°$, and
$\Omega = 4.11$ msr). The As signal extends from the surface energy position down
to the silicon signal. In the surface region, the spectrum height of the As
signal can be converted into values of $N_{As}$ by using Eq. (5.30) and the surface
energy approximation. At greater depths a correction must be applied.

Figure 5.11 shows $N_{As}(x)$ calculated from the data given in Fig. 5.10. The
depth scale in Fig. 5.11 was computed numerically by a simple computer
program (Chu and Lever, unpublished) which calculates the energy loss and
scattering cross section at each energy $E$ before scattering and the energy
loss after scattering. To evaluate the height $H_{As}(E_1)$ of the As signal at a

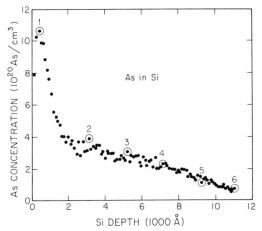

**Fig. 5.11**   Concentration of As in Si versus depth obtained from the data of Fig. 5.10 using Eq. (5.39) to determine $N_{As}$. The encircled points correspond to the data points in Fig. 5.10 and Table 5.5.

detected energy $E_1$, Eq. (5.4) is written in the form given by Eq. (3.50) for normal incidence ($\cos \theta_1 = 1$):

$$H_{As}(E_1) = \sigma_{As}(E)\Omega Q \frac{N_{As}(x)}{N_{Si}} \frac{\mathscr{E}}{[\varepsilon(E)]^{Si}_{As}} \frac{\varepsilon(K_{As}E)}{\varepsilon(E_1)}, \qquad (5.39)$$

where $[\varepsilon(E)]^{Si}_{As}$ is the stopping cross section factor evaluated at the energy $E$ before scattering and the ratio $\varepsilon(K_{As}E)/\varepsilon(E_1)$ corrects for the change with depth of the thickness of a slab corresponding to one channel.

Table 5.5 lists values for depth and $N_{As}$ calculated by the surface energy approximation [Eq. (5.30)] and by Eq. (5.39) for the labeled points in Fig. 5.10. Comparison of these values indicates that the errors incurred by using the

**TABLE 5.5**

Depth Distribution of As in Si Calculated from Circled Data Points in Fig. 5.10 Using the Surface Energy (SE) Approximation and Eq. (5.39)

| Labeled point | $\Delta E_{As}$ (keV) | $H_{As}$ (counts) | Depth (1000 Å) | | $N_{As}$ ($10^{20}/cm^3$) | |
|---|---|---|---|---|---|---|
| | | | SE[a] | Eq. (5.39) | SE[b] | Eq. (5.39) |
| 2 | 135 | 232 | 3.1 | 3.17 | 3.97 | 3.91 |
| 3 | 230 | 188 | 5.29 | 5.25 | 3.22 | 3.1 |
| 4 | 320 | 136 | 7.36 | 7.17 | 2.33 | 2.25 |
| 5 | 420 | 65 | 9.66 | 9.25 | 1.11 | 1.1 |
| 6 | 510 | 40 | 11.7 | 11.0 | 0.68 | 0.70 |

[a] Depth $= \Delta E/[S_0]^{Si}_{As}$, where $[S_0]^{Si}_{As} = 43.5$ eV/Å.
[b] From Eq. (5.30) with $N_{As} = (H_{As}/5.85) \times 10^{19}/cm^3$.

surface approximation are about 5% or less for the depth and arsenic concentration.

One can directly evaluate the difference in $N_{As}$ values by taking the ratio of Eqs. (5.39) to (5.4) as shown by

$$\frac{N_{As}(x)}{(N_{As}(x))_0} = \frac{[\varepsilon(E)]_{As}^{Si}}{[\varepsilon(E_0)]_{As}^{Si}} \frac{\sigma_{As}(E_0)}{\sigma_{As}(E)} \frac{\varepsilon(E_1)}{\varepsilon(K_{As}E)}. \tag{5.40}$$

For an energy $E_1$, the energy $E$ before scattering can be found from the mean energy approximation. For the example shown in Fig. 5.10, the correction introduced by the $\sigma_{As}$ ratio tends to be compensated by the $\varepsilon$ ratio. Again this indicates that the surface energy approximation is useful to obtain estimates of impurity distributions.

## 5.6   THICKNESS OF THIN FILMS

### 5.6.1   Elemental Films

A very thin film is usually deposited on a thick substrate. If elements in the substrate are of lighter mass than those in the film, the backscattering signal from the substrate does not interfere with the signal from the film. When a film is very thin, say from a fraction of a monolayer to a few hundred angstroms, it can be treated as a surface contamination of the substrate and analyzed by the methods demonstrated at the beginning of this chapter.

In this section, we treat thicker films ($t > 100$ Å) and consider different methods that can be used to find the number of atoms per square centimeter $Nt$ from backscattering spectra. Figure 5.12 shows seven backscattering

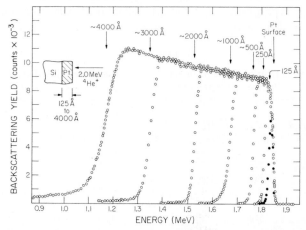

**Fig. 5.12**   Composite backscattering spectra for 2.0-MeV $^4$He ions incident on seven targets of Pt deposited on Si substrates. The Si signal is not shown in the composite figure. [From Chu *et al.* (1973a).]

spectra obtained by 2.0-MeV $^4$He ion backscattering from seven targets of platinum evaporated onto a silicon substrate (Chu *et al.*, 1973a). The thickness of the film, as determined by backscattering, ranges from 125 to 4000 Å. For simplicity, the contribution of the silicon substrate to the spectra at low energies is not plotted in this figure. The energy difference $\Delta E$ between particles scattered from the surface of the platinum and those scattered from the platinum–silicon interface is related to the thickness of the film by the energy loss; that is,

$$\Delta E = [\varepsilon]Nt \simeq [\varepsilon_0]Nt = (224.4 \times 10^{-15})Nt \quad \text{eV cm}^2 \tag{5.41}$$

where the stopping cross section factor $[\varepsilon_0]$ was taken from Table VIII.

If we assume that the density of the thin film is the same as that of bulk platinum, $N = 6.62 \times 10^{22}$ atoms/cm$^3$, the surface energy approximation provides a linear conversion between $\Delta E$ and $t$, as shown by the straight line (dashed) with a slope of 148.5 eV/Å in Fig. 5.13. The solid curve represents the nonlinear relation between $\Delta E$ and $t$, as obtained by the mean energy approximation

$$[\bar{\varepsilon}] = K_{\text{Pt}}\varepsilon(\bar{E}_{\text{in}}) + (1/\cos\theta_2)\varepsilon(\bar{E}_{\text{out}}). \tag{5.42}$$

Here $\bar{E}_{\text{in}}$ and $\bar{E}_{\text{out}}$ depend on the thickness of the platinum layer and are evaluated by the methods discussed in Section 3.2.2. The value of $[\bar{\varepsilon}]$ depends on $t$. We should emphasize that $[\bar{\varepsilon}]$ is used only in evaluating thickness, never in evaluating spectrum height. For the spectrum height, the energy $E$ at a given energy of the projectile immediately before scattering is needed.

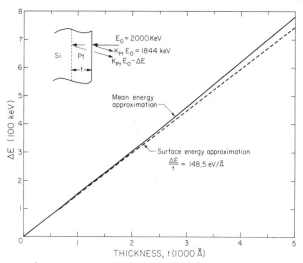

**Fig. 5.13**  Energy width $\Delta E$ of the Pt signal versus thickness of the Pt film for 2-MeV $^4$He ions at normal incidence, with $\theta = 170°$. The solid line is based on the mean energy approximation, and the dashed line on the surface energy approximation $\Delta E/t = 148.5$ eV/Å. A bulk density of $N_{\text{Pt}} = 6.62 \times 10^{22}$ atoms/cm$^3$ is assumed for the films.

Table 5.6 provides a comparison of thicknesses obtained from the energy width $\Delta E$ for the two different analytical approximations. Here $E_1$ is obtained by measuring the position of the trailing edge of each spectrum in Fig. 5.12. The trailing edge has a finite slope, which is due to the energy straggling of the helium ions, the system energy resolution, and the nonuniformities in the film. The position of the trailing edge is defined as the half-height of the step. At the lower end of the spectrum there is a finite background, which needs to be taken into account in determining the position of the half-height. From the data shown in Table 5.6, we can conclude that the two approximations differ by about 1% for each 1000 Å of film thickness.

**TABLE 5.6**

Comparison of Thickness Determinations by the
Surface Energy Approximation and the
Mean Energy Approximation from the Backscattering Spectra of
2.0-MeV He Ions Scattered from Pt Films (Fig. 5.12)

|  |  |  | | | | | |
|---|---|---|---|---|---|---|---|
| $E_1$ (keV): | | | 1170 | 1350 | 1526 | 1670 | 1765 |
| $\Delta E$ (keV): | | | 674 | 494 | 318 | 174 | 79 |
| Surface energy approximation | $t$ | (Å) | 4540 | 3330 | 2140 | 1170 | 530 |
| Mean energy approximation | $t$ | (Å) | 4320 | 3200 | 2100 | 1150 | 530 |
| Difference | | (%) | 5.0 | 3.9 | 1.9 | 1.7 | 0 |

The depth accessible with 2.0-MeV $^4$He depends on the energy loss of $^4$He in the target. For example, $\Delta E = 500$ keV will give $\Delta t = \frac{1}{3}$ $\mu$m for platinum, but about 1 $\mu$m for silicon or aluminum. For a platinum film, the maximum thickness that can be analyzed with 2.0-MeV $^4$He is about 1 $\mu$m. Since protons lose much less energy than $^4$He in the same material, thicker films can be measured with proton backscattering. For the spectra in Fig. 5.14, $^4$He ions and protons were backscattered from gold films deposited on a carbon substrate (Chu *et al.*, 1973b). Figure 5.14a gives the backscattering spectrum for 1.4-MeV $^4$He ions, and Fig. 5.14b indicates that 1.4-MeV protons can easily measure films 3 $\mu$m thick. In this part of the figure, signals from the carbon substrate can also be seen at lower energies. Backscattering with megaelectron volt $^4$He ions, then, is useful for analyzing layers about 1 $\mu$m thick, whereas a beam of protons is useful for analyzing layers from about 1 to 10 $\mu$m thick. This comparison is covered in more detail in Chapter 7 (Table 7.3).

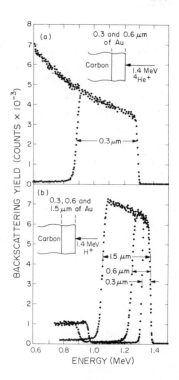

**Fig. 5.14** Composite backscattering spectra for 1.4-MeV $^4$He ions (a) and 1.4-MeV protons (b) incident normal to Au film deposited on carbon substrates. [From Chu *et al.* (1973b).]

Another method for obtaining the thickness of a thin film by backscattering is to calculate the film thickness from the area $A$ under a backscattering signal, as given in Eq. (5.1) and discussed in Section 4.2.2. If one takes into account the energy loss of the incident beam as it traverses the target, the scattering cross section will increase as the projectile loses energy. One method of including a correction for the energy dependence of the scattering cross section is described in Eq. (4.11). For a Pt film about 1000 Å thick, the correction is about 3.8% for 2.0-MeV He ions at normal incidence $(\cos\theta_1 = 1)$:

$$(Nt)_0 = A/\sigma(E_0)\Omega Q = 6.62 \times 10^{17} \quad \text{atoms/cm}^2 \tag{5.43a}$$

and

$$Nt = (Nt)_0\{1 - [\varepsilon(E_0)(Nt)_0/E_0]\} = (Nt)_0[1 - 0.038], \tag{5.43b}$$

where $\varepsilon(E_0) = 115 \times 10^{-15}$ eV cm$^2$ for Pt. For fixed incident energy, the amount of correction is directly proportional to $Nt$ and $\varepsilon$. For elements of low atomic number, $\varepsilon$ is smaller and so is the amount of correction calculated. For example, for a 1000-Å Si film, the amount of correction for 2-MeV helium backscattering is 1.2% rather than 3.8% as found for a Pt film.

### 5.6.2   Multielemental and Multilayered Films

For a multielemental thin film structure, we can obtain two types of information by backscattering spectrometry: composition and film thickness. Composition analysis has been discussed earlier for bulk samples. Here we discuss thickness measurement. Figure 5.15 shows an energy spectrum of 2-MeV $^4$He ions backscattered from an $SiO_2$ film thermally grown on a silicon substrate (Chu *et al.*, 1973a). When the film is thick enough, the energy shift is well defined and the thickness of the film can be readily calculated by using the stopping cross section factor:

$$Nt = \Delta E_{Si}/[\varepsilon_0]_{Si}^{SiO_2} = \Delta E_{Si}/226 \times 10^{-15} \quad \text{eV cm}^2,$$
$$Nt = \Delta E_O/[\varepsilon_0]_O^{SiO_2} = \Delta E_O/213 \times 10^{-15} \quad \text{eV cm}^2, \tag{5.44}$$

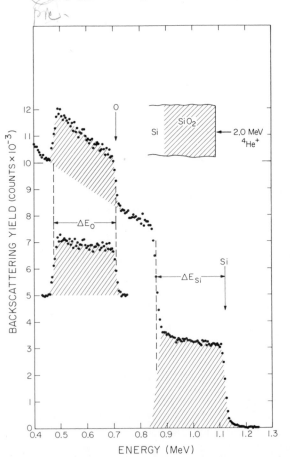

**Fig. 5.15**  Backscattering spectrum for 2.0-MeV $^4$He ions incident on a 5000 Å thick layer of $SiO_2$ thermally grown on a Si substrate. [From Chu *et al.* (1973a).]

where we use a molecule of $SiO_2$ as the unit of the compound stopping cross section, and adopt the surface energy approximation (Table 5.2). From Fig. 5.15 we have $\Delta E_{Si} = 262$ keV and $\Delta E_O = 238$ keV. By using both formulas in Eq. 5.44, we calculate $Nt$ to be $1.16 \times 10^{18}$ molecules/cm$^2$ from $\Delta E_{Si}$ and $1.12 \times 10^{18}$ molecules/cm$^2$ from $\Delta E_O$. The mean value, $1.14 \times 10^{18}$ SiO$_2$/cm$^2$, is equivalent to a 5000-Å $SiO_2$ film when a bulk density of $2.28 \times 10^{22}$ SiO$_2$ molecules/cm$^3$ is assumed for the oxide film.

If the mean energy approximation [Eq. (5.42)] is used, $\bar{E}_{in}$ and $\bar{E}_{out}$ depend on the thickness of the film, and in this particular case we have $\bar{E}_{in} = 1930$ keV, $\bar{E}_{out,Si} = 956$ keV, and $\bar{E}_{out,O} = 577$ keV. These energies give values of $[\bar{\varepsilon}]_{Si}^{SiO_2} = 234 \times 10^{-15}$ eV cm$^2$ and $[\bar{\varepsilon}]_O^{SiO_2} = 213 \times 10^{-15}$ eV cm$^2$, and an oxide thickness of 4910 Å. The mean energy and surface energy approximations differ by 0.4% for every 1000 Å of thickness.

Figure 5.16 gives backscattering energy widths for $SiO_2$, $Si_3N_4$, $Al_2O_3$, AlN, and $Ta_2O_5$ films obtained with 2-MeV $^4$He ions at normal incidence and $\theta = 170°$. The densities of the films are assumed to be identical with those of the bulk compounds. The value of $[\bar{\varepsilon}]$ is calculated by the mean energy approximation for scattering from the heavier of the two elements in the films.

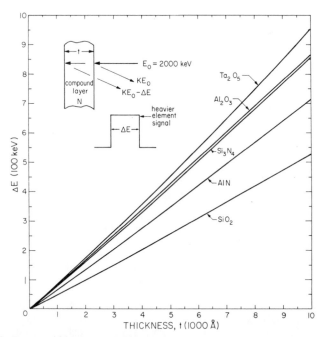

**Fig. 5.16**   Energy widths $\Delta E$ of the heavier element signal versus film thickness for different dielectric layers obtained for 2.0-MeV $^4$He ions at normal incidence, $\theta = 170°$, on the assumption of bulk density of the films.

The energy loss ratio method (see Section 3.3.1) provides an estimate of the energy $E$ before scattering. The method is based on the fact that the energy loss ratio ($\alpha = \Delta E_{out}/\Delta E_{in}$) does not change appreciably with depth. Values of $\alpha$ are listed in Table 5.7. For the layer of $SiO_2$ shown in Fig. 5.15, the value of $E$ at the $Si/SiO_2$ interface is found from Eq. (3.20):

$$E = (E_1 + \alpha E_0)/(K + \alpha) = 1.86 \quad MeV, \tag{5.45}$$

where $\alpha = 1.29$ for Si and $E_1 = 869$ keV for particles scattered from Si atoms at the $Si/SiO_2$ interface. As is shown by the values in Table 5.8, nearly the

TABLE 5.7

Values of $K$, $\alpha$, and $K + \alpha$ for
2-MeV $^4$He Ions Backscattering from a
Thin Film at Normal Incidence with
Scattering Angle at 170° [a]

| Target element | Target mass | $K$ | $\alpha$ | $K + \alpha$ |
|---|---|---|---|---|
| C  | 12   | 0.252 | 1.35 | 1.60 |
| O  | 16   | 0.362 | 1.35 | 1.71 |
| Al | 27   | 0.553 | 1.16 | 1.71 |
| Si | 28   | 0.566 | 1.29 | 1.86 |
| Cr | 52   | 0.736 | 1.11 | 1.85 |
| Cu | 63.5 | 0.78  | 1.06 | 1.84 |
| Ag | 108  | 0.86  | 1.07 | 1.93 |
| Ta | 181  | 0.92  | 1.03 | 1.95 |
| Au | 197  | 0.92  | 1.03 | 1.95 |
| U  | 238  | 0.92  | 1.01 | 1.95 |

[a] Values of $\alpha$ are determined from the surface energy approximation using Eq. (3.21): $\alpha = (\varepsilon(KE_0)/\varepsilon(E_0))\beta$, where $\beta = \cos\theta_1/\cos\theta_2 = (\cos 10°)^{-1}$.

TABLE 5.8

Calculation of the Energy $E$ before Scattering for the Spectrum
Shown in Fig. 5.15 Using the Energy Loss Ratio Method
[Eq. (3.20)] with $\alpha_{Si} = 1.29$ and $\alpha_O = 1.35$ (Table 5.7)

| Position | | $E_1$ (keV) | $E_1 + \alpha E_0$ (keV) | $E$ (keV) |
|---|---|---|---|---|
| Si at surface   | $K_{Si}E_0$              | 1131 | 3710 | 2000 |
| Si at interface | $K_{Si}E_0 - \Delta E_{Si}$ | 869  | 3449 | 1858 |
| O at surface    | $K_OE_0$                | 725  | 3425 | 2000 |
| O at interface  | $K_OE_0 - \Delta E_O$   | 487  | 3187 | 1860 |

same energy $E$ is obtained when the computation is made on the basis of scattering from O atoms at the interface.

The energy loss ratio calculated on the basis of the surface approximation can also be used with multilayer films to determine the energies before scattering at the various interfaces. Since the particle traverses exactly the same range of compositions on the outward path as on the inward path, the value of $\alpha$ will not change greatly even though the composition changes in the different layers. Figure 5.17 shows the spectrum for a sample with $Ni_2Si$ formed between Ni and the Si substrate (Tu et al., 1975). The presence of the silicide layer can be deduced from the step in the Ni and Si shoulders. The energies $E$ before scattering at the various interfaces are given in Table 5.9 for $\alpha_{Ni} = 1.11$ and the energy loss ratio method [Eq. (3.20)]. Rather good

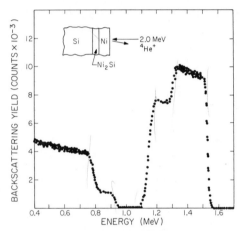

**Fig. 5.17**  Backscattering spectrum for 2.0-MeV $^4$He ions incident on a multilayer sample with $Ni_2Si$ formed between Ni and the Si substrate. [From Tu et al. (1975).]

**TABLE 5.9**

Calculation of the Energy $E$ before Scattering for the Spectrum
Shown in Fig. 5.17 Using the Energy Loss Ratio Method
[Eq. (3.21)] with $\alpha_{Ni} = 1.11$ and $\alpha_{Si} = 1.29$

| Position | Scattering element | $E_1$ (keV) | $E_1 + \alpha E_0$ (keV) | $E$ (keV) |
|---|---|---|---|---|
| Ni surface | Ni | 1525 | 3745 | 2000 |
| Ni/Ni$_2$Si interface | Ni | 1310 | 3530 | 1885 |
| Ni$_2$Si/Si interface | Ni | 1140 | 3360 | 1794 |
| Ni/Ni$_2$Si interface | Si | 930 | 3510 | 1896 |
| Ni$_2$Si/Si interface | Si | 780 | 3360 | 1810 |

agreement is found for the energies at the $Si/Ni_2Si$ interface as computed from scattering from Ni and Si atoms.

## REFERENCES

Chu, W. K., Mayer, J. W., Nicolet, M-A., Buck, T. M., Amsel, G., and Eisen, F. (1973a). *Thin Solid Films* **17**, 1.

Chu, W. K., Mayer, J. W., Nicolet, M-A., Buck, T. M., Amsel, G., and Eisen, F. (1973b). *in* "Semiconductor Silicon" (H. R. Huff and R. R. Burgess, eds.), p. 416. The Electrochemical Society, Princeton, New Jersey.

Howard, J. K., Chu, W. K., and Lever, R. F. (1976). *in* "Ion Beam Surface Layer Analysis" (O. Meyer, G. Linker, and F. Käppeler, eds.), p. 125. Plenum Press, New York.

Mayer, J. W., Ziegler, J. F., Chang, L. L., Tsu, R., and Esaki, L. (1973). *J. Appl. Phys.* **44**, 2322.

Nicolet, M-A., and Chu, W. K. (1975). *Am. Lab.* **7**, 22.

Sigmon, T. W., Chu, W. K., Müller, H., and Mayer, J. W. (1975). *Appl. Phys.* **5**, 347.

Tu, K. N., Chu, W. K., and Mayer, J. W. (1975). *Thin Solid Films* **25**, 403.

Chapter

# 6

# Instrumentation and Experimental Techniques

R. A. Langley[†]

## 6.1 INTRODUCTION

The main purpose of this chapter is to answer the question: What do I as an experimentalist need to get my job done? This chapter will concentrate on the technical aspects of the problem. It will be divided into three sections: (1) accelerator, (2) target chamber, and (3) energy analysis of the backscattered beam. To a large extent this chapter will be directed toward the research scientist who is attempting to set up or expand a program in backscattering spectrometry.

In essence, ion backscattering works in the following manner: A mono-energetic high-energy beam of ions (e.g., $H^+$ or $He^+$) impinges on a target from which some are backscattered; part of these backscattered ions are energy analyzed and counted, and the data stored. Figure 6.1 is a schematic drawing of an ion backscattering experiment divided into its essential components. In a sense, backscattering spectrometry has a beginning, a middle,

[†]Sandia Laboratories, Albuquerque, New Mexico. Presently at Oak Ridge National Laboratories, Oak Ridge, Tennessee.

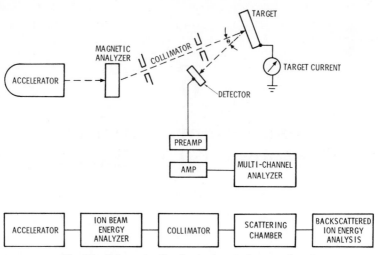

**Fig. 6.1**  Schematic of an ion backscattering experiment.

and an end: the birth, acceleration, and energy selection of an ion; its scattering; and its energy analysis. The organization of this chapter will reflect these three stages.

The accelerator consists of an ionization chamber for creation of the ions, followed by a column where the ions are accelerated. After passing through a short drift tube, the ions enter an analysis magnet, where the various species of ions are separated and only those energies are selected which are of use in the experiment. After passing through the magnet, the analyzed beam enters a drift tube and is collimated and directed onto the target. A few of the ions elastically backscatter from the target, but those which do scatter in a particular direction are detected by a particle detector; their energy is analyzed and stored in an appropriate data storage system. Since most experimenters will be using existing accelerators, not buying new ones, the first section on accelerators will be quite limited, but will include discussions relevant to backscattering spectrometry. References are included for those wishing further explanation and discussion.

## 6.2  ACCELERATOR

Many commercially available accelerators can be used for backscattering spectrometry. The one that best serves the experimenter depends to a large extent on his requirements. For industrial applications, where measurements are repetitive and must be accomplished quickly, the choice might be a Crockroft–Walton accelerator; for experimental applications that do not require large beam currents, it would probably be an electrostatic accelerator. Accelerators and problems related to them are discussed in the literature (Livingston and Blewett, 1962; Allen, 1974; Duggan, 1968; Duggan, 1970; Morgan and Duggan, 1974).

The most widely used and available electrostatic accelerator is the Van de Graaff generator (VDG). Its essential elements will be discussed, as will the Pelletron accelerator, which is similar to a VDG but has a different charging system. Finally, a tandem accelerator that has been proposed exclusively for backscattering spectrometry will be presented. Much of the material in the discussions of the different accelerators was gleaned from the companies product brochures and equipment manuals.

### 6.2.1   Van de Graaff Accelerator[†]

To produce an accelerated beam of positive ions requires an ion source, an accelerating voltage, an evacuated acceleration path, a vacuum system, and an adequate control system. The usual positive ion source uses a glass bottle to which rf energy is applied so that the gas introduced into the source bottle is ionized. This results in a plasma that is magnetically focused at the exit canal of the source bottle. Positive ions are initially expelled through the exit canal into the acceleration path by a potential applied to the anode of the source bottle (see Fig. 6.2). Further acceleration is provided by the voltage gradient developed along the column.

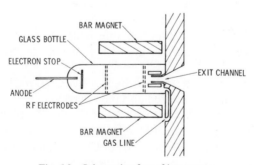

**Fig. 6.2**  Schematic of an rf ion source.

---

[†]High Voltage Engineering Corp., Product Brochure for AN-2500 Accelerator and Instruction Manual for AN-2000 Accelerator, Burlington, Massachusetts 01803.

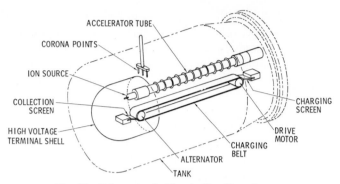

**Fig. 6.3** Schematic of a Van de Graaff accelerator.

A voltage is produced on the terminal shell (see Fig. 6.3) by continuously charging it by means of a rapidly moving belt that conveys charges between ground potential and the terminal. The generator operates in a compressed atmosphere (see the subsequent discussion for further details). A voltage-insulating column mechanically supports the terminal and the belt-charging system. This column provides the electrostatic environment for the accelerator tube, i.e., it holds the precision resistors that maintain the voltage gradient along the column. The drive motor moves the charging belt and runs the alternator that is built into the upper pulley for terminal power. The belt-charging current is supplied from a dc power source located outside the pressure vessel.

The acceleration path for the positive ion beam is provided by an accelerator tube, which is highly evacuated to minimize collisions between the accelerated particles and the extraneous gas molecules in the tube. The normal operating pressure of the accelerator tube system is about $10^{-6}$ Torr. The maximum permissible operating pressure is about $3 \times 10^{-5}$ Torr. Above this pressure, there is a serious possibility of voltage breakdown in the accelerator tube.

The accelerating column and the high-voltage terminal are surrounded by a tank, which is first evacuated and then filled with a combination of dry gases. The tank is generally filled with $N_2$, $CO_2$, or $SF_6$ to allow standoff of the high voltage on the terminal-from-ground potential. Some accelerators use only $SF_6$, some use a combination of $CO_2$ and nitrogen, and some use a combination of all three, depending on the manufacturer and the conditions under which the accelerator is used. The gases introduced into the tank must be extremely dry; dew point $< -55°C$. When filling the tank, it is necessary to fill first with the gas that has the lowest vapor pressure. This is usually $SF_6$, whose vapor pressure at room temperature is about 16 psi; some of it is condensed in liquid form. The liquid can be removed by placing a heating tape around the base of the supply tank that contains the gas, to help raise the temperature in the supply tank and vaporize the liquid $SF_6$. The gas with the next highest vapor pressure is $CO_2$, which has a vapor pressure of 120–150 psi. Again a heating tape around the base of the gas bottle allows use of the full contents of the supply bottle. Finally, the tank is topped off with $N_2$. Since $SF_6$ is an extremly good insulating gas, it is necessary to use extreme care when making electrical connections in the terminal area. A connection might have excellent conductivity in air, but extremely bad conductivity when exposed to $SF_6$. This can lead to problems that are difficult to analyze.

To stabilize the terminal voltage, the terminal load is varied by using corona discharge or, much less commonly, belt-spray stabilization. In a typical operation a discharge path is created between a series of sharp points (corona points) and the high-voltage terminal shell (Fig. 6.3). The flow of corona current imposes a controllable load on the high-voltage terminal, which can be used to compensate for tube loading or charge fluctuation. The corona current is usually controlled by one of the two following methods:

**a.   Manual Operation.**   With manual operation, the corona points can be driven toward or away from the tank by using an electrically operated motor. As the points are driven closer to the terminal shell, the corona load current increases, whereas when they are driven further from the terminal shell it decreases. The distance between the corona points and the terminal shell is set for the gross acceleration voltage consistent with the lowest controllable corona current (10–30 $\mu A$).

**b.   Automatic Operation.**   A signal proportional to the departure from the setpoint of the acceleration voltage is amplified and applied to the stabilizer tube of the corona circuit. This signal increases or decreases the bias of the tube that controls the voltage applied to the corona points, so that an increase in tube bias decreases the tube conduction and hence decreases the corona current. Conversely, a decrease in tube bias increases tube conduction and hence increases the corona load current. For the accelerator to work well, any deviation in the terminal voltage must be corrected by the effects of the feedback signal. The corona points must be sharp, straight, and of uniform length, and should be replaced periodically. The lifetime of these points is directly proportional to the current that passes through them, i.e., the corona load current. It is quite possible to run the corona current at a relatively low value ($\sim 10\ \mu A$) and thereby prolong the useful lifetime of the corona points. When an accelerator is installed, it is advantageous to run a plot of corona point position versus accelerating voltage for a specific corona load current. This gives the operator an idea of the maximum and minimum accelerating voltages for the corona point placement used by the manufacturer. The maximum or minimum energy can be increased or decreased by increasing or decreasing the distance from the extreme positions of the corona points to the terminal.

Many types of ion source have been used to provide ions for high-voltage accelerators. The most commonly used source is a radio-frequency (rf) ion source. These sources operate through the use of a rf oscillator circuit which applies rf power to the ion source proper (see Fig. 6.2). Each manufacturer has its own specifications for the ion source power supply and its tuning. Only two further comments will be made concerning the rf ion source. If there is a leak from the tank into the rf ion source, a sulfur deposit (brownish-yellow) from the breakdown of $SF_6$ may occur inside the ion source bottle and will in time shunt the rf coils and extinguish the plasma. If this condition occurs, one should carefully check all vacuum joints in the gas supply system. A combination of gases can be used in the gas supply bottle so that more than one type of ion can be obtained. Because the gases used have a wide range of ionization potentials, a judicious choice must be made when combining different gases. The following combinations have been found to work reasonably well: $H_2 + Ne$, $^3He + {}^4He$, and $Ar + Kr$ (Augustyniak, 1974). Most accelerators are supplied with at least two gas bottles, so that hydrogen or helium ions can be used interchangeably. A new technique has been devised by which these gas bottles can be replenished without venting the tank (Langley, 1976). This can considerably reduce downtime of the accelerator.

### 6.2.2   Pelletron Accelerator[†]

A Pelletron accelerator is very similar to a Van de Graaff accelerator. The main difference is that, whereas in the Van de Graaff device the charge is sprayed onto an insulating belt, the Pelletron has no belt, but instead a rugged chain consisting of metal cylinders joined by links of solid insulating plastic. The gaps between the metal cylinders serve as spark gaps, which provide excellent protection for the insulating links. Certain characteristics of the chain are said to make it superior to the belt: improved voltage stability, no dust or lint, relative insensitivity to moisture, high efficiency, no spark damage, dependable operation, and, finally, no tension adjustments are required.

### 6.2.3   Tandem Accelerator

A small tandem electrostatic accelerator has been proposed for back-scattering spectrometry.[‡] The approach used is to generate negative ions in an ion source. The negative ions from the source are attracted toward the

[†] National Electrostatics Corp., Middleton, Wisconsin 53562.
[‡] General Ionex Corp., Ipswich, Massachusetts 01938.

high-voltage terminal, where electrons are stripped from each ion in a gas-charge exchange cell producing positively charged ions. Since these particles are now positive, they are repelled by the high-voltage terminal and return to ground potential. This accelerator has been designed for a maximum voltage of $\sim 1$ MV (i.e., ion energy of 3.0 MeV for $He^{2+}$) and a capability of placing 200 nA of $He^{2+}$ on the target. The voltage is generated by a high-frequency voltage doubler power supply which has very good stability. This accelerator, a table-top device, is now in the first stage of construction. The advantage it offers is that it requires only a small area and a very small amount of radiation shielding, thus considerably decreasing the cost of the building. In addition, the ion source is near ground potential, so that various high-current ion sources can be used and various ions accelerated.

### 6.2.4  Safety Considerations

A main hazard in dealing with positive ion accelerators is *Bremsstrahlung* radiation induced by high-energy electrons and neutrons from (p, n) and (α, n) reactions. Electrons created by beam ionization of background gas and secondary electrons are accelerated into the terminal, creating high x-ray fluxes. The amount of radiation produced in the area of the terminal is directly proportional to the number of electrons created in the accelerator tube. These electrons are accelerated through the accelerator tube into the ion source area, causing *Bremsstrahlung* radiation. The radiation level they produce can be greatly reduced by increasing the pumping speed in the beam line system, thus reducing the scattering of the ion beam as it leaves the ion source, and reducing the amount of residual gas in the accelerator tube.

The other main hazard is neutrons from nuclear reactions between the accelerated ion and target material. These are well documented in the literature (Burrill, 1970). Some of the more troublesome reactions are listed in Table 6.1.

**TABLE 6.1**
Neutron-Producing Reactions

| Reaction | Reaction $Q$ (MeV) | Relative reaction strength | Reaction | Reaction $Q$ (MeV) | Relative reaction strength |
|---|---|---|---|---|---|
| $^3H(p, n)\,^3He$ | −0.764 | Weak | $^7Li(d, n)\,^8Be$ | 15.0 | Moderate |
| $^7Li(p, n)\,^7Be$ | −1.643 | Weak | $^9Be(d, n)\,^{10}B$ | 4.36 | Moderate |
| $^9Be(p, n)\,^9B$ | −1.85 | Weak | $^{10}B(d, n)\,^{11}C$ | 6.47 | Weak |
| $^2H(d, n)\,^3He$ | 3.3 | Moderate | $^{12}C(d, n)\,^{13}N$ | −0.28 | Weak |
| $^3H(d, n)\,^4He$ | 17.6 | Strong | $^{14}N(d, n)\,^{15}O$ | 5.1 | Weak |
| $^6Li(d, n)\,^7Be$ | 3.38 | Moderate | $^9Be(^4He, n)\,^{12}C$ | 1.26 | Strong |

Gamma and neutron detectors are needed to ensure that the radiation in the personnel area is below the established safe limits (NCRP, 1971). Gamma detectors made of scintillation crystals are more stable than gas chambers, and they are less responsive to electromagnetic fields. Of many gamma detection systems tried, one has been found to be quite satisfactory.[†] Also one neutron detector system has been found to be very stable and versatile.[‡]

### 6.3   ENERGY STABILIZATION SYSTEM

Backscattering spectrometry requires stable, nearly monoenergetic ($\pm 2$ keV) ion beams. These requirements can be met only if fluctuations in the terminal voltage are corrected in a fast negative feedback loop. The fluctuations can arise from many sources, e.g., variations in the belt-charging process and discharges along the insulating surfaces of drain resistors and voltage stand-off insulators. The feedback loop components differ with the design of the accelerator and with its application although there are two commonly used methods to provide the sensing for the negative feedback loop. The first involves the use of a generating voltmeter (Fig. 6.4a); the second involves the use of current sensing elements placed in the beam line after the analyzing magnet (Fig. 6.4b).

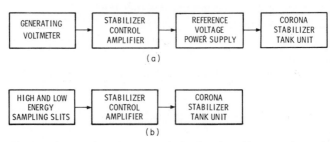

**Fig. 6.4**   Block diagram of energy control circuit using (a) generating voltmeter and (b) current sensing elements.

A generating voltmeter generates within its circuit a voltage proportional to the terminal voltage.[§] It can be used during electron or positive ion operation, and conversion is not necessary for a change from one polarity terminal operation to the other. The generating voltmeter has a motor-driven rotor and a fixed, insulated stator. The rotor, which has sectors cut out of it, revolves so that it alternately exposes and shields the high-voltage terminal to the

[†]Nuclear Measurements Corp., Gamma Alarm System GA-2TMO, Indianapolis, Indiana 46218.

[‡]Eberline Instrument Corp., Neutron Radiation Monitor RM-16, Santa Fe, New Mexico 87501.

[§]High Voltage Engineering Corp., The Generating Voltmeter, HVI-1015, Burlington, Massachusetts 01803.

stator plates. Essentially triangular wave ac voltages are electrostatically induced between adjacent sectors of the stator at a frequency proportional to the motor. These voltages are directly proportional to the terminal voltage. To be used as the sensing element for the negative feedback loop, the generating voltmeter must have an accuracy of at least 0.05%. It is sensitive to its geometric relationship with respect to the high-voltage terminal, and therefore must be calibrated after each removal of the tank as the generating voltmeter is coupled directly to the tank. Because the calibration curve for the generating voltmeter output is sufficiently linear, only one energy calibration point is needed.

The second sensing system consists of sampling the analyzed beam after it passes through the analysis magnet (Fig. 6.1). The beam is sensed by low-energy and high-energy slits. The current from these two slits is amplified by an amplifier, and the output is used in the negative feedback loop to determine the bias of the corona stabilizer circuit, discussed in Section 6.2.1.a. If the terminal voltage goes high, the beam is deflected less than normal, hits the high energy slit, and increases the corona load current, thus decreasing the terminal voltage. Conversely, if the terminal voltage is low, the beam energy is low and the beam is deflected more than normal, hits the low energy slit, decreases the corona load current, and thus increases the terminal voltage. This is the more commonly used feedback sensing system. It requires that the stability of the magnet and its power supply be better than $10^{-3}$.

Analyzing magnets that will easily satisfy the stability requirements of backscattering spectrometry are commercially available. An excellent discussion of various magnets and their properties has been given by Enge (1967).

Very precise energy control networks have been developed in which extremely fast variations in the terminal voltage are electronically sensed.[†] These networks are based on the use of a precision-generating voltmeter, a capacitive pickup, and an analyzing magnet followed by high- and low-energy slits. These precise systems are "automatic switching" in that the slow loop control is switched to the generating voltmeter in the event of loss of slit signals or large terminal fluctuations. The capacitive pickup is used to reduce short-term ripple. The ripple for this precision energy control network is 2 kV peak to peak, with a ±2-kV drift. This additional control network increases the expense of the accelerator by a few thousand dollars and is not required unless extremely high-precision energy analysis is required.

### 6.4  ENERGY CALIBRATION

The energy of the accelerator can be calibrated in numerous ways. The generally accepted procedure has been to use one or more suitably chosen

---

[†] High Voltage Engineering Corp., Super Stabilizing System, Product Procedure, Burlington, Massachusetts 01803.

nuclear resonance reactions, such as (p, n) or (p, $\gamma$), to establish an absolute energy calibration at a few specific energies and then rely on some secondary standard such as the magnetic field strength of the analyzing magnet to interpolate accurately to other energies. Many of these resonance reactions have been discussed previously in great detail (Marion, 1961; Bondelid and Kennedy, 1959; Bumiller et al., 1956) and will not be fully discussed here. Some of the reactions found most convenient for calibration are $^{19}F(p, \alpha\gamma) {}^{16}O$, $^{27}Al(p, \gamma) {}^{28}Si$, and $^{7}Li(p, n) {}^{7}Be$, where the emitted $\gamma$s are detected for the first two reactions and neutrons are detected for the last. For proton energies below 400 keV a method has been devised in which the $^{12}C(p, \gamma_0) {}^{13}N$ reaction is used to calibrate the proton beam with a precision of about $\pm 1.5$ keV (Switkowski and Parker, 1975). A thick target of $^{19}F$ is easily made by placing a drop of HF acid on almost any metal substrate. A thin target of LiF can be made by evaporating LiF onto a metal substrate. This target can be used for a large energy range, with both Li and F resonances. Thin-film targets of Al are easily made by vapor deposition and thin-film $^{7}Li$ targets which are $H_2O$-cooled may be purchased commercially.[†] For thick targets only step functions are observed in the output, whereas for thin targets the shape of the resonance is mimicked if the target is "thin" with respect to the resonance width. The detection of neutrons and $\gamma$s are fully discussed by Duseph (1975), Nicholson (1974), Cerny (1974), and ORTEC (1976a).

There are no nuclear reactions for He ions that can be easily used in the energy range of most backscattering spectrometry accelerators; therefore, no direct energy calibration of the accelerator can be made using He ions. There are indirect methods, however. One involves the use of He particles from a $^{241}Am$ source (Mitchell et al., 1976) to calibrate the detector and electronics. For energy calibration of the accelerator, it is necessary to backscatter He ions from a known target and compare the results to those obtained with the $^{241}Am$ source. If the radioactive source is thin, the energy spread can be as low as 6-keV FWHM. This arrangement makes possible not only a direct calibration of the detector system, but also a determination of specific quantities about the detector used in the backscattering setup. In order to transfer this calibration of the detector system to the accelerator voltage, the response of the detector to He must be evaluated and corrections made (see Section 6.7.1). If the accelerator is equipped with a precision-generating voltmeter, the energy calibration obtained with nuclear reactions for protons can be transferred directly to the He beam; however, calibration must be repeated whenever the accelerator tank is removed (see Section 6.3.)

The measurement of the magnetic field is extremely important, since this quantity is usually used for interpolating between the calibration energies.

---

[†] High Voltage Engineering Corp., Burlington, Massachusetts 01803.

The measurement can be made with sufficient accuracy by either of three methods. One is the use of a nuclear magnetic resonance (NMR) fluxmeter (Kinnard, 1956, pp. 290–292). This instrument is based on the principle that a nucleus, when placed in a magnetic field, precesses about an axis parallel to the direction of the magnetic field. The precessional rate (frequency) of the nucleus is directly proportional to the magnetic field strength. This nucleus sample, called the *probe*, is coupled to a variable frequency oscillator so that the resonance frequency of the system can easily be determined. The magnetic field strength is then established by reference to the gyromagnetic ratio for the particular nucleus used in the probe, nominally proton. The range of NMR Gaussmeters varies from a few hundred Gauss to tens of kilogauss. It may operate in field gradients up to 40 G/in., and should have a long-term stability of $10^4$ and a relative accuracy of 0.02 G.

A second method of measuring the magnetic field is to use a rotating coil Gaussmeter.[†] This unit is relatively inexpensive by comparison to an NMR fluxmeter, and has an absolute accuracy of a few gauss. The basic operation of this instrument consists of a rotating coil near the tip of a long probe. The coil is spun on one of its diameters so that it cuts the lines of magnetic field twice during each revolution, generating a relatively pure sine wave at half the frequency of the motor. The dimensions of the coil are carefully chosen to give the maximum output for a given volume of field occupied by the coil and also to give the best average reading in a highly nonuniform field. The coil acts as a simple ac generator, and the voltage output is a true measure of the field. Mounted on the same rotating shaft as the probe is a rotating magnet generator, which generates a constant ac voltage in phase with the signal from the pick-up coil. A precision voltage divider compares a fraction of this voltage against the pick-up coil voltage. The dials of the divider are adjusted until the two voltages are exactly equal, giving a null balance. In this mode the Gaussmeter accuracy is about 0.1% or $\pm 2$ G, whichever is larger. The range of measurement can vary from 0 to 80 G.

In the last few years the stability of Hall probes has been improved to the extent that they offer a viable method of measuring the magnetic field for backscattering spectrometry (Kinnard, 1956, pp. 288–290). Some Hall probes have stabilities quoted as $5 \times 10^{-5}$, quite sufficient for backscattering spectrometry.

The placement of all three types of probes in the magnetic field is critical, because the probe measures only the magnetic field strength in its active area, whereas the deflection of the ion beam is approximately proportional to the average strength of the magnetic field through which it passes. As the magnet saturates at higher fields, the correspondence between the measured

---

[†] Rawson–Lush Instrument Co., Instruction Manual for Type 820B Rotating Coil Gaussmeter Probe, Acton, Massachusetts 01720.

value and the average value, as seen by the ion beam, can become quite different. The sensing element should be placed as near the center of the pole pieces as possible, i.e., away from the fringing field, but with the restraint of keeping it out of the ion beam path.

## 6.5   THE VACUUM SYSTEM[†]

The vacuum system has three main parts—the accelerator, the beam line, and the experimental chamber. The gas load from the ion source must be accommodated by pumps near the exit of the accelerator. It is advantageous to place these pumps between the accelerator and the magnet, since the magnet chamber presents a fairly large pumping impedance. Ion pumps do not have enough pumping speed for $H_2$ or He to accommodate the source gas load, so that either diffusion pumps or turbomolecular pumps must be used. Either oil or Hg diffusion pumps may be used. In both cases the pumps need to be $LN_2$ trapped to prevent backflow of the pump vapor into the beam line and the accelerator tube. Special fluids have been developed for use in diffusion pumps placed on accelerators.[‡] Recent observations on the backflow of oil from turbomolecular pumps indicate that it is very small except under unusual conditions, e.g., venting the pump to the atmosphere. The forepump oil must be kept from migrating into the diffusion or turbo-molecular pump, since this oil has high vapor pressure and will easily migrate into the vacuum system. Migration can be prevented by placing a sorb trap or a cooled baffle in the foreline. It is appropriate to use ion pumps as holding pumps when the accelerator is not being operated. The pumping speed at the base of the accelerator should be at least 500 liter/sec. Accelerators are operated with the beam line at pressures from $10^{-7}$ to $1$–$2 \times 10^{-5}$ Torr; the lower the better, from both the background-radiation and beam-contamination standpoints (see Sections 6.2.4 and 6.6.3 for further detail).

The magnet chamber usually can have considerable volume and a substantial pumping impedance. We therefore recommend that it be pumped from both sides. Since a fairly long flight path from the magnet to the experimental chamber is desirable, a pumping station should be placed in the beam line. In most experimental setups more than one beam line is used, and it is suggested that each beam line be separately pumped. For this application ion pumps, diffusion pumps, or turbomolecular pumps have proved quite adequate. It is extremely useful to place pressure-sensing devices along the

[†]See, for example, Dushman (1962) and Power (1966).

[‡]These special fluids are of a silicon base rather than a petroleum base. They tend not to crack when exposed to air, and have very low vapor pressures at room temperature. See for example: Samtovac 5, Momsanto Co., St. Louis, Missouri; Dow Corning 705, Dow Corning Corp., Midland, Michigan.

vacuum envelope. They should be coupled to the accelerator so that, if the pressure goes above $10^{-5}$ Torr, the accelerator is automatically shut off, thus avoiding potential discharges down the accelerator tube. All thin-walled bellows should be protected by collimation of the beam before the bellows.

The requirements for the backscattering chamber are dictated by the anticipated experiments. Some *in situ* studies require ultra-high vacuum ($<10^{-9}$ Torr), and others are less stringent; a vacuum of $10^{-6}$ Torr is quite sufficient for most backscattering studies. The simplest chambers, and sometimes the most versatile, can be constructed from two double crosses, an example of which is shown in Fig. 6.5. This arrangement provides eight ports, which can be used for various instrumentation and sample holders. Other chambers have been designed and used for specific experiments, and these may be found throughout the literature.

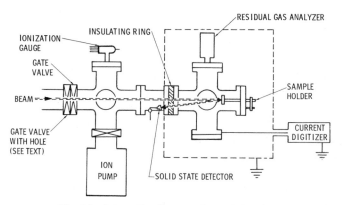

**Fig. 6.5**  Schematic of an experimental chamber.

The experimental chamber shown in Fig. 6.5 has a gate valve with a small hole ($\sim 3$ mm diameter) drilled in the valve. This serves both as the last aperture, so that most of any scattered beam is prevented from entering the experimental chamber, and as a constriction between the beam line and the experimental chamber for differential pumping if it is wanted. In addition, there is a gate valve for complete isolation when changing samples.

For experiments that require only high vacuum ($\sim 10^{-7}$ Torr), regular Viton O-rings suffice nicely for the gasket material. Unisex flanges, e.g., Dependex,[†] have been found to be quite successful. In situations requiring ultrahigh vacuum, metal gaskets are necessary. Flanging of the Conflat-type[‡]

---

[†] Aero Vac Corp., Burlington, Massachusetts.
[‡] Varian Associates, Palo Alto, California.

have proven quite adequate for ultrahigh vacuum conditions. It is also advisable to place a gate valve between the experimental chamber and the beam line (as shown in Fig. 6.5) so that the accelerator can remain running while samples are being changed.

In order to use a backscattering apparatus efficiently, the scattering chamber should be as small as possible and the pumping speed as large as possible. This arrangement minimizes the time taken for changing samples. The experimental chamber can be ion pumped, diffusion pumped, or turbomolecular pumped. In any case, with the use of a diffusion pump or a turbomolecular pump a liquid nitrogen trap should be placed between the pump and the sample chamber. This will reduce the hydrocarbon products in the vacuum chamber. The incident ion beam will break up these hydrocarbon molecules, and carbon will be deposited on the surface of the sample. This deposit can be extremely detrimental to backscattering analysis because it decreases the apparent resolution, and the carbon may physically combine with the sample thereby destroying the information sought. However, the carbon does allow an accurate determination of the beam spot size and position. Such deposits can be easily observed by breathing on the samples.

A simple solution to the problem of carbon contamination during analysis is to place a cryoshield around the target (Bottiger *et al.*, 1973). The shield completely surrounds the target except for small apertures ($\sim 10^{-3}$ Sr) for the incident beam and the scattered beam. The cryoshield temperature can be held as low as 20 K and an effective pressure of $\sim 5 \times 10^{-11}$ Torr maintained in the target region.

## 6.6  BEAM DEFINITION AND MEASUREMENT

This section deals with beam collimation, its measurement, and its contamination.

### 6.6.1  Collimation

Beam collimators of various designs are commercially available. Their function is to physically define the beam so that it has sharp edges, and to prevent beam particles which have scattered from the analyzing slits from entering the target chamber. The analyzing slits, discussed in Section 6.3, collimate the beam to a certain extent. They should not be used as the only collimation, however, since the beam is defined only in the horizontal plane and considerable scattering of the beam occurs from these slits. After the analyzing slits there should be at least one set of collimation slits, or preferably two sets. The analyzing slits are usually bulky, water cooled, and do not define the beam to sharp edges. It is necessary to water cool these slits because they intercept a large portion of the beam and must dissipate its

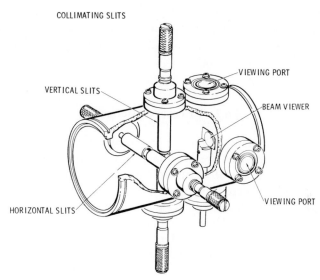

**Fig. 6.6**   Diagram of collimating slits and beam viewer.

energy. For convenience, the main collimation slits should be movable, and they should be driven by micrometers, either manually or electrically. These should have a reproducibility of at least ±0.001 in. The actual slit material can be made of stainless steel, but if high currents are anticipated, it is best that they be made of a material of high thermal conductivity. If a substantial portion of the beam is intercepted by the analyzing slits, the collimating slits need not be water cooled. When the system is being started, or after any major changes have been made, it is recommended that the micrometers and their respective slits be observed to make sure they are not overheating. A typical set of collimation slits is shown schematically in Fig. 6.6. If opposing slits are placed close together for very small beam definition, the beam can heat them, possibly causing enough expansion that the beam can be completely cut off and the slits welded together. It is advantageous to have viewing ports placed strategically along the beam path for observation of the beam. These are essential for diagnosing problems that may arise. It is suggested that beam viewers be placed on both sides of the analyzing slits and behind each set of collimating slits. This arrangement is sufficient for determining the character of the beam and analyzing most problems. Roughened quartz has been found to work well as the material for these viewers.

   It is necessary to be able to block the beam before it enters the experimental chamber, because in most cases one would like to limit the fluence on the target. The beam can be easily blocked by using the beam viewer that follows the last set of collimating slits. It may be advantageous to have this beam viewer remotely operated. The beam may also need to be blocked when gate valves in the beam line are being closed, since O-rings can be severely burned by the beam.

### 6.6.2   Fluence Measurement

To accurately measure the fluence placed on a sample, it is necessary to account for electrons that are moving with the beam and those that are emitted from the sample as secondary electrons. There are two generally accepted methods of correcting for the influence of secondary electrons. The first of these is by applying voltages to grids and plates so that secondary electrons are suppressed. The second method uses a Faraday cup in which the sample is placed. The flight path of the ions in the Faraday cup should be long with respect to the entrance aperture of the impinging beam and the exit aperture of the scattered beam. As the beam traverses the beam line between the magnet and the target chamber, it ionizes background gas atoms so that knock-on electrons can move with the ion beam and, in addition, the ions can pick up electrons to become neutralized. The effect increases as a function of decreasing energy, so that it might be insignificant at 2 MeV, but could become substantial at an energy of 1 MeV. The most efficient way to reduce this effect is to decrease the background gas pressure in the beam tube. Pressures in the low $10^{-7}$-Torr range have been found sufficient to reduce this effect to negligible proportions. Glancing collisions by the beam on the collimating slits can produce an abundance of electrons, but by placing a positive voltage of over 50 V relative to ground on the slits it was found that this source of electrons was greatly reduced (Khan and Potter, 1964).

A particular setup will be described in detail. The scattering chamber is electrically insulated and used as a Faraday cup. The criteria listed previously for the length of the flight path of the ions in the Faraday cup were met, in that the ion path was made long with respect to the entrance aperture of the incoming beam and the exit aperture of the scattered beam. It is not necessary in this case to apply any voltage to grids or plates in the scattering chamber. Using this configuration, both the sample and the sample chamber are floated with respect to ground and the beam current is measured directly with a current digitizer. Such a setup was shown schematically in Fig. 6.5. One of its advantages is that only a few millivolts of potential are applied to the Faraday cup, that being the input voltage of the electrometer. Voltages appled near the solid-state detector can have detrimental effects (see Section 6.7.1). A combination of the two techniques methods, application of voltages to slits and a floating Faraday cup, has been used where extremely accurate fluence measurements were necessary (Khan and Potter, 1964; Musket and Taatjes, 1973; and Singh, 1957).

For most of the backscattering spectrometry work, it is not necessary to measure the fluence absolutely, but only to determine it relatively. This can be done by using a rotating vane placed in the beam line before the sample chamber (Roth et al., 1974). The vane should rotate at a speed other than a multiple of the frequency of the power source of the accelerator. Ions are scattered periodically from the vane into a solid-state detector. Part of this signal is analyzed with a single-channel analyzer and a scaler. This accomplishes extremely accurate relative measurements of the beam fluence entering the target chamber. Since only the beam particles scatter and are detected by the solid-state detector, it makes no difference if the particles that impinge on the vane have been neutralized in the beam line, or even if electrons are being carried along with the beam. It is easy to make this method absolute by using either of the two methods previsouly described for calibrating the rotating vane.

### 6.6.3   Beam Contamination

The $^4He^+$ beam from an accelerator is often accompanied by a $^{16}O^+$ beam of the same energy. If, after acceleration and before magnetic analysis, one electron is stripped from the oxygen ion to form $^{16}O^{2+}$, these ions will not be separated from $^4He^+$ during magnetic analysis. Ion source conditions and the pressure in the beam lines strongly affect the $^{16}O^{2+}$ fraction. A formula for estimating the $^{16}O^{2+}$ beam intensity, based on charge exchange data and measurement of the primary $^{16}O^+$ beam intensity, is

$$N_{[O^{2+}]} = 6.5P[\text{Torr}]l[\text{cm}]N_{[O^+]},$$

where $P$ is the total gas pressure, $l$ the length of the beam line between the accelerator and switching magnet, and $N_{[O^{2+}]}$ and $N_{[O^+]}$ the beam intensities of the respective species (Picraux et al., 1973). The $O^{2+}$ contamination of the $He^+$ beam leads to spurious yields, which occur at lower energies in spectra than would be observed for a pure $He^+$ beam. Even if the contamination beam is only a small fraction of the $He^+$ beam, analysis of backscattering data can be extremely difficult since the Rutherford scattering cross section is about a factor of 16 larger for oxygen than for helium. Detrimental effects have alse been observed in studies of ion channeling, damage, and ion-beam-induced x-rays.

The contamination beam can be minimized by reducing the product $Pl$ (defined previously) and can be completely eliminated by electrostatic analysis after magnetic analysis.

## 6.7 BACKSCATTERING BEAM ENERGY ANALYSIS

Since all the experimental information is contained in the energy analysis of the backscattered beam, we will go into considerable detail and explanation in this section. The depth resolution of backscattering spectrometry is directly proportional to the energy resolution of the detection system; it is important, therefore, to have as good an energy resolution as possible. The accuracy of the elemental composition is determined by the statistics of the spectra; therefore, high count rates are desirable so that data can be accumulated in reasonable times. It is important to recognize the basic dichotomy between high resolution and high counting rates and to design experiments that minimize the inevitable compromises. Two methods have been used to energy-analyze the scattered beam: magnetic analysis and silicon barrier detector analysis. The magnetic analysis method provides a better energy resolution than does analysis with a silicon detector, but the data accumulation times are prohibitively long except in unusual circumstances. Magnetic analysis of scattered beams is discussed fully in the literature (Snyder *et al.*, 1956; Sippel, 1959; Rubin, 1967; Hirvonen and Hukler, 1976) and will not be discussed further here.

Since the energy of the backscattered beam particles is a function of the scattering angle, it is necessary to limit the angular acceptance of the detector. The angular effect of the elastic backscattering factor decreases as the scattering angle increases, up to 180°, so that large scattering angles are usually preferable, i.e., $\theta > 160°$ (see Fig. 2.2). The count rate should be as large as possible; moving the detector closer to the sample will increase it, but this will also increase the angle of acceptance. A distance of 10–15 cm between the detector and the target has been found quite sufficient for a 25-mm$^2$ detector. With this configuration the angular acceptance of the detector is about 2°, which corresponds to 30 mrad. The energy resolution of the detector system is based not only on the characteristics of the detector, but also on the characteristics of the preamplifier, the amplifier, and the multichannel analyzer system, each of which will be discussed separately. A typical electronic setup for use with a semiconductor detector is shown schematically in Fig. 6.7.

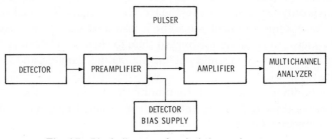

**Fig. 6.7** Block diagram of typical electronic setup.

Excellent discussions of nuclear instrumentation are contained in many books, articles, and product broachures; see for example, Duseph (1975), Nicholson (1974), Cerny (1974), and ORTEC (1974a). Some of the major manufacturers are listed in the footnote.[†] The electronic industry has standardized many of the components and signals under the broad heading of nuclear instrumentation modules (NIM). Most of the NIM components are interchangeable among different manufacturers so that systems with particular properties can be assembled with relative ease.

### 6.7.1   Detector

Silicon barrier detectors have been discussed in great detail in the literature (see for example: Duseph, 1975). Only factors affecting their resolution will be discussed here.

The energy resolution of the detector is governed by the straggling of the incoming ions in the gold barrier and the oxide layer on the surface of the detector, the uniformity in the thickness of these layers, the statistical spread in the number of the electron–hole pairs formed in the active volume of the detector, the efficiency of collecting electron–hole pairs, and the number of traps. The resolution of the detector system varies directly as the capacitance of the detector, the leakage current of the detector, and the capacitance between the detector and the input FET of the preamplifier. The detector capacitance can be reduced to a minimum by buying a quality detector with a small collection area and a large depletion depth. The detector capacitance increases linearly as the detector area increases, and decreases linearly as the depletion layer increases (see Fig. 6.8). Deterioration of the detector resolution has been observed to be caused by radiation damage resulting from the impinging ion beam and a pumping away of the absorbed oxygen in a passivating area on the detector. Marked broadening of the resolution has been observed for $10^8$ to $10^9$ He/cm$^2$, and complete failure at $10^{11}$ He/cm$^2$. This allows many months of constant analysis before the detector has to be replaced.

> The second problem is thought to be caused by pumping away of absorbed oxygen, which is passivating an area on the detector where undesirable surface impurities are present, or where the very thin oxide film responsible for the p-type nature of the surface has a flaw. If the microplasma breakdown in this region has not been too extensive or prolonged, the detector will frequently recover after several days or weeks of exposure to room air. It has been found that this problem is

[†] List of manufacturers: Camberra Industries, Meriden, Connecticut; Harshaw Chemical Co., Solon, Ohio; Northern Scientific Inc., Middleton, Wisconsin; Nuclear Data Inc., Schaumlurg, Illinois; ORTEC Inc., Oak Ridge, Tennessee; Princeton Gamma-Tech, Princeton, New Jersey; Tennecomp Systems Inc., Oak Ridge, Tennessee.

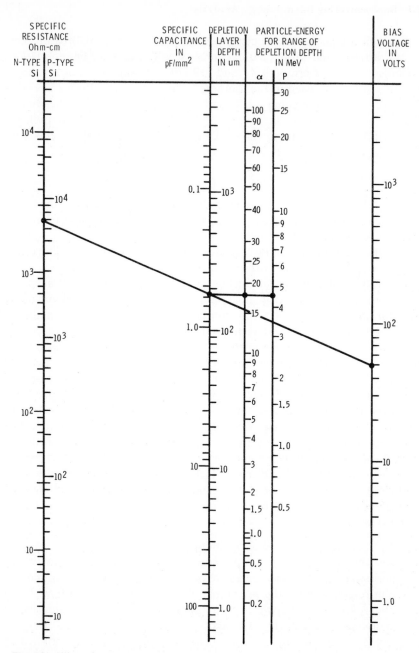

**Fig. 6.8** Silicon detector parameters nomogram. Use of the nomogram requires drawing a horizontal line which intersects the center vertical line at the required depletion depth using the maximum energy of the αs or protons expected in the experiment. From this intersection, combinations of resistivity and detector bias are obtained. [An example is given for a 4.5-MeV proton (18-MeV α) where the depletion depth is 180 μm. For a resistivity of $6 \times 10^3$ Ω-cm $p$-type ($17 \times 10^3$ Ω-cm $n$-type) a bias voltage of 50 V is required.]

significantly reduced by keeping a small nominal bias on the detector while it is in vacuum (ORTEC, 1976b).

This effect is much more observable where the pressure in the scattering chamber is below $10^{-6}$ torr. A detector with an area of about 25 mm$^2$ and a depletion depth of 200 to 1000 $\mu$m has been found to meet the requirements of backscattering spectrometry and is recommended for general use. In the construction of detectors the silicon wafer is expoxied into a ceramic ring. The inside diameter of this epoxy is usually not very uniform, so that some of the beam could pass through this nonuniform layer, causing degradation of the resolution. It is suggested that a cover be placed over the detector to aperture the beam to less than 80% of the active area of the detector.

Semiconductor detectors have been known for their remarkably uniform response to different types of ionizing radiation. After universal acceptance and many years of accumulated experience, second-order effects have been observed. These effects can introduce considerable experimental errors if not properly accounted for.

A simple expression has been proposed to account for these second-order effects (Langley, 1973):

$$E_{inc} = \Delta E_w + \mathscr{E}N + \Gamma, \tag{6.1}$$

where $E_{inc}$ is the energy of the particle incident on the detector, $\Delta E_w$ the energy lost in the window, $\mathscr{E}N$ the energy expended in producing electron–hole pairs, and $\Gamma$ the average energy that goes into atomic processes (Brice, 1970). The energy lost in the Au electrode can be determined by either of two methods: (1) experimentally measuring $E_{inc}$ and $N$, calculating $\Gamma$, and fitting the results to Eq. (6.1), where $N$ is the half-height channel of the leading edge of a backscattering spectrum or (2) tilting the detector with respect to the incoming beam. These two methods are discussed elsewhere in enough detail that no further explanation will be given here (Langley, 1973; Mitchell et al., 1976).

It has been experimentally observed that the average energy to create an electron–hole pair (usually called $\varepsilon$) in silicon is different for various incident ions. This effect is dependent on the incident ion atomic number, but not on its mass (Mitchell et al., 1976). The values of $\Gamma$ are obtained from calculations using the Thomas–Fermi elastic scattering cross section (Brice, 1975).

The constant $\mathscr{E}$ in Eq. (6.71) is the energy per channel and is directly proportional to $\varepsilon$. Previous experimental studies have shown this dependence of $\varepsilon$ on $Z$, $M$, and $E$, but were not in mutual agreement (Langley, 1973; Kemper and Fox, 1975). A recent study (Martini et al., 1975) has suggested that $\varepsilon$ is not constant with either the incident ion atomic number or its energy. This dependence is attributed to loss of ion-induced x-rays and electrons from the detector. Although this assertion is not conclusive, it is certainly indicative.

### 6.7.2 Preamplifier

A charge-sensitive preamplifer is used as the initial amplifying element. It is specifically designed to accept the signal from the detector and amplify that signal with some shaping so that it will preserve the maximum signal-to-noise ratio. In general, considerable attention must be focused on selecting and configuring the preamplifier's front-end components to preserve the maximum signal-to-noise ratio. This selecting and configuring of components is extremely dependent on the detail characteristics of the particular detector being used and sometimes on the types of signal processing that follow the preamplifier. Figure 6.9 shows two typical charge-sensitive preamplifier bias circuits for a semiconductor detector. Most preamplifiers contain a network of one of these types for biasing the detector. The bias circuit shown in Fig. 6.9a is ac-coupled, and that shown in Fig. 6.9b is dc-coupled. The dc-coupled configuration is preferred for very high-resolution systems, since the coupling capacitor in ac-coupling causes distortion in the output pulse shape for which there is no easy compensation. The ac-coupling configuration strikes a good balance between high resolution and high count rates. The charge-sensitive preamplifier noise is generally controlled by four components: the input FET, the input capacitance (the detector capacitance, the cabling capacitance, etc.), the resistance connected to the input, and the

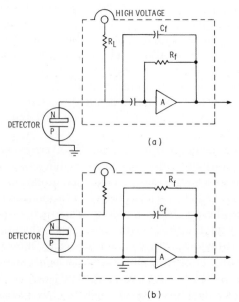

**Fig. 6.9** Detector–preamplifier coupling circuits: (a) ac-coupled charge-sensitive preamplifier; (b) dc-coupled charge-sensitive preamplifier.

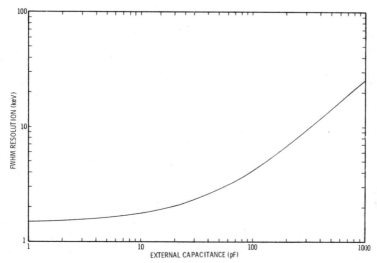

**Fig. 6.10** Resolution versus external capacitance.

detector leakage current. The leakage current usually is the dominating factor. The FET is selected for low noise performance. The preamplifier is designed with minimum internal circuit capacitance; in most applications, however, the user controls the major sources of input capacitance. A typical graph showing resolution versus external capacitance is shown in Fig. 6.10. The preamplifier should be selected to provide minimum noise for the external capacitance of the experiment. A preamplifier should be chosen only after consulting the literature of the major manufacturers.

### 6.7.3 Amplifier

Most preamplifiers have minimum shaping, which means that the output signal from the preamplifier is a step function of amplitude proportional to the input signal from the detector. The subsequent main amplifier must therefore create a suitable pulse shape optimizing resolution and count rate capability. There are two reasons for pulse shaping: (1) to avoid pulse pile up and (2) to enhance signal-to-noise ratio. Pulse pile up prevention can be explained most easily as an effect of each individual detector event which must be terminated in a time that is short compared to the average spacing of the pulses; otherwise, pulses will overlap and lead to erroneous amplitude measurements, as well as possible circuit difficulties when the piled up pulses exceed the available dynamic range. In high-resolution semiconductor spectrometry systems, the pulse-shaping method that would yield optimum signal-to-noise ratio enhancement is often in conflict with optimum methods for overlap prevention. This basic conflict requires some compromise in the design of an experiment; every effort must be made to keep the count rate within reasonably low limits by the proper selection of the experimental parameters. Many experimental situations demand high count rates and yet need the benefit of the highest possible resolution. In these circumstances, careful and empirical selection of the pulse-shaping method with an amplifier providing variable shaping is necessary. The use of pole-zero-canceled pre-amplifiers and amplifiers and a baseline restorer can improve resolution at high count rates. To reiterate, it is important to recognize the basic dichotomy between high resolution and high counting rates and to design experiments to minimize the inevitable compromises.

In backscattering spectrometry the useful information in the preamplifier output signal is the amplitude of each pulse. The pulse-shaping circuit of the subsequent main amplifier operates with time constants much shorter than the decay of the preamplifier signal and much longer than its rise time. The terms "clipping" and "differentiation" apply to these pulse-shorterning methods. This shaping effectively removes the slow component of the pre-amplifier signal and produces individual pulses whose amplitudes convey the quantity of interest, i.e., energy. The RC pulse-shaping applies to the use of resistors and capacitors as shaping networks and is generally used as the pulse-shortening method in the main amplifier. An RC integration filter affects the rise time of the pulse which attenuates the high-frequency components of the waveform. A CR differentiation filter affects the decay of the pulse and corresponds to a CR high-pass filter, which attenuates the low-frequency components of the waveform. Normally, CR differentiation and RC integration are used together, not separately. With CR–RC pulse shaping, in which the CR differentiation and RC integrator are cascaded, the resulting response removes both the low- and the high-frequency signal and noise component and significantly enhances the signal-to-noise ratio. For the majority of applications, equal CR and RC time constants optimize the results, but with high-noise detectors a deviation from equality should be tested empirically. Time constants of 0.2 to 1 $\mu$sec are often best for surface-barrier semiconductor detector systems used in backscattering spectrometry. For a more complete description of RC pulse shaping, see ORTEC (1976a).

### 6.7.4　Multichannel Analyzer

Multichannel analyzers (MCA) are special-purpose computers that pro-vide a variety of functions: data acquisition, storage, display, and interpreta-tion. MCAs are typically used in either of two distinct data analysis modes: the pulse-height-analysis (PHA) mode or the multichannel scaling (MCS) mode. The PHA mode is used almost exclusively for backscattering spectrometry. The MCS mode has been used for channeling measurements; it will be discussed at the end of this section, and its application in Section 6.8.2.

In the PHA mode, a spectrum (histogram) of the frequency distributions of the heights is accumulated from a sequence of input pulses. The desired spectrum is accumulated by measuring the amplitude of each input event, converting it to a number called the "channel address" or the "channel number" that is proportional to the pulse height, and storing the event as a count in a memory composed of individual channels. The number of counts in each channel at a given time is equal to the total number of pulses processed during the experiment up to that time whose amplitudes correspond to the channel address. A block diagram of the functional components of a typical multichannel analyzer system is shown in Fig. 6.11. In the PHA mode, the analog-to-digital converter (ADC) provides a channel address register. The number of counts contained in the appropriate channel is recalled from the memory and placed in the data register, and one count is added to the previous value. The updated number of counts is then returned to the memory. In pulse-height analysis, the ADC digitizes the input pulse amplitudes for acceptance by the memory system and thus establishes one of the accuracy limits of the information stored in the MCA memory. The ADC should be able to process the pulses from a detector and shaping amplifier without additional distortion of the spectral information. Figure 6.12 is a simplified block diagram of a typical ADC used in high-resolution spectroscopy applications. This type of ADC is essentially an amplitude-to-time converter. The dashed lines in Fig. 6.12 enclose the portion of the circuit that performs the conversion from a pulse height to a proportional digital address; the remaining function units control the range of analysis and acceptance of pulses of interest.

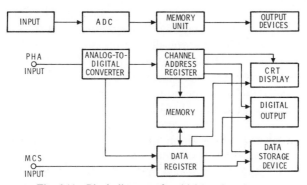

**Fig. 6.11**  Block diagram of multichannel analyzer.

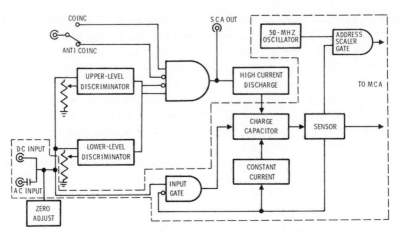

**Fig. 6.12**   Block diagram of typical analog-to-digital converter.

The input to the ADC is an analog pulse and should be in the NIM standard range of 0 to $+10$ V. The analog pulse may be either unipolar or bipolar with the positive portion leading. The unipolar pulse, which is singly differentiated, is used for optimum resolution; the bipolar pulse, which is doubly differentiated, is used for optimum count rate and timing requirements. A baseline restorer, not shown in Fig. 6.12, follows the ac input to minimize baseline shift when this input is used. An input pulse with an amplitude large enough to exceed the low-level discriminator charges the charge capacitor to a value directly proportional to the height of the incoming pulse. When the peak of the pulse is sensed, the voltage on the charge capacitor is discharged by a constant current. The time required to discharge the capacitor to the baseline, called the "run-down time," is proportional to the pulse height. When the discharge is begun, a gate is opened to allow the binary scaler to count clock pulses from the oscillator. The number of clock pulses counted during the run-down time represents the desired address information. At the end of the run-down time, the address and contents of the binary scaler are transferred into the memory address register for storage. An input gate prevents the acceptance of additional input pulses from time of the peak detection until the address transfer is complete. An oscillator frequency of no less than 50 MHz is recommended for the clock of the ADC.

The ADC should have, as a minimum, a zero-adjust, upper-level discriminator, a lower-level discriminator, and a coincidence–anticoincidence circuit. It should also have a variable conversion gain, which is the number of increments (channels) into which the 0- to $+10$-V input range is divided. This is determined by the magnitude of the current used to discharge the capacitor. The full-scale address should be at least 512 channels. It is preferable to have 4096 channels or even 8192 channels for memory storage. This allows from eight to sixteen 512-channel spectra to be stored at the same time. A majority of the backscattering spectrometry histograms are composed of 512 channels. For high-resolution systems, histograms of 1024 channels (or greater) are required. The ADC should have an indicator to present the percentage of dead time—that is, the percentage of the processing time required to process a pulse from the amplifier. Accurate dead-time correction is essential when the number of pulses processed in a given time period (a given fluence) is important in the analysis of the data. Dead-time correction compensates for the differing processing times, so that the number of counts in the spectra is directly proportional to the input fluence on the sample. A simple correction given by Nicholson (1974, p. 374) is

$$R = R'/(1 - R'\tau),$$

where $R$ is the mean rate of pulses at the input to the MCA, $R'$ the mean rate at the output, and $\tau$ the dead time.

The most important specification of an ADC intended for use in high-resolution spectroscopy is its nonlinearity. There are two contributions to the nonlinearity of an ADC: integral and differential. Integral nonlinearity is here defined as the maximum deviation of any address from its nominal position described by a straight-line plot of address versus input pulse amplitude. The integral nonlinearity should be less than $\pm 0.08\%$ of full scale over the top 99% of full scale. Differential nonlinearity describes the uniformity of address over the entire number of addresses of the ADC and is defined as the percent deviation of the maximum and minimum address widths to the average width of all addresses, divided by two. The factor of one-half is included so that the differential nonlinearity may be specified as a $\pm$ deviation. It should be less than or equal to $\pm 1\%$ of full scale over the top 99% of full scale.

The memory system of a multichannel analyzer should be a high-density data storage unit organized into a series of channels. Each channel should have a unique address, called a "channel number," capable of storing data values from 0 to at least 999,999. Two types of memory are currently available in MCA design: magnetic core and semiconductor. In both types of memory, the data are stored in binary digit (bit) form, where a data bit can have a value of 0 or 1. The advantage of the magnetic memory is that if power is lost to the instrument, the data stored are not lost, so that when power is reapplied the analysis can be continued. In a semiconductor memory, if power is interrupted, the data will be lost unless some power source is used during the power outage. The advantage of a semiconductor memory is that the memory cycle times are typically shorter than for those for magnetic core memory. Magnetic memory has been found preferable to the semiconductor memory in that the magnetic memory is sufficiently fast for backscattering spectrometry, while having the advantage that power can be interrupted without losing the data.

In most MCAs, the first channel in the memory, or in the active storage region, is used to store the elapsed time of analysis in the PHA mode. It usually records the lifetime of the analysis in one-second increments. Data manipulation and display are extremely important facets of the complete operation and can be very time-consuming if not used correctly. At the very minimum, the output of the MCA should have available an $x-y$ plotter, a hard-copy printer, and a storage device in which it is easy to get the data out of and back into the storage system of the multichannel analyzer. In addition, it is preferable for the MCA/computer system to have the capability of taking data and processing data, or outputting data at the same time. This can save considerable time during data processing.

For the multichannel scaling mode (MCS), individual channels of the memory act as a sequence of counters, with each channel counting the data for a predetermined "dwell time." The dwell time for each channel can be set by an internal clock or by an external channel advance signal. At the completion of each dwell time the counting operation is passed to the next channel; the result is a time histogram of the count rate data. The application of this mode to backscattering spectrometry is found in channeling and is discussed in Section 6.8.2.

### 6.7.5 Low-Noise, High-Resolution Detection System

For experiments that require the ultimate in depth resolution, i.e., energy resolution of the detector system, the input stage of the preamplifier can be placed inside the vacuum system next to the detector to reduce the input capacitance to the first stage of the preamplifier. In addition, all components placed inside the vacuum system can be cooled to reduce the bias current in the detector, the noise level in the detector, the input FET, and the feedback resistor. It has been observed experimentally that the noise of the system decreases as the temperature decreases, and is approximately constant in the region $-4$ to $-120°C$ (Ray and Barnett, 1969). As was mentioned in Section 6.7.2, a dc-coupling arrangement between the detector and the input stage of the FET is preferable for high-resolution systems. This dictates that the count rate must remain low, since baseline restoration methods will not maintain the good resolution for high count rates because of pulse pile up. Such a detector system is described by Ray and Barnett (1969), who observed a resolution of 1.65 keV for 100-keV protons, and about 4.5 keV for 100-keV $He^+$ ions. If one is going to use the energy resolution in the data analysis, the resolution should be covered by at least five channels; for example, if the full-width, half-maximum resolution is 5 keV, then the energy per channel should be 1 keV per channel or less. For a spectrum in which the resolution is not used, about three channels should cover the FWHM of the energy resolution.

## 6.8 SAMPLE HOLDERS

### 6.8.1 Standard Holders

For most experimental laboratories, it is desirable to have more than one type of holder. What is needed is one general-purpose sample holder with an experimental chamber in which numerous samples can be placed. The use of a manipulator of the type shown in Fig. 6.13 makes it possible to rotate different samples into the analysis beam and also provides two-axis positioning of the beam on the sample. The sample cover plate shown in Fig. 6.13 has been gold plated, first, to provide a nondeteriorating high-$Z$ material that can be used for relatively calibrating the mass scale by backscattering from it and, second, to allow absolute positioning of the beam on the sample. Special sample holders have been constructed for *in situ* film disposition (Langley and Blewer, 1973; Baglin and Hammer, 1976), sample annealing (Langley and Donhowe, 1976), surface reaction studies (Myers, 1974), and many other types of experiments. A quick search of the literature in any particular field yields many varied sample holders. The main limitations are imposed by imagination and funds.

**Fig. 6.13** Photograph of multiple-sample holder which provides two-axis positioning of the beam on the sample.

### 6.8.2   Channeling Goniometer

The primary additional piece of experimental equipment required for
channeling is a goniometer. This allows orientation of the sample crystal
with respect to the beam direction. An orientation accuracy of $\leq 0.05°$ is
required, whether the goniometer is two or three axes. One can use either an
x-ray goniometer that has been modified for operation in a vacuum (Behrisch
*et al.*, 1969), or a goniometer that specifically designed for channeling measurements, such as the one shown in Fig. 6.14.[†] The two-axis goniometer
shown in Fig. 6.14 allows tilt motion around the vertical axis of the entire
lower frame and 360° rotation of the crystal about an axis perpendicular to

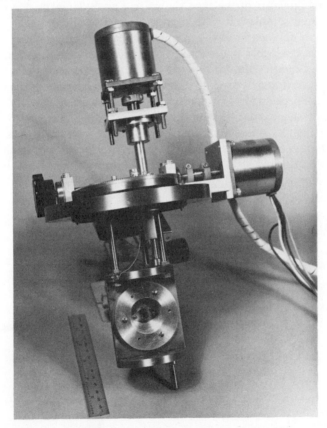

**Fig. 6.14**   Photograph of two-axis goniometer used for channeling measurements (Picraux,
1975).

[†] This goniometer was built from a design of W. Augustyniak (Bell Telephone Laboratories)
with modifications by J. Smalley (Sandia Laboratories).

the plane of the paper. This tilt axis should be perpendicular to the beam direction; a misalignment by $\Delta\theta$ will result in a solid angle of orientation which is inaccessible for all settings.

The first step in any channeling experiment is to orient the single crystal with respect to the beam direction. A procedure is given in Picraux (1975) for a two-axis goniometer. Further information on the experimental techniques of channeling is presented by Picraux (1975) and Morgan (1973), and in Chapter 8.

Automatic angular scanning can be used to facilitate data acquisition. Such devices have been designed and used (Abel et al., 1969; Borders and Picraux, 1970). Stepping motors fitted onto the goniometer are capable of driving any axis in small angular steps, typically 0.01°. Data acquisition is accomplished by amplifying the output pulses from the detector, as discussed in Sections 6.7.2 and 6.7.3, sampling the energy region of interest with a single-channel analyzer, and finally processing the signals with an MCA. The output of the single-channel analyzer is connected to the multiscale input of the MCA. The analyzer can be set to free run (predetermined dwell per channel) as the stepping motors drive the goniometer at a given rate (degrees per unit time), or it can be advanced by a signal from a current digitizer so that the fluence used is the same for each channel. With this arrangement, channeling data can be easily acquired and used.

### REFERENCES

Abel, F., Amsel, G., Bruneaux, M., and d'Artemane, E. (1969). *J. Phys. Chem. Solids* **30**, 687.

Allen, K. W. (1974). *In* "Electrostatic Accelerators, Nuclear Spectroscopy and Reactions" (J. Cerny, ed.), Part A, pp. 3–34. Academic Press, New York.

Augustyniak, W. (1974). Private communication, Bell Telephone Laboratories, Murray Hill, New Jersey.

Baglin, J. E. E., and Hammer, W. N. (1976). *In* "Ion Beam Surface Layer Analysis" (O. Meyer, G. Linker, and F. Käppeler, eds.), p. 447. Plenum Press, New York.

Behrisch, R., Muhlbauer, G., and Scherzer, B. M. U. (1969). *J. Phys.* E **2**, 381.

Bondelid, R. O., and Kennedy, C. A. (1959). *Phys. Rev.* **115**, 1601.

Borders, J. A., and Picraux, S. T. (1970). *Rev. Sci. Instrum.* **41**, 1230.

Bøttiger, J., Davis, A., Lozi, J., and Whitton, J. L. (1973). *Nucl. Inst. Meth.* **109**, 579.

Brice, D. K. (1970). *Radiat. Effects* **6**, 77.

Brice, D. K. (1975). *J. Appl. Phys.* **46**, 3385.

Bumiller, F., Staub, H. H., and Weaver, H. E. (1956). *Helv. Phys. Acta* **29**, 83.

Burrill, E. A. (1970). "Neutron Production and Protection." High Voltage Engineering Corp., Burlington, Massachusetts. (This publication has a list of many pertinent references.)

Cerny, J. (1974). "Nuclear Spectroscopy and Reactions," Part A, pp. 290–498. Academic Press, New York.

Duggan, J. L. (ed.) (1968). *Proc. 1st Conf. Appl. Small Accelerators* (CONF-680411). Available from the National Technical Information Service, U.S. Dept. of Commerce, Springfield, Virginia.

Duggan, J. L. (ed.) (1970). *Proc. 2nd Conf. Appl. Small Accelerators* (CONF-700322).

Duseph, P. J. (1975). "Introduction to Nuclear Radiation Detectors." Plenum Press, New York.

Dushman, S. (1962). "Scientific Foundations of Vacuum Technique." Wiley, New York.

Enge, H. A. (1967). In "Focusing of Charged Particles" (A. Septier, ed.). pp. 203–264. Academic Press, New York.

Hirvonen, J. K., and Hukler, G. K. (1976). In "Ion Beam Surface Layer Analysis" (O. Meyer, G. Linker, and F. Käppler, eds.), pp. 457–469. Plenum Press, New York.

Kemper, K. W., and Fox, I. D. (1975). Nucl. Instrum. Methods 105, 333.

Khan, J. M., and Potter, D. L. (1964). Phys. Rev. 133, A890.

Kinnard, I. F. (1956). "Applied Electrical Measurements." Wiley, New York.

Langley, R. A. (1973). Nucl. Instrum. Methods 113, 109.

Langley, R. A. (1976). For further information contact author at Oak Ridge National Laboratories, Oak Ridge, Tennessee.

Langley, R. A., and Blewer, R. S. (1973). Thin Solid Films 19, 187.

Langley, R. A., and Donhowe, J. M. (1976). J. Nucl. Mater. 63, 521.

Livingston, M. S., and Blewett, J. P. (1962). "Particle Accelerators." McGraw-Hill, New York.

Marion, J. B. (1961). Rev. Mod. Phys. 33, 139.

Martini, M., Raudorf, T. W., Scott, W. R., and Waddington, J. C. (1975). IEEE Trans. Nucl. Sci. NS-22, 145.

Mitchell, J. B., Agami, S., and Davies, J. A. (1976). Radiat. Effects 28, 133.

Morgan, D. V. (ed.) (1973). "Channeling." Wiley, New York.

Morgan, I. L., and Duggan, J. L. (eds.) (1974). Proc. 3rd Conf. Appl. Small Accelerators (CONF-7410400-P2).

Musket, R. G., and Taatjes, S. W. (1973). Rev. Sci. Instrum. 44, 1290.

Myers, S. M. (1974). J. Appl. Phys. 45, 4370.

NCRP (1971). Basic Radiation Protection Criteria, Report No. 39. NCRP Publ. Washington, D.C.

Nicholson, P. W. (1974). "Nuclear Electronics." Wiley, New York.

ORTEC (1976a). Radiation Spectroscopy and Analysis Instruments for Research and Industry, Catalog 1004, ORTEC, Oak Ridge, Tennessee.

ORTEC (1976b). Silicon Surface Barrier Radiation Detectors Instruction Manual. ORTEC, Inc., Oak Ridge, Tennessee.

Picraux, S. T. (1975). In "New Uses of Ion Accelerators" (J. F. Ziegler, ed.). p. 229. Plenum Press, New York.

Picraux, S. T., Borders, J. A., and Langley, R. A. (1973). Thin Solid Films 19, 371.

Power, B. D. (1966). "High Vacuum Pumping Equipment." Van Nostrand-Reinhold, Princeton, New Jersey.

Ray, J. A., and Barnett, C. F. (1969). IEEE Trans. Nucl. Sci. NS-16, 82.

Roth, J., Behrisch, R., and Scherzer, B. M. U. (1974). J. Nucl. Mater. 53, 147.

Rubin, S. (1967). "Ion Scattering Methods, Treatise on Analytical Chemistry" (I. M. Kolthoff and P. J. Elving, eds.), pp. 2075–2108. Wiley, New York.

Singh, B. (1957). Phys. Rev. 107, 711.

Sippel, R. F. (1959). Phys. Rev. 115, 1441.

Snyder, C. W., Rubin, S., Fowler, W. A., and Lauritsen, C. C. (1956). Rev. Sci. Instrum. 27, 899.

Switkowski, Z. E., and Parker, P. D. (1975). Nucl. Instr. Meth. 131, 263.

Chapter

# 7

# Influence of Beam Parameters

## 7.1 INTRODUCTION

In previous chapters we developed the formalism for backscattering analysis and gave examples based around the use of megaelectron volt $^4$He ions. There are several reasons for the choice of megaelectron volt $^4$He ions: most published data involves these beam parameters, data analysis is particularly simple in that conversion of energy to depth is given by a nearly constant factor, stopping cross sections have been measured, and scattering cross sections follow Rutherford's relation. Various other beam parameters have been used in the analysis of solids by ion beams. In some cases, other ion beams were chosen because of the availability of a particular accelerator. In other cases, the analytical problem under investigation dictated the choice of a particular analysis beam.

Although the formalism and experimental technique are the same for different particles and energies as for megaelectron volt $^4$He, the shape of the backscattering spectrum does depend on the choice of beam parameters.

For example, the magnitude and energy dependence of the stopping cross section differ for different projectiles. These differences and that due to the scattering cross section are reflected in the spectra. Mass and depth resolution, sensitivity to trace impurities, and accessible depth for analysis depend on the energy and mass of the projectile and on the scattering geometry. Resonances or nuclear reactions occur for projectile–target combinations at certain energies and can be used to detect light impurities in substrates. We do not treat the use of ion-induced x-rays, nuclear reactions, or resonant scattering. Treatments are given by Mayer and Rimini (1977), Ziegler (1975), Mayer and Ziegler (1974), and Meyer et al. (1976).

## 7.2  MASS RESOLUTION

In backscattering experiments, it is often desirable to increase the separation in energy between signals from different elements in the target. As pointed out in Chapter 2, backscattering spectrometry acquires its ability to sense the mass of an atom through the kinematic factor $K = E_1/E_0$, which depends on the ratio $M_1/M_2$ of the projectile and target masses, and on the scattering angle $\theta$ [Eq. (2.6)]. In Fig. 7.1, values of $K$ are plotted as a function of the target mass $M_2$ for $^1$H, $^4$He, $^{12}$C, $^{20}$Ne, and $^{40}$Ar for a scattering angle $\theta$ of 170°. These curves can all be derived from the one in Fig. 2.2 for $\theta = 170°$ by appropriately rescaling the abscissa. The figure shows that the energy after an elastic collision, which is proportional to $K$, differs little for hydrogen scattering from Si ($M_2 = 28$–30) and from Ge ($M_2 = 70$–76). That difference can be greatly increased by using a heavier projectile. When a target contains two types of atoms that differ in their masses by a small amount $\Delta M_2$, the difference $\Delta E_1$ in the projectile energy $E_1$ after collision is given by [Eq. (2.13)]:

$$\Delta E_1 = E_0(dK/dM_2)\Delta M_2. \tag{7.1}$$

For scattering at 180°, where $K = [(M_2 - M_1)/(M_2 + M_1)]^2$ [Eq. (2.8)], this expression can be written as [Eq. (2.14) with $\delta = 0$]:

$$\Delta M_2(E_0/\Delta E_1) = (M_1 + M_2)^3/4M_1(M_2 - M_1). \tag{7.2}$$

One is often interested in the attainable mass resolution for a specified incident energy $E_0$ and a given system energy resolution. As a measure of the *energy resolution*, we shall use here the FWHM of the Gaussian response function which characterizes the detection system, and we introduce the symbol $\delta E_1$ for this quantity. More generally, $\delta A$ will be defined in this chapter as the FWHM of the Gaussian which describes the fluctuations to which the quantity $A$ is subjected, and $\delta A$ will be referred to as the *resolution of* $A$. Numerically, $\delta A$ is equal to 2.355 times the standard deviation of $A$, and $\delta A$ corresponds to the difference in $A$ values between the 12% and 88% points of the corresponding step response function (Fig. 2.12).

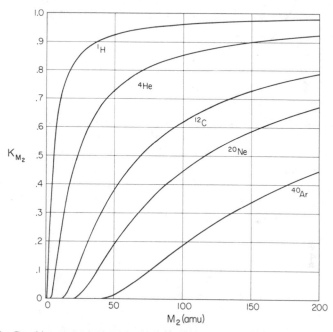

**Fig. 7.1** Graphic representation of the kinematic factor $K_{M_2}$ [Eq. (2.6)] for a scattering angle $\theta = 170°$ as a function of the target mass $M_2$ for $^1$H, $^4$He, $^{12}$C, $^{20}$Ne, and $^{40}$Ar. Each of these curves can be obtained from that of Fig. 2.2 for $\theta = 170°$ by appropriate rescaling of the abscissa.

If $\Delta E_1$ in the two previous equations is identified with the energy resolution $\delta E_1$ of the system, $\Delta M_2$ will be called the *mass resolution* $\delta M_2$. Figure 7.2 gives a plot for several projectiles of the dependence between these two quantities versus target mass $M_2$ as described by the right-hand side of Eq. (7.2). For $E_0 = 2$ MeV and $\delta E_1 = 0.02$ MeV, isotopic separation ($\delta M_2 \leq 1$) can be obtained for values of $\delta M_2 E_0 / \delta E_1 \lesssim 100$. The curves in Fig. 7.2 show that one can distinguish the isotopes of oxygen with hydrogen and the isotopes of Si with $^4$He.

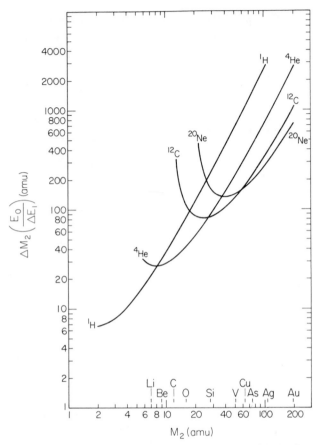

**Fig. 7.2**   A change of $\Delta M_2$ in the mass $M_2$ of the target atom produces a change $\Delta E_1$ in the detected energy whose relative value with respect to $E_0$ is given by the plots shown for four projectile atoms and a scattering angle $\theta = 180°$. (See also Fig. 7.3.)

The mass resolution curves also depend on the scattering angle, as is evident from Fig. 2.2. For example, for scattering at $\theta = 90°$, where $K = (M_2 - M_1)/(M_2 + M_1)$, the mass resolution equation [Eq. (7.1)] becomes

$$\Delta M_2(E_0/\Delta E_1) = (M_2 + M_1)^2/2M_1. \qquad (7.3)$$

This functional dependence is shown in Fig. 7.3 for five projectile atoms. The general trend is the same as in Fig. 7.2 for $\theta = 180°$, but there is a general worsening of the mass resolution for target atom $M_2 \gg M_1$. For example, with 2-MeV $^4$He and $\Delta E_1 = \delta E_1 = 0.02$ MeV, it would not be possible to resolve the isotopes of silicon as is possible at a scattering angle of $180°$.

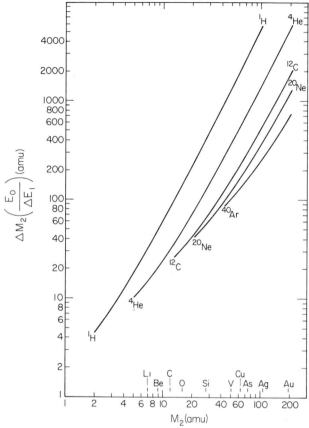

**Fig. 7.3** A change of $\Delta M_2$ in the mass $M_2$ of the target atom produces a change $\Delta E_1$ in the detected energy whose relative value with respect to $E_0$ is given by the plots shown for five projectile atoms and a scattering angle $\theta = 90°$. (See also Fig. 7.2.)

The value of $\delta E_1$ depends on the energy resolution of the system, which in turn depends on the detector and preamplifier used, as well as on the detected energy and the projectile. With conventional surface barrier detectors and no special precaution applied to the preamplifier, an energy resolution $\delta E_1$ of about 15 keV is obtained for 1- to 2-MeV $^4$He. With premium detectors and a cooled detector/preamplifier assembly to reduce electronic noise, one can achieve energy resolutions of about 10 keV. For hydrogen of energies from 0.1 to 0.3 MeV, which is a common energy range for low-energy accelerators, a system resolution of 5 keV can be achieved with a cooled detector and preamplifier. Figure 7.4 plots the mass

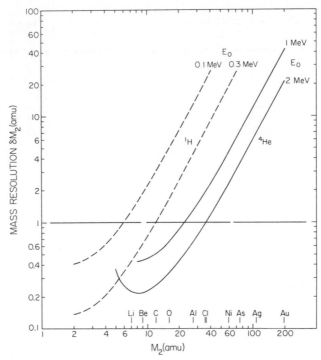

**Fig. 7.4**  Mass resolution $\delta M_2$ at a scattering angle of $\theta = 180°$ as a function of the mass $M_2$ of the target atom for $^1$H (dashed curves) at 0.1 and 0.3 MeV, assuming an energy resolution of 5 keV and for $^4$He (solid curves) at 1 and 2 MeV, assuming an energy resolution of 16 keV. Isotopic resolution is achieved for $\delta M_2 \leq 1$.

resolution $\delta M_2$ as a function of the mass $M_2$ of the target atom, where $\theta = 180°$ and where $\delta E_1 = 5$ keV for $^1$H and 16 keV for $^4$He. The horizontal line at $\delta M_2 = 1$ indicates the limit for isotopic resolution. The curves show that a $^1$H beam of 300 keV can resolve isotopes up to about carbon and that a 2.0-MeV $^4$He beam will resolve isotopes up to about chlorine. However, for 2-MeV $^4$He scattered from atoms of mass close to 200 amu, the mass resolution $\delta M_2$ is about 20. This means that with 2-MeV $^4$He one cannot distinguish among target atoms between $^{181}$Ta and $^{201}$Hg.

To obtain better mass resolution, it is clear that one should increase $E_0$ and $M_1$ and improve the energy resolution of the system. Such improvement can be obtained for $^1$H and $^4$He by cooling the silicon surface barrier detector and the preamplifier. However, the energy resolution of these detectors degrades for heavy ions such as $^{12}$C. The experimental studies by Bergstrom et al. (1968) and Peterson et al. (1973) show that the energy resolution of silicon surface barrier detectors depends on both the mass and the energy

**TABLE 7.1**

Mass Resolutions for Si Surface Barrier Detectors for Various Ions at Various Energies
for Target Masses around 100 and 200 amu

| Projectile | Detector resolution $\delta E$ (FWHM) (keV) | Incident energy $E_0$ (MeV) | $\delta E_1/E_0$ | Mass resolution $\delta M_2$ (FWHM) | |
|---|---|---|---|---|---|
| | | | | at $M_2 \simeq 100$ amu | at $M_2 \simeq 200$ amu |
| $^1$H | 5 | 0.5 | 0.0100 | 26 | 102 |
| | 5 | 1.0 | 0.0050 | 13 | 51 |
| | 5 | 1.5 | 0.0033 | 9 | 34 |
| | 5 | 2.0 | 0.0025 | 7 | 26 |
| $^4$He | 16 | 0.5 | 0.032 | 23 | 87 |
| | 16 | 1.0 | 0.016 | 12 | 43 |
| | 16 | 1.5 | 0.012 | 9 | 32 |
| | 16 | 2.0 | 0.008 | 6 | 22 |
| | 16 | 4.0 | 0.004 | 3 | 11 |
| $^{12}$C | 37 | 0.5 | 0.074 | 25 | 78 |
| | 48 | 1.0 | 0.048 | 16 | 51 |
| | 57 | 1.5 | 0.038 | 13 | 40 |
| | 61 | 2.0 | 0.031 | 10 | 33 |
| $^{20}$Ne | 49 | 0.5 | 0.098 | 26 | 72 |
| | 71 | 1.0 | 0.071 | 19 | 52 |
| | 86 | 1.5 | 0.057 | 15 | 42 |
| | 96 | 2.0 | 0.048 | 13 | 35 |

of the projectile. The mass resolution as a function of projectile energy listed in Table 7.1 is based on the resolution values given by Bergstrom *et al.* (1968), except that for hydrogen and helium the values of 5 and 16 keV have been assumed. An energy resolution $\delta E_1$ of 5 keV is typical for hydrogen and a cooled silicon surface barrier detector/preamplifier assembly; 16 keV is the resolution obtained with the same system for He without cooling the detector/preamplifier system. The last two columns in the table give the mass resolution $\delta M_2$ at target masses around 100 and 200 amu. For different energy resolutions than those assumed in the first column of the table, the proper mass resolution can be obtained by simply scaling the corresponding $\delta E_1$ and $\delta M_2$ values, since, by Eq. (7.3), $\delta M_2$ and $\delta E_1$ are linearly related to each other.

The tabulated values of $\delta M_2$ indicate that heavy ions do not offer substantial advantages, because of the poor energy resolution of the surface barrier detector. The work of Petersson *et al.* (1973) does indicate some gain in mass resolution for $^{16}$O at 20 MeV and a detector resolution of about 100 keV. A drawback of heavy ions is that the detector resolution degrades more rapidly than for H or He, because of radiation damage introduced by the incident ions. For heavy ions and detection rates of about $10^5$ counts/sec, a typical detector can be operated for only 5–20 hr, whereas with $^4$He it can be operated for several months without appreciable degradation of energy resolution.

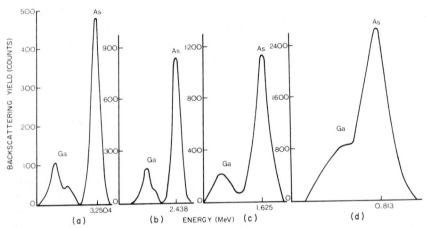

**Fig. 7.5**  Backscattering spectra for $^4$He incident along the $\langle 110 \rangle$ direction of a single-crystal target of GaAs and scattered through $\theta = 160°$. Values are: (a) 4 MeV; (b) 3 MeV; (c) 2 MeV; (d) 1 MeV. Each spectrum shows the surface peak of Ga ($M_2 = 69$ and 71) and of As ($M_2 = 75$) after subtraction of the background in the minimum yield (see Fig. 8.30). At a primary energy $E_0 = 4$ MeV, the two Ga isotopes ($\Delta M_2 = 2$) are distinguishable, but at 2 MeV, a mass difference of $\Delta M_2 = 5$ is barely resolved. [From Morgan and Wood (1973).]

Higher values of the incident energy $E_0$ for $^1$H and $^4$He can lead to better mass resolution, since the energy resolution of Si surface barrier detectors is approximately independent of energy in the megaelectron volt range. An example is shown in Fig. 7.5 for $^4$He incident along the $\langle 110 \rangle$ direction of a single-crystal GaAs sample. The spectra correspond to incident energies of 4, 3, 2, and 1 MeV and show the surface peak after subtraction of the background in the minimum yield (see Section 8.4.4). A mass difference of $\Delta M_2 = 5$ at $M_2 \approx 75$ between Ga and As can just barely be resolved at an energy of about 2 MeV, as could be deduced from the curves in Fig. 7.4. For 4-MeV $^4$He, the mass resolution is sufficient to distinguish the isotope $^{69}$Ga from $^{71}$Ga. The areas under the two peaks are consistent with the natural isotopic abundance ratio. That the mass resolution depends on $M_2$ can be deduced from this result and the fact that for $\Delta M_2 = 2$, $^{35}$Cl and $^{37}$Cl can be separated at 2 MeV (Mitchell et al., 1971) whereas 4 MeV is required to separate the two isotopes of Ga.

The discussion so far has been based on the use of silicon surface barrier detectors. If a high-resolution magnetic spectrometer is used, the system resolution can be improved dramatically to values of 1 to 2 keV for mega-electron volt $^4$He ions. The pronounced disadvantage of magnetic spectrometers is that their rate of data acquisition is very slow. This is so because their acceptance angle is small, because only a small part of the spectrum can be measured at a given magnetic field setting, and because the spectrometer accepts only particles of one charge state. Although this long data acquisition time is often intolerable for routine analysis, it is acceptable when high-resolution studies are necessary. However, radiation damage may set a definite upper limit to the acceptable dose of the primary beam (see, e.g., Section 8.6).

For a magnetic spectrometer of field intensity $B$, a nonrelativistic particle of momentum $p$ has a curved deflection of radius $R = p/Bq$. The variation of radius $\delta R$ with momentum $\delta p$ is given by $\delta R = \delta p/Bq$, and hence the variation $\delta E_1$ of $E_1$ is

$$E_1/\delta E_1 = \tfrac{1}{2}(p/\delta p) = \tfrac{1}{2}(R/\delta R) \qquad (7.4)$$

since $E_1 = p^2/2M_1$ and $\delta E_1 = p\ \delta p/M_1$. A typical value of $R/\delta R$ is about 2000, which gives a value for $E_1/\delta E_1$ of 1000. With this value, and with $E_1 \approx E_0$, the curves of Fig. 7.2 are again applicable and indicate that isotopic mass resolution ($\delta M_2 = 1$) is achieved at target masses around that of $^{118}$Sn. This mass resolution is sufficient to separate the signals from adjacent heavy elements.

Electrostatic analyzers can also be used to obtain improved system resolution. As an example, Feuerstein et al. (1976) report 0.7% energy resolution for 100 to 250 keV protons and $^4$He ions.

For low-energy particles, typically below a few hundred kiloelectron volts, electrostatic analyzers are most commonly used (Honig and Harrington, 1973; Buck, 1975; Chu et al., 1973). Such systems have typical energy resolutions of about 1 keV. This corresponding mass resolution can again be estimated from Figs. 7.2 and 7.3. The use of electrostatic analyzers in material analysis with ion beams is described by van Wijngaarden et al. (1971).

## 7.3  ACCESSIBLE DEPTH

### 7.3.1  Elemental Samples and Thin Films

One factor that enters into the choice of a given beam parameter is the depth below the surface of the sample that one desires to reach, or the film thickness that one can measure. The depth that is accessible in backscattering analysis is in the general region of 1 to 10 $\mu$m. The actual depth, however, depends on the target elements, projectile energy, and mass.

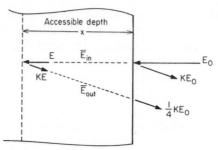

**Fig. 7.6**  An empirical way to define the depth accessible to a backscattering analysis is to require that particles scattered at that depth be detected with an energy that is $\frac{1}{4}$ of the energy possessed by particles scattered from the surface.

A general consideration in evaluating the accessible depth is that the particle must emerge from the target with sufficient energy $E_1$ to be detected. On the basis of empirical data we impose the somewhat arbitrary criterion that $E_1 \geq \frac{1}{4}KE_0$, as shown schematically in Fig. 7.6. The total energy interval $\Delta E$ between the energy of a particle scattered at the surface and its energy at the accessible depth $x$ is

$$\Delta E = KE_0 - \tfrac{1}{4}KE_0 = \tfrac{3}{4}KE_0. \tag{7.5}$$

This energy interval can be related to the depth $x$ through Eqs. (3.7) and (3.13) which we repeat here for normal incidence ($\cos\theta_1 = 1$ and $\cos\theta_2 = -\cos\theta$):

$$\Delta E = [\bar{S}]x \tag{7.6}$$

and

$$[\bar{S}] = K\, dE/dx|_{\bar{E}_{in}} + (1/\cos\theta_2)\, dE/dx|_{\bar{E}_{out}}. \tag{7.7}$$

Using the symmetrical mean energy approximation [Eqs. (3.17) and (3.18)] to estimate $\bar{E}_{in}$ and $\bar{E}_{out}$ gives

$$\bar{E}_{in} = E_0 - \tfrac{3}{16}KE_0 \tag{7.8}$$

and

$$\bar{E}_{out} = \tfrac{1}{4}KE_0 + \tfrac{3}{16}KE_0. \tag{7.9}$$

Table 7.2 lists values for the energy loss factor $[\bar{S}]$ for $^1H$, $^4He$, and $^{16}O$ incident on various targets for which bulk densities are assumed. The targets were chosen to span the range from medium to heavy mass elements, and oxygen was chosen as representative of a heavy mass ion. With these values of $[\bar{S}]$ and the equation for the accessible depth

$$x = \tfrac{3}{4}KE_0/[\bar{S}], \tag{7.10}$$

one derives the values for the accessible depth given in Table 7.3. These values show that for $E_0 = 2$ MeV, the accessible depth for $^4He$ is about

**TABLE 7.2**

Energy Loss Factor $[\bar{S}]$ for $^1$H, $^4$He, and $^{16}$O in
Al, Ni, Ag, and Au for $\Delta E = \frac{3}{4}KE_0$ ($\theta = 170°$)

| Projectile | Projectile energy $E_0$ (MeV) | $[\bar{S}]$ (eV/Å) for | | | |
|---|---|---|---|---|---|
| | | Al | Ni | Ag | Au |
| $^1$H[a] | 0.5 | 14 | 31 | 29 | 33 |
| | 1.0 | 9.2 | 22 | 20 | 25 |
| | 2.0 | 5.9 | 15 | 14 | 18 |
| $^4$He | 0.5 | 46 | 112 | 114 | 132 |
| | 1.0 | 51 | 126 | 121 | 148 |
| | 2.0 | 46 | 119 | 105 | 135 |
| $^{16}$O[a] | 1.0 | — | 191 | 256 | 288 |
| | 2.0 | — | 273 | 363 | 433 |
| | 20.0 | 147 | 575 | 633 | 845 |

[a] Stopping cross sections for hydrogen and oxygen taken
from Northcliffe and Schilling (Appendix D).

**TABLE 7.3**

Accessible Depth for Backscattering Analysis with
$^1$H, $^4$He, and $^{16}$O in Al, Ni, Ag, and Au for
$\Delta E = \frac{3}{4}KE_0$ ($\theta = 170°$)

| Projectile | Projectile energy $E_0$ (MeV) | Accessible depth ($\mu$m) in | | | |
|---|---|---|---|---|---|
| | | Al | Ni | Ag | Au |
| $^1$H[a] | 0.5 | 1.8 | 1 | 1 | 1 |
| | 1.0 | 4.5 | 2.4 | 2.7 | 2.4 |
| | 2.0 | 15 | 6.9 | 7.7 | 6.4 |
| $^4$He | 0.5 | 0.6 | 0.3 | 0.3 | 0.3 |
| | 1.0 | 1 | 0.5 | 0.6 | 0.5 |
| | 2.0 | 1.7 | 1 | 1.1 | 1 |
| $^{16}$O[a] | 2.0 | 0.3 | 0.2 | 0.3 | 0.3 |
| | 20.0 | 1 | 1 | 1.4 | 1.4 |

[a] Stopping cross sections for hydrogen and oxygen taken
from Northcliffe and Schilling (Appendix D).

one-seventh that of $^1$H and that of $^{16}$O is one-fourth that of $^4$He. As a rule
of thumb, the accessible depths are about 10 $\mu$m for $^1$H, 1 $\mu$m for $^4$He, and
0.3 $\mu$m for $^{16}$O at 2 MeV. In the analysis of a thin film, the thickest measurable film can be equated with the accessible depth. An example of the increase in film thickness that can be analyzed with $^1$H as compared to $^4$He
is shown in Fig. 5.14. One should also note that the values of $[\bar{S}]$ given in

Table 7.2 for $^4$He at 1.0 and 2.0 MeV differ by about 1% or less from the values of $[S_0]$ given in Table IX.

### 7.3.2  Impurities in Elemental Samples

The depth $x$ over which impurity distributions can be measured in an elemental sample is determined by the atomic mass of the impurity and host atoms as well as by the beam parameters. The impurity generates a detectable signal only if its mass exceeds that of the host atom. In this case, the cutoff toward low energies for the impurity signal is determined by the energy of particles scattered from surface atoms of the host. Consequently, the total energy interval $\Delta E_{imp}$ of the impurity signal is given by

$$\Delta E_{imp} = (K_{imp} - K_{sub})E_0, \qquad (7.11)$$

where the subscript "imp" refers to the impurity atom, and "sub" to the atom in the host substrate. The accessible depth $x_{imp}$ is given by

$$x_{imp} = \Delta E_{imp}/[\overline{S}]_{imp}^{sub}, \qquad (7.12)$$

where the energy loss factor $[\overline{S}]$ is calculated on the basis of the energy loss in the substrate incurred by particles scattered from impurity atoms (Section 3.6).

Using the assumption that the impurity concentration is low enough that the elemental energy loss values can be used, we have calculated in Table 7.4 the accessible depth for different impurities in three host targets.

**TABLE 7.4**

Accessible Depth for Detecting Impurities in Si, Ge, and Sn with
$^1$H, $^4$He, and $^{16}$O at Various Energies ($\theta = 170°$)

| Projectile | Projectile energy (MeV) | Accessible depth ($\mu$m) in | | | | | |
|---|---|---|---|---|---|---|---|
| | | Si for | | | Ge for | | Sn for |
| | | Ni | Ag | Au | Ag | Au | Au |
| $^1$H[a] | 0.5 | 0.27 | 0.40 | 0.47 | 0.05 | 0.097 | 0.033 |
| | 1.0 | 0.83 | 1.2 | 1.4 | 1.14 | 0.27 | 0.092 |
| | 2.0 | 2.5 | 3.7 | 4.3 | 0.40 | 0.78 | 0.27 |
| $^4$He | 0.5 | 0.16 | 0.23 | 0.26 | 0.022 | 0.059 | 0.022 |
| | 1.0 | 0.33 | 0.47 | 0.55 | 0.042 | 0.12 | 0.042 |
| | 2.0 | 0.83 | 1.2 | 1.5 | 0.097 | 0.27 | 0.098 |
| $^{16}$O[a] | 2.0 | 0.40 | 0.50 | 0.60 | 0.13 | 0.24 | 0.17 |
| | 20.0 | 2.3 | 3.7 | 4.7 | 0.74 | 1.5 | 0.88 |

[a] Stopping cross sections for hydrogen and oxygen taken from Northcliffe and Schilling (Appendix D). For Si and Sn, the stopping cross sections of Al and Ag were substituted as the nearest known values. The kinematic factor for oxygen was calculated from Eq. (2.11). The average atomic mass of the impurity was used for $M_2$.

Since the table is meant to provide estimates only, $[\bar{S}]$ has been approximated by $[S_0]$ in the calculations.

### 7.3.3  Thin Compound Films

To determine the combination of composition and thickness of a compound film as in Section 4.4, it is generally desirable to avoid overlap of the signals from the elements in the film (Section 4.4.4). Figure 7.7 shows schematically the backscattering spectrum for a thin film of compound AB. The notation $E_{1A}$ and $E_{1B}$ is used for simplicity to denote the detected energies of particles scattered from the rear interface at thickness $t$. From Eq. (4.51), the thickness at which the signals of A and B overlap is given by

$$t = (K_A - K_B)E_0/[\bar{S}]_A^{AB}. \tag{7.13}$$

This relation is analogous to the one obtained for the accessible depth for analysis of impurities in an elemental target.

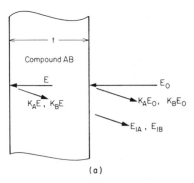

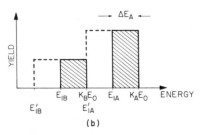

**Fig. 7.7**  (a) Schematic for and (b) backscattering spectrum of a thin compound film. The atomic mass of element A is larger than that of element B.

An example of signal overlap is shown in Fig. 7.8 for backscattering spectra of 400-keV $^4$He ions incident on $Cr_3Pt$ films on $SiO_2$ on Si substrates. In Fig. 7.8a, for 130- and 270-Å-thick films, the Pt and Cr signals are separated. In Fig. 7.8b, for the 530-Å-thick film, the signals overlap. In this case, the overlap could be prevented by raising the primary beam energy $E_0$.

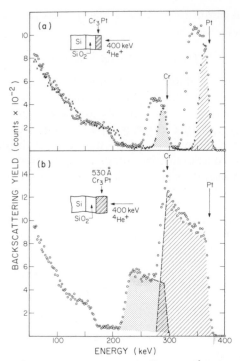

**Fig. 7.8** (a) For a sufficiently thin compound film [here 130 Å ( × ) and 270 Å (○) Cr$_3$Pt], the individual signals of the elements are distinct. (b) As the film thickness increases, the signals begin to overlap. This overlap could be prevented by raising the primary beam energy. [From Chu *et al.* (1973).]

## 7.4  DEPTH RESOLUTION AT NORMAL INCIDENCE

Along with mass resolution and accessible depth, the depth resolution is an important factor in the application of backscattering spectrometry to the analysis of materials. Depth resolution refers to the ability to sense composition changes with depth or variations in impurity distributions with depth. Just as the energy scale of detected particles is translated into a depth scale through $[\varepsilon]$ or $[S]$ factors, so the lowest resolvable energy width $\delta E$ can be translated into smallest resolvable depth interval $\delta t$. Following the convention adopted at the beginning of this chapter (Section 7.2), we define the *depth resolution* $\delta x$ by the energy width $\delta E_1$ between the positions at 12% and 88% of the full height of a signal that corresponds to an abrupt change in sample composition, as shown in Fig. 2.12 and the relationship

$$\delta x = \delta E_1 / [\overline{S}] \tag{7.14}$$

(see Section 3.2). The energy resolution $\delta E_1$ is normally determined by two contributions, one stemming from the system resolution $\delta E_r$ and the other from energy straggling $\delta E_s$. In the Gaussian approximation, they add up in quadrature:

$$(\delta E_1)^2 = (\delta E_r)^2 + (\delta E_s)^2. \tag{7.15}$$

The influence of the system resolution $\delta E_r$ and of the energy straggling $\delta E_s$ can be seen in Fig. 7.9 for the high- and low-energy edges of the energy spectrum of 2.0-MeV $^4$He ions backscattered from a 2000-Å-thick Pt film. The derivatives of the edges of the Pt signal are shown in the lower part of

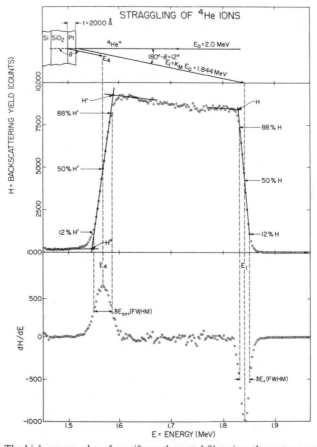

**Fig. 7.9**  The high-energy edge of a uniform elemental film gives the system resolution $\delta E_r$. The low-energy edge has a width $\delta E_{s+r}$ which is wider because of the added effect of energy straggling in particles that have passed in and out of the film. The bottom part of the figure gives a numerically determined derivative of the spectrum. [From Harris *et al.* (1973).]

the figure. The derivatives yield Gaussian-like functions whose full width at half maximum $\delta E$ is equivalent to the energy width between 12% and 88% of the signal height. The energy $\delta E_1$ of the high-energy edge gives the depth resolution $\delta x$ at the surface. For this edge, $\delta E_1 = 18$ keV and hence $\delta E_r = 18$ keV. The energy loss factor $[S_0]$ in Pt is 148 eV/Å, so that the depth resolution at the surface is $\delta x = 121$ Å.

The width $\delta E_1$ of the low-energy edge gives the depth resolution at the thickness $t$ in the film. From Fig. 7.9, $\delta E_1 = 34$ keV and hence $\delta x = 230$ Å, where for $[\bar{S}]$ we use the value $[S_0] = 148$ eV/Å in the surface energy approximation. The error incurred by this simplification is minor (see Section 5.6). The depth resolution for 2-MeV $^4$He on Pt thus is roughly 200 Å at a depth of 2000 Å when a surface barrier detector is used. At this depth, the contribution of energy straggling to $\delta E_1$, as determined by use of Eq. (7.15), is $\delta E_s = 28.9$ keV.

From the knowledge of the energy loss factor $[S_0]$ and the choice of a system resolution $\delta E_r$, one can plot depth resolution $\delta x$ at the surface versus atomic number, as displayed in Fig. 7.10 for $\delta E_r = 15$ keV. The large variation in depth resolution between adjacent elements is due to differences in

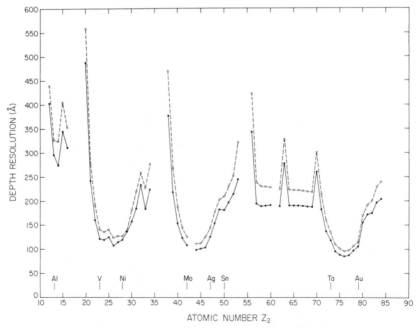

**Fig. 7.10** Depth resolution at the surface of an elemental sample versus its atomic number $Z_2$ for a detector resolution $\delta E_r$ of 15 keV and $^4$He of 1.0 MeV (●) and 2.0 MeV (○). Bulk densities and scattering through an angle $\theta$ of 170° are assumed. The large variation in depth resolution is due to differences in bulk densities.

bulk densities. For example, the alkali metals have low densities and the layer required for a given energy loss is thicker than for transition metals. As a rule of thumb, the depth resolution at the surface is between 100 and 200 Å for most elements with 1.0- to 2.0-MeV $^4$He.

Similarly, one can determine the depth resolution at the surface for other analyzing particles from calculated values of $[S_0]$ and measured values of the system resolution. In Table 7.5 we give such calculated values of the depth resolution, assuming $\delta E_r = 5$ keV for $^1$H, 15 keV for $^4$He, and 60, 80, and 100 keV for 1.0-, 2.0-, and 20.0-MeV oxygen. For convenience, we have utilized the values of $[\bar{S}]$ from Table 7.2 rather than $[S_0]$. This substitution is of little consequence, since the table is meant to provide estimates only. Tables of this nature, which rely on specific values of the system resolution, serve as guidelines. Improved values of depth resolution can be obtained with coolable surface barrier detectors or with magnetic or electrostatic analyzers.

**TABLE 7.5**

Depth Resolution near the Surface for $^1$H, $^4$He, and $^{16}$O Backscattered through $\theta = 170°$ from Al, Ni, Ag, and Au for Energy Resolutions $\delta E_r$ and Primary Energies $E_0$ as Indicated

| Projectile | Projectile energy $E_0$ (MeV) | Depth resolution (Å) in | | | |
|---|---|---|---|---|---|
| | | Al | Ni | Ag | Au |
| $^1$H | 0.5 | 360 | 160 | 170 | 150 |
| ($\delta E_r = 5$ keV) | 1.0 | 540 | 230 | 250 | 200 |
| | 2.0 | 850 | 330 | 360 | 280 |
| $^4$He | 0.5 | 330 | 130 | 130 | 110 |
| ($\delta E_r = 15$ keV) | 1.0 | 290 | 120 | 120 | 100 |
| | 2.0 | 330 | 130 | 140 | 110 |
| $^{16}$O | | | | | |
| 60 keV) | 1.0 | — | 310 | 230 | 210 |
| ($\delta E_r = $ 80 keV) | 2.0 | — | 290 | 220 | 180 |
| 100 keV) | 20.0 | 680 | 170 | 160 | 120 |

To estimate the depth resolution for scattering at depth $x$ in the sample, one must also consider energy straggling. In accordance with the treatment in Section 4.6.1, based on Bohr's theory [Eq. (4.65)], the variance for normal incidence is given by

$$\Omega_s^2 = 4\pi(Z_1 e^2)^2 Z_2 Nx[K^2 + (1/\cos\theta_2)]. \tag{7.16}$$

Since energy straggling in Bohr's theory is independent of energy, it is possible to construct tables of $\Omega_s^2/Nx$ and of $\Omega_s$ for scattering at a given

**TABLE 7.6**

Calculation of $Q_s^2$ and $Q_s$ at $t = 1$ $\mu$m for $^1$H, $^4$He, and $^{16}$O
Backscattered through $\theta = 170°$ from Al, Ni, Ag, and Au [from Eq. (7.16)][a]

| Projectile | $Q_s^2/Nt$ $[10^{-17}$ $(\text{keV cm})^2]$ | | | | $Q_s$ at $x = 1$ $\mu$m[b] (keV) | | | |
|---|---|---|---|---|---|---|---|---|
| | Al | Ni | Ag | Au | Al | Ni | Ag | Au |
| $^1$H | 0.6 | 1.4 | 2.3 | 4.1 | 6 | 11 | 12 | 16 |
| $^4$He | 1.8 | 4.6 | 8.6 | 15.3 | 10 | 20 | 22 | 30 |
| $^{16}$O | 22 | 52 | 103 | 202 | 36 | 69 | 79 | 109 |

[a] Values of $Q_s$ (right half of table) for thicknesses other than those given are obtained by scaling the numbers with $\sqrt{t}$.

[b] Bulk densities of 6.023, 9.126, 5.848, and 5.905 $\times$ $10^{22}$/cm$^3$ are assumed for Al, Ni, Ag, and Au, respectively.

depth and any given combination of projectile and target atom. Such a tabulation is given in Table 7.6. Again, the table provides mere estimates. Since $Q_s \propto \sqrt{t}$, the value of energy straggling for scattering at depths other than those assumed in the table are obtained by scaling with $\sqrt{t}$.

The energy width $\delta E_s$ for energy straggling is equal to $2.355 Q_s$. Combining this width with the system resolution $\delta E_r$ as shown in Eq. (7.15), one can obtain the energy resolution $\delta E_1$ and hence, from Eq. (7.14), the depth resolution. Values of the depth resolution are given in Table 7.7 for the

**TABLE 7.7**

Depth Resolution of Backscattering Analysis for a Thick Target Derived
from the Data of Tables 7.2 and 7.6 with Eqs. (7.14) and (7.15)

| Projectile | Projectile energy $E_0$ (MeV) | Depth resolution $\delta x$ (Å) for a given depth $x$ ($\mu$m) | | | | | | | |
|---|---|---|---|---|---|---|---|---|---|
| | | Al | | Ni | | Ag | | Au | |
| | | $\delta x$ | $x$ | $\delta x$ | $x$ | $\delta x$ | $x$ | $\delta x$ | $x$ |
| $^1$H | 0.5 | 1000 | 1 | 800 | 1 | 900 | 1 | 1000 | 1 |
| ($\delta E_r = 5$ keV) | 1.0 | 2300 | 4 | 1500 | 2 | 1700 | 2 | 1900 | 2 |
| | 2.0 | 7000 | 15 | 4000 | 7 | 5000 | 8 | 5000 | 7 |
| $^4$He | 0.5 | 500 | 0.5 | 350 | 0.3 | 400 | 0.3 | 400 | 0.3 |
| ($\delta E_r = 15$ keV) | 1.0 | 500 | 1 | 350 | 0.5 | 350 | 0.5 | 400 | 0.5 |
| | 2.0 | 700 | 1.7 | 400 | 1 | 450 | 1 | 550 | 1 |
| $^{16}$O | | | | | | | | | |
| ($\delta E_r = $ 80 keV) | 2.0 | | | 500 | 0.2 | 500 | 0.3 | 500 | 0.3 |
| ($\delta E_r = $ 100 keV) | 20.0 | 1000 | 1 | 400 | 1 | 650 | 1.4 | 500 | 1.4 |

energy loss factors $[\bar{S}]$ of Table 7.2 and the values of detector resolution used in Table 7.5.

The values in Table 7.7, like those in Table 7.5, are intended only as an indication of approximate depth resolution for analysis with different projectiles. Data on energy straggling in solid targets are scarce, and deviations from Bohr's theory have been found (Harris *et al.*, 1973; Luomajarvi *et al.*, 1976; Hoffman and Powers, 1976; Chu, 1976).

### 7.5  DEPTH RESOLUTION AT GLANCING INCIDENCE

#### 7.5.1  General Scattering Geometry

In the examples discussed so far, we have treated the incident beam direction, the target normal, and the emergent path of the scattered particle as coplanar. A general scattering geometry is shown in Fig. 7.11. Here the plane defined by the incident beam direction and the target normal is at an angle $\phi$ with respect to the plane defined by the target normal and emergent path.

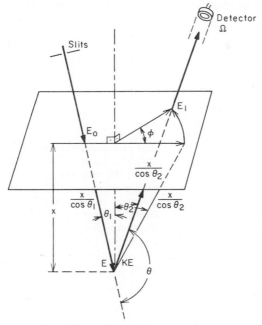

**Fig. 7.11**  In the most general arrangement for a backscattering measurement, the detector is not positioned in the plane defined by the incident beam and the sample normal, but is placed outside of it, at an angle $\phi$. The scattering angle $\theta$ is then given no longer simply by $180° - \theta_1 - \theta_2$, but by Eq. (7.17). The angles $\theta_1$, $\theta_2$, and $\phi$ are positive regardless of their orientation.

We maintain the notation for incident and emergent angles as $\theta_1$ and $\theta_2$, so that for scattering at depth $x$ the incident and outgoing path lengths remain $x/\cos\theta_1$ and $x/\cos\theta_2$. With this notation there is no change in the formalism developed in previous chapters except that the scattering angle $\theta$ is now given by

$$\theta = \cos^{-1}(\sin\theta_1\,\sin\theta_2\,\cos\phi - \cos\theta_1\,\cos\theta_2), \qquad (7.17)$$

where $\theta_1, \theta_2$, and $\phi$ are positive regardless of orientation.

### 7.5.2  Target Tilting and Glancing Incidence

One method of increasing the depth resolution was not mentioned in Section 7.4. By tilting the target with respect to the direction of the incident beam, one obtains a geometrical increase in the total path length required to reach a given point below the surface; this increase, in turn, produces an increase in the effective depth resolution. This, of course, reduces the actual accessible depth for analysis, but can result in dramatic improvements in depth resolution for near-surface analysis.

An example is shown in Fig. 7.12a, depicting a view of an arrangement in which the detector is mounted directly below the incident beam and the

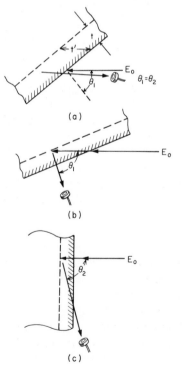

**Fig. 7.12**  Three different ways to increase the total path length that a particle must traverse in the sample to reach the detector after scattering at a fixed depth $x$ below the surface. The increase in total path length can greatly improve the effective depth resolution, provided the sample is large enough and laterally uniform. Incident beam, sample normal, and direction of detection are coplanar, and the tilting axis is normal to that plane. (a) $t' = t/\cos\alpha$; (b) $\theta_1 = 180° - \theta = \alpha$ and $\theta_2 = 0°$; (c) $\theta_1 = 0°$ and $\theta_2 = 180° - \theta = \alpha$.

sample is tilted by an angle $\theta_1$. The effective film thickness is then increased from $t$ to $t/\cos\theta_1$. For $\theta_1 = 60°$ the film thickness perceived by the beam is doubled.

We note in passing that sample tilting is an effective method to quickly verify that an impurity lies on the sample surface. The signal of such an impurity does not shift in energy with tilt angle, though the signal of one below the surface does. Tilting also increases the energy width $\Delta E$ of the signal from a very thin film, so that $\Delta E$ can become greater than the detector resolution. This circumvents the difficulties depicted in Fig. 4.10 (Section 4.6) for the analysis of the signal heights of very thin films.

The tilting method must be applied with caution. The sample must be large and flat enough, since the beam spot becomes larger than the actual cross section of the beam. The axis of rotation should coincide with both the sample surface and the incident beam spot, or else the beam spot will wander across the sample surface. For glancing angles of incidence and/or exit, it is also most important to measure accurately the value of the angles, because small uncertainties will produce large errors in calculated path lengths. For the same reason, it is necessary to reduce the acceptance angle of the detector when the detector is positioned at glancing angles of exit.

Other geometries can be employed, as sketched in Fig. 7.12. In the arrangement shown in Fig. 7.12b, the incident path is increased by a factor $1/\cos\theta_1$, and in the setup shown in Fig. 7.12c, it is the outgoing path length that is increased by the factor $1/\cos\theta_2$. These unequal path lengths are sometimes used also to change the segments of the stopping cross section curve that are traced along the inward and outward paths. Both cases lead to an improvement in depth resolution. The arrangement in Figure 7.12c has the advantage that with a beam of normal incidence, channeling can be combined with the increased depth resolution obtained from the glancing exit path.

Williams (1975, 1976) used low-angle glancing incidence and obtained depth resolutions of 25 Å in silicon with 2-MeV $^4$He ions and surface barrier detectors of a resolution of 15 keV (FWHM). The optimum geometry in his work was $\theta_1 = 85°$ ($1/\cos\theta_1 = 11.5$) and $\theta_2 = 78°$ ($1/\cos\theta_2 = 4.8$). This choice gave a high depth resolution for analysis of very thin films and implanted layers. Feuerstein *et al.* (1976) combined target tilting with electrostatic analysis of the backscattered particles; for $^4$H of 250 keV incident on a Au film and a tilt angle of 83.5°, the surface resolution was about 5 Å. Pabst (1976) showed that the geometrical effect with megaelectron volt $^4$He at glancing angles was more effective in improving depth resolution than the use of $^{14}$N ions at normal incidence. At these glancing angles of incidence one must be aware of surface topography, beam alignment with the sample surface, charge integration, and accelerated degradation of depth resolution below the surface due to energy straggling and multiple scattering.

Tilting the target not only increases the depth scale, but also affects the height of a spectrum. As shown in Chapter 3 [Eq. (3.50)] the height of a spectrum is proportional to

$$H \propto \sigma(\theta, E)/[\varepsilon(\theta_1, \theta_2, E)] \cos \theta_1, \tag{7.18}$$

where $\theta$ has been introduced in the argument of $\sigma$ to indicate that the scattering cross section depends on the scattering angle $\theta$; similarly, the argument of $[\varepsilon]$ indicates that the stopping cross section factor depends on $\theta_1$ and $\theta_2$ (see Fig. 7.11 or Figs. 3.6 and 7.12 for the coplanar arrangement). In contrast, the total number of counts $A$ of a signal such as that produced by a surface impurity is proportional to [Eq. (4.7)]:

$$A \propto \sigma(\theta, E)/\cos \theta_1. \tag{7.19}$$

For example, assume that $\theta_1 = \theta_2$ as shown in Fig. 7.12a. Going from a system with normal arrangement ($\theta_1 = \theta_2 = 0$) to one with $\theta_1 = \theta_2 = 60°$ does not change the height of the spectrum for scattering from the surface region of the sample, because the number of atoms per unit area normal to the beam is doubled at 60°. However, the energy lost per unit length normal to the sample surface is doubled, and this reduces by half the width $\tau_0$ of a slab that corresponds to one channel of the analyzer. The net result is no change in the number of scattering events per channel. In other words, the factor $\cos \theta_1$ in the denominator of Eq. (7.18) cancels the factor $1/\cos \theta_1$ in $[\varepsilon_0]$ [Eq. (3.12)]. On the other hand, the total number of counts of a signal from a surface impurity on that same sample (e.g., atoms of a heavy element, as discussed in Fig. 5.1) will double for the same tilt because the number of surface atoms illuminated by the beam doubles. Whether the shape of the impurity signal changes in the process or not depends on a number of factors, such as the detector resolution, the actual thickness of the impurity layer, and the energy per channel $\mathscr{E}$; but the total number of counts in the signal will double, as stated by Eq. (7.19).

As another example, imagine that the direction of detection is normal to the sample surface and fixed ($\theta_2 = 0$), but that the direction of the incident beam changes from normal ($\theta_1 = 0$) to $\theta_1 = 60°$, as sketched in Fig. 7.12b. The contribution to the height of the signal due to scattering from the near-surface region of the sample is again doubled as a consequence of the doubling of the atoms traversed by the incident beam on reaching a given depth measured normally below the sample surface. Now, however, the energy lost along only the incident path is doubled and that dissipated along the outward path is unchanged. The stopping cross section factor thus increases by a factor of $\frac{3}{2}$, so that the surface height $H_0$ of the sample signal rises by $2/\frac{3}{2} = \frac{4}{3}$. The total number of counts in the signal of the surface impurity doubles as before. These changes have to be modified further by the increase in the value of the scattering cross section when we go from a backscattering angle of $\theta = 180°$ for $\theta_1 = 0$ to a more forward angle of $\theta = 120°$ for $\theta_1 = 60°$. The effect adds another factor of approximately $\sin^{-4} 60°/\sin^{-4} 90° = (\sqrt{3}/2)^{-4} = 16/9 = 1.78$ [Eq. (2.24) and Fig. 2.5] both to the surface height $H_0$ of the substrate signal and the total number of counts A of the impurity. Finally, there is also a shift in the position of the signals on the energy axis because the kinematic factor $K$ varies with the backscattering angle $\theta$ also [Eq. (2.6) and Fig. 2.2]. One additional complication can arise. Since the incident beam is tilted away from the surface normal, the area of the sample surface illuminated by the beam increases in size. This expansion broadens the range of angles $\theta$ by which particles can backscatter and still reach the detector. That effect also modifies the signals. One remedy is to maintain an adequate distance between the beam spot on the sample and the detector, but the cure comes at the cost of a reduction of the counting rate.

The first example demonstrates the advantage of the arrangement in which $\theta_1$ and $\theta_2$ are the same. In this configuration, tilting the sample away from normal incidence magnifies the equivalent depth scale on the energy axis and leaves the surface height of the signal unchanged. When the tilt axis of the sample is normal to the sample surface and falls within the plane defined by the direction of the beam and the detector, the change in the size of the beam spot has a limited effect on the backscattering angles accepted by the detector. The effect vanishes altogether when the direction of detection and the incident beam are collinear, as is the case for annular detectors or any configuration where the detected particles are backscattered by nearly 180°.

## 7.6 SENSITIVITY TO DETECTION OF SURFACE IMPURITIES

In Section 5.2, we discussed the detection of surface impurities with a heavy mass on a lower-mass substrate and presented an empirical formula [Eq. (5.3)] for the minimum amount of surface impurities that can be detected by 2-MeV $^4$He ion backscattering. This sensitivity limit is based on standard operating conditions and predicts a limit of $3 \times 10^{12}$ Au/cm$^2$ on a silicon substrate.

A major factor limiting the detection of trace amounts of impurities on the sample surface is that the background counts extend to higher energies above the high-energy edge of the signal from the substrate. The background counts are generally caused by pulse pile up in the detection system; that is, they arise when more than one single-counting event occurs within the time resolution of the electronic system. This leads to counts being recorded in the multichannel analyzer at higher channel numbers than would be the case if the individual events had occurred at sufficiently large time intervals. Pulse pile up leads to background counts extending to energies above the high-energy edge of the substrate signal. Although the height of the background is small compared to the height of the substrate signal (typically the height ratio is $10^{-3}$), it is sufficient to limit the detection of impurities.

An example [quoted by Chu et al. (1973)] of this background is shown in Fig. 7.13a for detecting $6 \times 10^{12}$ Au atoms/cm² on an $SiO_2$ substrate. The high-energy edge of the substrate signal (not shown) occurs at channel 9 in this spectrum. The pulse pile up background can be reduced by lowering the primary beam current. This reduces the average count rate and increases

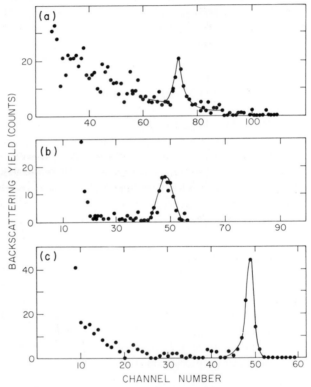

**Fig. 7.13**   (a) Backscattering spectrum registered with slow electronics of 1.6-MeV $^4$He incident on a Si sample with $6 \times 10^{12}$ Au atoms/cm² on the surface. The Si edge (8.8 keV/channel) is located at channel 9. Only the portion of the spectrum above the Si edge is reproduced, showing the Au signal and the background due to pile up. (b) The spectrum registered with fast electronics for an experiment otherwise identical with that shown in (a) for an Si edge (13.5 keV/channel) located at channel 7. The background is cleaner, at some cost in resolution. (c) Backscattering spectrum of 1.6-MeV $^{12}$C registered with slow electronics for a sample with only one-twelfth of the Au used for the spectrum in (a). The energy separation between the Au signal and the Si edge has increased, shifting the Au signal away from the pile-up region of the spectrum and increasing the sensitivity to heavy surface impurities. The Si edge (20.3 keV/channel) is in channel 0. [From Chu et al. (1973).]

the average time interval between individual counts. This reduction in pile up is obtained at the expense of longer exposure time. Generally, however, fast electronics and pulse-gating techniques are used. Although this leads to some loss in the energy resolution of the system, the background can be reduced substantially, as is shown by the spectrum given in Fig. 7.13b.

An alternative approach is to use heavy ions for the analysis beam (Hart *et al.*, 1973). This decreases the energy of particles scattered from the substrate and the heavy impurity, but the energy separation between the two edges increases, with the result that the background due to pulse pile up is reduced. Again this reduction is achieved at the expense of some loss in energy resolution. The improvement in sensitivity obtained with 1.6-MeV $^{12}$C ions in detecting $5 \times 10^{11}$ Au atoms/cm$^2$ is evident in Fig. 7.13c.

As was mentioned in Section 7.5, trace impurity sensitivity can be improved by tilting the target. In this case, however, near-glancing incidence is required to produce a substantial decrease in the sensitivity limit.

### 7.7  LOW-ENERGY TAILS

In the backscattering spectra of thin films, such as those shown in Fig. 5.12, one notes the presence of signals appearing in energy below the low-energy edge of the thin-film signal. These counts are often referred to as the low-energy *tail of a backscattering signal.* A more pronounced example of this phenomenon is shown in Fig. 7.14 for 0.56-MeV $^4$He incident on a Pt film

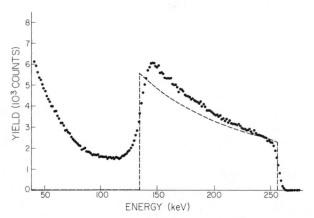

**Fig. 7.14**  Large discrepancies are observed between a computer-generated spectrum (dashed line) and an actually measured spectrum of a thin Pt film at energies below the low-energy edge of the Pt signal (0.56-MeV $^4$He$^{++}$ → 1130-Å Pt/SiO$_2$). These counts, referred to as the *tail* of a thin-film spectrum, are of uncertain origin. They severely limit the application of backscattering spectrometry, particularly at low incident energies, where the tail is observed to be quite substantial. [From Scherzer *et al.* (1976).]

on an $SiO_2$ substrate (Scherzer et al., 1976). In this example, the height of the tail is nearly 50% of the signal height from the film.

The physical origin of the low-energy tail is not understood at present. It may arise from low-energy particles in the incident beam due to glancing angle collisions on the collimating slits, or from deficiencies of the detection system.

The height of the tail is approximately proportional to the total number of counts in the signal of the thin film. A reduction of the tail height is also observed with increasing incident beam energy. This reduction can also be obtained by use of protons rather than $^4He$ ions.

The presence of this tail does impose limitations on thin-film analysis beyond that specified by the accessible depth of analysis. It also restricts use of sub-megaelectron volt $^4He$ ions to film thicknesses less than a few thousand angstroms.

### 7.8  NON-RUTHERFORD SCATTERING, NUCLEAR REACTIONS, AND DETECTION OF LOW-MASS IMPURITIES

Trace amounts of an impurity with low atomic mass contained within a substrate with higher atomic mass are usually quite difficult to detect by backscattering techniques. The yield from the substrate is generally so high that the impurity signal cannot be distinguished from that of the substrate. Some special situations, however, do allow detection of impurities with low atomic mass. With single-crystal substrates, channeling can be used to reduce the yield from the substrate so that surface impurities in monolayer amounts can be detected (Fig. 8.21). For example, the signal from surface oxide layers on Si containing $10^{15}$ oxygen atoms/cm$^2$ can be extracted that way from the signal of the Si substrate (Chu et al., 1973). Another situation occurs when the impurities are present in concentrations high enough, typically greater than several atomic percent, that the yield from the substrate is reduced because of the decrease in the number of substrate atoms in an incremental slab whose thickness $\tau$ corresponds to one channel width $\mathscr{E}$. By comparing the yield from an impurity-free substrate to that from the substrate under investigation, the amount of impurity can be determined (Mayer et al., 1973; Gamo et al., 1976). This method has been used routinely to investigate the incorporation of impurities in thin films (Kräutle et al., 1974; Mayer et al., 1973).

Two methods, nuclear reaction and resonant scattering, can be used to improve the detection sensitivity for low-mass impurities. Both are highly selective but require some care in calibration.

With resonant scattering, the scattering cross section is greater than for pure Rutherford scattering. An example of the energy dependence of the scattering cross section for $^4$He incident on $^{16}$O is shown in Fig. 7.15 (Cameron, 1953). A pronounced increase in yield is found, for example, at 3.05 MeV. This resonance energy has been used by Mezey *et al.* (1976) to investigate thin oxide layers. The resulting yield was greater by a factor of 30 than that calculated on the basis of Rutherford scattering. For quantitative

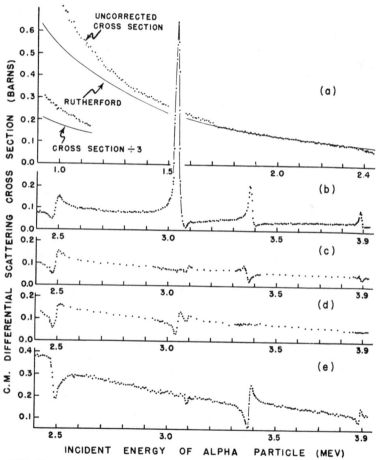

**Fig. 7.15**  Energy dependence of the elastic scattering cross section of $^4$He on $^{16}$O for (a) $\theta = 168.0°$ in the energy range 0.94–2.4 MeV, and at (b) 168.0°, (c) 140.1°, (d) 124.6°, (e)124.6°, and (f) 90.0° in the 2.4–4.0 MeV energy range. The value of $\theta$ is given in the center-of-mass coordinate system ($\theta_c$ in Chapter II, $\theta_{cm}$ here). [From Cameron (1953).]

measurements in order to maintain the conditions of a thin-layer approxima-tion, it was necessary that the oxide layer remain thin enough that its signal would not exceed the half-width of the resonance. With thicker layers, the $^{16}O$ yield is increased only in the narrow portion of the spectrum which corresponds to the resonance energy, as shown in Fig. 7.16 (Chu et al., 1973). Under these conditions, quantitative measurements of impurity concentra-tions become more difficult.

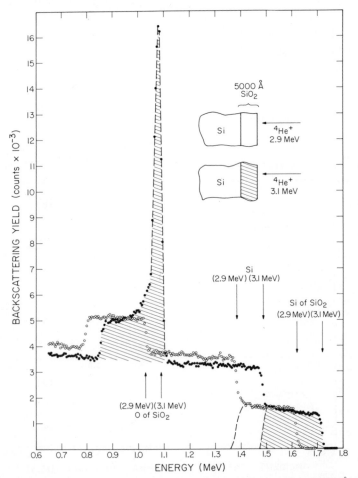

**Fig. 7.16**  Energy spectra for 2.9- and 3.1-MeV $^4He$ backscattered from 5000 Å of $SiO_2$ coated on a Si substrate. The contributions of Si and oxygen from the $SiO_2$ layer to the 3.1-MeV spectrum are shaded. The abnormally high yield of backscattering from oxygen is due to an elastic nuclear resonance at 3.05 MeV. [From Chu et al. (1973).]

Nuclear reactions have also been used in studies of oxidation phenomena (Amsel *et al.*, 1971). Table 7.8 (Feldman and Picraux, 1977) lists some of the more commonly used reactions. Since the energy released in the reactions is high, the impurity signal appears at higher energies than that of the substrate. The backscattering signal from the substrate is normally undesirable, so that usually the detector is shielded from the substrate signals by a thin absorbing film. Because the reaction cross sections are smaller than those for Rutherford scattering, high beam currents are needed. Recent work by L'Ecuyer *et al.* (to be published) has shown that a $^6$Li beam of 5 to 10 MeV can be used quite generally to profile quantitatively light elements ($5 < M_2 < 20$) by nuclear reactions induced by $^6$Li bombardment.

**TABLE 7.8**

Nuclear Reactions Used for the Detection of Light Elements[a]

| Detected element | Reaction | Incident energy $E_0$ (MeV) | Emitted particle energy[b] (MeV) | Mylar absorber thickness ($\mu$m) | Yield (counts/$\mu$C) |
|---|---|---|---|---|---|
| $^2$H | $^2$H(d, p)$^3$H | 1.0 | 2.3 | 14 | 30 |
| $^2$H | $^2$H($^3$He, p)$^4$He | 0.7 | 13.0 | 6 | 380 |
| $^3$He | $^3$He(d, p)$^4$He | 0.45 | 13.6 | 8 | 400 |
| $^6$Li | $^6$Li(d, $\alpha$)$^4$He | 0.7 | 9.7 | 8 | 35 |
| $^7$Li | $^7$Li(p, $\alpha$)$^4$He | 1.5 | 7.7 | 35 | 9 |
| $^9$Be | $^9$Be(d, $\alpha$)$^7$Li | 0.6[c] | 4.1 | 6 | 6 |
| $^{10}$B | $^{10}$B(n, $\alpha$)$^7$Li* | thermal | 1.78 | — | — |
| $^{11}$B | $^{11}$B(p, $\alpha$)$^8$Be | 0.65 | 5.57 | 10 | 0.7 |
| $^{12}$C | $^{12}$C(d, p)$^{13}$C | 1.20 | 3.1 | 16 | 210 |
| $^{13}$C | $^{13}$C(d, p)$^{14}$C | 0.64 | 5.8 | 6 | 2 |
| $^{14}$N | $^{14}$N(d, $\alpha$)$^{12}$C | 1.5 | 9.9 | 23 | 3.6 |
| $^{15}$N | $^{15}$N(p, $\alpha$)$^{12}$C | 0.8[d] | 3.9 | 12 | 90 |
| $^{16}$O | $^{16}$O(d, p)$^{17}$O* | 0.90 | 2.4 | 12 | 25 |
| $^{18}$O | $^{18}$O(p, $\alpha$)$^{15}$N | 0.730[d] | 3.4 | 11 | 90 |
| $^{19}$F | $^{19}$F(p, $\alpha$)$^{16}$O | 1.25 | 6.9 | 25 | 3 |
| $^{23}$Na | $^{23}$Na(p, $\alpha$)$^{20}$Ne | 0.592 | 2.238 | 6 | 25 |
| $^{27}$Al | $^{27}$Al(p, $\gamma$)$^{28}$Si | 0.992 | 1.77 | — | 80[e] |
| $^{31}$P | $^{31}$P(p, $\alpha$)$^{28}$Si | 1.514 | 2.734 | [f] | 100 |

[a] From Feldman and Picraux (1977).

[b] For laboratory emission angle of 150° with recoil nucleus in ground state.

[c] 0.6 MeV is optimum for Be in a light $Z$ matrix and 1.6 MeV for Be in a high $Z$ matrix.

[d] Maximum energy for Mylar to stop backscattered proton.

[e] With 3-in. × 3-in. NaI(T1) detector, 1 cm from the target; $\gamma$ energy window = 9–13 MeV.

[f] Range of $\alpha$ < range of p.

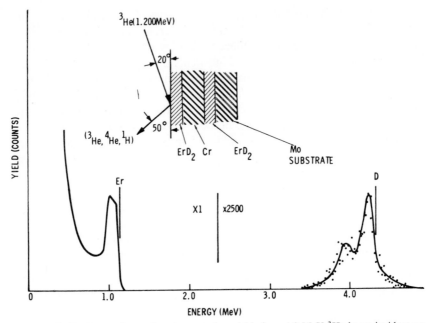

**Fig. 7.17** Backscattering and nuclear reaction yields for a 1.2-MeV $^3$He beam incident on a multilayered $ErD_2$ target. The angle of observation is 70° away from the direction of incidence. [From Langley *et al.* (1974).]

Although it is not the intent of this section to discuss nuclear reactions and non-Rutherford scattering in any detail, we will illustrate the use of these methods to detect hydrogen isotopes in solids. Langley *et al.* (1974) used megaelectron volt beams of $^4$He ions and detected $^4$He ions generated in the reactions $d(^3He, p)^4He$. For this reaction, they found that the best depth resolution was obtained in forward detection geometries. The concepts were tested with multilayer deuterided films and 1.2-MeV $^3$He ions as shown in Fig. 7.17. The angle of incidence of the $^3$He ions with respect to the surface was 20°, and the detector was positioned at an angle of 70° away from the direction of the incident beam. The counts at energies below 1.2 MeV correspond to Rutherford scattering of $^3$He; the position of the particles scattered from Er atoms at the surface is indicated by a vertical bar. The much lower yield of particles at energies around 4.0 MeV is due to $^4$He particles from the nuclear reaction. The vertical bar corresponds to $^4$He emanating from deuterons at the surface. The two peaks indicate the presence of the two $ErD_2$ layers.

Nuclear reactions induced by heavy ions are also useful to profile hydrogen. For example, Leich and Tombrello (1973) have used 16–18-MeV $^{19}$F and the resonant nuclear reaction $^1\text{H}(^{19}\text{F}, \alpha\gamma)\,^{16}\text{O}$ to obtain direct hydrogen depth profiles on lunar samples. Ligeon and Guivarc'h (1976) have used a 2-MeV $^{11}\text{B}$ beam and the reaction $^1\text{H}(^{11}\text{B}, \alpha)\alpha\alpha$, and Lanford *et al.* (1976) have used a 7-MeV $^{15}\text{N}$ beam and the reaction $^1\text{H}(^{15}\text{N}, \alpha\gamma)\,^{12}\text{C}$ to profile hydrogen implanted at low energies into Si. The references cited in Section 10.9 give other examples of the application of nuclear reactions to the detection of hydrogen and helium.

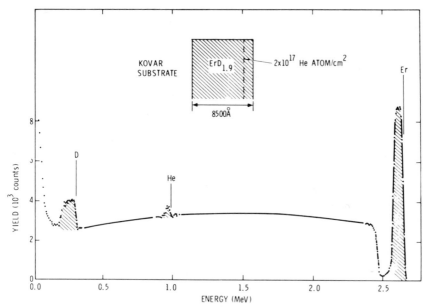

**Fig. 7.18**  Energy spectrum for 2.8-MeV $^1\text{H}$ incident on an ErD film implanted with 50-keV $^4\text{He}$ with a mean range of 1500 Å. Because the elastic scattering cross section for $^1\text{H}$ on $^2\text{D}$ and $^4\text{He}$ is several hundred times as great as that of pure Rutherford scattering, the signals of $^4\text{He}$ and $^2\text{D}$ stand out from the background signal from the substrate.

The alternative approach is to exploit the two orders of magnitude increase in scattering cross section obtained with 2.0- to 3.0-MeV protons incident on deuterons (Blewer, 1973). Figure 7.18 shows the energy spectrum for 2.8-MeV $^1\text{H}$ scattered from an erbium deuteride film implanted with $2 \times 10^{17}\,^4\text{He}$ atoms/cm$^2$. Since the enhancement in scattering cross section for 2.8-MeV protons is about 250 for deuterons and about 300 for $^4\text{He}$ (Langley, 1976), the signals from these elements emerge quite distinctly from the background of protons scattered from the Kovar substrate.

The use of enhanced cross sections or nuclear reactions adds significantly to the capability of ion beam analysis. A description of these techniques, as well as of ion-induced x-rays, is given, for example, by Mayer and Rimini (1977).

A modification of the standard Rutherford scattering technique that works particularly well for profiling $^1$H in self-supported films makes use of coincidence measurements. The concept is explained in Fig. 7.19. When an energetic incident proton is scattered by 45° in the laboratory frame of reference by a proton (i.e., the hydrogen nucleus) in the film, this target proton must itself be scattered to take up the recoil and will emerge at 45° on the other side of the transmitted beam and in the same scattering plane. By symmetry, the energies of both protons are the same. The sum of these energies will be equal to $E_0 - \Delta E$, where $\Delta E$ is the energy lost by the two particles as they traverse the material. If two identical detectors are positioned at 90° to each other to intercept these particles, their signals will exactly coincide in both time and amplitude. With the help of a fast coincidence circuit, only those events are counted that occur simultaneously, to ensure a specific detection of events caused by scattering only from $^1$H in the film. Accidental coincidences can be eliminated by retaining only events that produce signals of equal magnitude in both detectors. The number of such events versus the sum of their amplitudes produces an image of the hydrogen profile in the film. This method is absolutely specific to hydrogen and has a sensitivity of 1 ppm.

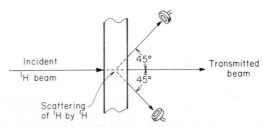

**Fig. 7.19**  Energetic protons scattered through 45° by a hydrogen nucleus in a film will impart a recoil to the target proton which ejects it at 45° on the other side of the transmitted beam. The two protons generate simultaneous signals of equal magnitude in the two detectors. This coincidence in time and amplitude is unique to scattering from protons in the target and offers the opportunity of $^1$H-specific detection and profiling.

A hydrogen profile obtained in this manner is shown in Fig. 7.20 (Cohen *et al.*, 1972). It gives the spectrum of three 17-$\mu$m-thick Fe foils interleaved with four 4-$\mu$m-thick Mylar sheets, recorded with two surface barrier detectors and a primary proton beam of 17 MeV. The high energy ensures a clean transmission of the beam through such a thick target. The sensitivity of the system is such that about one monolayer of hydrogen on the surface

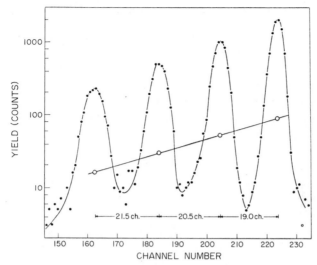

**Fig. 7.20**  Coincidence spectrum of an Fe–Mylar sample consisting of three 17-$\mu$m-thick Fe foils interleaved with four 4-$\mu$m-thick Mylar sheets. The plot gives the number of coincidences versus the total energy of both detected particles. The open circles give the total number of counts contained in each peak ( $\times$ 1/1000). They show that counting losses vary exponentially with energy (channel number). [From Cohen (1972).]

should still be detectable, but the resolution in depth is clearly quite poor. The reduction of total counts per Mylar layer with decreasing energy—the lowest energy corresponds to the point of first impact of the primary beam on the sample—is caused by coincidences that are lost because one of the two particles has undergone excessive low-angle multiple scattering on its traverse through the target and has missed its detector.

The method can be applied to the selective profiling of any element by using that same element as the beam particle. Moore *et al.* (1975) have used an $^{16}$O beam to profile oxygen in Ni foils about 2200 Å thick. The same authors have generalized the method to detect elements other than those of the beam. The two detectors then have to be placed at different angles, and the coincidence circuitry must introduce a delay in one of the two signals because the recoiled target atom emerges from the sample at an angle and a velocity different from those of the scattered projectile. Mass resolution and depth perception are optimum when $M_1 \simeq M_2$. For foils of the order of 2000 Å, energies are typically 1 MeV/amu. Coincidence measurements in transmission thus normally necessitate beams of much higher energy than is typical for backscattering spectrometry with $^1$H or $^4$He (Artemov *et al.*, 1973; Jarvis and Sherwood, 1974).

### 7.9   MICROSCOPIC BEAM AND NONUNIFORM LAYERS

Thus far we have been concerned with laterally uniform films or layered structures. Often, however, nonuniform film thicknesses are present, in particular after a heat treatment of the sample. For example, if the adhesion between a metal film and the substrate is poor, the metal can agglomerate into balls or islands. Another example occurs in thin-film reactions when the reaction process does not progress within a uniform reaction front but where crystallites grow within the film.

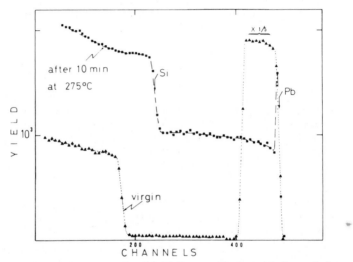

**Fig. 7.21**   Backscattering spectra for a thin Pb film on a Si substrate before and after annealing at 275°C for 10 min. After annealing, the Pb yield has decreased by a factor of 10, and the high-energy edge of the Si spectrum has shifted to the energy corresponding to backscattering from the surface. This change is the consequence of the lateral nonuniformity of the sample (see Fig. 7.22). [From Campisano *et al.* (1975).]

The heat treatment of a Pb layer on a silicon substrate shows the influence of island formation (Campisano *et al.*, 1975). The backscattering spectrum in Fig. 7.21 for the deposited film shows a well defined flat-topped signal from the Pb film. After heating at 275°C, the signal height from the Pb film decreases and the Pb signal extends down to low energies. In fact, the silicon signal now rides on the Pb signal. From these spectra alone it is difficult to ascertain whether the Pb is penetrating uniformly into the Si crystal or whether Pb islands form on the surface. The scanning electron micrograph of the sample after the heat treatment (Fig. 7.22) shows that islands have formed.

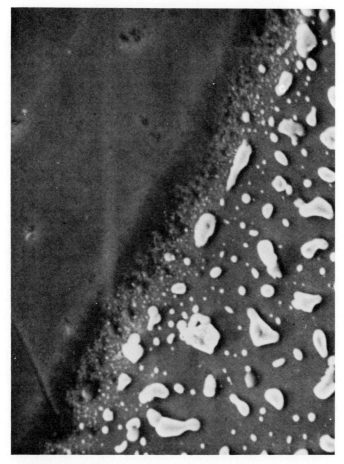

**Fig. 7.22** Scanning electron micrograph of the sample after annealing (Fig. 7.21). The upper right half is the surface of the Si substrate. The lower left half shows that the Pb film has agglomerated into islands of an average dimension of 2 $\mu$m during annealing at 275°C for 10 min. [From Campisano *et al.* (1975).]

These results, as well as others (for example, Nakamura *et al.*, 1975), demonstrate that it is incumbent upon the investigator to verify the lateral uniformity of the samples. Frequently, lateral nonuniformities can be seen readily with optical microscopes. For finer nonuniformities, electron microscopy must be used.

It is possible to gain some insight into the irregularities of the sample surface by tilting the sample with respect to the incident beam. As sketched

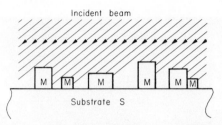

**Fig. 7.23** The presence of small islands of a metal M on a substrate S can reveal themselves by tilting the sample with respect to the incident beam. The island shields the substrate from the incident beam. There is no scattering from the substrate surface, and the substrate signal is below the edge $K_s F_0$ in the spectrum.

in Fig. 7.23, tilting the sample can shadow the substrate when islands are present on the surface. This would shift the high-energy edge of the substrate signal to lower energies and would cause the height and width of the signal from the islands to increase. Nonuniform reaction fronts in thin-film structures produce sloping edges in the signals, as do uniform interdiffusion zones. Sample tilting does not produce a marked difference in the spectra for these two cases, since both the diffusion zone and the irregular front show an increase in effective thickness as the length of the beam path is increased.

Lateral nonuniformities in the sample can be investigated with microbeams, i.e., beams of very small cross sections. Examples are listed in Section 10.11. Beam spot widths of about 10 $\mu$m and less have been achieved (Pierce, 1974; Cookson and Pilling, 1973). Scanning of such a microbeam across the sample surface will produce changes in the spectra if the nonuniformities of the sample are of about the size of the beam spot or larger. Microbeams could also be useful to analyze local areas of integrated circuits or devices directly on samples of the production-line scale, but the system would have to meet stringent requirements in the accuracy of the positioning of the beam and the sample.

**REFERENCES**

Amsel, G., Nadai, J. P., Artemare, E. D., David, D., Girard, E., and Moulin, J. (1971) *Nucl. Instrum. Methods* **92**, 481.
Artemov, K. P., Gol'dberg, V. Z., Petrov, I. N., Rudakov, V. P., Serikov, I. N., and Timofeev, V. A., (1973). *Nucl. Sci. Abstr.* **28**, 1419.
Bergstrom, I., Bjorkqvist, K., Domeij, B., Fladda, G., and Andersen, S., (1968). *Can. J. Phys.* **46**, 2679.
Blewer, R. S. (1973). *Appl. Phys. Lett.* **23**, 593.
Buck, T. M. (1975). *in* "Methods of Surface Analysis" (A. W. Czanderna, ed.), p. 75. Elsevier, Amsterdam.

Cameron, J. R. (1953). *Phys. Rev.* **90**, 839.
Campisano, S. U., Foti, G., Grasso, F., and Rimini, E. (1975). *Thin Solid Films* **25**, 431.
Chu, W. K. (1976). *Phys. Rev. A* **13**, 2057.
Chu, W. K., Mayer, J. W., Nicolet, M-A., Buck, T. M., Amsel, G., and Eisen, F. H. (1973). *Thin Solid Films* **17**, 1.
Cohen, B. L., Fink, C. L., and Degnan, J. H. (1972). *J. Appl. Phys.* **43**, 19.
Cookson, J. A., and Pilling, F. D. (1973). *Thin Solid Films* **19**, 381.
Feldman, L. C., and Picraux, S. T. (1977). *In* "Ion Beam Handbook for Material Analysis" (J. W. Mayer and E. Rimini, eds.), Chapter 4. (Academic Press, New York.)
Feuerstein, A., Grahmann, H., Kalbitzer, S., and Oetzmann, H. (1976). *In* "Ion Beam Surface Layer Analysis" (O. Meyer, G. Linker, and F. Käppeler, eds.), p. 471. Plenum Press, New York.
Gamo, K., Inada, T., Samid, I., Lee, C. P., and Mayer, J. W. (1976). *In* "Ion Beam Surface Layer Analysis" (O. Meyer, G. Linker, and F. Käppeler, eds.), p. 375. Plenum Press, New York.
Harris, J. M., Chu, W. K., and Nicolet, M-A. (1973). *Thin Solid Films* **19**, 259.
Hart, R. R., Dunlap, H. L., Mohr, A. J., and Marsh, O. J. (1973). *Thin Solid Films* **19**, 137.
Hoffman, G. E., and Powers, D. (1976). *Phys. Rev. A* **13**, 2042.
Honig, R. E., and Harrington, W. L. (1974). *In* "Ion Beam Surface Layer Analysis" (J. W. Mayer and J. F. Ziegler, eds.), p. 43. Elsevier, Amsterdam.
Jarvis, O. N., and Sherwood, A. C. (1974). *Nucl. Instrum. Methods* **115**, 271.
Kräutle, H., Nicolet, M-A., and Mayer, J. W. (1974). *J. Appl. Phys.* **45**, 3304.
Lanford, W. A., Trautvetter, H. P., Ziegler, J. F., and Keller, J. (1976). *Appl. Phys. Lett.* **28**, 566.
Langley, R. A. (1976). *In* "Ion Beam Surface Layer Analysis" (O. Meyer, G. Linker, and F. Käppeler, eds.), p. 201. Plenum Press, New York.
Langley, R. A., Picraux, S. T., and Vook, F. L. (1974). *J. Nucl. Mater.* **53**, 257.
L'Ecuyer, J., Brassard, C., Cardinal, C., Deschênes, L., Jutras, Y., and Labrie, J. P. (to be published).
Leich, D. A., and Tombrello, T. A. (1973). *Nucl. Instrum. Methods* **108**, 67.
Ligeon, E., and Guivarc'h, A. (1976). *Radiat. Effects* **27**, 129.
Luomajarvi, M., Fontell, A., and Bister, M. (1976). *In* "Ion Beam Surface Layer Analysis" (O. Meyer, G. Linker, and F. Käppeler, eds.), p. 75. Plenum Press, New York.
Mayer, J. W., and Rimini, E. (eds.) (1977). "Ion Beam Handbook for Material Analysis." Academic Press, New York.
Mayer, J. W., and Ziegler, J. F. (eds.) (1974). "Ion Beam Surface Layer Analysis." Elsevier Sequoia, Lausanne.
Mayer, J. W., Ziegler, J. F., Chang, L. L., Tsu, R., and Esaki, L. (1973). *J. Appl. Phys.* **44**, 2322.
Meyer, O., Linker, G., and Käppeler, F. (eds.) (1976). "Ion Beam Surface Layer Analysis." Plenum Press, New York.
Mezey, G., Gyulai, J., Nagy, T., Kotai, E., and Manuaba, A. *In* "Ion Beam Surface Layer Analysis" (O. Meyer, G. Linker, and F. Käppeler, eds.), p. 303. Plenum Press, New York.
Mitchell, I. V., Kamoshida, M., and Mayer, J. W. (1971). *J. Appl. Phys.* **42**, 4378.
Moore, J. A., Mitchell, I. V., Hollis, M. J., Davies, J. A., and Howe, L. M. (1975). *J. Appl. Phys.* **46**, 52.
Morgan, D. V., and Wood, D. R. (1973). *Proc. Roy. Soc. London A* **335**, 509.
Nakamura, K., Nicolet, M-A., Mayer, J. W., Blattner, R. J., and Evans, C. A. (1975). *J. Appl. Phys.* **46**, 4678.
Pabst, W. (1976). *In* "Ion Beam Surface Layer Analysis" (O. Meyer, G. Linker, and F. Käppeler, eds.), p. 211. Plenum Press, New York.

Petersson, S., Tove, P. A., Meyer, O., Sundqvist, B., and Johansson, A. (1973). *Thin Solid Films* **19**, 157.

Pierce, T. B. (1974). *In* "Characterization of Solid Surfaces" (P. F. Kane and G. B. Larrabee, eds.), p. 419. Plenum Press, New York.

Sherzer, B. M. U., Børgesen, P., Nicolet, M-A., and Mayer, J. W. (1976). *In* "Ion Beam Surface Layer Analysis" (O. Meyer, G. Linker, and K. Käppeler, eds.), p. 33. Plenum Press, New York.

van Wijngaarden, A., Miremadi, B., and Baylis, W. E. (1971). *Can. J. Phys.* **49**, 2440.

Williams, J. S. (1975). *Nucl. Instrum. Methods* **126**, 205.

Williams, J. S. (1976). *In* "Ion Beam Surface Layer Analysis" (O. Meyer, G. Linker, and F. Käppeler, eds.), p. 223. Plenum Press, New York.

Ziegler, J. F. (ed.) (1975). "New Uses of Low Energy Accelerators." Plenum Press, New York.

Chapter

# *8*

# Use of Channeling Techniques

## 8.1 INTRODUCTION

In the preceding chapters, we treated backscattering analysis on the basis that the targets were either amorphous or composed of randomly oriented polycrystallites. That approach ignores one of the important effects available in particle–solid interactions namely, the perception of structural and crystalline order by use of channeling effects. The channeling effect arises because rows or planes of atoms can "steer" energetic ions by means of a correlated series of gentle, small-angle collisions. In terms of backscattering spectrometry, channeling effects produce strikingly large changes in the yield of backscattered particles as the orientation of the single-crystalline target is changed with respect to the incident beam. Channeling effect measurements have had three major applications in backscattering analysis: (1) amount and depth distribution of lattice disorder, (2) location of impurity atoms in the lattice sites, and (3) composition and thickness of amorphous surface layers.

In this chapter, we shall concentrate on these applications rather than develop the theory and basic experimental data that have led to the present understanding of channeling phenomena. Channeling of energetic ions is well described (Morgan, 1973; Gemmell, 1974; Dearnaley et al., 1973; Picraux, 1975; Mayer et al., 1970), and even a cursory glance at a model of lattice atoms, such as that in Fig. 8.1, would suggest that pronounced effects would occur when the crystal orientation is shifted with respect to the direction of the incident beam. Indeed, there can be a hundred-fold decrease in the number of backscattered particles when the crystal is rotated so that the beam is incident along axial directions rather than viewing the crystal as a random collection of atoms. Aligning the beam with planar directions also causes a decrease in yield. In both cases particles that are steered along the "channels" in the crystal do not approach the lattice atoms in the axial rows and planes closely enough to undergo the wide-angle elastic scattering processes.

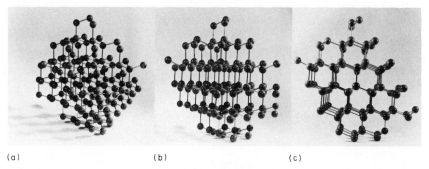

(a)                                    (b)                                    (c)

**Fig. 8.1**   Model of lattice atoms showing the atomic configuration in the diamond-type lattice viewed along (a) random, (b) planar, or (c) axial directions.

From an analytical viewpoint, then, the channeled component of the beam acts as a probe to detect atoms, either host or impurity, that have been displaced from substitutional lattice sites by distances exceeding about 0.1 to 0.2 Å. Using the same analytical concepts developed in earlier chapters, one can determine the number of displaced host atoms (the amount of disorder), the number of nonsubstitutional impurity atoms, and their depth distribution.

Several effects can complicate the analysis. We have assumed either that the atoms are on well-defined substitutional lattice sites or they are displaced well away from these sites. In a number of cases, however, the atoms are displaced only by small distances from their normal sites. Further, we have implicitly assumed that the channeled beam is distributed uniformly across the channels between the axial rows or planes. This is not true, as it has been shown that the beam can be concentrated or focused near the center of the channels. This can lead to substantial changes in the distribution of beam particles (called "flux peaking") across the channels. We have also assumed that the radiation damage produced by the analysis beam does not influence the lattice location of either host or impurity atoms. Again, it has been found that this assumption does not necessarily hold.

In spite of these possible complications, the channeling effect is a remarkably powerful analytical technique. It provides simple and direct values for the amount of lattice disorder and the number of nonsubstitutional atoms in many situations. There are also standard methods that allow one to determine if any of the difficulties previously listed play a role in the analysis.

Our approach in the remainder of the chapter shall be to start with the simplest system, the perfect crystal with substitutional impurity atoms, and give without derivation the equations used to characterize axial channeling. Then we shall describe the other simple case of an amorphous layer on a single crystal. From these examples we shall progress to more complex cases where, for example, dechanneling and flux peaking must be considered in some detail. Our intention is to discuss topics that involve evaluation of crystal structures rather than those that deal with channeling as a subject in itself.

## 8.2   CRYSTAL ALIGNMENT PROCEDURES

The channeling effect itself provides a simple method to orient the crystal axes or planes with respect to the beam. This alignment procedure is always the first step in experiments designed to determine either lattice site location of impurities or the amount of lattice disorder.

To develop the concept, we use the models of the crystal lattice in Fig. 8.1 and visualize the samples shown in Fig. 8.2 as being composed of either (a) a random arrangement of atoms (an amorphous solid), (b) a crystal composed of a set of planes randomly occupied by atoms and parallel to each other (planar sample), or (c) a crystal composed of two sets of planes at right angles to each other forming a cubic crystal with rows of atoms along the lines of intersection of the planes (axial sample). As shown in Fig. 8.2a, if a beam of energetic ions is incident on the amorphous sample, the yield of scattered particles will remain constant as the orientation of the sample is tilted and rotated with respect to the beam. In this figure we show schematically the backscattered yield from the surface layers of the sample. The scattering yield can be calculated directly from the equations given in Chapter 3.

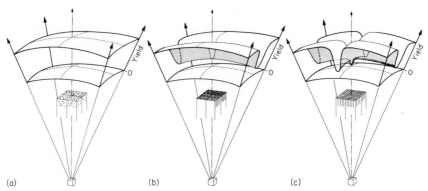

**Fig. 8.2** Backscattering yield from the near-surface layers of (a) amorphous, (b) planar, and (c) cubic crystals as a function of the orientation of the sample with respect to a collimated beam of megaelectron volt $^4$He ions. The inserts under the angular yield profiles give a magnified view of the crystal structure.

For the "crystal" of stacked planes, the scattering yield decreases markedly as the crystal is tilted so that the beam is aligned with the planes (Fig. 8.2b). In an effect similar to the optical view shown in Fig. 8.1b, the particles in the beam can make close impact collisions only with the outermost atoms in the planes. The atoms lying below the surface are shadowed by the outermost atoms. This planar channeling effect produces a decrease in the yield of backscattered particles. In the angular yield profile shown in Fig. 8.2b, the presence of planar channeling is evidenced by the trough that lies parallel to the intersection of planes with the sample surface. Where there are two sets of planes at right angles to each other, the two troughs in the backscattering yield are at right angles to each other (Fig. 8.2c). When the beam is aligned with the crystal at the intersection of the two troughs, the yield decreases further as the crystal appears as rows of atoms with only the uppermost atoms in the row "visible" to the beam. At this alignment, one refers to axial channeling as distinct from planar channeling.

The alignment of a crystal with respect to the beam is based on the existence of well-defined planar minima in the yield of backscattered particles at beam-to-substrate directions that correspond directly with the planes in the crystal. One method for carrying out the alignment procedure is to tilt the crystal with respect to the beam direction. In Fig. 8.3 the beam is incident on a crystal whose axis is tilted $\theta$ degrees away from the beam direction. As the crystal is rotated around its axis there are minima in the backscattering yield at angular positions of the rotation where the planes are aligned with the beam. For a crystal with two planes (Fig. 8.3), there will be four planar minima in 360° of rotation. The direction of the beam with respect to the crystal axis can be found by plotting the angular positions of the minima

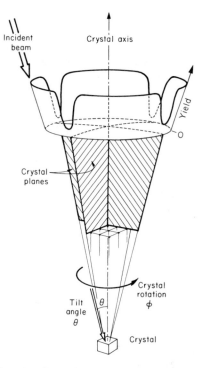

**Fig. 8.3**  Backscattering yield as a function of rotation angle for a crystal whose axis is tilted $\theta$ degrees from the direction of the incident beam of megaelectron volt He ions. The minima in the backscattering yield occur at rotation angles $\phi$ where the beam is aligned with planar directions in the crystal.

on polar coordinate graph paper and finding the tilt and rotation coordinates ($\theta$ and $\phi$) at the intersection of lines connecting the minima.

We next describe how to align a wafer of single-crystal silicon (cubic diamond structure) that is cut so that the $\langle 110 \rangle$ axis[†] is nearly normal to the sample surface. The crystal is mounted on a goniometer whose axis of rotation is tilted by 6° from incident beam direction as shown schematically in Fig. 8.4a. As the sample is rotated, the backscattering yield is recorded. Part of the yield versus rotation angle plot is shown in Fig. 8.4b. The angular positions of the most pronounced minima, near 40, 90, and 120°, are recorded on polar coordinates as in Fig. 8.4c. Around the $\langle 110 \rangle$, there will be eight minima. The rotation is then made at a tilt angle of 5°, and the positions of the eight minima on the polar coordinates are recorded again. The lines connecting the minima then correspond to the $\{100\}$, $\{110\}$, and two $\{111\}$ planes. The intersection of these lines correspond to the $\langle 110 \rangle$ axial direction. The coordinates of the point of intersection give the goniometer position that will line up the $\langle 110 \rangle$ axis with the beam. In this case the coordinates are $\theta = 0.9$ and $\phi = 60°$.

[†] We follow the convention that the symbols $\langle hkl \rangle$ and $\{hkl\}$ refer to the family of axes and planes designated by the Miller indices $h$, $k$, and $l$.

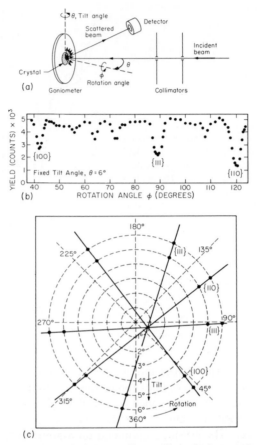

**Fig. 8.4**   Use of channeling techniques to align a silicon single crystal with respect to the direction of the incident beam of 2-MeV $^4$He ions: (a) schematic drawing of the scattering geometry; the tilt axis is perpendicular to the incident beam, and the detector lies in the plane defined by these two intersecting lines; (b) scattering yield from the surface of the crystal as a function of rotation angle $\phi$ for part of the 360° of rotation; and (c) location of the planar minima on polar coordinates for tilt angles of 5 and 6°. The solid lines in (c) correspond to the crystal planes of the sample, which in this case is prepared so that the $\langle 110 \rangle$ axis is nearly normal to the sample surface.

The number and orientation of the planes around each axis in such a polar diagram are determined by the identity of the axis. The three parts of Fig. 8.5 show the major planes around the three principal axes in cubic crystal structures. The most pronounced planar minima for the diamond structure are the $\{110\}$, as shown in Fig. 8.4b; for the face-centered cubic, the $\{111\}$ planar minima are the most pronounced. Table 8.1 lists the planar and axial minima in order of decreasing magnitude for the major orientations in cubic structures, i.e., from strong to weak decreases in the yield. Sterographic

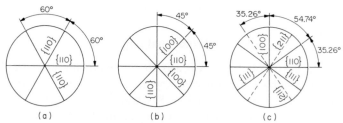

**Fig. 8.5** Orientation of the major planes around the three principal low-index axes in cubic crystals: (a) ⟨111⟩; (b) ⟨100⟩; and (c) ⟨110⟩. Around the ⟨110⟩ axis, the {211} planar minima are pronounced in body-centered cubic crystals but are not as great as {111} minima in diamond or face-centered cubic structures. [From Appleton and Foti, (1977).]

TABLE 8.1

Planar and Axial Indices for Cubic Structures, Listed in Order of Decreasing Magnitude of the Amount of Reduction in Yield as the Crystal is Tilted from "Random" to "Aligned" Orientations[a]

|  | Diamond | FCC | BCC |  | Diamond | FCC | BCC |
|---|---|---|---|---|---|---|---|
| Planes {hkl} | 110 | 111 | 110 | Axes ⟨hkl⟩ | 110 | 110 | 111 |
|  | 111 | 100 | 100 |  | 111 | 100 | 100 |
|  | 100 | 110 | 112 |  | 100 | 111 | 110 |

[a] Data taken from blocking patterns shown by Barrett *et al.* (1968).

projections and tables of angles between planes are given by Appleton and Foti (1977).

At this point we note that although orienting the crystal to obtain spectra for axial–aligned crystals is a straightforward matter orienting it so that channeling effects are excluded is not a trivial task. Backscattering spectra that coincide with those from an amorphous sample are hard to obtain. They can be obtained by orienting the crystal so that the incident beam is incident along a direction that does not coincide with major crystallographic axes or planes. The choice of such a direction depends on the crystal symmetry around each axis or plane. An alternative procedure that is frequently used is to tilt the crystal so that the incident beam is well away from the crystallographic axis and then continuously rotate the crystal while the backscattering spectrum is acquired. Spectra obtained by either orienting the crystal so that channeling effects are minimized or rotating the crystal are called "random" since the beam "sees" a random arrangement of atoms (Fig. 8.1a).

In this chapter we follow the common practice of referring to "aligned" and "random" spectra rather than, for example, "spectra obtained with the incident beam aligned with a given crystallographic axis." This notation is a convenient short form for descriptions of channeling measurements.

## 8.3   PERFECT CRYSTAL

### 8.3.1   Substitutional Impurities

One of the most straightforward aspects of channeling measurements is the analysis of single-crystal substrates with substitutional impurity atoms. The basic concept is outlined in Fig. 8.6, which shows schematically the interactions that take place when an incident beam of megaelectron volt $^4$He ions is aligned with a low-index axial direction of the crystal. When the beam is aligned with an axial direction of a single-crystal substrate, between 95 and 98% of the incident particles can be steered or channeled after entering the crystal. Of course a small fraction (2–5%) of the particles will be incident close enough to the outermost atoms in the axial rows that they undergo wide-angle scattering collisions and hence travel through the crystal without being steered or channeled. The magnitude of this "random" component of

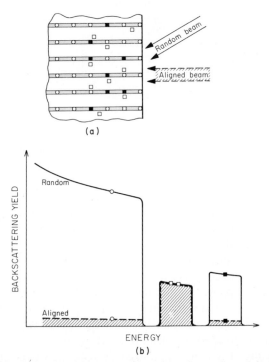

**Fig. 8.6** (a) Schematic and (b) backscattering spectra for random and aligned beams for megaelectron volt $^4$He ions incident on a crystal containing heavy impurities in the near-surface region that are on substitutional sites (■) and interstitial sites (□) at random positions in the lattice (○, host lattice atoms). The shaded part represents the magnitude of the yield in the aligned spectrum.

the aligned beam can be estimated for single crystals (Section 8.3.2) but does depend on surface conditions. The presence of a thin oxide or hydrocarbon layer or surface disorder can cause an increase in the random component of the aligned beam. Conversely, of course, channeling measurements can be used to evaluate the condition of the surface.

The aligned or channeled component of the beam represents the condition in which the particles in the beam are steered without approaching the axial rows closer than about 0.1–0.2 Å. This distance is orders of magnitude greater than the impact parameter for wide-angle elastic scattering. Consequently, one can consider the crystal atoms (open circles in Fig. 8.6) on lattice sites below the surface as "invisible" to the channeled particles from the standpoint of backscattering spectrometry. Therefore, as the beam-to-crystal orientation is changed from random to aligned directions, the yield of backscattered particles from the host crystal atoms will decrease by nearly two orders of magnitude, as is shown schematically by the dashed line in Fig. 8.6b.

Substitutional impurity atoms (filled squares in Fig. 8.6) too are shielded from direct interactions with the channeled component of the beam. The steering of the channeled particles is established by the host crystal lattice atoms, and hence the host atoms in effect shield the substitutional impurity atoms. Consequently, the backscattering yield from substitutional impurity atoms will exhibit the same decrease in yield as that from the host atoms when the crystal is shifted from random to aligned orientations with respect to the beam. The signals for substitutional impurities located in the near-surface region under random and aligned beam conditions are shown in Fig. 8.6b.

Not all the impurity atoms will be found in well-defined substitutional lattice positions. For the extreme case, all the impurities (denoted by the open squares in Fig. 8.6a and b) are displaced well away from lattice sites, at random positions within the lattice structure. For this case, there is no change in the yield from impurity atoms as the beam-to-substrate orientation is changed. As is shown in Fig. 8.6b, the spectra for such impurities coincide for both aligned and random beam-to-crystal orientations.

There is a dangerous oversimplification inherent in this description. We have implicitly assumed that the impurities are substitutional or else are located well away from lattice sites. However, there are well-documented cases in which the impurity atoms are displaced slightly (0.1–0.3 Å) from a substitutional lattice site, or occupy well-defined interstitial sites.

To determine if the impurities are truly on substitutional or random lattice sites, it is generally necessary to measure the backscattering yield along different axial directions and as a function of tilt angle through certain axial directions (Section 8.5).

## 8.3.2   Axial Half-Angle $\psi_{1/2}$ and Minimum Yield

In this section, we rely on the extensive literature on channeling measurements (Lindhard, 1965; Gemmell, 1974; Morgan, 1973) and list the major equations without derivation. Calculated values of critical angles for channeling and the magnitude of the axial minima are given so that they can be compared with experimental values for evaluation of the quality of the crystal. Equations and numerical values are also given by Appleton and Foti (1977).

Schematic backscattering yields as a function of tilt angle are shown in Fig. 8.7a for megaelectron volt $^4$He ions incident on a single crystal. Backscattering spectra for a beam incident along a low-index crystallographic axis (aligned spectrum) and along a nonchanneling direction (random spectrum) are shown in Fig. 8.7b. The ratio $H_A/H$ of the heights of two spectra taken in the near-surface region for aligned and random orientation is referred to as the *minimum yield* $\chi_{min}$ (Fig. 8.7c). The heights are measured

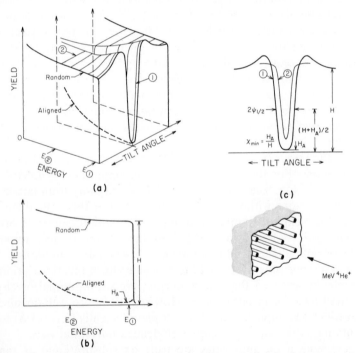

**Fig. 8.7**  Channeling measurements with megaelectron volt $^4$He ions incident on a crystal along a major crystallographic axial direction: (a) Schematic representation of backscattering spectra measured as a function of tilt angle, (b) random and aligned spectra, and (c) angular yield profiles measured in the near-surface region and at depth.

in a narrow energy region at energy $E_①$, just below the edge of the substrate signal. In the near-surface region of a perfect crystal, the values for the minimum yield lie in the range of 1 to $3 \times 10^{-2}$ for low-index axial directions. Below the surface, the aligned yield (dashed line, Fig. 8.7b) increases with increasing depth within the crystal as the channeled particles are scattered into the random component of the beam (dechanneled). To estimate the substitutional concentration of impurity atoms near the surface, one can generally neglect the influence of dechanneling.

Aside from the minimum yield, the other parameter that can be measured is the *half-angle* $\psi_{1/2}$ for axial channeling. To obtain the value of $\psi_{1/2}$, one measures the yield in a narrow energy window as a function of tilt angle from an axial direction, but taking care to avoid a tilt along a planar channel. The angular yield profile for such a series of measurements is shown in Fig. 8.7c for energy windows at $E_①$ and $E_②$. As the crystal orientation is tilted toward the axial alignment, the yield rises to a maximum value and then decreases to its minimum value. The maximum yield, in the shoulders of the angular yield profile, occurs when the beam is oriented at an angle just slightly larger than that required to steer or channel the particles. Under this geometrical configuration, the incident particles can undergo more close-encounter wide-angle scattering interactions with lattice atoms than under conditions of random incidence. The half-angle $\psi_{1/2}$ is the angular half-width of the yield profile at the yield value halfway between the minimum yield $\chi_{min}$ and the yield for random incidence. For megaelectron volt $^4$He ions incident along low-index crystallographic directions, the axial half-angle values lie in the range of 0.4 to 1.2°. As in the case of the minimum yield, the half-angle is depth-dependent and is usually determined for energies corresponding to near-surface regions (see Fig. 8.7b and 8.7c).

The values of the axial half-angle can be calculated directly by the procedure of Barrett (1971), or estimated rather closely by using the continuum model of Lindhard (1965). The basic concept is that when the angle of incidence of an energetic particle with an axial row exceeds a certain critical angle, the particle is no longer steered by a series of correlated collisions, but rather "sees" the lattice atoms as individual scattering centers. Lindhard introduced the *characteristic angle* $\psi_1$ to describe the case in which the distance of closest approach of a particle of energy $E$ to the center of an axial row of atoms with spacing $d$ is approximately the Thomas–Fermi screening distance $a$. The expression for $\psi_1$ is

$$\psi_1 = (2Z_1 Z_2 e^2 / Ed)^{1/2} = 0.307(Z_1 Z_2 / Ed)^{1/2} \quad \text{degrees}, \quad (8.1)$$

where $E$ is the incident energy in megaelectron volts $d$ the atomic spacing in angstroms along the axial direction, and $Z_1$ and $Z_2$ the atomic numbers of the projectile and target atoms, respectively. Table 8.2 gives values by which

**TABLE 8.2**

Values by Which the Lattice Constant $d_0$ Must Be Multiplied
to Compute the Interatomic Spacings $d$ in Axial Directions[a]

| Structure | Atoms per unit call | Axial directions | | |
|---|---|---|---|---|
| | | $\langle 100 \rangle$ | $\langle 110 \rangle$ | $\langle 111 \rangle$ |
| bcc | 2 | 1 | $\sqrt{2}$ | $\sqrt{3}/2$ |
| fcc | 4 | 1 | $1/\sqrt{2}$ | $\sqrt{3}$ |
| fcc (diamond) | 8 | 1 | $1/\sqrt{2}$ | $\sqrt{3}/4, 3\sqrt{3}/4$ |

[a] Taken from Gemmel (1974).

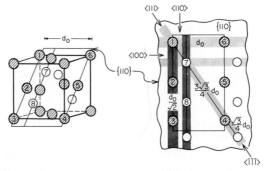

**Fig. 8.8**  Configuration for the diatomic lattice of the zincblende structure (fcc, diamond) showing the atomic positions on the {110} plane. The shading along the rows of atoms indicates the three different principal axial directions. The numbers indicate the atomic spacing in terms of the lattice constant $d_0$.

the lattice constant $d_0$ of cubic crystals must be multiplied to compute $d$. For diatomic lattices (Fig. 8.8 for the fcc, diamond structure) one uses average values for $Z_2$ and $d$ along rows of mixed atoms such as $\langle 111 \rangle$ in Fig. 8.8. Along the $\langle 100 \rangle$ and $\langle 110 \rangle$, each of the monoatomic rows has a separate critical angle. The *critical angle* $\psi_{1/2}$ for channeling is related to the *characteristic angle* $\psi_1$ by

$$\psi_{1/2} = \alpha\psi_1,  \tag{8.2}$$

where values of $\alpha$ are generally between 0.8 and 1.2, depending on the vibrational amplitude of the lattice atoms. In experimental work, the critical angle $\psi_{1/2}$ is also referred to as the *half-angle* or the *half-width* $\psi_{1/2}$.

In the Monte Carlo calculations of Barrett (1971) one considers the average interatomic potential $V_{RS}(\rho)$ for an ion moving at a distance $\rho$ with a trajectory of angle $\psi_0$ to an isolated atomic row. As the value of $\psi_0$ increases, the distance of closest approach decreases, and at a certain angle $\psi_c$ it equals the critical distance $\rho_c$ for sustaining a stable channeling trajectory:

$$E\psi_c{}^2 = V_{RS}(\rho_c).  \tag{8.3}$$

Aside from numerical constants of order unity, the problem is then to determine $V_{RS}$ and $\rho_c$.

In Barrett's treatment the half-width $\psi_{1/2}$ of the axial dip taken halfway between the minimum and random yield is given in radians by

$$\psi_{1/2} = 0.80[V_{RS}(1.2u_1)/E]^{1/2},  \tag{8.4}$$

where the minimum distance $\rho_c$ of closest approach for the trajectory of channeled particle is taken at $1.2u_1$ and $u_1$ is the one-dimensional rms thermal amplitude. The factor 0.80 is the ratio of the half-width $\psi_{1/2}$ to the angle $\psi_c$.

The expression for the half-angle can be written in degrees as

$$\psi_{1/2} = 0.8F_{RS}(\xi)\psi_1  \tag{8.5}$$

where values of $F_{RS}$ versus $\xi$ are given in Fig. 8.9, and the normalized distance $\xi$ of closest approach is given by

$$\xi = 1.2u_1/a,  \tag{8.6}$$

where the Thomas–Fermi screening radius $a$ is given by[†]

$$a = 0.8853a_0(Z_1^{1/2} + Z_2^{1/2})^{-2/3}  \tag{8.7}$$

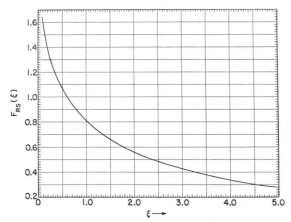

**Fig. 8.9**  The value of $F_{RS}$ as a function of $\xi$, the normalized distance of closest approach. [From Barrett (1971).]

---

[†] We follow the usage of Barrett and Gemmell. Alternative functional forms (Lindhard) give $a = 0.885a_0(Z_1^{2/3} + Z_2^{2/3})^{-1/2}$.

for the Bohr radius, $a_0 = 0.528$ Å. Hence

$$a = 0.4685 \text{ Å}/(Z_1^{1/2} + Z_2^{1/2})^{2/3} \approx 0.47/Z_2^{1/3} \quad \text{Å}. \tag{8.8}$$

Values typically range between 0.1 and 0.2.

The value of the one-dimensional rms thermal vibration amplitude $u_1$ is computed from the Debye theory of thermal vibrations:

$$u_1 = 12.1 \left[ \left( \frac{\phi(x)}{x} + \frac{1}{4} \right) \Big/ M_2 \theta_D \right]^{1/2} \quad \text{Å}, \tag{8.9}$$

where $M_2$ is the atomic weight in atomic mass units, $\theta_D$ the Debye temperature (in degrees Kelvin), $T$ the crystal temperature (in degrees Kelvin), $x = \theta_D/T$, and $\phi(x)$ the Debye function $x^{-1} \int_0^x t(e^t - 1)^{-1} dt$ which is tabulated by Appleton and Foti (1977). Values of $u_1$ generally range from 0.05 to 0.1. Table 8.3 lists values of $u_1$ for some selected crystals along with a comparison of calculated and measured values of $\psi_{1/2}$ for axial channeling. As can be seen, the agreement between the two values is quite good.

The value for the minimum yield $\chi_{\min}$ is determined by particles entering the crystal close enough to the atomic row to be scattered by the outermost atoms to angles greater than $\psi_{1/2}$. Lindhard has estimated the value of the minimum yield as

$$\chi_{\min} = N d\pi (2u_1^2 + a^2), \tag{8.10}$$

**TABLE 8.3**

Some Lattice Constants and Measured Critical Angles[a]

| Name structure | $Z_2$ | $M_2$ | $N$ (atom/cm³) $\times 10^{22}$ | $a$ (Å) ($=0.47Z_2^{-1/3}$) | $\theta_D$ (K) | $u_1$ (Å) (293 K) | $d_0$ (Å) | $\psi_{1/2}$ (Å)[b] Calculated [Eq. (8.5)] | Measured |
|---|---|---|---|---|---|---|---|---|---|
| C fcc (dia) | 6 | 12.01 | 1.13 | 0.258 | 2000 | 0.04 | 3.567 | 0.75 | 0.75 |
| Al fcc | 13 | 26.98 | 6.02 | 0.199 | 390 | 0.105 | 4.050 | 0.45[c] | 0.4[c] |
| Si fcc (dia) | 14 | 28.09 | 4.99 | 0.194 | 543 | 0.075 | 5.431 | 0.73 | 0.75 |
| Ge fcc (dia) | 32 | 72.59 | 4.42 | 0.148 | 290 | 0.085 | 5.657 | 0.93 | 0.95 |
| W bcc | 74 | 183.85 | 6.32 | 0.112 | 310 | 0.050 | 3.165 | 0.83[d] | 0.85[d] |

[a] Taken from Gemmel (1974) and Mayer and Rimini (1977).
[b] Values are for 1.0-MeV He along ⟨110⟩ unless otherwise specified.
[c] For 1.4-MeV He along ⟨110⟩.
[d] For 3.0-MeV H along ⟨111⟩.

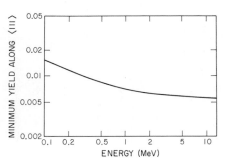

**Fig. 8.10**   Calculated value of the minimum yield from Eq. (8.11) for protons in W with $T = 298$ K. [Adapted from Barrett, (1971).]

where $N$ is the number of atoms per unit volume. A more accurate value, obtained by an empirical fit to computer calculation (Barrett), gives an energy dependence as

$$\chi_{min} = 18.8 N d u_1^{2}(1 + \zeta^{-2})^{1/2} \tag{8.11}$$

where $\zeta$ can be expressed as

$$\zeta \simeq 126 u_1/\psi_{1/2} d \tag{8.12}$$

and $\psi_{1/2}$ is expressed in degrees. The dependence of $\chi_{min}$ on incident energy $E_0$ is shown in Fig. 8.10 for protons in W. As the particle energy increases, the minimum yield decreases. Because the measured value of $\chi_{min}$ depends on surface preparation, the variation of $\chi_{min}$ with energy for a wide variety of crystals, has not been systematically studied.

If channeling measurements were performed on a perfect crystal at 0 K, the spectra would exhibit a peak at energies corresponding to scattering from the surface atoms as shown in Fig. 8.11. The outermost atoms are visible to the beam, and the area of the peak is equivalent to $Nd$ atoms/cm² (the number of atoms per square centimeter on the outer row). Even in the perfect crystal, the surface peak does not arise from scattering only from the outermost atoms. Thermal vibrations of the lattice atoms lead to contributions from the underlying atoms—contributions that can be larger than those of just the surface atoms (Fig. 8.11). The effective number $L$ of surface layers contributing to the surface peak will equal unity in the absence of thermal vibrations at 0 K and will depend on the thermal vibration amplitude $u_1$ at higher temperatures. To calculate the effective number $L$ of surface layers, we follow the prescription of Barrett (1971):

$$L = (1 + \zeta^{2})^{1/2}, \tag{8.13}$$

where $\zeta^2$ is proportional to $E/d$ from Eq. (8.12). A value for the case of He in Si at 20°C, with parameters given in Table 8.3, is $L = (1 + 43.9E/d)^{1/2}$, where

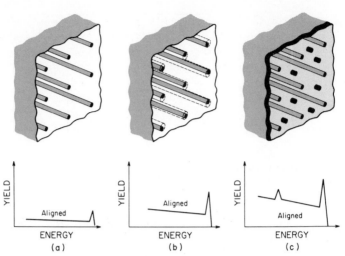

**Fig. 8.11**   Aligned spectra showing surface peaks for (a) perfect crystals at 0 K, (b) an elevated temperature, and (c) with a thin oxide layer with $T = 0$ K. In (b), when thermal vibration of the lattice is present, the surface peak arises from scattering from outermost atoms as well as contributions from the next nearest underlying atoms. The cylinders represent the shadowing of the underlying row of atoms by the surface atoms. The dashed cylinder shows the extent of the thermal vibration of individual atoms in the rows.

$E$ is the energy in megaelectron volts and $d$ the spacing in angstroms. For 2-MeV He on $\langle 100 \rangle$ Si, $L$ has a value of 4.15. This means, in effect, that there are 3.15 atoms below the surface layer that would contribute to the surface peak. One surface atom per row on $\langle 100 \rangle$–oriented Si corresponds to $2.72 \times 10^{15}$ Si atoms/cm$^2$.

There is some difficulty in obtaining the predicted values in Eq. (8.13) in real crystals. The value of $L$ is very sensitive to the presence of oxide layers or surface disorder as shown in Fig. 8.11. At this stage, we can only suggest that one use Barrett's formulation as a guideline. It is certainly possible to determine relative changes in the number of surface atoms. This procedure has been followed in evaluating the surface condition of compound semiconductors (Morgan and Wood, 1973; Morgan and Bøgh, 1972).

### 8.3.3   Planar Channeling

In comparison with axial channeling, planar channeling is characterized by narrower critical angles and higher minimum yields (Gemmell, 1974; Morgan, 1973; Appleton and Foti, 1977). Planar channels have not been used extensively to evaluate disorder profiles or to determine the lattice location of impurities. In special cases in which impurities are located on well-defined interstitial sites, scans across planes can give information on the site location (Davies, 1973; Picraux, 1975). In routine evaluation of samples, however, planar channeling effects are used primarily in the alignment procedures to determine the axial directions as explained in Section 8.2. Consequently, we have not emphasized planar channeling.

### 8.4   LATTICE DISORDER, AMORPHOUS LAYERS, AND POLYCRYSTALLINE FILMS

### 8.4.1   Introduction

Channeling effect measurements can also be used to determine the number of host atoms displaced from their crystal lattice sites. Under channeling conditions the aligned component of the beam can interact with displaced atoms in both wide-angle collisions and forward elastic scattering events, as is shown schematically in Fig. 8.12. The wide-angle, close-encounter collisions are those leading to the direct detection of displaced atoms through backscattering analysis. The small-angle, forward-scattering collisions can cause the channeled particles to be scattered at angles greater than the critical angle for channeling. This process is called *dechanneling*. The dechanneled particles can then interact with the nondisplaced host lattice atoms and give a much higher backscattering yield than that of the channeled particles, which are backscattered from the displaced atoms. Consequently, a proper treatment of dechanneling processes is necessary in order to extract the number of displaced atoms from the spectra. In some cases this is a simple procedure; in others it is very difficult unless one makes assumptions about the microscopic nature of the disordered region or has electron microscopy data or other detailed information about the defects.

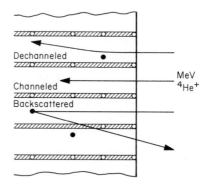

**Fig. 8.12**   Schematic representation of particles in an aligned beam interacting with displaced atoms to cause either backscattering events or forward scattering at angles outside the critical angle (dechanneling).

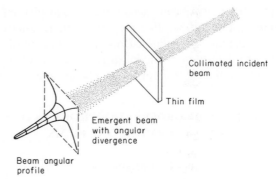

**Fig. 8.13**  Angular divergence introduced into a collimated beam by a thin film as indicated by the contour lines on the angular profile of the emergent beam.

The influence of dechanneling can be described by first considering the angular divergence in the beam caused by interposing a thin film in the path of a well collimated beam. Figure 8.13 shows the angular profile of the beam after it traverses such a film. For films a few thousand angstroms thick, most of the particles undergo some deflection in their trajectory, but are essentially moving along paths closely aligned with their initial direction.

Channeling phenomena are characterized by critical angles $\psi_{1/2}$, and we can estimate the cross section $\sigma_D(\psi_{1/2})$ for atoms in the film to deflect particles through angles equal to or greater than the critical angle. As a first approximation we can consider the film to be thin enough so that the particles, on the average, make only one forward-scattering collision (single-scattering approximation). One uses the Rutherford differential scattering cross section $d\sigma/d\Omega$ and integrates from $\psi_{1/2}$:

$$\sigma_D(\psi_{1/2}) = \int_{\psi_{1/2}}^{\infty} (d\sigma/d\Omega)\, d\Omega \qquad (8.14)$$

to obtain (Bøgh, 1968; Eisen, 1973)

$$\sigma_D(\psi_{1/2}) = \pi Z_1{}^2 Z_2{}^2 e^4 / E^2 \psi_{1/2}^2. \qquad (8.15)$$

This expression can be related to the Lindhard characteristic angle $\psi_1$ [Eq. (8.1)] to give

$$\sigma_D(\psi_{1/2}) = \tfrac{1}{4}\pi(\psi_1{}^4/\psi_{1/2}^2)\, d^2 \qquad (8.16)$$

when the film is composed of the same material as in the crystal with critical angle $\psi_{1/2}$. Since $\psi_{1/2} = 0.8 F_{RS}\psi_1$ [Eq. (8.5)], this expression can be given in numerical terms as

$$\sigma_D(\psi_{1/2}) = 3.53 \times 10^{-21}(Z_1 Z_2 d / E F_{RS}^2)\quad \text{cm}^2, \qquad (8.17)$$

where the value of the atomic spacing $d$ along an axial direction is given in angstroms and the incident energy $E$ in megaelectron volts. For megaelectron volt He ions, the value of $F_{RS}$ in crystals is about 0.9, and for lattice spacings $d$ of 4 Å, the factor $Z_1 d / F_{RS}^2$ is about 10; so the expression can be estimated by

$$\sigma_D(\psi_{1/2}) \approx 3.5 \times 10^{-20} Z_2/E \quad cm^2. \qquad (8.18)$$

The probability $P(\psi_{1/2})$ that a particle will be scattered through angles greater than $\psi_{1/2}$ in a single collision when traversing a film of $Nt$ atoms/cm$^2$ is given by

$$P(\psi_{1/2}) = \sigma_D(\psi_{1/2})Nt. \qquad (8.19)$$

If we consider a Ge film ($Z_2 = 32$) with $10^{17}$ atoms/cm$^2$ ($t \approx 200$ Å) with 1-MeV He atoms incident we obtain the values of $\sigma_D \approx 10^{-18}$ cm$^2$ and $P \approx 0.1$. In this case, 10% of the incident particles would be deflected beyond a critical angle characteristic of megaelectron volt He ions in Ge.

For channeling, the yield is extremely sensitive to small changes in the angular spread in the beam. For megaelectron volt He ions in most crystals the value of $\psi_{1/2}$ lies between 0.5 and 1.0°, and we are concerned with an angular spread in the beam of even less than $\psi_{1/2}$. We can evaluate the influence of the beam divergence by using a collimated beam and tilting the crystal as shown in the upper portion of Fig. 8.14. The lower part of Fig. 8.14 shows backscattering spectra for particles incident in random alignment and

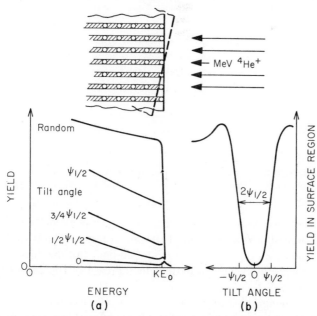

**Fig. 8.14**  Random spectrum and aligned spectra at different tilt angles (a) and the angular yield profile (b) measured from the scattering yield in the surface region of the crystal as a function of tilt angle.

for tilt angles between normal incidence (tilt angle $= 0$) and at the critical angle (tilt angle $= \psi_{1/2}$). From the backscattering spectra one can see that even if the crystal is tilted by one-half the value of $\psi_{1/2}$, there is a marked influence on the yield.

Figure 8.14b shows the yield measured at energies just below $KE_0$ (surface region of the crystal) as a function of tilt angle. This yield versus tilt angle curve is often called the *angular yield profile*. Knowing the profile of a divergent beam such as that in Fig. 8.13), one can obtain the aligned yield at normal incidence by convolution of the angular yield profile with the beam profile (Rimini *et al.*, 1972a; Lugujjo and Mayer, 1973).

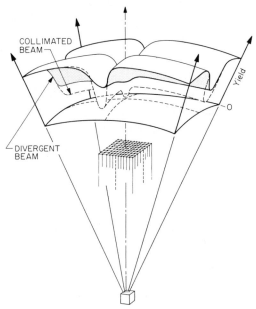

**Fig. 8.15** Backscattering yield from the near-surface region of a crystal as a function of tilt and rotation for a collimated beam (dashed line) and a beam with angular divergence (solid lines).

In Fig. 8.15, we show the influence of beam divergence on the surface yield from a crystal as a function of tilt and rotation. The main features of the planar and axial dips are preserved (solid line), but the dips are less pronounced and more smeared out than in the case of the collimated beam (dashed line).

In the following subsections we shall first discuss the simplest case of evaluation of dechanneling in a crystal covered with an amorphous layer. Then we shall treat the case of a polycrystalline film, and finally the evaluation of lattice disorder in a crystal.

### 8.4.2   Amorphous Layers on Single Crystals

When a well collimated beam of particles traverses a thin, amorphous film (or a film composed of randomly oriented polycrystallites), scattering events within the film cause a spread or divergence in the beam, as is shown schematically in Fig. 8.16. If the film is thick enough, some of the particles will undergo a sufficient number of forward scattering events so that they are deflected from the original beam directions by angles greater than $0.5-1°$, typical values of the critical angle for channeling. Consequently, if the film is deposited on a single crystal with an axial half-angle $\psi_{1/2}$, particles denoted by the shaded area in Fig. 8.16 will lie outside the critical angle for channeling. Therefore the aligned yield from the near-surface region of underlying crystal will be increased.

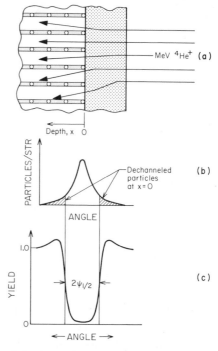

**Fig. 8.16** (a) Well collimated beam traversing an amorphous film becomes divergent. (b) Particles are dechanneled when entering the crystal at angles greater than $\psi_{1/2}$, where $\psi_{1/2}$ is determined from the angular yield profile of an uncovered single crystal. (c) Angular yield profile for megaelectron volt He ions incident on a single crystal without an amorphous layer.

Greater insight into the broadening of the beam is obtained from angular distribution profiles as shown in Fig. 8.17 (Rimini *et al.*, 1972a). The dashed lines in Fig. 8.17c are axial (left side) and planar (right side) angular distributions obtained on single-crystal Si with 1.8-MeV $^4$He ions. However, when the crystal is covered with an Al layer the normal incidence (tilt angle = 0) yield from the silicon crystal is increased and the angular distribution (solid line in Fig. 8.17c) is broadened. For the same thickness of film, the planar yield

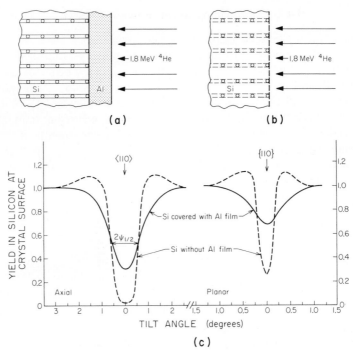

**Fig. 8.17**   Schematic representation of 1.8 MeV $^4$He incident on (a) Si crystal covered with on Al film and (b) uncovered Si crystal. (c) Axial and planar angular yield profiles obtained from the scattering yield measured in the near-surface region of Si crystals covered with an Al film (solid line) and without the film (dashed line). (Adapted from Rimini *et al.* (1972a).)

is influenced more than the axial yield because the channeling critical angle is smaller (compare axial and planar dashed lines). Both angular yield profiles are broadened because some particles are scattered into channeling directions even if the crystal is tilted by angles of more than the critical angle.

The increase in minimum yield in the crystal region immediately adjacent to an amorphous layer can be calculated to a fair degree of accuracy by determining only the number of particles scattered outside the axial half-angle $\psi_{1/2}$. This procedure, in effect, assumes a step function or square well approximation to the angular yield profile. With this approximation, particles scattered through angles greater than $\psi_{1/2}$ are dechanneled on entering the crystal. Experimentally it was found (Rimini *et al.*, 1972a; Lugujjo 1974) that axial minimum-yield values agreed within 10% with step function approximation calculations for measurements of 0.4 to 1.8-MeV $^4$He and H ions incident on Si single crystals covered with Al or Au films.

To find the value for the minimum yield, a first approximation is to use single scattering [Eq. (8.19)] to estimate the number of particles scattered at angles greater than $\psi_{1/2}$. A better procedure for films several thousand angstroms thick is to use multiple scattering calculations. One first determines the value $m$ of the reduced film thickness:

$$m = \pi a^2 Nt, \tag{8.20}$$

where $a$ is the Thomas–Fermi screening radius [Eq. (8.8)] and $Nt$ is the number of atoms per square centimeter in the film. Values of $Nt$ can be determined directly from backscattering measurements as outlined in earlier chapters. The values for the Thomas–Fermi screening radius range between 0.1 and 0.2 Å for He ions: for example, $a = 0.105$ Å for Au and $a = 0.176$ Å for Al. Physically, $m$ is the mean value of the number of collisions of the particles with the atoms in the thin film for a cross section of $\pi a^2$. A value of $m = 10$ is equivalent to an Al film thickness of 1550 Å if we use the relation that $6.02 \times 10^{17}$ Al atoms/cm$^2$ is equivalent to 1000 Å. We note that the single-scattering approximation holds for $m < 0.2$.

Next one determines the number of particles scattered by angles greater than the critical angle $\psi_{1/2}$. In this treatment we follow the approach given by Meyer (1971) and use reduced angles $\tilde{\theta}$ where

$$\tilde{\theta} = Y\theta \tag{8.21}$$

with values of the real angle $\theta$ given in radians (or degrees) and

$$Y = aE/(2Z_1 Z_2 e^2). \tag{8.22}$$

For 1.8-MeV $^4$He incident on Al, $Y = 420$ and on Au, $Y = 41.6$. Consequently, for 1.8-MeV $^4$He on $\langle 110 \rangle$ Si where $\psi_{1/2} = 0.01$ rad, the *reduced critical angle* $\tilde{\theta}_c$ equivalent to scattering at angles greater than $\psi_{1/2}$ is given by

$$\tilde{\theta}_c = Y\psi_{1/2} \tag{8.23}$$

and has values $\tilde{\theta}_c = 4.2$ for Al and 0.42 for Au films for $\psi_{1/2} = 0.01$ rad.

The value of $\tilde{\theta}_c$ can also be given by

$$\tilde{\theta}_c = 1.49 \times 10^2 F_{RS} a(E/Z_1 Z_2 d)^{1/2}, \tag{8.24}$$

where $a$ and $d$ are in angstroms and $E$ is in megaelectron volts.

To determine the number of particles scattered outside the $\psi_{1/2}$ value, one uses data from Meyer (1971) similar to that shown in Fig. 8.18. For a given value of $\tilde{\theta}_c$ and film thickness $m$, the minimum yield $\chi$ is given by

$$\chi = P(\tilde{\theta}_c, m). \tag{8.25}$$

For example, for $\tilde{\theta}_c = 1.0$ and $m = 1.0$, the value of the minimum yield is 0.2. This approach holds reasonably well for $0.2 < m < 20$.

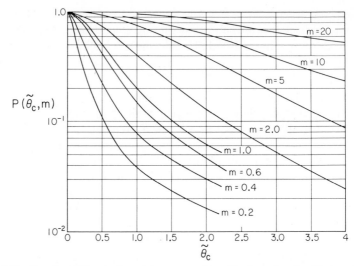

**Fig. 8.18**   Fraction $P(\theta_c, m)$ of particles scattered outside the reduced critical angle $\theta_c$ for channeling versus $\theta_c$ for various values of the reduced film thickness $m$. [From Lugujjo and Mayer (1973).]

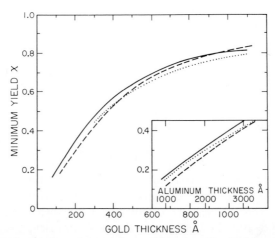

**Fig. 8.19**   Calculated values of the minimum yield versus film thickness for 1.8-MeV $^4$He ions incident along the $\langle 110 \rangle$ direction in Si covered with Al and Au. The three curves refer to different methods of determing the probability that the particles are dechanneled:—, axial scan (tilt only); – – –, step function approximation; $\cdots$ azimuthally averaged. Experimental values fall within the three curves. [From Lugujjo and Mayer (1973).]

In Fig. 8.19, values of the minimum yield at the outermost surface layer of the crystal are shown for 1.8-MeV $^4$He incident on $\langle 110 \rangle$ Si covered with Au or Al films. The dashed line refers to the step function or square well approach where the value of $\chi_{min}$ is given by the number of particles scattered beyond $\psi_{1/2}$. Shown for comparison are minimum yield values calculated from the convolution of the beam profile with the angular yield profile determined from axial scans obtained by tilting only (solid line) as shown in Fig. 8.14 and by tilting and rotating (dotted lines). The experimental yield values lie within the three curves.

As in the case of the single-scattering approximation, the minimum yield decreases with an increase in beam energy. The energy dependence of the yield arises from two factors—the critical angle $\psi_{1/2}$ and the number of particles scattered outside of a fixed angle. As the energy is increased, the beam angular distribution narrows more than the critical angle decreases. This leads to a decrease in the minimum yield.

In evaluating ion-implanted semiconductors, a common example, shown in Fig. 8.20, is the formation of an amorphous layer by the implanted ions.

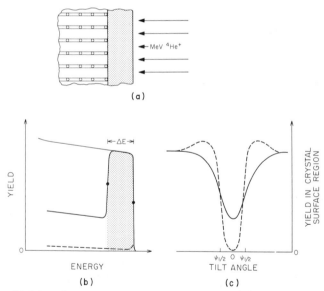

**Fig. 8.20** (a) Schematic representation of megaelectron volt $^4$He ions incident on a silicon crystal in which the outer layer has been rendered amorphous by ion implantation. (b) Random and aligned spectra for silicon containing an amorphous layer (shaded region). The thickness of the amorphous layer is taken from $\Delta E$ measured at the half-heights (●). The dashed line is for the virgin crystal before implantation. The angular yield profile (c) is measured in the surface region of the crystal (beneath the amorphous layer in the implanted case.)

The aligned spectrum (heavy line) coincides with the random one over some energy interval, in this case the near-surface region. The thickness of the amorphous layer can be found from the energy width $\Delta E$ taken from the full width at half-height (filled circles in Fig. 8.20) of the signal from the amorphous layer. The minimum yield $\chi_{min}$ in the underlying undamaged crystal is given by $P(\psi_{1/2})$ or $P(\bar{\theta}_c, m)$ in Eqs. (8.19) and (8.25), respectively. The angular yield profile (Fig. 8.20c) is similar to those shown in Fig. 8.17c.

In this discussion we have only considered the calculation of the yield $\chi_{min}$ in the outermost layers of the single crystal. It is, of course, possible to calculate the dechanneling that occurs at deeper depths within the crystal. One procedure is to assume that the beam profile is unchanged after the beam enters the crystal. The yield at any depth can be calculated from the convolution of the angular yield profile taken at that depth on an uncovered crystal with the beam profile. In effect this procedure is equivalent to that used to find the aligned yield in the surface region.

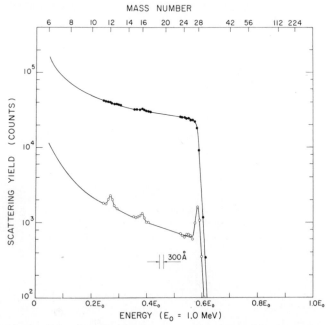

**Fig. 8.21**  Random (●) and ⟨111⟩-aligned (○) backscattering spectra for a silicon crystal with the scattering yield plotted logarithmically. A mass number scale for surface impurities is shown. The three peaks in the aligned spectrum—at masses 12, 16, and 28—indicate, respectively, the amounts of carbon, oxygen, and silicon in the surface region. [From Davies *et al.* (1967).]

One of the earliest applications of channeling effect measurements to surface analysis was the reduction of the scattering yield from the substrate. This makes it possible to detect lower amounts of surface impurities—for example, the yield from about two monolayers of oxygen in an oxide layer on Si. An example of this early work in 1967 is shown in Fig. 8.21, which gives spectra for aligned and random orientations of 1.0-MeV He ions incident on a Si single crystal. The peaks due to scattering from carbon and oxygen on the surface are clearly visible in the aligned spectrum but are buried in the random spectrum. This same technique of background reduction has also been extensively applied (Meek and Gibbon, 1974, Hart *et al.*, 1973) to improve the sensitivity for the detection of impurities with higher mass than that of the substrate.

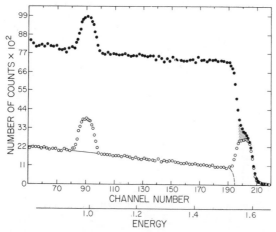

**Fig. 8.22**  Random (●) and ⟨111⟩-aligned (○) spectra for 2.8-MeV $^4$He ions incident on a Si sample covered with a 1300-Å-thick silicon oxide layer. The shaded area is the contribution of the mass 29 and 30 Si isotopes to the leading edge of the Si substrate signal. The dashed line represents the signal from the Si substrate normalized to the aligned yield at the Si/SiO$_2$ interface. [From Linker *et al.* (1973).]

The reduction of the yield from the substrate is also useful in determining the thickness and composition of surface layers (Meyer *et al.*, 1970; Mitchell *et al.*, 1971; Della Mea *et al.*, 1975a, b). An example is shown in Fig. 8.22 (Linker *et al.*, 1973) for analysis of an SiO$_2$ layer on Si with 2.8-MeV He ions. The leading edge of the random spectrum does not have a sharp step but is smeared because of the contribution (shaded area) to the leading edge from the isotopes $^{29}$Si (4.7 at.%) and $^{30}$Si (3.1 at.%). This interference makes it difficult to determine the composition of the oxide layer from the random spectrum alone. In the aligned spectrum, however, the yield from the Si in the $SiO_2$ stands out quite clearly. The composition can then be determined from the ratio of spectrum areas of the Si and O signals in the aligned spectrum. The major problem is to correct for the contribution from the aligned yield of the Si substrate.

One can extract the profile of the amorphous surface component by subtracting a normalized random spectrum from the aligned spectrum (Mitchell et al., 1971). This assumes sharp interfaces so that the intersecting yield profiles from substrate and amorphous layers are images of each other except for a scaling factor that relates to the atomic concentrations in the two media. The procedure is shown for an oxide layer on Si in Fig. 8.23. The random spectrum is normalized to the height of the aligned spectrum in the Si

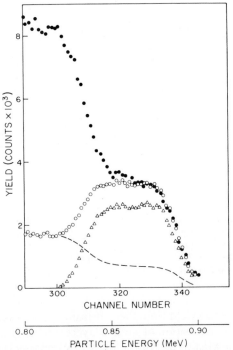

**Fig. 8.23**   Random (●) and ⟨111⟩-aligned ( ○ ) spectra for a 1000-Å-thick silicon oxide film on silicon. The dashed line represents the random spectrum normalized to the aligned spectrum at the Si/$SiO_2$ interface. The extracted spectrum (Δ) of the Si signal in the $SiO_2$ layer is obtained by subtracting the normalized random spectrum from the aligned spectrum. [From Mitchell et al. (1971).]

substrate region just below the $Si/SiO_2$ interface. The normalized random spectrum is shown by a dashed curve in Fig. 8.23. When this dashed curve is substrated from the aligned spectrum (open circles) to give the reduced yield profile of Si in $SiO_2$ (open triangles), the correct yield of Si in $SiO_2$ can be obtained by renormalizing the extracted profile.

The normalization technique can also be applied to thin layers, as was shown by Della Mea et al. (1975a). An alternative procedure, basically a variation of the method just described, is to use two aligned spectra. For example, an aligned spectrum taken with the beam parallel to a major crystallographic axis can be compared to a spectrum obtained by tilting the sample slightly ($0.2–0.4°$) off alignment (Chu et al., 1973). For thin amorphous layers, however, this method gives not only the number of atoms in the amorphous layer, but also a contribution from the uppermost atoms in the underlying crystal substrate. This is the same problem encountered in analysis of the surface peak (Section 8.3.2). The extracted number of atoms per square centimeter can be an overestimate of the number of atoms in the amorphous layer due to the contribution from the substrate. Although there are uncertainties in determining the absolute number of surface atoms, the composition of the surface layers can be determined by analyzing different thicknesses of layers prepared in the same fashion (Poate et al., 1973). By this procedure it was found that thermally grown oxide layers on Si were stoichiometric in the region away from the interface (Sigmon et al., 1974).

### 8.4.3  Polycrystalline Layers

The crystalline nature of a layer can be investigated in certain cases by channeling measurements. The schematic diagrams in Fig. 8.24 show samples and spectra for three different structures of films of the same elemental composition as the underlying crystal substrate. In Fig. 8.24a the spectra for an amorphous film is shown for comparison with the polycrystalline layers. The analysis of a highly polycrystalline film (Fig. 8.24b) with randomly oriented polycrystallites follows the treatment for an amorphous layer. If the polycrystallite size is much smaller than the beam spot size, the randomly oriented grains should lead to a yield of backscattered particles identical to that from an amorphous target. These examples are in contrast to the case (Fig. 8.24c) in which the crystallites are highly oriented with respect to the substrate and the aligned yield falls between that for a single crystal and that for a layer composed of random polycrystallites. The aligned yield for oriented crystallites varies with the energy of the incident beam in a manner different from that for dechanneling by randomly displaced host atoms. This provides a strong guide for the interpretation of the data.

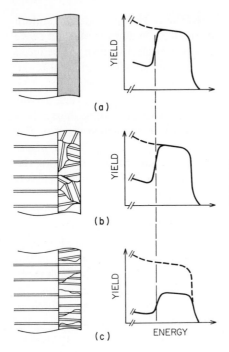

**Fig. 8.24**  Schematic diagrams for an amorphous layer (a) and polycrystalline layers [(b) random; (c) $<2°$ off axis] of Si on a single-crystal Si with the associated random (dashed lines) and aligned (solid lines) spectra.

The phrase "highly oriented crystallites" refers to the more general cases of mosaic spread or misoriented twins in single crystals as well as near-epitaxial layers containing some misoriented regions. This comparison has been made in ion-implanted semiconductors (Csepregi *et al.*, 1976; Johansen, *et al.*, 1976; Foti *et al.*, 1977, 1978), in silicide layers (Sigurd *et al.*, 1973b, Tu *et al.*, 1974), and in solid–phase crystallization of Ge and Si in metal films (Sigurd *et al.*, 1974). In general, the use of channeling techniques are useful for cases in which the layer is preferentially oriented parallel to the crystallographic axis of the underlying single-crystal substrates and the spread in orientation in small, within 1 or 2°. That is, the misorientation is comparable to or less than the critical half-angle $\psi_{1/2}$ for channeling.

The minimum yield in the portion of the spectrum corresponding to the polycrystalline region is dependent on the energy of the incident particles. This is illustrated in Fig. 8.25 with spectra taken with three different beam energies for a $Pd_2Si$ layer formed on $\langle 111 \rangle$ Si. It can be seen that the minimum yield of the Pd signal decreases with decreasing energy of the incident beam. For example, near the high-energy edge of the Pd signal, the minimum yield ranges from about 60% at 2 MeV (Fig. 8.25a) to about 20% at 0.4-MeV beam energy (Fig. 8.25c).

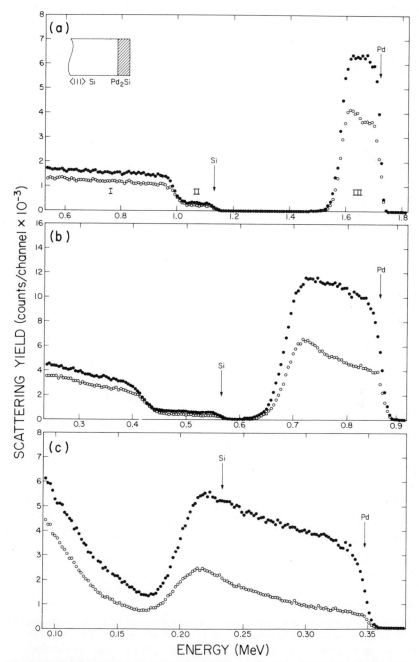

**Fig. 8.25** Random (●) and aligned (○) spectra for (a) 2-, (b) 1-, and (c) 0.4-MeV $^4$He ions backscattered from a 1350-Å-thick Pd$_2$Si layer on $\langle 111 \rangle$ Si. Vertical arrows indicate energies corresponding to scattering from surface atoms. [From Sigurd *et al.* (1973b).]

Insight into this behavior can be gained by making the simplifying assumption that the distribution of crystallite orientations is Gaussian characterized by one angular coordinate $\theta$. The distribution $g(\theta)$ of crystallite orientations with standard deviation $\sigma$ is then

$$g(\theta) = (1/\sigma^2)\exp(-\theta^2/2\sigma^2). \tag{8.26}$$

Similarly, one can ascribe a Gaussian form to the angular yield profile $f(\theta)$ in the near-surface region:

$$f(\theta) = 1 - (1 - \chi_{min})\exp\left(-\frac{\theta^2}{2(\psi_{1/2}^2/\ln 4)}\right). \tag{8.27}$$

The constants in the expression for $f(\theta)$ were chosen to give a minimum yield $\chi_{min}$ at normal incidence, a value of 0.5 at $\theta = \psi_{1/2}$, and unity for $\theta \gg \psi_{1/2}$. Another expression for the angular yield profile is given by the square well approximation $f(\theta) = \chi_{min}$ for $\theta < \psi_{1/2}$ and $f(\theta) = 1$ for $\theta > \psi_{1/2}$ discussed for dechanneling by amorphous layers. More exact angular yield profiles can be derived from experimental measurement or calculation.

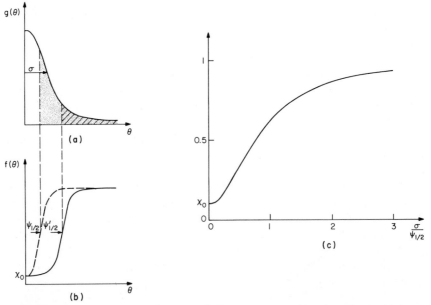

**Fig. 8.26**  Schematic diagrams [(a) and (b)] of the functions used in calculating the minimum yields for particles scattered in the near-surface regions of a polycrystalline layer. Part (a) indicates the distribution of crystalline orientations $g(\theta)$, and (b) indicates the channeling angular yield for two critical angles of single-crystal materials. The shaded area indicates crystallites whose orientation lies outside the critical angles. Part (c) shows how the calculated minimum yield depends on the parameter $\sigma/\psi_{1/2}$. [From Sigurd, *et al.* (1973b).]

Convolution of $f(\theta)$ and $g(\theta)$ [Eqs. (8.26) and (8.27)] yields a minimum yield related directly to the ratio of the crystallite standard deviation to the axial half-width as given by

$$\chi = [(\sigma/\psi_{1/2})^2 \ln 4 + \chi_{\min}]/[(\sigma/\psi_{1/2})^2 \ln 4 + 1]. \tag{8.28}$$

The variation of the minimum yield $\chi$ as a function of the parameter $\sigma/\psi_{1/2}$ is shown in Fig. 8.26c. If the spread of the crystallite orientations is large compared with the critical angle for channeling, the minimum yield approaches unity. In effect this is the limit for randomly oriented crystallites, for which the aligned yield is that for an amorphous layer. At the other extreme, for $\sigma/\psi_{1/2} \ll 1$, the minimum yield approaches the value $\chi_{\min}$ for a single crystal. The strongest variation in the minimum yield occurs in the region where the standard deviation $\sigma$ is comparable to the half-angle $\psi_{1/2}$.

The physical origin of this behavior is shown in Figs. 8.26a and 8.26b. For the crystallite distribution and angular profiles shown by the solid lines, the minimum yield is determined mainly by the crystallites with orientations in the shaded part under the $g(\theta)$ curve. This is equivalent to the square well approximation where all crystallites oriented at angles more than $\psi_{1/2}$ have unity yield. If the critical angle is decreased to the value shown by the dashed curve, more crystallites will lie outside the $\psi_{1/2}$ value and the minimum yield value will be higher.

Experimental measurements of the energy dependence of the minimum yield of the Pd signal for $Pd_2Si$ layers indicate agreement with the behavior predicted from Eq. (8.28). The minimum yield values decreased with a decrease in $\sigma/\psi_{1/2}$ and the values of $\sigma$ deduced from channeling measurements were close to those deduced from x-ray diffraction (Sigurd et al., 1973b). More recently, Ishiwara and Furukawa (1976) measured the angular yield profiles from a $Pd_2Si$ layer and found that these profiles were in reasonable agreement with calculated values.

In this treatment we have been concerned with the minimum yield in the surface region of crystalline layers. The focus was on the energy dependence of the minimum yield, since it differs from that predicted for scattering in an amorphous layer. We have not treated the depth dependence of the aligned yield, nor the aligned yield in the underlying crystal substrate. A proper treatment of these topics would require detailed assumptions about the defect structure in the film.

### 8.4.4  Crystals Containing Disorder

In this section we treat the analysis of aligned backscattering spectra of crystals containing host atoms displaced from their lattice sites. We will use the generic term "disorder" to include all displaced host atoms including

point defects, dislocation loops, stacking faults, and twins. In the discussion
we will initially assume that the disorder can be treated as randomly displaced
atoms, in effect a dilute concentration of atoms in amorphous zones, that
are located within a perfect crystal that has a well defined value of $\psi_{1/2}$. Of
course, this assumption does not hold in the general case.

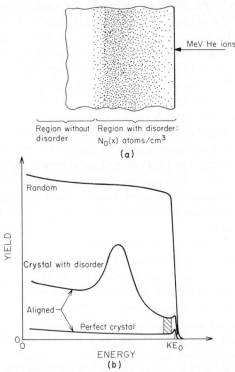

**Fig. 8.27**    (a) Schematic and (b) random and aligned spectra for megaelectron volt $^4$He ions
incident on a crystal containing disorder. The aligned spectrum for a perfect crystal without
disorder is shown for comparison. The difference (shaded portion) in the aligned spectra between
disordered and perfect crystals can be used to determine the concentration $N_D(0)$ of displaced
atoms at the surface.

The axially aligned and random spectra for megaelectron volt He ions
incident on a crystal containing disorder near the surface region are shown
in Fig. 8.27. For comparison, the aligned spectra for the same crystal without
disorder, a perfect crystal, is shown. This latter spectrum is often referred
to as virgin or perfect. The aligned spectra for the disordered crystal lie
above the virgin spectrum even at depths below the disordered region.

The yield at depths below the disordered region is increased because some of the channeled ions have become dechanneled as a result of the scattering of the particles as they pass through the disordered area. These dechanneled particles may then interact with and be scattered from all the atoms of the crystal. This is the same problem considered in a previous section where the dechanneling in a perfect crystal overlaid by an amorphous layer was treated. In this case, however, the disorder is contained in the region of the crystal in which dechanneling occurs. The problem is to separate the two contributions: backscattering of the channeled particles from displaced atoms, and backscattering of dechanneled particles from all the atoms of the crystal.

Since the amount of dechanneling is proportional to the number of displaced atoms the beam traverses, the contribution of dechanneling to the aligned beam in the surface region is minimal. Consequently a very good approximation to the concentration $N_D(0)$ of displaced atoms at the surface is the difference between aligned and virgin spectra at $E_1 = KE_0$. Since there are often oxide or thin amorphous layers on the crystal surface, the disorder is evaluated just behind the surface peak as shown by the shaded area in Fig. 8.27.

The disorder concentration $N_D(0)$ is given by

$$N_D(0) = N[\chi(0) - \chi_V(0)]/[1 - \chi_V(0)], \tag{8.29}$$

where $N$ is the bulk density in atoms per cubic centimeter and $\chi$ the ratio of aligned to random yields at the energy $E$ where the disorder is evaluated. The yield ratio $\chi_V$ for the virgin crystal has a value $\chi_V(0)$ at the surface that is equal to the minimum yield $\chi_{min}$. In Eq. (8.29), $1 - \chi_V(0)$ is the fraction of the incident beam that is channeled (typically 0.95–0.98), and the term $\chi(0) - \chi_V(0)$, which equals $[1 - \chi_V(0)] - [1 - \chi(0)]$, represents the difference between the channeled fractions in the virgin crystal and the damaged one.

Having determined the amount of disorder in the surface layer, we then must evaluate the situation at depth $x$. One approach is to remove successive layers and measure the surface disorder. An example is shown in Fig. 8.28a for a sample containing a disorder distribution $N_D(x)$. The normalized yield $\chi$ of the sample before layer removal is shown by the solid line in Fig. 8.28b. The solid point and the shaded area represent the amount of disorder at the surface. After a certain thickness of material $t$ (for this example) is removed, the aligned spectrum is again measured (dashed line in Fig. 8.28b). The open circle represents the disorder measurement at the new surface and hence gives the value of $N_D$ at $x = t$. By a series of such measurements one can determine $N_D(x)$.

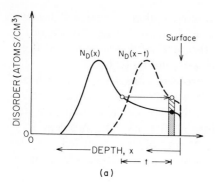

(a)

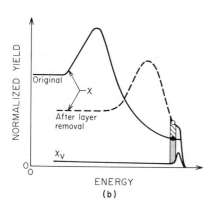

ENERGY
(b)

**Fig. 8.28**  Use of layer removal techniques to determine the distribution of lattice disorder. Part (a) shows the disorder distributions on the original sample (solid line) and after (dashed line) a layer of thickness of $t$ was removed. (b) The aligned spectra give the normalized yield in the original and stripped sample along with the yield from a perfect (virgin) crystal.

Another approach is to use an iterative procedure to determine the amount of dechanneling. This technique is based on several assumptions (Eisen, 1973; Eisen and Mayer, 1976): (1) the critical angle $\psi_{1/2}$ is unchanged by the introduction of disorder; (2) the flux of channeled particles is distributed uniformly within the channels; (3) *all* the displaced atoms can interact with the channeled particles; and (4) dechanneled particles are not scattered back into channels. The problem is to determine the dechanneled fraction $\chi_R(x)$ in the damaged crystal as a function of depth.

The procedure is shown in Fig. 8.29 for a sample containing a disorder distribution $N_D(x)$. The normalized yield is given by (Eisen, 1973)

$$\chi(x) = \chi_R(x) + [1 - \chi_R(x)][N_D(x)/N], \qquad (8.30)$$

where $\chi_R(x)$ is the fraction of dechanneled particles (these can interact with all the crystal atoms) and $[1 - \chi_R(x)]$ the fraction of channeled particles (these can interact wih $N_D(x)$ displaced atoms). The dechanneled fraction $\chi_R$ is often referred to as the random component of the aligned beam.

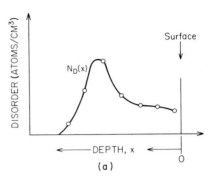

(a)

**Fig. 8.29**   Use of iterative procedures to determine the amount of dechanneling in a sample in which (a) the disorder distribution as shown by the solid line with open circles and (b) the aligned spectrum is given by the solid line labeled $\chi$. In (b) the value $\chi_R(1)$ is the amount of dechanneling caused by $N_D(0)$ displaced atoms in the surface layer of thickness $\Delta t$. The number of displaced atoms $N_D(1)$ in the next layer is determined from the shaded area between $\chi$ and $\chi_R(1)$. The dashed curve gives the random component of the aligned beam (dechanneled fraction) and the shaded area represents the concentration $N_D(x)$ of displaced atoms.

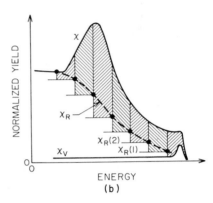

(b)

The dechanneled fraction is usually approximated by (Eisen, 1973)

$$\chi_R(x) = \chi_V(x) + [1 - \chi_V(x)]P(x, \tilde{\theta}_c), \qquad (8.31)$$

where $\chi_V(x)$ is the aligned yield from a virgin crystal, and where $P(x, \tilde{\theta}_c)$ is the probability that channeled particles are dechanneled by defects between the surface and depth $x$. This assumes that the dechanneling due to the disorder is linearly added to the dechanneling that occurs in the crystal in the absence of disorder. The value of $P(x, \tilde{\theta}_c)$ can be found from the procedure used to determine the amount of dechanneling in a perfect crystal overlaid by an amorphous layer [Eq. (8.19) or (8.25)]. As an approximation, $P(x, \tilde{\theta}_c)$ is directly proportional to the number of displaced atoms the particles traverse if the amount of disorder is small and the single-scattering approximation holds.

As sketched schematically in Fig. 8.29b, the aligned spectrum is divided into equivalent thickness increments $\Delta t$, and it is assumed that $N_D(x)$ is constant in each increment. One determines the number of displaced atoms

$N_D(0)\Delta t$ in the surface layer from Eq. (8.29). From the value of $N_D(0)\Delta t$ one calculates the dechanneling probability $P$ and, from Eq. (8.31), the amount of dechanneling $\chi_R(1)$ in the next layer caused by the displaced atoms in the surface layer. From the difference in the heights of the curves between $\chi$ and $\chi_R(1)$ one determines $N_D(1)$ and hence $\chi_R(2)$. The procedure is iterated to determine $N_D(2)$ in the next layer. The values of $N_D(i)$ are plotted as open circles in the disorder distribution curve in Fig. 8.29a.

For the example shown in Fig. 8.29, the disorder does not extend deep into the crystal. Consequently, it is possible to test the consistency of the calculation procedure in that $\chi_R$ should equal $\chi$ at depths below the disordered region. This test is commonly used in evaluating disorder in ion-implanted semi-conductors.

When there is a well-defined disorder peak in the aligned spectrum and the amount of dechanneling is not too large, as shown in Fig. 8.30a, it is possible to make a simple estimate of the total number of displaced atoms per square centimeter. One uses a straight–line approximation for $\chi_R$ and

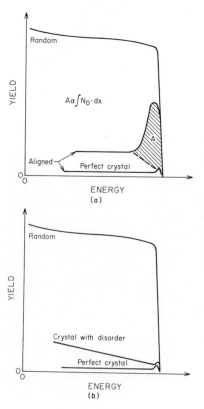

**Fig. 8.30**   Use of approximations to determine the amount of disorder. (a) An implanted sample with a large amount of disorder near the surface. The shaded area $A$ of the peak in aligned spectrum is proportional to the number of displaced atoms per square centimeter, and the dechanneling is estimated by the dashed line. (b) A crystal containing a small concentration of displaced atoms distributed uniformly in depth. The slope of the aligned spectrum is proportional to the concentration of displaced atoms.

determines the area $A$ of the shaded portion of the peak in the aligned spectrum. As pointed out in earlier chapters, the number of displaced atoms per square centimeter is found from the ratio of $A$ to the height of the random spectrum in the surface energy approximation.

A different set of approximations can be used when there are relatively small concentrations (of the order of 1 at.%) of displaced atoms in the crystal. The aligned spectra, as shown in Fig. 8.30b, often exhibit a linear increase with depth; the increase is more rapid with increased amounts of disorder. If it is assumed that the concentration of disorder $N_D$ is constant with depth and that the single-scattering model can be used to calculate $P$, then the disorder can be found from the slope of the aligned yield:

$$d\chi/dx \approx \sigma_D N_D. \tag{8.32}$$

This estimate is very crude as it neglects the contribution due to dechanneling $\chi_V$ by the host lattice itself. A number of studies (Merkle et al., 1973; Swanson et al., 1975; and Pronko and Merkle, 1974) have treated the contribution from the host lattice with the assumption that the dechanneling from the host atoms is independent of the presence of defects. Although this treatment is clearly an improvement over the zero-order estimate of Eq. (8.32), more detailed measurements are required to show that the dechanneling processes can be treated independently. A similar approach has been used to evaluate disorder in metals, where the dechanneling cross section has been evaluated for more complex defects such as dislocations or stacking faults (Mory and Quéré, 1972; Quéré, 1974).

Throughout this section, we have treated the conversion of the energy-to-depth scale as a simple procedure that could be handled as outlined in Chapter 3. In fact, this is not so. The aligned beam has a significantly lower stopping cross section than the random beam for particles confined to the center of the channel. The energy-to-depth conversion for these well-channeled particles can differ by 20 to 30% between random and aligned spectra. However, the particles in the aligned beam that are dechanneled have trajectories close to the axial rows and hence have higher stopping cross sections than the well-channeled particles. Bøttiger and Eisen (1973) have shown that the stopping cross section for dechanneled particles is close to that of the random beam. The approach in disorder analysis is often to use the stopping cross section in a random media for the energy-to-depth conversion in aligned spectra.

### A NOTE OF CAUTION

The use of channeling techniques to evaluate disorder in crystals has some pitfalls. In the best cases, the aligned spectra do give accurate values

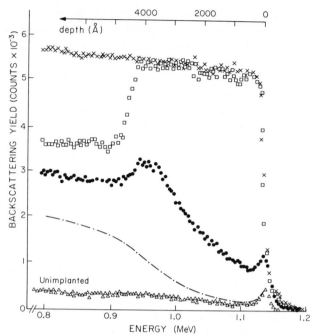

**Fig. 8.31** Backscattering spectra for 2.0-MeV $^4$He ions incident in the $\langle 111 \rangle$ direction on ion implanted with $^{28}$Si before ($\square$) and after ($\bullet$) annealing at 600°C for 100 min. The values shown are: the amount of dechanneling calculated from the method shown in Fig. 8.29 ($\cdot - \cdot -$); random ($\times$); and unimplanted ($\triangle$). [From Pronko *et al.* (1977).]

of the depth distribution $N_D(x)$ of displaced atoms. On the other hand, there are clear cases for which channeling gives misleading information. The spectra in Fig. 8.31 are for a crystal with an amorphous layer formed by implantation and then annealed. The open squares show the aligned Si yield for the as-implanted case. The amount of dechanneling in the crystal under the amorphous layer (about 5000 Å below the surface) agrees with calculated values. After the annealing, however, the dechanneling in the underlying crystal layer is substantially above that (dot–dashed line) calculated from the procedure shown in Fig. 8.29. The studies of Pronko *et al.* (1977) demonstrated that the conventional analysis procedure (i.e., Fig. 8.29) gave an incorrect disorder distribution and that the increased amount of dechanneling is due to the presence of twins. (Foti *et al.*, 1977, 1978). The presence of dislocations can also lead to an increase in dechanneling (Tseng *et al.*, 1978).

It is becoming increasingly evident that the assumptions of randomly displaced atoms and unique critical angles do not hold generally. The defect complexes in crystals can involve dislocations, stacking faults, small atom displacements, and other configurations. These configurations can give different contributions to the aligned spectra than randomly displaced atoms.

One can test the disorder analysis procedure by comparing results obtained with different projectiles, different axial directions, and different incident beam energies. One can also use both layer removal and iterative techniques on the same sample. If the disorder distributions are consistent with each other, then one can place some reliance on the analysis procedure and quantitative statements can be made. However, even if the interpretation of the aligned spectra is ambiguous, the channeling techniques can indicate the presence of disorder and relative changes due to heat treatment or other process steps.

Our concern here is to warn the reader to be cautious when using channeling techniques to extract quantitative data on disorder. We do not wish to imply that channeling is not a powerful technique to obtain qualitative estimates of disorder distribution. It has been used successfully in many laboratories. The techniques are often used to categorize samples that are later to be examined by transmission electron microscopy to characterize the detailed nature of the disordered regions.

### 8.4.5  Summary of Disorder Analyses

In channeling effect measurements of lattice disorder, the yield in the aligned spectrum is due to channeled particles backscattered from displaced atoms and dechanneled particles backscattered from lattice site atoms. The key to the analysis lies in the procedure for predicting the amount of dechanneling as the particles penetrate deeper into the crystal. The approach followed here, as well as in most of the published work, is based on the assumption that the displaced atoms are at random positions within an otherwise perfect crystal lattice. Dechanneling arises from particles scattered through angles greater than the axial half-angle $\psi_{1/2}$. This approach is not valid in all cases and can lead to erroneous interpretations of the aligned spectra. Certain self-consistency checks can be applied to determine if the analysis procedure is in serious trouble. However, the concept of displaced atoms in random positions is very useful in evaluating crystals containing amorphous regions.

The advantage of the channeling technique is that it provides a fast and simple evaluation of the crystalline quality of the sample. The technique has been applied successfully in many cases discussed in this chapter and in the chapter on applications. The difficulty is that the microscopic nature of the defect configuration can play a dramatic role. Other measurements, such as transmission electron microscopy, should also be used, as has been done in some cases (Picraux and Thomas, 1973; Merkle *et al.*, 1973, Johansen *et al.*, 1976, Rechtin *et al.*, 1978) to provide information about the defect configuration.

We should point out that even if the nature of the defects is known, the amount of dechanneling can be difficult to predict. The influence of thermal vibrations of the host lattice is not, in general, a simply additive contribution to dechanneling, and the crystal lattice may be distorted by the presence of defects. Further, we have assumed that the flux of channeled particles is uniformly distributed across the channels.

We believe that the application of channeling effect measurements to the evaluation of disorder will continue as an active field. Measurements will be made routinely, as they are at present, in well defined situations such as those for ion-implanted semiconductors or for epitaxial layers. In addition, more sophisticated analysis involving the correlation between defect configuration and dechanneling rates will be made in special situations. The fundamental processes underlying dechanneling phenomena in both perfect and disordered crystals will be studied by experiment, calculation, and computer simulation.

## 8.5 FLUX PEAKING AND LATTICE SITE LOCATION OF IMPURITIES

### 8.5.1 Introduction

In the earlier discussions on the use of channeling to determine the fraction of impurity atoms on substitutional lattice sites and the amount of lattice disorder, it was tacitly assumed that the channeled particles were uniformly distributed across the channel. For truly substitutional impurities this assumption is not very crucial, since both impurity and host atoms would exhibit the same minimum yields and angular yield profiles. For interstitial atoms on well-defined interstitial sites, for example in the center of a channel, the aligned yield depends strongly on the distribution of channeled particles across the channel. For example, if the particles moved predominantly along the center of the channel where the impurities were located, the backscattering yield from the impurities would be much higher than if the flux of channeled particles were uniform across the channel.

The possibility of a concentrated flux ("flux peaking") near the center of the channel was largely ignored until two nearly simultaneous investigations (Andersen *et al.*, 1971; Domeij *et al.*, 1970) of ion-implanted silicon gave evidence for flux peaking. One of these early sets of angular yield profiles is given in Fig. 8.32. The dashed lines are the yield from the implanted ytterbium (Yb) atoms, and the solid lines are the yield from the host Si atoms in the

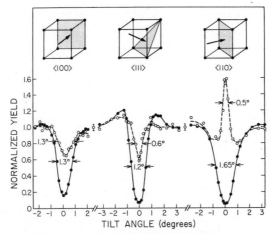

**Fig. 8.32**   Angular yield profiles around the three principal axes for 1.0-MeV $^4$He ions incident on an Yb-implanted silicon crystal. The solid lines are the Si signal and the dashed lines the Yb signal yield profiles. [From Andersen *et al.*, (1971).]

implanted region. The axial scans along the $\langle 100 \rangle$ and $\langle 111 \rangle$ orientations show a dip in the yield from Yb, but the $\langle 110 \rangle$ axial scan shows a pronounced increase. This increase was attributed to a strong flux peaking in the mid-channel region, along with preferential location of the Yb atoms near the center of the $\langle 110 \rangle$ channel.

The presence of flux-peaking effects is both a help and a hindrance in the analysis of solids. It is a help in that a well focused beam in the center of the channel can be used to probe well defined interstitial positions and can provide a characteristic peak in the angular yield profile of the interstitial impurity signal. Flux peaking is a hindrance, in a sense, because the amount of flux peaking is sensitive to the initial perfection of the crystal, to the beam divergence, and to the amount of disorder produced during analysis. Consequently, it is difficult to specify the extent of the flux concentration in any given situation. Further, there are often several nearly equivalent interstitial sites, so that even with measurements along different axial and planar directions, it may be difficult to make an unambiguous site assignment.

### 8.5.2   Flux Peaking

An ion incident parallel to a low-index axial direction in a crystal will tend to be deflected by the atoms in the axial rows. The amount of deflection depends on the position of entry into the channels relative to the rows. If the position of entry is close to one of the rows, the deflection or transverse

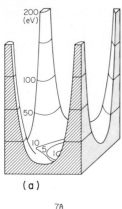

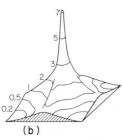

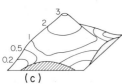

(a)

(b)

(c)

**Fig. 8.33**   Representation of (a) the continuum potential across a ⟨100⟩ channel in Cu, and the particle flux distribution for (b) a well collimated beam and (c) a slightly divergent (±0.23°) beam of 1-MeV $^4$He ions. [Adapted from the contour profiles in Morgan and Van Vliet, (1972).]

momentum is large; if the position of entry is near the center, the deflection will be small. If the interaction between the energetic particle and the atomic rows is represented by a continuum potential (Van Vliet, 1973), such as that shown in Fig. 8.33a for a Cu crystal, then the transverse energy is determined solely by the potential at the point of entry. The particle will then be confined to the region within the center of the channel where the potential is equal to or less than that at the point of entry.

The confinement would mean that all particles entering the Cu crystal at positions corresponding to the 10-eV equipotential contour in Fig. 8.33a would be constrained within the contour. After penetrating deep enough (≈ 1000 Å) to achieve an equilibrium distribution, the particles that started on the 10-eV contour line can be found anywhere within the portion of the channel bounded by the 10-eV contour line. This same argument for other contours would lead to a substantially larger flux of particles in the central region of the channel than at the edges along the atomic rows. For the same

Cu crystal, the intensity contours for the flux of 1-MeV He ions are shown in Figs. 8.33b and 8.33c (Morgan and Van Vliet, 1972).

Figure 8.33b represents the ideal case of a perfectly rigid lattice and perfect beam alignment. In that case there are no scattering mechanisms; so the particles would remain confined inside their equipotential contour. The maximum flux in the center of the channel is 7.1 times that for particles distributed uniformly over the channel (Alexander *et al.*, 1974).

The assumptions of a perfect crystal and perfectly aligned beam are not realistic. If one adds thermal vibrations and also multiple scattering from electrons, the maximum flux in the central region is reduced. The flux peak is also reduced if the incident beam has some divergence, as is shown in Fig. 8.33c. In real crystals, of course, surface layers, thermal vibrations of the host lattice, lattice disorder, and beam divergence will all cause a decrease in the maximum flux concentration. From examination of the experimental evidence, however, it seems reasonable to estimate that a flux enhancement of a factor of 2 to 4 could be achieved in practical situations. The pronounced decrease in the magnitude of the flux peak caused by beam divergence indicates that flux-peaking effects do not play a large role in disordered crystals in which dechanneling effects are large.

The flux distribution within the channel is influenced, of course, by the incident angle of the collimated beam. In Fig. 8.34 we depict 1.5-MeV $^4$He ions incident at an angle $\theta$ to the $\langle 100 \rangle$ row in Cu. The lower portion of the

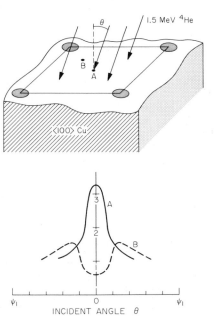

**Fig. 8.34**   Variation as a function of incident angle of the flux of particles below the surface at two positions $A$ and $B$ for 1.5-MeV $^4$He ions incident on a $\langle 100 \rangle$ Cu crystal. The characteristic angle $\psi_1$ for channeling is indicated for comparison of the angular widths. [Adapted from Morgan and Van Vliet (1972).]

figure represents the flux experienced at two positions within the channel, as computed by Morgan and Van Vliet (1972). Although shown on the surface, the positions are deep enough within the crystal so that the flux peaking can be established. For the central position *A*, the flux shows a pronounced drop for small tilts from the direction of perfect alignment. For position *B*, off-center on the diagonal, the flux is small for perfect alignment, since most of the particle trajectories are in the central region around *A*. For increasing tilt angle, the flux first increases to a maximum and then decreases toward the value (unity) expected if flux-peaking effects were absent.

### 8.5.3 Substitutional Impurities

The analysis is particularly simple for the lattice site location of impurities positioned on substitutional lattice sites. The ratio of aligned to random yields for the impurity signal matches that of the host lattice. Angular yield profiles of impurity and host lattice also match. This equivalence between the angular yield profiles of the signals from host and impurity atom has been demonstrated in both early and more recent studies of impurities in Cu lattices (Alexander and Poate, 1972; Borders and Poate, 1976).

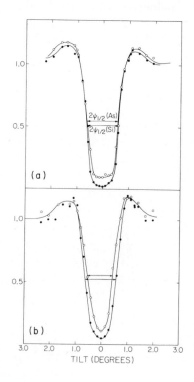

**Fig. 8.35** Angular yield profiles for tilts through the $\langle 100 \rangle$ axis for 1.8-MeV $^4$He ions incident on two crystals of Si (●) doped with different concentrations of arsenic (○): (a) $6 \times 10^{19}$ As/cm$^3$; (b) $1.5 \times 10^{21}$ As/cm$^3$. At the higher As concentrations, the As signal half-width is narrower than that of the Si host crystal. [From Haskell *et al.* (1972).]

Angular yield profiles are shown in Fig. 8.35 for 1.8-MeV $^4$He ions incident across the $\langle 110 \rangle$ axis of two silicon crystals doped with different concentrations of arsenic. For the sample with the smaller As concentration (Fig. 8.35a), the value of the silicon minimum yield is 0.03 and the axial half-angle $\psi_{1/2}$ is 0.62°. Both values are close to those predicted in Section 8.3.2, indicating that the silicon crystal is of good quality. The angular yield profile for the As signal has a half-width only slightly smaller (0.6°) than the silicon signal, showing that the As atoms are predominantly on substitutional lattice sites. From the value of the As signal minimum yield, $\chi_{min}$(impurity) = 0.08, the fraction $f$ of As on sites along the $\langle 110 \rangle$ rows is given by

$$f = [1 - \chi_{min}(\text{impurity})]/(1 - \chi_{min}) = (1 - 0.08)/(1 - 0.03) = 0.95, \quad (8.33)$$

where the term $1 - \chi_{min}$ gives the fraction of the incident aligned beam that is channeled.

At the higher As concentration, the As signal half-width is appreciably narrower (0.5° compared to 0.6°) than that of the Si host lattice (Fig. 8.35b). The narrowing of the impurity signal half-width is an indication that the As atoms are displaced somewhat from the well defined, crystallographic lattice sites.

If the impurity atoms are displaced by only small amounts, about 0.2 Å, from substitutional sites, then they lie within regions of high potential (within the potential contours of 50 to 100 eV in Fig. 8.33) within the channel. In order for the incident channeled particle of energy $E$ to make a close impact collision, it must have an angle of incidence $\theta_{in}$ high enough that the transverse energy $E\theta_{in}^2$ at least equals the value of the potential. One can obtain a relationship between the displacement $r$ of an atom from a row and the angle $\theta_{in}$ a particle must have to interact with the atom by using Lindhard's continuum potential,

$$U(r) = \psi_1^2 E[\ln(Ca/r)^2 + 1], \quad (8.34)$$

where $C \approx \sqrt{3}$, $a$ is the Thomas–Fermi screening radius ($a \approx 0.1$–$0.2$ Å), and $\psi_1$ is the characteristic angle [eq. (8.1)]. For the conditions $E\theta_{in}^2 = U(r)$ and $r = a_{TF}$, we have $\theta_{in} = \psi_1(\ln 4/2)^{1/2} = 0.83\psi_1$. Since the axial half-angle $\psi_{1/2}$ for channeling is close to that of the characteristic angle $\psi_1$, these relations imply that the beam must enter at angles slightly less than the critical angle in order to interact with atoms displaced 0.1–0.2 Å from their substitutional positions.

This argument suggests that one can distinguish between substitutional and slightly displaced impurities by measuring the width of the impurity angular yield profile. This concept can be misleading because channeling

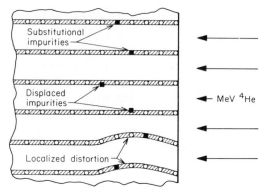

**Fig. 8.36**   Schematic diagram of a crystal lattice containing substitutional impurities and displaced impurities in a perfect region of the lattice and substitutional impurities in a region containing localized distortion of the lattice. The angular yield profiles of the impurity signal for displaced impurities or substitutional impurities in a distorted region would be narrower than those of the host lattice.

conditions are established by averages along rows (or planes) of atoms. Displacements are measured from the center of the row not from the position of the local site. If localized strain displaces the sites, in effect putting a "kink" in the row as in Fig. 8.36, even substitutional impurities in that region will appear displaced. This difference in interpretation is all the more obvious in analyzing ion-implanted semiconductors in which the ion comes to rest in a locally disordered region. For high concentrations of implanted atoms the impurity angular scans are nearly always somewhat narrower than those from the host lattice (Picraux et al., 1972, Sigurd and Bjorkqvist, 1973). The data do not distinguish between substitutional impurities in locally perturbed lattices or impurities displaced from substitutional sites.

More detailed calculations have been carried out by several groups (Picraux et al., 1972; Sigurd and Bjorkqvist, 1973). The calculated angular distributions for 1-MeV $^4$He ions as a function of displacement distance of impurities in $\langle 110 \rangle$ Si are shown in Fig. 8.37. For 0.1-Å displacements, narrowing of the half-angle of the impurity angular yield profile is apparent, and for 0.2-Å displacements there is also an increase in the minimum yield.

The assignments of site locations cannot be made solely on the basis of angular scans across one axial direction. Scans through other axes and planar directions must also be made. Another refinement is the use of double alignment techniques, by which both the incident and outgoing (detected) beams are aligned simultaneously with the major crystal axis (Gemmell, 1974).

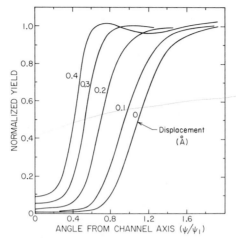

**Fig. 8.37**  Calculated angular yield profiles of the signal from an impurity atom as a function of the displacement of the impurity from the $\langle 110 \rangle$ row for 1-MeV $^4$He ions incident on Si at 296 K. The angles are normalized to the characteristic angle $\psi_1$. [Picraux *et al.* (1972).]

The point we are trying to emphasize here and in the following section is that a precise determination of nonsubstitutional site location requires more than a routine measurement. To determine that some fraction of the impurity atoms is on well defined substitutional sites or is displaced by small increments from substitutional sites is fairly easy. To specify the site location for atoms that are not on well defined substitutional sites is quite another matter.

### 8.5.4  Well Defined Interstitial Sites

One of the common signatures of an impurity on a well defined interstitial site is an increased aligned yield above the random value and a narrow width in the angular yield profiles. Such effects give evidence for both an enhanced flux concentration and a preferred site location in the central region of the channel.

To determine the exact interstitial sites, one makes angular scans about various axial and planar directions. As a first step, one generally considers what are the possible preferred interstitial positions, and next determines what angular scans are required to distinguish between the different possible locations. For example, if the impurity was located in the center of the cubic lattice (circle in Fig. 8.38), then it would be visible along the $\langle 100 \rangle$ and $\langle 110 \rangle$ directions, but shadowed by the axial rows in the $\langle 111 \rangle$ direction. Lattice configurations for specific interstitial sites are given by Appleton and Foti (1977).

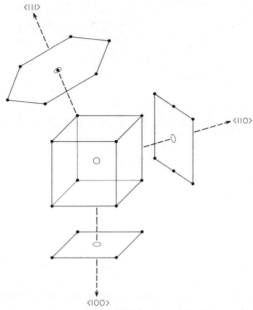

**Fig. 8.38**  Projection along the three principal axes of the position of an impurity (circle) located at body-center of a cubic lattice. The impurity would be visible along the ⟨100⟩ and ⟨110⟩ axes, shielded by the atomic rows when viewed along the ⟨111⟩ axis.

It is often difficult to make unamibuous site locations for interstitial species. The influence of thermal vibrations or small displacements due to local disorder can cloud the picture. Further, since there are often a number of possible sites, there may not be simple channeling directions where pronounced dips or peaks will be observed. Computer-synthesized, angular-scan calculations for interstitial positions in Si have been presented, and a survey of lattice location measurements has been given, in several review papers (Davies, 1973; Picraux, 1975).

### 8.6  INFLUENCE OF THE ANALYSIS BEAM

Bombardment of crystalline materials with megaelectron volt He ions is not a gentle process. There are changes in the composition of surface layers, movement of dopant atoms off lattice sites, and changes in the host lattice itself. In fact, megaelectron volt ion beams can be used to both introduce and analyze lattice disorder. It is important to recognize that the properties of the sample can change during analysis.

The fact that the process of analysis can influence the properties of the crystal under study has been recognized and treated in a number of review articles (Davies, 1973; Eisen and Mayer, 1976; Picraux, 1975). Rather than discuss the matter in detail, we shall only cite a few pertinent examples. However, it is incumbent on the users of channeling techniques to determine the influence of the analysis beam. Two techniques commonly used are: (1) to look for changes in the crystal as a function of the total dose used during analysis and (2) to move the beam to fresh spots on the crystal during data acquisition.

Beam-induced movement of impurity atoms off lattice sites has been observed in silicon for a number of cases (Fladda et al., 1969, Allen and Bicknell, 1974; Rimini et al., 1972b; Fujimoto et al., 1972). The movement of dopant atoms off lattice sites is not found for all species in silicon, and no general rules for predicting the magnitude of the effect have been formulated. However, it has been found that the dose of $^4$He ions required to align the crystal is often sufficient to cause movement of atoms off lattice sites. Consequently, after the crystal is aligned with the beam, it is desirable to move the beam spot to an unbombarded region before taking an aligned spectrum.

The beam can also affect the site location of impurities in metals. Impurity–defect interactions in Al have been studied by Howe and Swanson (1976).

Irradiation-induced defects in the host crystal have been investigated in semiconductors (Baeri et al., 1975; Pabst and Palmer, 1973), in ionic crystals (Matske, 1971; Hollis, 1973), and in metals (Merkle et al., 1973; Pronko 1974). In these cases, one often uses a random alignment to produce the defects and a channeled alignment to investigate the defects. A random alignment tends to produce several orders of magnitude more defects than can be produced in the channeled alignment. The analysis beam can also cause a decrease in the amount of disorder. Westmoreland et al. (1970) observed that in boron-implanted samples, the analysis beam caused a reduction in the disorder peak.

The composition of a surface layer can change during analysis. Buck and Wheatley (1972) found an increase in the silicon surface peak as a function of ion dose. These results indicate that evaluation of surface layers requires movement of the beam to ensure that no major changes occur.

## REFERENCES

Alexander, R. B., and Poate, J. M. (1972). *Radiat. Effects* **12**, 211.
Alexander, R. B., Callaghan, P. T., and Poate, J. M. (1974). *Phys. Rev. B* **9**, 3022.
Andersen, J. U., Andreason, O., Davies, J. A., and Uggerhøj, E. (1971). *Radiat. Effects* **7**, 25.
Allen, C. R., and Bicknell, R. W. (1974). *Phil. Mag.* **30**, 483.
Appleton, B. R., and Foti, G. *In* "Ion Beam Handbook for Material Analysis" (1977). (J. W. Mayer and E. Rimini, eds.). Academic Press, New York.
Baeri, P., Campisano, S. U., Foti, G., Rimini, E., and Davies, J. A. (1975). *Appl. Phys. Lett.* **26**, 424.

Barret, C. S., Mueller, R. M., and White, W. (1968). *J. Appl. Phys.* **39**, 4694.
Barrett, J. H. (1971). *Phys. Rev. B* **3**, 1527.
Bøgh, E. (1968). *Can. J. Phys.* **46**, 653.
Borders, J. A., and Poate, J. M. (1976). *Phys. Rev. B* **13**, 969.
Bøttiger, J., and Eisen, F. H. (1973). *Thin Solid Films* **19**, 239.
Buck, T. M., and Wheatley, G. H. (1972). *Surf. Sci.* **33**, 35.
Chu, W. K., Lugujjo, E., Mayer, J. W., and Sigmon, T. W. (1973) *Thin Solid Films* **19**, 329.
Csepregi, L., Chu, W. K., Muller, H., Mayer, J. W., and Sigmon, T. W. (1976). *Radiat Effects* **28**, 227.
Davies, J. A. (1973). *In* "Channeling" (D. V. Morgan, ed.) Chapter 13. Wiley, New York.
Davies, J. A., Denhartog, J., Eriksson, L., and Mayer, J. W. (1967). *Can. J. Phys.* **45**, 4053.
Dearnaley, G., Freeman, J. H., Nelson, R. S., and Stephen, J. (1973). "Ion Implantation." North Holland Publ., Amsterdam.
Della Mea, G., Drigo, A. V., Lo Russo, S., Mazzoldi, P., Yamaguchi, S., and Bentini, G. G. (1975a). *Appl. Phys. Lett.* **26**, 147.
Della Mea, G., Drigo, A. V., Lo Russo, S., Mazzoldi, P., Cornara, G., and Yamaguchi, S. (1975b). *In* "Atomic Collisions in Solids" (S. Datz, B. R. Appleton, and C. D. Moak, eds.), Vol. 2, p. 811. Plenum Press, New York.
Domeij, B., Fladda, G., and Johansson, N. G. E. (1970). *Radiat. Effects* **6**, 155.
Eisen, F. H. (1973). *In* "Channeling" (D. V. Morgan, ed.), Chapter 14. Wiley, New York, 1973.
Eisen, F. H., and Mayer, J. W. (1976). *In* "Treatise on Solid State Chemistry" (N. B. Hannay, ed.), Vol. 6B, Surfaces II, Chapter 2. Plenum Press, New York.
Fladda, G., Mazzoldi, P., Rimini, E., Sigurd, D., and Eriksson, L., (1969). *Radiat. Effects* **1**, 243.
Foti, G., Csepregi, L., Kennedy, E. F., Pronko, P. P., and Mayer, J. W. (1977). *Phys. Lett* **64A**, 265.
Foti, G., Csepregi, L., Kennedy, E. F., Mayer, J. W., Pronko, P. P., and Rechtin, M. D., (1978). *Phil. Mag.*
Fujimoto, F., Komaki, K., Watanabe, M., and Yonezawa, T. (1972). *Appl. Phys. Lett.* **20**, 248.
Gemmell, D. S. (1974). *Rev. Mod. Phys.* **46**, 129.
Hart, R. R., Dunlap, H. L., Mohr, A. J., and Marsh, O. J. (1973). *Thin Solid Films* **19**, 137.
Haskell, J., Rimini, E., and Mayer, J. W. (1972). *J. Appl. Phys.* **43**, 3425.
Hollis, M. J. (1973). *Phys. Rev. B* **8**, 931
Howe, L. M., and Swanson, M. L. (1976). *Inst. Phys. Conf. Ser. No.* **28**, 273.
Ishiwara, H., and Furukawa, S. (1976). *J. Appl. Phys.* **47**, 1686.
Johansen, A., Svenningsen, B., Chadderton, L. T., and Whitton, J. L. (1976). *Inst. Phys. Conf. Ser. No.* **28**, 267,
Lindhard, J. (1965). *Mat. Fys. Medd. Dan. Vid. Selsk.* **34** (14), 1.
Linker, G., Meyer, O., and Scherber, W. (1973). *Phys. Status Solidi (a)* **16**, 377.
Lugujjo, E. (1974). Ph.D. Thesis, California Inst. of Technol.
Lugujjo, E., and Mayer, J. W. (1973). *Phys. Rev. B* **7**, 1782.
Matske, Hj. (1971). *Phys. Status Solidi (a)* **8**, 99.
Mayer, J. W., and Rimini, E. (eds.) (1977). "Ion Beam Handbook for Material Analysis." Academic Press, New York.
Mayer, J. W., Eriksson, L., and Davies, J. A. (1970). "Ion Implantation in Semiconductors." Academic Press, New York.
Meek, R. L., and Gibbon, C. F. (1974). *J. Electrochem. Soc.* **121**, 444.
Merkle, K. L., Pronko, P. P., Gemmell, D. S., Mikkelson, R. D., and Wrobel, J. R. (1973). *Phys. Rev. B* **8**, 1002.

Meyer, L. (1971). *Phys. Status Solidi* **44**, 253.

Meyer, O., Gyulai, J., and Mayer, J. W. (1970). *Surf. Sci.* **22**, 263.

Mitchell, I. V., Kamoshida, M., and Mayer, J. W. (1971). *J. Appl. Phys.* **42**, 4378.

Morgan, D. V. (ed.) (1973). "Channeling." Wiley, New York.

Morgan, D. V., and Bøgh, E. (1972). *Surf. Sci.* **32**, 278.

Morgan, D. V., and Van Vliet, D. (1972). *Radiat. Effects* **12**, 203.

Morgan, D. V., and Wood, D. R. (1973). *Proc. Roy. Soc. London* **A335**, 509.

Mory, J., and Quéré, Y. (1972). *In* "Atomic Collisions in Solids" (S. Andersen, K. Bjorkqvist, B. Domeij and N. G. E. Johansson, eds.), Vol. IV, p. 303. Gordon and Breach, New York.

Picraux, S. T. (1975). *In* "New Uses of Ion Accelerators" (J. F. Ziegler, ed.), Chapter 4. Plenum Press, New York.

Picraux, S. T., and Thomas, G. J. (1973). *J. Appl. Phys.* **44**, 594.

Pircraux, S. T., Brown, W. L., and Gibson, W. M. (1972). *Phys. Rev. B* **6**, 1382.

Poate, J. M., Buck, T. M., and Schwartz, B. (1973). *J. Phys. Chem. Solids* **34**, 779.

Pabst, H. J., and Palmer, D. W. (1973). *Inst. Phys. Conf. Ser. No. 16* 438.

Pronko, P. P., and Merkle, K. L. (1974). *In* "Applications of Ion Beams to Metals" (S. T. Picraux, E. P. EerNisse, and F. L. Vook, eds.), p. 481. Plenum, New York.

Pronko, P. P., Rechtin, M. D., Foti, G., Csepregi, L., Kennedy, E. F., and Mayer, J. W. (1977). *In* "Ion Implantation in Semiconductors and Other Materials," (F. Chernow, J. Borders, and D. K.Brice, eds.), p. 503. Plenum Press, New York.

Quére, Y. (1974). *J. Nucl. Mater.* **53**, 262.

Rechtin, M. D., Pronko, P. P., Foti, G., Csepregi, L., Kennedy, E. F., and Mayer, J. W. (1978). *Phil. Mag.*

Rimini, E., Lugujjo, E., and Mayer, J. W. (1972a). *Phys. Rev. B* **6**, 716.

Rimini, E., Haskell, J., and Mayer, J. W. (1972b). *Appl. Phys. Lett.* **20**, 234.

Sigmon, T. W., Chu, W. K., Lugujjo, E., and Mayer J. W. (1974). *Appl. Phys. Lett.* **24**, 105.

Sigurd, D., and Bjorkqvist, K. (1973a). *Radiat. Effects* **17**, 209.

Sigurd, D., Bower, R. W., van der Weg, W. F., and Mayer, J. W. (1973b). *Thin Solid Films* **19**, 319.

Sigurd, D., Ottaviani, G., Arnal, H., Mayer, J. W. (1974). *J. Appl. Phys.* **45**, 1740.

Swanson, M. L., Howe, L. M., and Quenneville, A. F. (1975). *Radiat. Effects* **25**, 61.

Tseng, W. F., Gyulai, J., Koji, T., Lau, S. S., Roth, J., and Mayer, J. W. (1978). *Nucl. Instr. and Meth.* **149**, 615.

Tu, K. N., Alessandrini, E. I., Chu, W. K., Krautle, H., and Mayer, J. W. (1974). *Jpn. J. Appl. Phys. Suppl. 2* Part 1, 669.

Van Vliet, D. (1973). *In* "Channeling" (D. V. Morgan, ed.), Chapter 2. Wiley, New York.

Westmoreland, J. E., Mayer, J. W., Eisen, F. H., and Welch, B. (1970). *Radiat. Effects* **6**, 161.

Chapter

# 9

# Energy-Loss Measurements

## 9.1 INTRODUCTION

In the preceding chapters, we assumed knowledge of the stopping cross sections $\varepsilon$ and used values of $\varepsilon$ in backscattering analyses to determine depth scales and sample composition. In this chapter we shall reverse the procedure and use knowledge of the sample composition to determine stopping cross section values from energy widths or signal heights of backscattering spectra.

Backscattering is just one of the three general methods that have been used to determine $\varepsilon$ values. The measurement of energy lost by particles which traverse thin self-supported foils was one of the first methods used, and the measurement of Doppler shift attenuation represents the most recent method. In this chapter we concern ourselves with the backscattering method, since it represents one of the applications of backscattering spectrometry. Transmission and Doppler-shift attenuation measurements have been reviewed by Chu (1979).

The energy of backscattered particles depends on the energy loss along both the incoming and outgoing paths. Consequently, from the backscattering measurements one obtains values for the stopping cross section factor $[\varepsilon]$ rather than the stopping cross section $\varepsilon$. We show in the next section how one can extract values of $\varepsilon$ from $[\varepsilon]$.

276

One can determine values of the stopping cross section factor $[\varepsilon]$ from measurements of either the energy width or the height of a backscattering signal. In the former case, measurements are made on thin films and the parameter that has to be known is the number of atoms per unit area. To determine $[\varepsilon]$ from the signal height requires, in turn, an absolute calibration of the solid angle of detection $\Omega$ (i.e., of the effective area of the detector and of the distance between the detector and the target) as well as an accurate measurement of the total number $Q$ of incident particles. In the first method, then, the main experimental effort resides in the preparation of the target, whereas in the second method the emphasis is on the detector setup and the charge collection system. The two methods will be discussed further in the last two sections of the chapter.

### 9.2  EXTRACTION OF ε VALUES FROM [ε] MEASUREMENTS

#### 9.2.1  Mean Energy Approximation and Expansion of ε

Backscattering measurements give information on the stopping cross section factor

$$[\bar{\varepsilon}] = \frac{K}{\cos\theta_1}\,\varepsilon(\bar{E}_{in}) + \frac{1}{\cos\theta_2}\,\varepsilon(\bar{E}_{out}),\qquad(9.1)$$

where $\bar{E}_{in}$ and $\bar{E}_{out}$ are energies taken somewhere along the incoming and outgoing paths. Assume that values for $[\varepsilon]$ have been experimentally determined. The problem then is how to extract $\varepsilon$ from such measurements.

One approach is to derive one value of the stopping cross section $\varepsilon$ from each measurement of $[\varepsilon]$ and to assign this stopping cross section value to an energy $E_x$ that is intermediate to $\bar{E}_{in}$ and $\bar{E}_{out}$. In general, the value of $E_x$ will lie in the energy interval between the energy region spanned along the incoming path and that along the outgoing path (i.e. between $E$ and $KE$). Since one measurement of $[\varepsilon]$ contains two unknowns, $\varepsilon(\bar{E}_{in})$ and $\varepsilon(\bar{E}_{out})$, a value of $\varepsilon(E_x)$ can be obtained only with the help of additional assumptions. Warters (1953) assumes that $\varepsilon(E)$ can be expanded in a Taylor series about the intermediate energy $E_x$ and in this way obtains the needed relationship between $(\bar{E}_{in})$ and $(\bar{E}_{out})$:

$$\varepsilon(\bar{E}_{in}) = \varepsilon(E_x) + \frac{d\varepsilon}{dE}\bigg|_{E=E_x}(\bar{E}_{in} - E_x) + \cdots,\qquad(9.2)$$

$$\varepsilon(\bar{E}_{out}) = \varepsilon(E_x) + \frac{d\varepsilon}{dE}\bigg|_{E=E_x}(\bar{E}_{out} - E_x) + \cdots.\qquad(9.3)$$

Substituting these expansions for $\varepsilon(\bar{E}_{in})$ and $\varepsilon(\bar{E}_{out})$ in the expression for $[\varepsilon]$ gives

$$[\varepsilon] = \left(\frac{K}{\cos\theta_1} + \frac{1}{\cos\theta_2}\right)\varepsilon(E_x)$$

$$+ \left[\frac{K}{\cos\theta_1}(\bar{E}_{in} - E_x) + \frac{1}{\cos\theta_2}(\bar{E}_{out} - E_x)\right]\frac{d\varepsilon}{dE}\bigg|_{E=E_x} + \cdots . \quad (9.4)$$

The energy $E_x$ is now chosen in such a way that the second term in this expansion vanishes. If we neglect the terms of higher order, both $E_x$ and $\varepsilon(E_x)$ can be specified as

$$\varepsilon(E_x) = \frac{[\varepsilon]}{(K/\cos\theta_1) + (1/\cos\theta_2)} \quad (9.5)$$

and

$$E_x = \frac{(K/\cos\theta_1)\bar{E}_{in} + (1/\cos\theta_2)\bar{E}_{out}}{(K/\cos\theta_1) + (1/\cos\theta_2)}. \quad (9.6)$$

To relate $\bar{E}_{in}$ or $\bar{E}_{out}$ to the energy loss $\Delta E_{in}$ and $\Delta E_{out}$, the simplest procedure is to use the mean energy approximation discussed in Section 3.2.2 and set $\bar{E}_{in} = E_0 - \frac{1}{2}\Delta E_{in}$ and $\bar{E}_{out} = E_1 + \frac{1}{2}\Delta E_{out}$; this gives

$$E_x = \frac{(K/\cos\theta_1)E_0 + (1/\cos\theta_2)E_1}{(K/\cos\theta_1) + (1/\cos\theta_2)} + \frac{1}{2}\frac{(-K/\cos\theta_1)\Delta E_{in} + (1/\cos\theta_2)\Delta E_{out}}{(K/\cos\theta_1) + (1/\cos\theta_2)}.$$

$$(9.7)$$

The second term can be expressed in terms of $\Delta E$ since

$$\Delta E = K\,\Delta E_{in} + \Delta E_{out} = \Delta E_{in}(K + \Delta E_{out}/\Delta E_{in}),$$

leading to

$$E_x = \frac{(K/\cos\theta_1)E_0 + (1/\cos\theta_2)E_1}{(K/\cos\theta_1) + (1/\cos\theta_2)}$$

$$+ \frac{\Delta E/2}{(K/\cos\theta_1) + (1/\cos\theta_2)}\frac{(-K/\cos\theta_1) + (n/\cos\theta_2)}{K + n}, \quad (9.8)$$

where $n = \Delta E_{out}/\Delta E_{in}$. For $\theta_1 = \theta_2$, this expression reduces to

$$E_x = \left(\frac{KE_0 + E_1}{K + 1}\right) + \frac{1}{2}\left(\frac{KE_0 - E_1}{(K + 1)}\right)\left(\frac{n - K}{n + K}\right), \quad (9.9)$$

which is the formula given by Warters (1953). The second term is usually small, and the ratio $n = \Delta E_{out}/\Delta E_{in}$ can be approximated without much error by the stopping cross section ratio $\varepsilon(E_1)/\varepsilon(E_0)$ (see Fig. 3.7). As a first approximation, one takes $n$ as unity.

### 9.2.2  Surface Energy Approximation and Expansion of $\varepsilon$

Energy-loss measurements are often made under conditions that meet the surface energy approximation; that is, the value of $\bar{E}_{\text{in}}$ and $\bar{E}_{\text{out}}$ are taken as $E_0$ and $KE_0$, respectively. This means that $\Delta E_{\text{in}}$ and $\Delta E_{\text{out}}$ are negligibly small. The second term in Eq. (9.7) for $E_x$ vanishes and $E_1 \simeq KE_0$, so that in this case

$$\varepsilon(E_x) = \frac{[\varepsilon_0]}{(K/\cos\theta_1) + (1/\cos\theta_2)} \tag{9.10}$$

and

$$E_x = \frac{(1/\cos\theta_1) + (1/\cos\theta_2)}{(K/\cos\theta_1) + (1/\cos\theta_2)} KE_0. \tag{9.11}$$

### 9.2.3  Surface Energy Approximation and Ratio of $\varepsilon$

Another approach that is particularly useful in the surface energy approximation is to rewrite the stopping cross section factor $[\varepsilon_0]$ in the form [see (Eq. 3.12)]

$$[\varepsilon_0] = \varepsilon(E_0)\left[\frac{K}{\cos\theta_1} + \frac{1}{\cos\theta_2}\frac{\varepsilon(KE_0)}{\varepsilon(E_0)}\right]. \tag{9.12}$$

Often the shape of the $\varepsilon$ versus $E$ curve is known either from measurements on neighboring elements or from theoretical predictions. The ratio $\varepsilon(KE_0)/\varepsilon(E_0)$ is then known, and $\varepsilon(E_0)$ can be determined.

### 9.3  MEASUREMENT OF [ε] FROM THIN-FILM DATA

As was stated initially, one way to determine the stopping cross section factor $[\varepsilon]$ is to measure the energy width $\Delta E$ of a backscattering signal for a thin film with a known value $Nt$ of atoms per unit area. In these measurements, the most difficult task lies in preparing the film and measuring the thickness. The film must be free of impurities to the extent that the ratio of the product of concentration times the stopping cross section for impurity to host $N_{\text{imp}}\varepsilon^{\text{imp}}(E_0)/N\varepsilon(E_0)$ should be less than a few percent. The film should be amorphous or polycrystalline and with no preferred orientation of the crystallites that might influence the measurement of $\Delta E$ through channeling effects.

The measurement of the film weight divided by the film area gives the mass per unit area $\rho t$, where $\rho$ is the volume density in grams per cubic centimeter. The value $Nt$ is given by the product $\rho t \cdot N_0/M$, where $N_0$ is Avogadro's number and $M$ the atomic weight of the sample [Eq. (2.38)]. The film preparation and the weight measurements are difficult since the films should be thin enough so that $\Delta E \ll E_0$.

A last consideration is that the kinematic factor $K$ should be as large as possible. This places the energy $\bar{E}_{out}$ of the outward track close to that of $\bar{E}_{in}$ of the inward track and makes it easy to extract $\varepsilon$ values from $[\varepsilon]$. With a high $Z$ film, this condition poses no problem, but with a low $Z$ film it is indicated to deposit the film on a high $Z$ substrate. We will illustrate the latter case with a numerical example.

For measurements of the stopping cross section of vanadium (Chu and Powers, 1969), the target was prepared by vacuum deposition of the low-atomic-mass element vanadium ($M_2 = 51$) on the high-atomic-mass substrate of tantalum ($M_2 = 181$). The difference in energy $\Delta E$ between particles scattered from covered and uncovered Ta gives a value of $[\bar{\varepsilon}]_{Ta}^V$. Figure 9.1

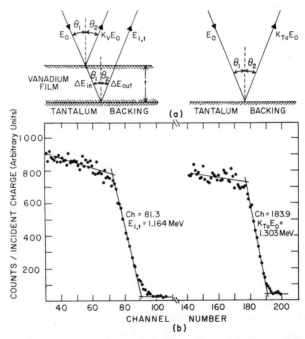

**Fig. 9.1** Typical example of (a) the scattering kinematics and (b) the corresponding spectra for $^4$He backscattered through an angle of $\theta = 130°$ ($\theta_1 = \theta_2 = 25°$). The incident energy is $E_0 = 1.402$ MeV. The detected energy $E_1$ is 1.303 MeV for scattering from the bare Ta backing, and 1.164 MeV for scattering from the Ta backing covered with a vanadium film of 59.6 $\mu$gm/cm$^2$ [From Chu and Powers (1969).]

shows backscattering spectra for 1.402-MeV $^4$He scattered from the samples at $\theta_1 = \theta_2 = 25°$ ($\theta = 130°$). Particles scattered from the uncovered Ta are detected at an energy $K_{Ta}E_0 = 1.303$ MeV. Particles scattered from the Ta backing covered with the V layer have an energy $E_{1,t} = 1.164$ MeV. The value of $\Delta E$ is given by the difference $\Delta E = 139$ keV.

The sample shown in Fig. 9.1 has a mass per unit area $\rho t = 59.6$ $\mu$gm/cm$^2$. From the values of $N_0$ (6.025 × 10$^{23}$ atom/mole) and the atomic weight of vanadium ($M_V = 50.9$), the number $Nt$ of V atoms per square centimeter is $7.05 × 10^{17}$. The value of $[\bar{\varepsilon}]$ is then given by

$$[\bar{\varepsilon}] = \frac{\Delta E}{Nt} = \frac{139 \quad \text{keV}}{7.05 × 10^{17} \quad \text{atom/cm}^2} = 197 × 10^{-15} \quad \frac{\text{eV cm}^2}{\text{atom}} \quad (9.13)$$

and, from Eq. (9.5),

$$\varepsilon(E_x) = \frac{\cos \theta_1}{(1 + K_{Ta})} \frac{\Delta E}{Nt} = 92.6 × 10^{-15} \quad \frac{\text{eV cm}^2}{\text{atom}}. \quad (9.14)$$

To determine the value of the intermediate energy $E_x$, we substitute in Eq. (9.9) values of $K_{Ta}$ ($\theta = 130°$) = 0.9299, $n = 1.028$ (although a value of unity could have been chosen as a first guess), $E_0 = 1.402$ MeV, and the measured value of $E_1 = 1.164$ MeV. Using the first two terms of Eq. (9.9) gives

$$E_x = (1278.3 + 1.8) \quad \text{keV} = 1280 \quad \text{keV}. \quad (9.15)$$

To determine an entire $\varepsilon$ versus energy curve, one measures the energy difference $\Delta E$ for a sequence of beam energies $E_0$ and for different target thicknesses (Chu and Powers, 1969).

This example was for a low-mass element on a high-mass substrate. For high-mass elements on a low-mass substrate, the spectra are as described in Chapter 4 with some additional examples given in Chapter 5. The extraction of $\varepsilon(E_x)$ values from measurements of $\Delta E$ for samples of known $Nt$ follows the procedure previously described.

## 9.4   DETERMINATION OF [ε] FROM SIGNAL HEIGHT

In the preceding section, the values of the stopping cross section factor $[\bar{\varepsilon}]$ were obtained from measurements of an energy difference in the backscattering spectra of thin films. The major uncertainty in the result arises from difficulties in sample preparation and characterization: thickness, uniformity, contaminants, texture, etc.

Another approach is to measure signal heights. This technique was first used by Wenzel and Whaling (1952) in their measurements of the proton stopping cross section of ice. Later measurements in more conventional samples have been made by Bethge and Sandner (1965), Chu *et al.* (1973), Leminen (1972), Lin *et al.* (1974), Feng *et al.* (1973), Behrisch and Scherzer (1973), and others. The concept is based on the fact that the signal height $H(E_1)$ is proportional to $[\varepsilon]^{-1}$, as described in Chapter 3 [Eqs. (3.38) and (3.50)]. Experimentally, the major difficulty with this method lies in the accurate measurement of the incident dose $Q$ and the solid angle $\Omega$ of the detector. These problems were discussed in Chapter 6. In the following, we illustrate some of the ways in which signal heights have been used to determine stopping cross sections.

### 9.4.1   Height of the Surface Yield

Figure 9.2 (Chu *et al.*, 1973) shows the energy spectrum of 2-MeV $^4$He scattered from normal incidence through $\theta = 170°$ from an indium sample. In this spectrum, the integrated number $Q$ of particles is $3.084 \times 10^{13}$ $^4$He$^+$ ions (4.94 $\mu$C), $\Omega = 4.306 \times 10^{-3}$ sr, $\mathscr{E} = 4.88$ keV/channel, and $\sigma(E_0) =$

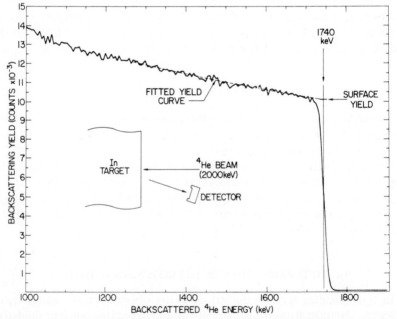

**Fig. 9.2**   Spectrum of 2-MeV $^4$He backscattered through $\theta = 170°$ from a thick In sample. The total incident dose of $^4$He$^+$ is 4.94 $\mu$C; the detector solid angle $\Omega$ is 4.306 msr; the energy per channel $\mathscr{E}$ is 4.88 keV/channel. The height of the yield at the edge of the signal, which corresponds to scattering from the surface layer, is obtained by extrapolating the signal height to the energy position of the edge. [From Chu *et al.* (1973).]

$3.152 \times 10^{-24}$ cm². The height $H(E_0)$ is found by extrapolating the yield to the energy $K_{In}E_0 = 1.74$ MeV. The height in Fig. 9.2 has a value of 10,200 counts. From Eq. (3.38) and the values just given,

$$[\varepsilon_0] = \sigma(E_0)\Omega Q(\mathscr{E}/H_0) = 200.3 \times 10^{-15} \quad \text{eV cm}^2/\text{atom}, \quad (9.16)$$

and, from Eq. (9.10),

$$\varepsilon(E_x) = \frac{[\varepsilon_0]}{(K + (1/\cos\theta_2))} = 106.2 \times 10^{-15} \quad \text{eV cm}^2/\text{atom}, \quad (9.17)$$

where, from Eq. (9.11), $E_x = 1861$ keV.

Another method of extracting $\varepsilon$ values from $[\varepsilon_0]$ measurements is to use the $\varepsilon$ ratio method of Section 9.2.3. In that case the ratio $\varepsilon(KE_0)/\varepsilon(E_0)$ needs to be known. One can use the ratio of experimental values of the stopping cross sections of neighboring elements, of theoretical calculations, or of semiempirical tabulations. In the latter case (Table VI),

$$\varepsilon(1.74 \quad \text{MeV})/\varepsilon(2.00 \quad \text{MeV}) = 1.035, \quad (9.18)$$

so that by Eq. (9.12),

$$\varepsilon(2.0 \quad \text{MeV}) = [\varepsilon_0]\frac{1}{K_{In} + (1.035/\cos\theta_2)} = 104.3 \times 10^{-15} \quad \frac{\text{eV cm}^2}{\text{atom}}. \quad (9.19)$$

### 9.4.2 Ratio of Surface Yields

One method of removing the experimental uncertainties associated with the determination of the values for $\Omega$ and $Q$ is to compare the surface yields for two different elements A and B; Eq. (3.38) gives

$$[\varepsilon_0]_A^A = \frac{\sigma_A}{\sigma_B}\frac{H_{0B}}{H_{0A}}[\varepsilon_0]_B^B. \quad (9.20)$$

If $[\varepsilon_0]_B^B$ is known from previous measurements of $\varepsilon^B$, then $[\varepsilon_0]_A^A$ can be determined and values of $\varepsilon^A$ extracted as discussed previously. The only experimental difficulty in such measurements is to ensure the reproducibility of the incident dose. It is often simpler, though, to perform reproducible measurements than an absolute measurement.

Leminen (1972) has used this comparison technique to determine stopping cross sections for hydrogen in various metallic elements. Figure 9.3 shows the energy spectra of 0.5-MeV $^1$H scattered from thick targets of Ti, Cu, Mo, Ag, W, and Au. He used previously established $\varepsilon$ values for Cu, Ag, and Au as standards and as checks on the internal consistency of the data.

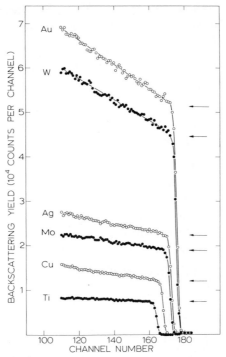

**Fig. 9.3** Energy spectra of 3-$\mu$C-dosed, 0.5-MeV $^1$H backscattered from thick samples of Ti, Cu, Mo, Ag, W, and Au. The Cu, Ag, and Au spectra serve as standards against which the others are calibrated and as checks on the consistency of the data. The arrows indicate the values of the extrapolated surface yields. [From Leminen (1972).]

### 9.4.3  Relative Yields

Another way to circumvent absolute measurements of $Q$ and $\Omega$ is to use layered samples. One determines the yield of particles scattered from the interface between a deposited film and the underlying element in the substrate or another film underneath. In this case, the ratio of the yield from the elements on either side of the interface specifies a ratio of the stopping cross section factors $[\varepsilon]$. However, as was shown in Chapter 4, the analytical formulation is more complicated than for measurements of surface yields, since one has to account for the effect of the energy loss in the overlying film on the yields. One correction arises from the difference between the energy width $\mathscr{E}$ of a channel and the corresponding energy width $\mathscr{E}'$ for particles scattered at the interface below the surface. Another point is that the values of the stopping cross section factors $[\varepsilon]$ are obtained for the energy $E$ of the particles before scattering at the interface rather than for the incident energy $E_0$.

In the example shown in Fig. 9.4, the ratio of the heights at the interface is given by (see Section 4.3.3)

$$\frac{H_{Au}}{H_{Al}} = \frac{\sigma_{Au}(E)}{\sigma_{Al}(E)} \frac{[\varepsilon(E)]^{Al}_{Al}}{[\varepsilon(E)]^{Au}_{Au}} \frac{\mathscr{E}'_{Au}}{\mathscr{E}'_{Al}}. \tag{9.21}$$

The last ratio can be expressed as

$$\frac{\mathscr{E}'_{Au}}{\mathscr{E}'_{Al}} = \frac{\varepsilon^{Au}(K_{Au}E)}{\varepsilon^{Au}(E_{1Au,t_{Au}})} \frac{\varepsilon^{Au}(E_{1Al,t_{Au}})}{\varepsilon^{Au}(K_{Al}E)} \tag{9.22}$$

or approximated by

$$\frac{\mathscr{E}'_{Au}}{\mathscr{E}'_{Al}} = 1 + \frac{N^{Au}t_{Au}}{\cos\theta_2}\left[\frac{d\varepsilon^{Au}}{dE}(K_{Au}E) - \frac{d\varepsilon^{Au}}{dE}(K_{Al}E)\right] \tag{9.23}$$

as obtained from Eq. (3.49). From their results, Feng et al. (1973) established that the stopping cross sections commonly accepted for Au and Al were at variance with the observed yield ratio. By measuring the interface yields of an Al film on a Si film and relying on the fact that the stopping cross sections of both Au and Si had been measured a number of times by independent investigators with consistent results, they inferred that the Al stopping cross section was in error. Later measurements have borne this out (Feng, 1975; Harris and Nicolet, 1975; Luomajärvi et al., 1976). The same technique was applied to test the internal consistency of the stopping cross sections

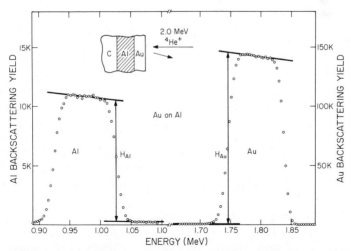

**Fig. 9.4**   Backscattering spectrum for 2.0-MeV $^4$He incident on a two-layered sample of a Au film on top of an Al film on a carbon substrate, demonstrating the method of obtaining the interface yields of two elements. The ratio of these yields determines the ratio of the stopping cross section factors of these two elements. [From Feng et al. (1973)].

of Au, Ag, and Cu by measuring the interface yields of all possible pairs of films deposited on each other (Feng *et al.*, 1973).

### 9.4.4 Thick-Target Yields

As is shown in Eq. (3.50), the height of the backscattering yield of a spectrum for a thick target depends on the stopping cross section factor $[\varepsilon(E)]$. If the spectrum is taken with an absolutely calibrated system, so that the total number $Q$ of incident particles and the solid angle of detection $\Omega$ are known, and if the Rutherford formula for the cross section is valid, one can determine $[\varepsilon(E)]$ from one spectrum for a range of energies $E$. The spectrum, however, is recorded as a function of the detected energy $E_1$, not as a function of the energy $E$ before scattering. The relationship between these two energies also depends on the stopping cross section factor and is therefore initially unknown as well. The extraction of $[\varepsilon(E)]$ from a thick-target spectrum thus requires some initial assumptions, followed later by an iterative procedure or some checks to verify the assumptions or to arrive at a self-consistent solution.

Lin *et al.* (1974) use a formula of Brice (1972) that describes the shape of the stopping cross section in parametric form. They express the thick-target yield as a function of Brice's stopping power parameters and then use a numerical method of iteration to extract these parameters from the thick-target spectrum. Their method also takes into account the influence of the finite resolution of the detection system. They have determined the $^4$He stopping cross section in Au and Ag by this technique and obtained good agreement with earlier measurements.

Another approach has been taken by Behrisch and Scherzer (1973). They assume that the energy loss of the moving particle in the target has the energy dependence $dE/dx = A_v E^v$, where the exponent $v$ is constant over some energy range and has values of $v = \frac{1}{2}$, 0, or $-1$ depending on the energy region considered, and where $A_v$ is a constant (cf. Section 3.3.3b). They can then analytically express the stopping cross section $\varepsilon$ at the detected energy $E_1$ in terms of the backscattering yield $H(E_1)$ at that energy, the energies $E_1$ and $E_0$, and $v$. Scherzer *et al.* (1976) have demonstrated the consistency between this method and that based on thin-films measurements described in Section 9.3 by applying them both to artificial spectra generated by a computer program (Børgesen *et al.*, 1976). They also investigated how the shape of the assumed $\varepsilon(E)$ function (i.e., the value of $v$) influenced the results. The computer-generated spectra were further compared to measured backscattering spectra, with generally good agreement. The advantage of this method is that it is rather simple in execution and provides a point-by-point transposition of $H(E_1)$ to $\varepsilon(E_1)$ once the shape of $\varepsilon$ versus $E$ is fixed. When the material under investigation contains a light element, however, the thin-film method of Section 9.3 is preferable to the thick-target method.

## REFERENCES

Behrisch, R., and Scherzer, B. M. U. (1973). *Thin Solid Films* **19**, 247.
Bethge, K., and Sandner, P. (1965). *Phys. Lett.* **19**, 241.
Børgesen, P., Harris, J. M., and Scherzer, B. M. U. (1976). Interim Tech. Rep. RADC-TR-76-182, available from Rome Air Development Center, Air Force Systems Command, Griffiss Air Force Base, New York.
Brice, D. K. (1972). *Phys. Rev. A* **6**, 1791.
Chu, W. K. (1979). *In* "Accelerators in Atomic Physics" (P. Richard, ed.). A volume in *Methods of Experimental Physics*. Academic Press, New York. (To be published.)
Chu, W. K., and Powers, D. (1969). *Phys. Rev.* **187**, 478.
Chu, W. K., Ziegler, J. F., Mitchell, I. V., and Mackintosh, W. D. (1973). *Appl. Phys. Lett.* **22**, 437.
Feng, J. S.-Y. (1975). *J. Appl. Phys.* **46**, 444.
Feng, J. S.-Y., Chu, W. K., Nicolet, M-A., and Mayer, J. W. (1973). *Thin Solid Films* **19**, 195.
Harris, J. M., and Nicolet, M-A. (1975). *Phys. Rev. B* **11**, 1013.
Leminen, E. (1972). *Ann. Acad. Sci. Fenn. A VI* No. 386.
Lin, W. K., Matteson, S., and Powers, D. (1974). *Phys. Rev. B* **10**, 3746.
Luomajärvi, M., Fontell, A., and Bister, M. (1976). *In* "Ion Beam Surface Layer Analysis" (O. Meyer, G. Linker, and F. Käppeler, eds.), p. 75. Plenum Press, New York.
Scherzer, B. M. U., Børgesen, P., Nicolet, M-A., and Mayer, J. W. (1976). *In* "Ion Beam Surface Layer Analysis" (O. Meyer, G. Linker, and F. Käppeler, eds.), p. 33. Plenum Press, New York.
Warters, W. D. (1953). Thesis, California Inst. of Technol.
Wenzel, W. A., and Whaling, W. (1952). *Phys. Rev.* **87**, 499.

Chapter

# *10*

# Bibliography on Applications of Backscattering Spectrometry

## 10.1 INTRODUCTION

Backscattering spectrometry has been applied to a wide variety of subjects that require knowledge of the properties of near-surface regions. The analytical methods used in these applications of backscattering spectrometry have been covered in previous chapters. In this chapter, we cite publications dealing with applications of the techniques. We have not included references to studies of atomic collision phenomena, channeling, energy loss, scattering cross sections, or other basic phenomena, but rather concentrate on papers in which backscattering is used as an analytical tool, with some examples of applications of ion-induced x-rays and nuclear reactions. As a practical measure due to the large number of publications, we cite only late 1975–1976 references to ion implantation in semiconductors. The cutoff date on the citations is August 1976. Except for Section 10.1.3, the references are listed in alphabetical order of the last name of the first author.

### 10.1.1 Early Papers

Surface Analysis by Means of Nuclear Particle Scattering
  W. W. Buechner and J. Robertshaw, *Trans. Am. Nucl. Soc.* **5**, 197 (1962).

Rutherford Scattering and Channeling—A Useful Combination for Chemical Analysis of Surfaces
  W. D. Mackintosh and J. A. Davies, *Anal. Chem.* **41**, 26A (1969).
Chemical Analysis of Surfaces Using Alpha Particles
  J. H. Patterson, A. L. Turkevich, and E. Franzgrote, *J. Geophys. Res.* **70**, 1311 (1965).
Analysis of Surfaces by Scattering of Accelerated Alpha Particles
  M. Peisach and D. O. Poole, *Anal. Chem.* **38**, 1345 (1966).
Microanalysis of Surfaces by Ion Beam Scattering
  M. Peisach and D. O. Poole, *in* "Electron and Ion Beam Technology" (R. Bakish, ed.), Vol. 2, p. 1195. Gordon and Breach, New York, 1969.
Surface Analysis by Charged Particle Spectroscopy
  S. Rubin, *Nucl. Instrum. Methods* **5**, 177 (1959).
Chemical Analysis of Surfaces by Nuclear Methods
  S. Rubin, T. O. Passell, and L. E. Bailey, *Anal. Chem.* **29**, 736 (1957).
Ion-Scattering Methods
  S. Rubin, *in* "Treatise on Analytical Chemistry" (I. M. Kolthoff, P. J. Elving, and E. B. Sandell, eds.), Vol. 4, Chapter 41, p. 2075. Wiley, New York, 1963.
Diffusion Measurements in the System Cu–Au by Elastic Scattering
  R. F. Sippel, *Phys. Rev.* **115**, 1441 (1959).
Chemical Analysis of Surfaces by Use of Large-Angle Scattering of Heavy Charged Particles
  A. Turkevich, *Science* **134**, 672 (1961).
Application of the Coulomb Backscattering of the Heavy Charged Particles.
  I—Thickness Measurements of Thin Foils and Surface Layers, A. Turos and Z. Wilhelmi, *Nukleonika* **13**, 975 (1968).

  II—Chemical Analysis of Surface Layers, A. Turos and Z. Wilhelmi, *Nukleonika* **14**, 319 (1969).

## 10.1.2 Review Articles

Microanalysis by the Direct Observation of Nuclear Reactions Using a 2 MeV van de Graaff
  G. Amsel, J. P. Nadai, E. D'Artemare, D. David, E. Girard, and J. Moulin, *Nucl. Instrum. Methods* **92**, 481 (1971).
Depth Distribution of Damage Obtained by Rutherford Backscattering Combined with Channeling
  R. Behrisch and J. Roth, *in* "Ion Beam Surface Layer Analysis" (O. Meyer, G. Linker, and F. Käppeler, eds.), Vol. 2, p. 539. Plenum Press, New York, 1976.
Characteristic X-Ray Production by Heavy Ion Bombardment as a Technique for the Examination of Solid Surfaces
  J. A. Cairns, *Surf. Sci.* **34**, 638 (1973).
Principles and Applications of Ion Beam Techniques for the Analysis of Solids and Thin Films
  W. K. Chu, J. W. Mayer, M-A. Nicolet, T. M. Buck, G. Amsel, and F. Eisen, *Thin Solid Films* **17**, 1 (1973).
Ion Beam Analysis Techniques in Corrosion Science
  G. Dearnaley, *in* "Ion Beam Surface Layer Analysis" (O. Meyer, G. Linker, and F. Käppeler, eds.), Vol. 2, p. 885. Plenum Press, New York, 1976.
Ion Implantation and Channeling
  F. H. Eisen and J. W. Mayer, *In* "Treatise on Solid State Chemistry" (B. Hannay, ed.), Vol. 6B, Surfaces II, Chapter 2. Plenum Press, New York, 1976.
Surface and Thin Film Analysis of Semiconductor Materials
  R. E. Honig, *Thin Solid Films* **31**, 89 (1976).

Rutherford Scattering
    W. D. Mackintosh, *in* "Characterization of Solid Surfaces" (P. F. Kane and G. B. Larrabee, eds.), Chapter 16, p. 403. Plenum Press, New York, 1974.
Comparison of Surface Layer Analysis Techniques
    J. W. Mayer and A. Turos, *Thin Solid Films* **19**, 1 (1973).
Thin Films and Solid-Phase Reactions
    J. W. Mayer, J. M. Poate, and K. N. Tu, *Science* **190**, 228 (1975).
Ion Implantation in Superconductors
    O. Meyer, *in* "New Uses of Ion Accelerators" (J. F. Ziegler, ed.), Chapter 6. Plenum Press, New York, 1975.
Principles and Applications of Elastic Scattering with High-Energy Protons and Heavy ions for Quantitative Microanalysis of Metal Surfaces, Thin Layers and Sandwich Structures
    P. Muller and G. Ischenko, *J. Appl. Phys.* **47**, 2811 (1976).
Microanalysis of Materials by Backscattering Spectrometry
    M-A. Nicolet, I. V. Mitchell, and J. W. Mayer, *Science* **177**, 841 (1972).
Accelerator Microbeam Techniques
    T. B. Pierce, *in* "Characterization of Solid Surfaces" (P. F. Kane and G. B. Larrabee, eds.), Chapter 17, p. 419. Plenum Press, New York, 1974.
Lattice Location of Impurities in Metals and Semiconductors
    S. T. Picraux, *in* "New Uses of Ion Accelerators" (J. F. Ziegler, ed.), Chapter 4. Plenum Press, New York, 1975.
Ion Scattering Spectroscopy
    J. M. Poate and T. M. Buck, *Exp. Methods Catal. Res.* **3**, 175 (1976).
Charged Particle Elastic Scattering and Channeling Techniques for Surface and Near Surface Analyses
    E. A. Wolicki, J. K. Hirvonen, J. E. Westmoreland, A. G. Pieper, and W. H. Weisenberger, Rep. of NRL Progr., p. 1 (November 1972).
Material Analysis by Nuclear Backscattering
    J. F. Ziegler, J. W. Mayer, B. M. Ullrich, and W. K. Chu, *in* "New Uses of Ion Accelerators" (J. F. Ziegler, ed.), Chapter 2, p. 75. Plenum Press, New York, 1975.

### 10.1.3  Books and Conference Proceedings

Atomic Collision and Penetration Studies
    Univ. of Toronto Press, Toronto, Canada, 1967 (*Conf. Proc., Chalk River, Canada, 1967*).
Radiation Effects in Semiconductors
    F. L. Vook (ed.), Plenum Press, New York, 1968 (*Conf. Proc., Santa Fe, New Mexico, 1967*).
Ion Implantation in Semiconductors
    J. W. Mayer, L. Eriksson, and J. A. Davis. Academic Press, New York, 1970.
Atomic Collision Phenomena in Solids
    D. W. Palmer, M. W. Thompson, and P. D. Townsend (eds.). North Holland Publ., Amsterdam, 1970 (*Conf. Proc., Univ. of Sussex, England, 1969*).
Ion Implantation
    F. H. Eisen and L. T. Chadderton (eds.), Gordon and Breach, New York, 1971 (*Conf. Proc., Thousand Oaks, California, 1970*).
Radiation Effects in Semiconductors
    J. W. Corbett and G. K. Watkins (eds.). Gordon and Breach, New York, 1971 (*Conf. Proc., State Univ. of New York, Albany, New York, 1970*).

Ion Implantation in Semiconductors
  I. Ruge and J. Graul (eds.). Springer-Verlag, New York, 1971 (*Conf. Proc., Garmisch-Partenkirchen, Germany, 1971*).
Ion Implantation in Semiconductors and Other Materials
  B. L. Crowder (ed.). Plenum Press, New York, 1973 (*Conf. Proc., Yorktown Heights, New York, 1973*).
Channeling
  D. V. Morgan (ed.). Wiley, New York, 1973.
Ion Implantation
  G. Dearnaley, J. H. Freeman, R. S. Nelson, and J. Stephan. North-Holland Publ. Amsterdam, 1973.
Ion Beam Surface Layer Analysis
  J. W. Mayer and J. F. Ziegler (eds.). American Elsevier, Amsterdam, 1974; *Thin Solid Films, Vol.* **19** (*Conf. Proc., Yorktown Heights, New York, 1973*).
Atomic Collisions in Solids
  S. Datz, B. R. Appleton, and C. D. Moak. Plenum Press, New York, 1975 (*Conf. Proc., Gatlinburg, Tennessee 1973*).
Applications of Ion Beams to Metals
  S. T. Picraux, E. P. EerNisse, and F. L. Vook (eds.). Plenum Press, New York, 1974 (*Conf. Proc. Albuquerque, New Mexico, 1973*).
Lattice Defects in Semiconductors 1974
  *Inst. Phys. Conf. Ser.* **23**, (1975) (*Conf. Proc., Freiburg, Germany, 1974*).
Characterization of Solid Surfaces
  P. F. Kane and G. B. Larrabee (eds.). Plenum Press, New York, 1974.
Ion Implantation in Semiconductors
  S. Namba (ed.). Plenum Press, New York, 1975 (*Conf. Proc., Osaka, Japan, 1974*).
New Uses of Ion Accelerators
  J. F. Ziegler (ed.). Plenum Press, New York, 1975.
Applications of Ion Beams to Materials 1975
  G. Carter, J. S. Colligon, and W. A. Grant (eds.), *Inst. Phys. Conf. Ser.* **28** (1976) (*Conf. Proc., Univ. of Warwick, 1975*).
Ion Implantation, Sputtering and Their Applications
  P. D. Townsend, J. C. Kelly, and N. E. W. Hartley (eds.). Academic Press, New York, 1976.
Low Temperature Diffusion and Applications to Thin Films
  A. Gangulee, P. S. Ho, and K. N. Tu (eds.). American Elsevier, Amsterdam, 1975. Journal Publication: *Thin Solid Films* **25** (1975) (*Conf. Proc., Yorktown Heights, New York, 1974*).
Methods of Surface Analysis
  A. W. Czanderna (ed.). Elsevier, Amsterdam, 1975.
Ion Beam Surface Layer Analysis
  O. Meyer, G. Linker, and F. Käppeler (eds.). Plenum Press, New York, 1976 (*Conf. Proc., Karlsruhe, Germany, 1975*).
Atomic Collisions in Solids
  F. W. Saris and W. F. Van der Weg (eds.). North Holland Publ., Amsterdam, 1976 (*Conf. Proc., Amsterdam, Holland, 1975*); Journal Publication: *Nucl. Instrum. Methods* 132 (1976).
Radiation Effects on Solid Surfaces
  M. Kaminsky (ed.). Advances in Chemistry Series 158 (Amer. Chem. Soc. Washington, D.C. 1976).
Ion Beam Handbook for Material Analysis
  J. W. Mayer and E. Rimini (eds.). Academic Press, New York, 1977.

Material Characterization Using Ion Beams
  A. Cachard and J. P. Thomas (eds.). Plenum Press, New York, 1977.
Ion Implantation of Semiconductors
  G. Carter and W. A. Grant. Wiley, New York, 1977.
Ion Implantation in Semiconductors and Other Materials
  F. Chernow, J. A. Borders, and D. K. Brice (eds.). Plenum Press, New York, 1977 (*Conf.*
  *Proc., Boulder, Colorado, 1976*).
Ion Beam Analysis
  E. A. Wolicki. J. W. Bulter, and P. A. Treado (eds.). North Holland Publ., (*Conf. Proc.*
  *Washington., D.C., 1977*); Journal Publication: *Nucl. Instrum. Methods* (1978).
Thin Films-Interdiffusion and Reactions
  J. M. Poate, K. N. Tu, and J. W. Mayer (eds.). Wiley, New York, 1978.

## 10.2  SURFACES

### 10.2.1  General Analysis

Application to Surface Studies
  E. Bøgh, *in* "Channeling" (D. V. Morgan, ed.), Chapter 15, p. 435. Wiley, New York, 1973.
Ion Scattering for Analysis of Surfaces and Surface Layers
  T. M. Buck and J. M. Poate, *J. Vac. Sci. Technol.* **11**, 289 (1974).
Studies of Solid Surfaces with 100 keV $^4$He$^+$ and H$^+$ Ion Beams
  T. M. Buck and G. H. Wheatley, *Surf. Sci.* **33**, 35 (1972).
The Use of Ion Beams in Surface Physics Studies
  G. Carter, *J. Vac. Sci. Technol.* **10**, 95 (1973).
Scattering, Energy Loss and X-Ray Production of Protons of 150–175 KeV at Grazing Inci-
dence on Tungsten Crystals
  G. S. Harbinson, B. W. Farmery, H. J. Pabst, and M. W. Thompson, *Radiat. Effects* **27**, 97
  (1975).
The Detection Sensitivity of Heavy Impurities in Si Using 280 keV He$^{2+}$ and C$^{2+}$ Backscattering
  R. R. Hart, H. L. Dunlap, A. J. Mohr, and O. J. Marsh, *Thin Solid Films* **19**, 137 (1973).
Application of a High Resolution Magnetic Spectrometer to Near-Surface Materials Analysis
  J. K. Hirvonen and G. K. Hubler, *in* "Ion Beam Surface Layer Analysis" (O. Meyer, G.
  Linker, and F. Käppeler, eds.), Vol. I, p. 457. Plenum Press, New York, 1976.
Rutherford Scattering and Channeling—A Useful Combination for Chemical Analysis of
Surfaces
  W. D. Mackintosh and J. A. Davies, *Anal. Chem.* **41**, 26A (1969).
Near-Surface Investigation by Backscattering of N$^+$ Ions and Grazing Angle Beam Incidence
  W. Pabst, *in* "Ion Beam Surface Layer Analysis" (O. Meyer, G. Linker, and F. Käppeler,
  eds.), Vol. I, p. 211. Plenum Press, New York, 1976.
An Investigation of RF Sputter Etched Silicon Surfaces Using Helium Ion Backscattering
  G. W. Sachse, W. E. Miller, and C. Gross, *Solid State Electron.* **18** 431 (1975).
The Application of Low Angle Rutherford Backscattering to Surface Layer Analysis
  J. S. Williams, *in* "Ion Beam Surface Layer Analysis" (O. Meyer, G. Linker, and F. Käppeler,
  eds.), Vol. I, p. 223. Plenum Press, New York, 1976.

### 10.2.2  Impurities on Semiconductor Surfaces

The Detection Sensitivity of Heavy Impurities in Si using 280 keV He$^{2+}$ and C$^{2+}$ Backscattering
  R. R. Hart, H. L. Dunlap, A. J. Mohr, and O. J. Marsh, *Thin Solid Films* **19**, 137 (1973).

Implantation of Argon into $SiO_2$ Films Due to Backsputter Cleaning
  F. B. Koch, R. L. Meek, and D. V. McCaughan, *J. Electrochem. Soc.* **121**, 558 (1974).
Determination of Lighter Impurities on Silicon by 90° Forward Ion Scattering through Thin
Targets
  R. L. Meek and W. M. Gibson, *Nucl. Instrum. Methods* **98**, 375 (1972).
High Sensitivity Detection of Solution Contaminants by MeV Ion Scattering
  R. L. Meek, *J. Electrochem. Soc.* **121**, 172 (1974).
Silicon Surface Contamination: Polishing and Cleaning
  R. L. Meek, T. M. Buck, and C. F. Gibbon, *J. Electrochem. Soc.* **120**, 1241 (1973).
The Determination of Surface Contamination on Silicon by Large Angle Ion Scattering
  D. A. Thompson, H. D. Barber, and W. D. Mackintosh, *Appl. Phys. Lett.* **14**, 102 (1969).

### 10.2.3   Surface Structure and Channeling

Line-Shape Extraction Analysis of Silicon Oxide Layers on Silicon by Channeling Effect
Measurements
  W. K. Chu, E. Lugujjo, J. W. Mayer, and T. W. Sigmon, *Thin Solid Films* **19**, 329 (1973).
Measurement of Surface Relaxation by MeV He Ion Backscattering and Channeling
  J. A. Davies, D. P. Jackson, J. P. Mitchell, P. R. Norton, and R. L. Tapping, *Phys. Lett.* **54A**,
  239 (1975).
Surface Relaxation Effects in (111) Platinum Measured by Backscattering and Channeling
  J. A. Davies, D. P. Jackson, J. B. Mitchell, P. R. Norton, and R. L. Tapping, *Nucl. Instrum.
  Methods* **132**, 609 (1976).
Analysis of Surface Layers by the Channeling Technique: Beam Energy Dependence
  G. Della Mea, A. V. Drigo, S. Lo Russo, P. Mazzoldi, S. Yamaguchi, and G. G. Bentini,
  *Appl. Phys. Lett.* **26**, 147 (1975).
Energy Dependence of the Surface Minimum Yield for Axial Channeling
  G. Della Mea *et al.*, *in* "Atomic Collisions in Solids" (S. Datz, B. R. Appleton, and C. D.
  Moak, eds.), p. 811. Plenum Press, New York, 1976.
On the Analysis of the Backscattering Spectra of Compound Semiconductors
  J. J. Grob, A. Ghitescu, and P. Siffert, *Radiat. Effects* **18**, 139 (1973).
Surface Preparation of Single Crystal *n*-Type GaAs Substrates Studied by the Channeling
Technique
  J. O. Hvalgård, S. L. Andersen, and T. Olsen, *Phys. Status Solidi (a)* **5**, K83 (1971).
On the Application of Rutherford Scattering and Channeling Techniques to Study Semi-
conductor Surfaces
  D. V. Morgan and E. Bøgh, *Surf. Sci.* **32**, 278 (1972).
Surface Studies by Means of α Particles of High Energy (2–4 MeV)
  D. V. Morgan and D. R. Wood, *Proc. R. Soc. London* **A335**, 509 (1973).
Heat Cycling of Un-Passivated GaAs Surfaces
  D. V. Morgan and D. R. Wood, *Phys. Status Solidi (a)* **23**, 325 (1974).
Surface Channeling
  R. Sizmann and C. Varelas, *Nucl. Instrum. Methods* **132**, 633 (1976).
Surface Structure Analysis by Means of Rutherford Scattering: Methods to Study Surface
Relaxation
  W. C. Turkenburg, W. Soszka, F. W. Saris, H. H. Kersten, and B. G. Colenbrander, *Nucl.
  Instrum. Methods* **132**, 587 (1976).
The Characterization of GaAs Surfaces
  D. R. Wood and D. V. Morgan, *J. Electrochem. Soc.* **122**, 773 (1975).
Damage in the Surface Region of Silicon Produced by Sputter-Etching
  Y. Yamamoto, K. Shinada, T. Itoh, and K. Yada, *Jpn. J. Appl. Phys.* **13**, 551 (1974).

Characterization of Reordered ⟨001⟩ Au Surfaces by Positive-Ion-Channeling Spectroscopy, LEED and AES
  D. M. Zehner, *et al., J. Vac. Sci. Technol.* **12**, 454 (1975).

## 10.2.4   Gettering on Silicon Surfaces

Gettering Rates of Various Fast-Diffusing Metal Impurities at Ion-Damaged Layers on Silicon
  T. M. Buck, K. A. Pickar, J. M. Poate, and C-M. Hsieh, *Appl. Phys. Lett.* **21**, 485 (1972).
A Rutherford Scattering Study of the Diffusion of Heavy Metal Impurities in Silicon to Ion-Damaged Surface Layers
  T. M. Buck, J. M. Poate, K. A. Pickar, and C-M. Hsieh, *Surf. Sci.* **35**, 362 (1973).
Preliminary Results of an Ion Scattering Study of Phosphosilicate Glass Gettering
  R. L. Meek and C. F. Gibbon, *J. Electrochem. Soc.* **121**, 444 (1974).
The Diffusion of Cu through Si and Gettering at Ion Damaged Surface Layers in the Presence of O
  J. M. Poate and T. E. Seidel, *in* "Ion Implantation in Semiconductors and Other Materials" (B. L. Crowder, ed.), p. 317. Plenum Press, New York, 1973.
Ion Implantation Damage Gettering and Phosphorus Diffusion Gettering of Cu and Au in Silicon
  T. E. Seidel and R. L. Meek, *in* "Ion Implantation in Semiconductors and Other Materials" (B. L. Crowder, ed.), p. 305. Plenum Press, New York, 1973.
Ion Implantation Gettering of Gold in Silicon
  T. W. Sigmon, L. Csepregi, and J. W. Mayer, *J. Electrochem. Soc.* **123**, 1116 (1976).

## 10.3   BULK PROPERTIES

## 10.3.1   Composition and Structure

Diffusion and Aggregation of Implanted Ag and Au in a Lithia–Alumina–Silica Glass
  G. W. Arnold and J. A. Borders, *Inst. Phys. Conf. Ser.* **28**, 121 (1976).
Applications of Ion Beam Analysis to Insulators
  J. A. Borders and G. W. Arnold, *in* "Ion Beam Surface Layer Analysis" (O. Meyer, G. Linker, and F. Käppeler, eds.), Vol. I, p. 415. Plenum Press, New York, 1976.
Backscattering of 1 MeV He$^+$ on SiC
  M-A. Nicolet, H. R. Bilger, and O. Meyer, *Phys. Status Solidi (a)* **3**, 1019 (1970).
Elemental Analysis by Differential Backscattering Spectrometry
  M. Peisach, *Thin Solid Films* **19**, 297 (1973).
Analysis of Glass by Simultaneous Spectrometry of Scattered Alpha Particles and Prompt Protons
  M. Peisach and R. Pretorius, *J. Radioanal. Chem.* **16**, 559 (1973).
The Niobium–Hydrogen System: A Study of the $\alpha \to \alpha + \beta$ Phase Change by Channeling of 1 MeV He ions
  J. L. Whitton, J. B. Mitchell, T. Schober, and H. Wenzl, *Scripta Metall.* **9**, 851 (1975).
A Rutherford Backscattering Study of Precipitation Phenomena in Dilute Alloys of Nb–H and Nb–D
  J. L. Whitton, J. B. Mitchell, T. Schober, and H. Wenzl, *Acta Metall.* **24**, 483 (1976).
Determination of Surface Impurity Concentration Profiles by Nuclear Backscattering
  J. F. Ziegler and J. E. E. Baglin, *J. Appl. Phys.* **42**, 2031 (1971).

Defining the "Random" Spectrum as Used in the Channeling Technique of Nuclear Back-scattering
  J. F. Ziegler and B. L. Crowder, *Appl. Phys. Lett.* **20**, 178 (1972); **22**, 347 (1973) (erratum).

## 10.3.2   Diffusion and Solubility

Study of Tin Diffusion into Silicon by Backscattering Analysis
  Y. Akasaka, K. Horie, G. Nakamura, K. Tsukamoto, and Y. Yukimoto, *Jpn. J. Appl. Phys.* **13**, 1533 (1974).
Channeling and Electrical Investigations of Au-Doped CdTe
  W. Akutagawa, D. Turnbull, W. K. Chu, and J. W. Mayer, *Solid State Commun.* **15**, 1919 (1974).
Solubility and Lattice Location of Au in CdTe by Backscattering Techniques
  W. Akutagawa, D. Turnbull, W. K. Chu, and J. W. Mayer, *J. Phys. Chem. Solids* **36**, 521 (1975).
A Study of 2 MeV He-Irradiated Phosphorus Diffused Si
  C. R. Allen and R. W. Bicknell, *Phil. Mag.* **30**, 483 (1974).
Investigation of Interstitial Zn Concentrations in Additively Colored ZnO Using the Uni-Directional Channeling and Blocking Technique
  B. R. Appleton and L. C. Feldman, *J. Phys. Chem. Solids* **33**, 507 (1972).
A Study of the Diffusion of Oxygen in Zirconium by Means of the Nuclear Reaction $^{18}O(p, \alpha)^{15}N$
  G. B. Fedorov, N. A. Skakun, O. N. Khar'kov, G. V. Fetisov, and A. Ye. Kissil', *Fiz. Met. Metalloved.* **35**, 978 (1973).
Application of the Backscattering Method for the Measurement of Diffusion of Zinc in Aluminum
  A. Fontell, E. Arminen, and M. Turunen, *Phys. Status Solidi (a)* **15**, 113 (1973).
Effect of Atmospheres on Arsenic Diffusion into Silicon from the Doped Oxide Layer
  T. Itoh, K. Shinada, Y. Ohmura, and K. Kirita, *J. Appl. Phys.* **46**, 1943 (1975).
Measurement of Oxygen-Diffusion-Profiles in Silicon-Single-Crystals by Means of the Nuclear Reaction $^{18}O(p, \alpha)^{15}N$
  J. E. Gass, H. H. Müller, H. Schmied, L. Jörissen, and G. Ziffermayer, *Nucl. Instrum. Methods* **106**, 109 (1973).
Grain Boundary Segregation and Desegregation of Antimony in Temper-Brittle Steel, Identified and Measured by the Elastic Scattering of Energetic Ions
  M. Guttmann, P. R. Krahe, F. Abel, G. Amsel, M. Bruneaux, and C. Cohen, *Scripta Metall.* **5**, 479 (1971).
Heavy Elements Diffusion in Cadmium Telluride
  M. Hage-Ali, I. V. Mitchell, J. J. Grob, and P. Siffert, *Thin Solid Films* **19**, 409 (1973).
Measurement of Diffusion of Gold in Copper by Elastic Scattering of Deuteron
  S. Kawasaki and E. Sakai, *J. Nucl. Sci. Technol.* **4**, 273 (1967).
Diffusion Processes at the Cu–CdTe Interface for Evaporated and Chemically Plated Cu Layers
  H. Mann, G. Linker, and O. Meyer, *Solid State Commun.* **11**, 475 (1972).
Equilibrium Positions of Au Atoms in Sn Crystals as Determined by a Channeling Technique
  J. W. Miller, D. S. Gemmell, R. E. Holland, J. C. Poizat, J. N. Worthington, and R. E. Loess, *Phys. Rev. B* **11**, 990 (1975).
Study of Cu Diffusion in Be Using Ion Backscattering
  S. M. Myers, S. T. Picraux, and T. S. Prevender, *Phys. Rev. B* **9**, 3953 (1974).
Measurements of $^{18}O$ Concentration Profiles Using Resonant Nuclear Reactions
  D. J. Neild, P. J. Wise, and D. G. Barnes, *J. Phys. D.: Appl. Phys.* **5**, 2292 (1972).

Diffusion Profiles of Arsenic in Silicon Observed by Backscattering Method and by Electrical Measurement
   S. Ohkawa, Y. Nakajima, T. Sakurai, H. Nishi, and Y. Fukukawa, *Jpn. J. Appl. Phys.* **13**, 361 (1974).
Effects of Atmosphere During Arsenic Diffusion in Silicon from Doped Oxide
   T. Sakurai, H. Nishi, T. Furuya, H. Hashimoto, and H. Shibayama, *Appl. Phys. Lett.* **22**, 219 (1973).
Diffusion Measurements in the System Cu–Au by Elastic Scattering
   R. F. Sippel, *Phys. Rev.* **115**, 1441 (1959).

### 10.3.3    Lattice Location of Foreign Atoms

Precipitation of Boron Atoms Implanted in Silicon as Detected By Channeling Analysis
   Y. Akasaka and K. Horie, *J. Appl. Phys.* **44**, 3372 (1973).
Use of the Channeling Technique and Calculated Angular Distributions to Locate Br Implanted into Fe Single Crystals
   R. B. Alexander, P. T. Callaghan, and J. M. Poate, *Phys. Rev. B* **9**, 3022 (1974).
The Use of Channeling Effect Techniques to Locate Interstitial Foreign Atoms in Silicon
   J. U. Andersen, O. Andreasen, J. A. Davies, and E. Uggerhoj, *Radiat. Effects* **7**, 25 (1971).
Use of the Channeling Technique to Locate Interstitial Impurities
   J. U. Andersen, E. Laesgaard, and L. C. Feldman, *Radiat. Effects* **12**, 219 (1972).
Lattice Site Location of Ion-Implanted Impurities in Cu and Other fcc Metals
   J. A. Borders and J. M. Poate, *Phys. Rev. B* **13**, 969 (1976).
Lattice Location of Low-Z Impurities in Medium-Z Targets Using Ion-Induced X-Rays. I. Analytical Technique
   J. F. Chemin, I. V. Mitchell, and F. W. Saris, *J. Appl. Phys.* **45**, 532 (1974).
Lattice Location of Low-Z Impurities in Medium-Z Targets Using Ion-Induced X-Rays. II. Phosphorus and Sulfur Implants in Germanium Single Crystals
   J. F. Chemin, I. V. Mitchell, and F. W. Saris, *J. Appl. Phys.* **45**, 537 (1974).
Channeling Flux in Single Crystals With Interstitial Atoms: Impurity Concentration Dependence
   G. Della Mea, A. V. Drigo, S. Lo Russo, P. Mazzoldi, S. Yamaguchi, G. G. Bentini, A. Desalvo and R. Rosa, *Phys. Rev. B* **10**, 1836 (1974).
Channeling and Related Effects in the Motion of Charged Particles through Crystals
   D. S. Gemmel, *Rev. Mod. Phys.* **46**, 129 (1974).
Channeling Measurements in As-Doped Si
   J. Haskell, E. Lugujjo, and J. W. Mayer, *J. Appl. Phys.* **43**, 3425 (1972).
The Combined Effects of Lattice Vibration and Irradiation-Produced Defects on Dechanneling
   L. M. Howe, M. L. Swanson, and A. F. Quenville, *Nucl. Instrum. Methods* **132**, 241 (1976).
Lattice-Location Study of Hf Implanted in Ni
   E. N. Kaufmann, J. M. Poate, and W. M. Augustynaik, *Phys. Rev. B* **7**, 951 (1973).
Equilibrium Positions of Au Atoms in Sn Crystals as Determined by a Channeling Technique
   J. W. Miller, D. S. Gemmell, R. E. Holland, J. C. Poizat, J. N. Worthington, and R. E. Loess, *Phys. Rev. B* **11**, 990 (1975).
Investigation of Te-Doped GaAs Annealing Effects by Optical and Channeling-Effect Measurements
   I. V. Mitchell, J. W. Mayer, J. K. Kung, and W. G. Spitzer, *J. Appl. Phys.* **42**, 3982 (1971).
Lattice Location of Impurities in Metals and Semiconductors
   S. T. Picraux, *in* "New Uses of Ion Accelerators" (J. F. Ziegler, ed.), Chapter 4, p. 229. Plenum Press, New York, 1975.

Lattice Location by Channeling Angular Distributions: Bi Implanted in Si
  S. T. Picraux, W. L. Brown, and W. M. Gibson, *Phys. Rev. B* **6**, 1382 (1972).
Channeling Evidence for a Shallow Trapping Configuration of Copper Interstitials at Gold
Atoms in a Cu–0.05 at% Au Crystal
  M. L. Swanson, A. F. Quenneville, and F. Maury, *Phys. Status Solidi (a)* **27**, 281 (1975).
The Application of Channeling for Locating Impurity Atoms and Investigating Impurity–Defect
Interactions in Zirconium
  M. L. Swanson and L. M. Howe, *J. Nucl. Mater.* **54**, 155 (1974).
The Location of Displaced Manganese and Silver Atoms in Irradiated Aluminum Crystals by
Backscattering
  M. L. Swanson, F. Maury, and A. F. Quenneville, *in* "Applications of Ion Beams to Metals"
  (S. T. Picraux, E. P. EerNisse, and F. L. Vook, eds.), p. 393. Plenum Press, New York, 1974.
Trapping Configuration of Al Interstitial Atoms at Zn Solute Atoms in Al–0.1 at% Zn Crystals
  M. L. Swanson, L. M. Howe, and A. F. Quenneville, *Phys. Status Solidi (a)* **31**, 675 (1975).
The Combined Effects of Lattice Vibrations and Irradiation-Induced Defects on Dechanneling
in Copper
  M. L. Swanson, L. M. Howe, and A. F. Quenneville, *Phys. Status Solidi (a)* **33**, 265 (1976).
Location of Impurities in Alkali-Halide Crystals by Channeling Studies
  K. Tachibana, K. Morita, and N. Itoh, *Solid State Commun.* **9**, 1425 (1971).
Lattice Location of Impurities Implanted into Metals
  H. de Waard and L. C. Feldman, *in* "Applications of Ion Beams to Metals" (S. T. Picraux,
  E. P. EerNisse, and F. L. Vook, eds.), p. 317. Plenum Press, New York, 1974.

## 10.3.4  Channeling and Hyperfine Interactions

Combined Lattice Location and Hyperfine Field Study of Yb Implanted into Fe
  R. B. Alexander, E. J. Ansaldo, B. I. Deutch, J. Gellert, and L. C. Feldman, *in* "Applications
  of Ion Beams to Metals" (S. T. Picraux, E. P. EerNisse, and F. L. Vook, eds.), p. 365. Plenum
  Press, New York, 1974.
Use of the Channeling Technique and Calculated Angular Distributions to Locate Br Implanted
into Fe
  R. B. Alexander, P. T. Callaghan, and J. M. Poate, *Phys. Rev. B* **9**, 3022 (1974).
Lattice Location of I Implanted into Fe Single Crystals
  P. T. Callaghan, P. K. James, and N. J. Stone, *Phys. Rev. B* **12**, 3553 (1975).
Impurity Site Distribution of Implanted Bi in Iron and Nickel Studied by Channeling and
Nuclear Orientation
  P. T. Callaghan, P. Kittel, N. J. Stone, and P. D. Johnston, *Phys. Rev. B* **14**, 3722 (1976).
Lattice Location and Hyperfine Fields of Rare Earth Ions Implanted into Iron
  R. L. Cohen, G. Beyer, and B. Deutch, *Phys. Rev. Lett.* **33**, 518 (1974).
Channeling in Iron and Lattice Location of Implanted Xenon
  L. C. Feldman and D. E. Murnick, *Phys. Rev. B* **5**, 1 (1972).
Mossbauer Studies of Implanted $I^{129}$ Ions in Semiconductors and Alkali Halides
  D. W. Hafemeister and H. de Waard, *Phys. Rev. B* **7**, 3014 (1973).
Combined Lattice-Location–Hyperfine-Interaction Experiments on Hg Implanted in Fe
  P. K. James, P. Herzog, N. J. Stone, and K. Freitag, *Phys. Rev. B* **13**, 59 (1976).
Lattice-Location Study of Hf Implanted in Ni
  E. N. Kaufmann, J. M. Poate, and W. M. Augustyniak, *Phys. Rev. B* **7**, 951 (1973).
Preferred Sites of Impurities Implanted in Be: Lattice Location and Quadrapole Interactions
  E. N. Kaufmann, P. Raghaven, R. S. Raghaven, E. J. Ansaldo, and R. A. Naumann, *Phys.
  Rev. Lett.* **34**, 1558 (1975).

Study of the Rare-Earth–Oxygen Interaction in Iron by Lattice Location and Perturbed-Angular-Correlation Experiments
  L. Thome, H. Hernas, F. Abel, M. Bruneaux, C. Cohen, and J. Chaumont, *Phys. Rev. B* **14**, 2787 (1976).
Lattice Location of Impurities Implanted into Metals
  H. de Waard and L. C. Feldman, *in* "Applications of Ion Beams to Metals" (S. T. Picraux, E. P. EerNisse, and F. L. Vook, eds.), p. 317. Plenum Press, New York, 1974.
Direct Comparison of Mossbauer and Channeling Studies of Implanted $^{119}$Sn in Si Single Crystals
  G. Weyer, J. U. Andersen, B. I. Deutch, J. A. Golovchenko, and A. Nylandsted-Larsen, *Radiat. Effects* **24**, 117 (1975).

## 10.3.5  Defects Introduced by the Analysis Beam

A Study of 2 MeV Helium-Irradiated Phosphorus-Diffused Silicon
  C. R. Allen and R. W. Bicknell, *Phil. Mag.* **30**, 483 (1974).
Channeling-Effect Study of Deuteron-Induced Damage in Si and Ge Crystals
  P. Baeri, S. U. Campisano, G. Foti, E. Rimini, and J. A. Davies, *Appl. Phys. Lett.* **26**, 424 (1975).
Effects of Ion Bombardment on Na and Cl Motion in $SiO_2$ Thin Films
  W. Beezhold, *IEEE Trans. Nucl. Sci.* **NS-21**, 62 (1974).
Beam-Induced Lattice Disorder in Channeling Experiments on Si and Ge
  S. U. Campisano, G. Foti, F. Grasso, and E. Rimini, *Appl. Phys. Lett.* **21**, 425 (1972).
Displacement of Arsenic Atoms in Silicon Crystals During Irradiation
  F. Fujimoto, K. Komaki, M. Ishii, H. Nakayama, and K. Hisatake, *Phys. Status Solidi (a)* **12**, K7 (1972).
Channeling of 1-MeV He$^+$ ions in NaCl Damage and Temperature Effects
  M. J. Hollis, *Phys. Rev. B* **8**, 931 (1973).
Channeling Investigations of the Interaction between Solute Atoms and Irradiation Produced Defects in Metals
  L. M. Howe and M. L. Swanson, *Inst. Phys. Conf. Ser. No. 28* 273 (1976).
The Combined Effects of Lattice Vibrations and Irradiation Produced Defects on Dechanneling
  L. M. Howe, M. L. Swanson, and A. F. Quenneville, *Nucl. Instrum. Methods* **132**, 241 (1976).
Damage Production and Arsenic Displacement in Silicon by Proton and Helium Irradiation
  W. H. Kool, H. E. Roosendaal, L. W. Wiggers, and F. W. Saris, *Nucl. Instrum. Methods* **132**, 285 (1976).
Channeling of MeV Projectiles in Diatomic Ionic Crystals
  Hj. Matske, *Phys. Status Solidi (a)* **8**, 99 (1971).
Dechanneling from 2 MeV He$^+$ Damage in Gold
  K. L. Merkle, P. P. Pronko, D. S. Gemmell, R. C. Mikkelson, and J. R. Wrobel, *Phys. Rev. B* **8**, 1002 (1973).
Channeling Studies of Radiation Damage in Sodium Chloride Crystals
  C. S. Newton, R. B. Alexander, G. J. Clark, H. J. Hay, and P. B. Treacy, *Nucl. Instrum. Methods* **132**, 213 (1976).
Irradiation-Induced Defects Studied by the Channeling Method in $p$- and $n$-type Silicon
  H. J. Pabst and D. W. Palmer, *Inst. Phys. Conf. Ser. No. 16* 438 (1973).
Beam Effects in the Analysis of As-Doped Silicon by Channeling Measurements
  E. Rimini, J. Haskell, and J. W. Mayer, *Appl. Phys. Lett.* **20**, 237 (1972).
The Application of Channeling for Locating Impurity Atoms and Investigating Impurity–Defect Interactions in Zirconium
  M. L. Swanson and L. M. Howe, *J. Nucl. Mater.* **54**, 155 (1974).

Investigation of the Interaction between Irradiation-Induced Defects and Solute Atoms in Dilute Copper Alloys by Ion Channeling
   M. L. Swanson, L. M. Howe, and A. F. Quenneville, *Radiat. Effects* **28**, 205 (1976).
Identification of Irradiation-Induced Al–Mn Dumbells in Al Crystals by Backscattering
   M. L. Swanson, F. Maury, and A. F. Quenneville, *Phys. Rev. Lett.* **31**, 1057 (1973).


## 10.4  OXIDE AND NITRIDE LAYERS

### 10.4.1  Composition and Thickness

Effect of Arsenic on the Composition and Optical Constants of Iron Passive Film Reduced at 100% Current Efficiency
   B. Agius and J. Siejka, *J. Electrochem. Soc.* **122**, 723 (1975).
Microanalysis of the Stable Isotopes of Oxygen by Means of Nuclear Reactions
   G. Amsel and D. Samuel, *Anal. Chem.* **39**, 1689 (1967).
Analysis of Silicon Oxynitride Layers by the Complementary Use of Elastic Backscattering and Nuclear Reactions
   A. Barcz, A. Turos, L. Wielunski, and I. Skrzynecka, *Phys. Status Solidi (a)* **28**, 293 (1975).
Study of Aluminum Oxide Films by Ion-Induced X-Rays and Rutherford Backscattering
   W. Bauer and R. G. Musket, *J. Vac. Sci. Technol.* **10**, 273 (1973).
Study of Aluminum Oxide Films by Ion-Induced X-Rays and Rutherford Backscattering
   W. Bauer and R. G. Musket, *J. Appl. Phys.* **44**, 2606 (1973).
Analysis of Evaporated Silicon Oxide Films by Means of $(d, p)$ Nuclear Reactions and Infrared Spectrophotometry
   A. Cachard, J. A. Roger, J. Pivot, and C. H. S. Dupuy, *Phys. Status Solidi (a)* **5**, 637 (1971).
Comparison of Backscattering Spectrometry and Secondary Ion Mass Spectrometry by Analysis of Tantalum Pentoxide Layers
   W. K. Chu, M-A. Nicolet, J. W. Mayer, and C. A. Evans, *Anal. Chem.* **46**, 2136 (1974).
Investigation of the Composition of Sputtered Silicon Nitride Films by Nuclear Microanalysis
   M. Croset, S. Rigo, and G. Amsel, *Appl. Phys. Lett.* **19**, 33 (1971).
A Model for Filament Growth and Switching in Amorphous Oxide Films
   G. Dearnaley, D. V. Morgan, and A. M. Stoneham, *J. Non-Crystall. Solids* **4**, 593 (1970).
Analysis of Silicon Nitride Layers on Silicon by Backscattering and Channeling Effect Measurements
   J. Gyulai, O. Meyer, J. W. Mayer, and V. Rodriguez, *Appl. Phys. Lett.* **16**, 232 (1970).
Evaluation of Silicon Nitride Layers of Various Compositions by Backscattering and Channeling Effect Measurements
   J. Gyulai, O. Meyer, J. W. Mayer, and V. Rodriguez, *J. Appl. Phys.* **42**, 451 (1971).
Studies of Tantalum Nitride Thin Film Resistors
   R. A. Langley, *in* "Ion Beam Surface Layer Analysis" (O. Meyer, G. Linker, and F. Käppeler, eds.), Vol. I, p. 337. Plenum Press, New York, 1976.
Influence of Heat Treatment on Aluminum Oxide Films on Silicon
   M. Kamoshida, I. V. Mitchell, and J. W. Mayer, *J. Appl. Phys.* **43**, 1717 (1972).
Residual Chlorine in $O_2$ : HCl Grown $SiO_2$
   R. L. Meek, *J. Electrochem. Soc.* **120**, 308 (1973).
Properties of $SiO_2$ Grown in the Presence of HCl or $Cl_2$
   Y. J. van der Meulen, C. M. Osburn, and J. F. Ziegler, *J. Electrochem. Soc.* **122**, 284 (1975).
Analysis of Amorphous Layers on Silicon by Backscattering and Channeling Effect Measurements
   O. Meyer, J. Gyulai, and J. W. Mayer, *Surf. Sci.* **22**, 263 (1970).

Analysis of Silicon Nitride Layers Deposited from $SiH_4$ and $N_2$ on Silicon
  O. Meyer and W. Scherber, *J. Phys. Chem. Solids* **32**, 1909 (1971).
Enhanced Sensitivity of Oxygen Detection by the 3.05 MeV ($\alpha$, $\alpha$) Elastic Scattering
  G. Mezey, J. Gyulai, T. Nagy, E. Kotai, and A. Manuaba, *in* "Ion Beam Surface Layer Analysis" (O. Meyer, G. Linder, and F. Käppeler, eds.), Vol. I, p. 303. Plenum Press, New York, 1976.
Channeling-Effect Analysis of Thin Films on Silicon: Aluminum Oxide
  I. V. Mitchell, M. Kamoshida, and J. W. Mayer, *J. Appl. Phys.* **42**, 4378 (1971).
Backscattering Analyses of the Composition of Silicon–Nitride Films Deposited by rf Reactive Sputtering
  C. J. Mogab and E. Lugujjo, *J. Appl. Phys.* **47**, 1302 (1976).
An Investigation of the Stoichiometry and Impurity Content of Thin Silicon Oxide Films Using Rutherford Scattering of MeV $\alpha$-Particles
  D. V. Morgan and R. P. Gittins, *Phys. Status Solidi (a)* **13**, 517 (1972).
Oxide Thickness Determination by Proton-Induced X-Ray Fluorescence
  R. G. Musket and W. Bauer, *J. Appl. Phys.* **43**, 4786 (1972).
Effect of B, C, N and Ne Ion Implantation on the Oxidation of Polycrystalline Cu
  H. M. Naguib, R. J. Kriegler, J. A. Davies, and J. B. Mitchell, *J. Vac. Sci. Technol.* **13**, 396 (1976).
Determination of Alumina Film Thickness by Alpha Particle Scattering
  M. Peisach, D. O. Poole, and H. F. Röhm, *Talanta* **14**, 187 (1967).
Impurity Incorporation During rf Sputtering of Silicon Oxide Layers
  S. Petersson, G. Linker, and O. Meyer, *Phys. Status Solidi (a)* **14**, 605 (1972).
Oxygen Implantation in Stoichiometric Silicon Monoxide
  J. Pivot, D. Morelli, A. Cachard, J. Tardy, J. A. Roger, J. P. Thomas, and C. H. S. Dupuy, *Inst. Phys. Conf. Ser. No. 28* 117 (1976).
Ion Implantation and Backscattering from Oxidized Single Crystal Copper
  J. Rickards and G. Dearnaley, *in* "Applications of Ion Beams to Metals" (S. T. Picraux, E. P. EerNisse, and F. L. Vook, eds.), p. 101. Plenum Press, New York, 1974.
Investigation of Reactively Sputtered Silicon Nitride Films by Complementary Use of Backscattering and Nuclear-Reaction Microanalysis
  S. Rigo, G. Amsel, and M. Croset, *J. Appl. Phys.* **47**, 2800 (1976).
Stoichiometry of Thin Silicon Oxide Layers on Silicon
  T. W. Sigmon, W. K. Chu, E. Lugujjo, and J. W. Mayer, *Appl. Phys. Lett.* **24**, 105 (1974).

## 10.4.2   Native and Thermally Grown Oxide Layers

Study of Aluminum Oxide Films by Ion-Induced X-Rays and Rutherford Backscattering
  W. Bauer and R. G. Musket, *J. Appl. Phys.* **94**, 2606 (1973).
Line Shape Extraction Analysis of Silicon Oxide Layers on Silicon by Channeling Effect Measurements
  W. K. Chu, E. Lugujjo, J. W. Mayer, and T. W. Sigmon, *Thin Solid Films* **19**, 329 (1973).
The Influence of Ion Implantation on the Oxidation of Nickel
  P. D. Goode, *Inst. Phys. Conf. Ser. No. 28* 154 (1976).
Rutherford Scattering Investigation of Thermally Oxidized Tantalum on Silicon
  J. Hirvonen, A. G. Revesz, and T. K. Kirkendall, *Thin Solid Films* **33**, 315 (1976).
Residual Chlorine in $O_2$:HCl Grown $SiO_2$
  R. L. Meek, *J. Electrochem. Soc.* **120**, 308 (1973).
Properties of $SiO_2$ Grown in the Presence of HCl or $Cl_2$
  Y. J. van der Meulen, C. M. Osburn, and J. F. Ziegler, *J. Electrochem. Soc.* **122**, 284 (1975).

Analysis of Amorphous Layers on Silicon by Backscattering and Channeling Effect Measurements
  O. Meyer, J. Gyulai, and J. W. Mayer, *Surf. Sci.* **22**, 263 (1970).
Enhanced Oxidation on Ion-Implanted Silicon
  O. Meyer and J. W. Mayer, *Radiat. Effects* **3**, 139 (1970).
The Effect of Ion-Implanted Impurities on the Oxidation of Chromium
  S. Muhl, R. A. Collins, and G. Dearnaley, *Inst. Phys. Conf. Ser. No. 28* 147 (1976).
Surface Analysis Using Proton Beams
  R. G. Musket and W. Bauer, *Thin Solid Films* **19**, 69 (1973).
Oxide Thickness Determination by Proton-Induced X-Ray Fluorescence
  R. G. Musket and W. Bauer, *J. Appl. Phys.* **43**, 4786 (1972).
A Rutherford Scattering Study of the Chemical Composition of Native Oxides on GaP
  J. M. Poate, T. M. Buck, and B. Schwartz, *J. Phys. Chem. Solids* **34**, 779 (1973).
Lithium Ion Backscattering as a Novel Tool for the Characterization of Oxidized Phases of Aluminum Obtained from Industrial Anodization Procedures
  J. P. Thomas, A. Cachard, M. Fallavier, J. Tardy, and S. Marsaud, *in* "Ion Beam Surface Layer Analysis" (O. Meyer, G. Linker, and F. Käppeler, eds.), Vol. I, p. 425. Plenum Press, New York, 1976.
Determination of the Total Amount of Oxygen Atoms in Silicon Oxide Surface Layers by the Nuclear Reactions $^{16}O(d, p_1)^{17}O*$ and $^{16}O(d, \alpha)^{14}N$
  A. Turos, L. Wieluński, and J. Oleński, *Phys. Status Solidi (a)* **16**, 211 (1973).
Use of the Nuclear Reaction $^{16}O(d, \alpha)^{14}N$ in the Microanalysis of Oxide Surface Layers
  A. Turos, L. Wieluński, and A. Barcz, *Nucl. Instrum. Methods* **111**, 605 (1973).

## 10.4.3  Anodic Oxidation

The Influence of the Electrolyte on the Composition of Anodic Oxide Films on Tantalum
  G. Amsel, C. Cherki, G. Feuillade, and J. P. Nadai, *J. Phys. Chem. Solids* **30**, 2117 (1969).
The Use of Rutherford Backscattering to Study the Behavior of Ion-Implanted Atoms During Anodic Oxidation of Aluminum: Ar, Kr, Xe, K, Rb, Cs, Cl, Br and I
  F. Brown and W. D. Mackintosh, *J. Electrochem. Soc.* **120**, 1096 (1973).
Comparison of Backscattering Spectrometry and Secondary Ion Mass Spectrometry by Analysis of Tantalum Pentoxide Layers
  W. K. Chu, M-A. Nicolet, J. W. Mayer, and C. A. Evans, *Anal. Chem.* **46**, 2136 (1974).
An $O^{18}$ Study of the Source of Oxygen in the Anodic Oxidation of Silicon and Tantalum in Some Organic Solvents
  M. Croset, E. Petreanu, D. Samuel, G. Amsel, and J. P. Nadai, *J. Electrochem. Soc.* **118**, 717 (1971).
The Combined Use of He-Backscattering and He Induced X-Rays in the Study of Anodically Grown Oxide Films on GaAs
  L. C. Feldman, J. M. Poate, F. Ermanis, and B. Schwartz, *Thin Solid Films* **19**, 81 (1973).
On the Anodic Oxide Growth on Silicon Carbide
  M. Huez, G. Restelli, and A. Manara, *Thin Solid Films* **23**, S33 (1974).
Backscattering Studies of Anodization of Aluminum Oxide and Silicon Nitride on Silicon
  M. Kamoshida and J. W. Mayer, *J. Electrochem. Soc.* **119**, 1084 (1972).
Analysis of Plasma-Grown GaAs Oxide Films
  R. L. Kauffman, L. C. Feldman, J. M. Poate, and R. P. H. Chang, *Appl. Phys. Lett.* **30**, 319 (1977).
Mobility of Metallic Foreign Atoms During the Anodic Oxidation of Aluminum
  W. D. Mackintosh, F. Brown, and H. H. Plattner, *J. Electrochem. Soc.* **121**, 1281 (1974).

The Anodic Oxidation of Vanadium: Transport Numbers of Metal and Oxygen and the Metal/Oxygen Ratio in the Oxide Films
   W. D. Mackintosh and H. H. Plattner, *J. Electrochem. Soc.* **123**, 523 (1976).
Anodic Oxide Films on GaP
   J. M. Poate, P. J. Silverman, and J. Yahalom, *J. Electrochem. Soc.* **120**, 844 (1973).
The Growth and Composition of Anodic Films on GaP
   J. M. Poate, P. J. Silverman, and J. Yahalom, *J. Phys. Chem. Solids* **34**, 1847 (1973).
A Rutherford Backscattering Analysis of Anodic Tantalum–Titanium Oxides
   R. L. Ruth and N. Schwartz, *J. Electrochem. Soc.* **123**, 1860 (1976).
A Study of the Oxygen Growth Laws of Anodic Oxide Films on Aluminum and Tantalum Using Nuclear Microanalysis of $O^{16}$ and $O^{18}$
   J. Siejka, J. P. Nadai, and G. Amsel, *J. Electrochem. Soc.* **118**, 727 (1971).
A Study of Passivity Phenomena by Using $O^{18}$ Tracer Techniques
   J. Siejka, C. Cherki, and J. Yahalom, *J. Electrochem. Soc.* **119**, 991 (1972).
A Rutherford Scattering Analysis of Anodic Tantalum–Silicon Oxides
   P. J. Silverman and N. Schwartz, *J. Electrochem. Soc.* **121**, 550 (1974).
Anodic Processing for Multilevel LSI
   G. C. Schwartz and V. Platter, *J. Electrochem. Soc.* **123**, 34 (1976).

## 10.5  DEPOSITED AND GROWN LAYERS

### 10.5.1  Composition and Thickness

Preparation and Characterization of Au–$SiO_2$ Radio Frequency Co-Sputtered Thin Films
   A. Armiglia, G. G. Bentini, S. Guerra, P. Ostoja, and L. Morettini, *Thin Solid Films* **33**, 355 (1976).
Analysis of Surface Layers by Light Ion Backscattering and Sputtering Combined with Auger Electron Spectroscopy (AES)
   R. Behrisch, B. M. U. Scherzer, and P. Staib, *Thin Solid Films* **19**, 57 (1973).
The Origin of Non-Gaussian Profiles in Phosphorus Implanted Si
   P. Blood, G. Dearnaley, and M. A. Wilkins, *J. Appl. Phys.* **45**, 5123 (1974).
Ion Backscattering Study of $Cu_2Si$ Formation on Single Crystal CdS
   J. A. Borders, *J. Electrochem. Soc.* **123**, 37 (1976).
Densities of Amorphous Si Films by Nuclear Backscattering
   M. H. Brodsky, D. Kaplan, and J. F. Ziegler, *Appl. Phys. Lett.* **21**, 305 (1972).
Structural Influence on Electrical Properties of Metal-Oxide–Metal Devices
   A. Cachard, J. A. Roger, J. Pivot, C. Diaine, and C. H. S. Dupuy, *Thin Solid Films* **13**, 231 (1972).
Analysis for Impurities by Nuclear Scattering
   B. L. Cohen and R. A. Moyer, *Anal. Chem.* **43**, 123 (1971).
Deposition of Thin Films by Retardation of an Isotope Separation Beam
   J. S. Colligon, W. A. Grant, J. S. Williams, and R. P. W. Lawson, *Inst. Phys. Conf. Ser.* **28**, 357 (1976).
Structure and Composition of Sputtered Tantalum Thin Films on Silicon Studied by Nuclear and X-Ray Analysis
   M. Croset and G. Velasco, *J. Appl. Phys.* **43**, 1444 (1972).
Characterization of Evaporated Gold–Indium Films on Semiconductors
   T. G. Finstad, T. Andreassen, and T. Olsen, *Thin Solid Films* **29**, 145 (1975).

Analysis of $Ga_{1-x}Al_xAs$–GaAs Heteroepitaxial Layers by Proton Backscattering
  K. Gamo, T. Inada, I. Samid, C. P. Lee, and J. W. Mayer, *in* "Ion Beam Surface Layer
  Analysis" (O. Meyer, G. Linker, and F. Käppeler, eds.), Vol. I, p. 375. Plenum Press, New
  York, 1976.
Profiling of SiGe Superlattices by He Backscattering
  E. Kasper and W. Pabst, *Thin Solid Films* **37**, L5 (1976).
Analysis of Thin Films on Glass by Nuclear Techniques
  V. Gottardi *et al.*, *Glass Technol.* **17**, 26 (1975).
Formation Process of MnBi Thin Films
  Y. Iwama, U. Mizutani, and F. B. Humphrey, *IEEE Trans. Magn.* 487 (1972).
Detection of Chlorine on Aluminum by Means of Nuclear Reactions
  A. R. Knudson and K. L. Dunning, *Anal. Chem.* **44**, 1053 (1972).
Ion Backscattering Study of Tantalum Nitride Thin Film Resistors
  R. A. Langely and D. J. Sharp, *J. Vac. Sci. Technol.* **12**, 155 (1975).
Measurements of the Stopping Cross Sections for Protons and $^4$He Ions in Erbium and Erbium
Oxide: A Test of Bragg's Rule
  R. A. Langley and R. S. Blewer, *Nucl. Instrum. Methods* **132**, 109 (1976).
Analysis of Phosphosilicate Glass Layers by Backscattering and Channeling Effect Mea-
surements
  G. Linker, O. Meyer, and W. Scherber, *Phys. Status Solidi (a)* **16**, 377 (1973).
Backscattering Energy Loss Parameter Measurements in Thin Metal Films
  G. Linker, O. Meyer, and M. Gettings, *Thin Solid Films* **19**, 177 (1973).
Profiling of Periodic Structures (GaAs–GaAlAs) by Nuclear Backscattering
  J. W. Mayer, J. F. Ziegler, L. L. Chang, R. Tsu, and L. Esaki, *J. Appl. Phys.* **44**, 2322 (1973).
Observation of Film Growth Process by Means of Backscattering Technique
  O. Meyer, H. Mann, and G. Linker, *Appl. Phys. Lett.* **20**, 259 (1972).
Detection of Low-Mass Impurities in Thin Films Using MeV Heavy-Ion Elastic Scattering and
Coincidence Detection Techniques
  J. A. Moore, I. V. Mitchell, M. J. Hollis, J. A. Davies, and L. M. Howe, *J. Appl. Phys.* **46**,
  52 (1975).
Thin Film Analysis Using Rutherford Scattering
  D. V. Morgan, *J. Phys. D.: Appl. Phys.* **7**, 653 (1974).
Effects of Deposition Parameters on Properties of rf Sputtered Molybdenum Films
  R. S. Nowicki, W. D. Buckley, W. D. Mackintosh, and I. V. Mitchell, *J. Vac. Sci. Technol.*
  **11**, 675 (1974).
The Determination of Gold Coating Thicknesses of Proton Scattering
  M. Peisach and D. O. Poole, *J. S. Afr. Chem. Inst.* **18**, 61 (1965).
Multilayer Thin-Film Analysis by Ion Backscattering
  S. T. Picraux and F. L. Vook, *Appl. Phys. Lett.* **18**, 191 (1971).
Predeposition in Silicon as Affected by the Formation of Orthorhombic SiP and Cubic
$SiO_2 \cdot P_2O_5$ at the PSG–Si Interface
  S. Solmi, G. Celotti, D. Nobili, and P. Negrini, *J. Electrochem. Soc.* **123**, 654 (1976).
Ion-Beam-Deposited Polycrystalline Diamond-Like Films
  E. G. Spencer, P. H. Schmidt, D. C. Joy, and F. J. Sansalone, *Appl. Phys. Lett.* **29**, 118 (1976).
Ionized-Cluster Beam Deposition
  T. Takagi, I. Yamada, and A. Sasaki, *J. Vac. Sci. Technol.* **12**, 1128 (1975).
Application of the Coulomb Backscattering of the Heavy Charged Particles. I. Thickness
Measurements of Thin Foils and Surface Layers
  A. Turos and Z. Wilhelmi, *Nukleonika* **13**, 975 (1968).

Application of the Coulomb Backscattering of the Heavy Charged Particles. II. Chemical
Analysis of Surface Layers
  A. Turos and Z. Wilhelmi, *Nukleonika* **14**, 1 (1969).
Backscattering of Light Atomic Projectiles from Au Films in the Energy Range 50–110 keV
  A. van Wijngaarden, E. J. Brimner, and W. E. Baylis, *Can. J. Phys.* **48**, 1835 (1970).
Energy Spectra of keV Backscattered Protons as a Probe for Surface-Region Studies
  A. van Wijngaarden, B. Miremadi, and W. E. Baylis, *Can. J. Phys.* **49**, 2440 (1971).
Analysis of Lead Azide Films by Rutherford Backscattering
  H. M. Windawi, S. P. Varma, C. B. Cooper, and F. Williams, *J. Appl. Phys.* **47**, 3418 (1976).
A Nuclear-Detection Method for Sulfur in Thin Films
  E. A. Wolicki and A. R. Knudson, *Int. J. Appl. Radiat. Isotopes* **18**, 429 (1967).
Backscattering Measurements on Ag Photodoping Effect in $As_2S_3$ Glass
  Y. Yamamoto, T. Itoh, Y. Hirose, and H. Hirose, *J. Appl. Phys.* **47**, 3603 (1976).
Thermal Stability of a Proposed Magnetic Bubble Metallurgy
  J. F. Ziegler, J. E. E. Baglin, and A. Gangulee, *Appl. Phys. Lett.* **24**, 36 (1974).
Nuclear Backscattering Analysis of $Nb-Nb_2O_5-Bi$ Structure
  J. F. Ziegler, M. Berkenblit, T. B. Light, K. C. Park, and A. Reisman, *IBM J. Res. Develop.*
  **16**, 530 (1972).

## 10.5.2   Epitaxial Layers

Ion Beam Analysis of Aluminum Profiles in Heteroepitaxial $Ga_{1-x}Al_xAs$-Layers
  P. Bayerl, W. Pabst, and P. Eichinger, *in* "Ion Beam Surface Layer Analysis" (O. Meyer,
  G. Linker, and F. Käppeler, eds.), Vol. I, p. 363. Plenum Press, New York, 1976.
Studies of Compound Thin Film Semiconductors by Ion Beam and Electron Microscopy
Techniques
  S. U. Campisano, G. Foti, E. Rimini, G. Vitali, and C. Corsi, *in* "Ion Beam Surface Layer
  Analysis" (O. Meyer, G. Linker, and F. Käppeler, eds.), Vol. II, p. 585. Plenum Press,
  New York, 1976.
Ion Backscattering and Electron Microscopy Analyses of Multi-Layer Metal-$Pb_xSn_{1-x}Te$
Epitaxial Film Structures Obtained by Radio Frequency Sputtering
  C. Corsi, S. U. Campisano, G. Foti, E. Rimini, and G. Vitali, *Thin Solid Films* **32**, 315 (1976).
Single-Crystal Heteroepitaxial Growth of $Pb_xSn_{1-x}Te$ Films on Germanium Substrates by
Rutherford Sputtering
  C. Corsi, E. Fainelli, G. Petrocco, G. Vitali, S. U. Campisano, G. Foti, and E. Rimini, *Thin
  Solid Films* **33**, 135 (1976).
Rutherford Scattering from Ultra-Thin Epitaxial Films
  E. B. Dale, G. C. Poole, S. Özkök, and H. Goldberg, *Radiat. Effects* **13**, 3 (1972).
Epitaxial Silicon Layers Grown on Ion-Implanted Silicon Nitride Layers
  R. J. Dexter, S. B. Watelski, and S. T. Picraux, *Appl. Phys. Lett.* **23**, 455 (1973).
Separate Estimate of Crystallite Orientations and Scattering Centers in Polycrystals by Back-
scattering Techniques
  H. Ishiwara and S. Furakawa, *J. Appl. Phys.* **47**, 1686 (1976).
Epitaxial Growth of Silicon Assisted by Ion Implantation
  T. Itoh and T. Nakamura, *Radiat. Effects* **9**, 1 (1971).
Nuclear Microanalysis of Oxygen Concentration in Liquid–Phase Epitaxial Gallium Phosphide
  E. C. Lightowlers, J. C. North, A. S. Jordan, L. Derick, and J. L. Merz, *J. Appl. Phys.* **44**,
  4758 (1973).
Some Aspects of Ge Epitaxial Growth by Solid Solution
  G. Ottaviani, C. Canali, and G. Majni, *J. Appl. Phys.* **47**, 627 (1976).

Ion-Channeling Studies of Epitaxial Layers
S. T. Picraux, *Appl. Phys. Lett.* **20**, 91 (1972).
Ion Channeling Studies of the Crystalline Perfection of Epitaxial Layers
S. T. Picraux, *J. Appl. Phys.* **44**, 587 (1973).
Correlation of Ion Channeling and Electron Microscopy Results in the Evaluation of Hetero-
epitaxial Silicon
S. T. Picraux and G. J. Thomas, *J. Appl. Phys.* **44**, 594 (1973).
Characterization of Polycrystalline Layers by Channeling Measurements
D. Sigurd, R. W. Bower, W. F. van der Weg, and J. W. Mayer, *Thin Solid Films* **19**, 319 (1973).
Crystallization of Ge and Si in Metal Films. II.
D. Sigurd, G. Ottaviani, H. J. Arnal, and J. W. Mayer, *J. Appl. Phys.* **45**, 1740 (1974).
The Analysis of Thin Epitaxial Layers of GaAs Using MeV α-Particles
D. R. Wood and D. V. Morgan, *Phys. Status Solidi (a)* **17**, K143 (1973).

## 10.5.3 Superconducting Films

Backscattering Energy Loss Parameter Measurements in Thin Metal Films
G. Linker, O. Meyer, and M. Gettings, *Thin Solid Films* **19**, 177 (1973).
Superconducting Properties and Structural Transformations of Nitrogen Implanted Molyb-
denum Films
G. Linker and O. Meyer, *Solid State Commun.* **20**, 695 (1976).
Validity of Bragg's Rule in Sputtered Superconducting NbN and NbC Films of Various
Compositions
O. Meyer, G. Linker, and B. Kraeft, *Thin Solid Films* **19**, 217 (1973).
Oxygen Distribution in Sputtered Nb–Ge Films
J. R. Gavaler, J. W. Miller, and B. R. Appleton, *Appl. Phys. Lett.* **28**, 237 (1976).
Studies of Surface Contaminations, Composition and Formation of Superconducting Layers
of V, $Nb_3Sn$ and of Tunneling Elements Using High Energy Protons Combined with Heavy Ions
P. Muller, G. Ischenko, and F. Gabler, *in* "Ion Beam Surface Layer Analysis" (O. Meyer,
G. Linker, and F. Käppeler, eds.), Vol. I, p. 265. Plenum Press, New York, 1976.

## 10.6 THIN-FILM REACTIONS

### 10.6.1 Metal–Semiconductor

Meausrement of the Solubility of Germanium in Aluminum Using MeV $^4He^+$ Backscattering
J. Caywood, *Metall. Trans.* **4**, 735 (1973).
Ion-Induced Migration of Cu into Si
R. R. Hart, H. L. Dunlap, O. J. Marsh, *J. Appl. Phys.* **46**, 1947 (1975).
Enhanced Migration of Implanted Sb and In in Si Covered with Evaporated Al
R. R. Hart, D. H. Lee, and O. J. Marsh, *Appl. Phys. Lett.* **20**, 76 (1972).
Formation of Silicon-Oxide over Gold Layers on Silicon Substrates
A. Hiraki, E. Lugujjo, and J. W. Mayer, *J. Appl. Phys.* **43**, 3643 (1972).
Low Temperature Migration of Silicon through Metal Films: Importance of Silicon–Metal
Interface
A. Hiraki, E. Lugujjo, M-A. Nicolet, and J. W. Mayer, *Phys. Status Solidi* **7**, 401 (1971).
Low Temperature Migration of Silicon in Thin Layers of Gold and Platinum
A. Hiraki, M-A. Nicolet, and J. W. Mayer, *Appl. Phys. Lett.* **18**, 178 (1970).

Effects of Al Films on Ion-Implanted Si
  D. H. Lee, R. R. Hart, and O. J. Marsh, *Appl. Phys. Lett.* **20**, 73 (1972).
Diffusion Processes at the Cu–CdTe Interface for Evaporated and Chemically Plated Cu Layers
  H. Mann, G. Linker, and O. Meyer, *Solid State Commun.* **11**, 475 (1972).
Interaction of Al Layers with Polycrystalline Silicon
  K. Nakamura, M-A. Nicolet, J. W. Mayer, R. J. Blattner, and C. A. Evans, Jr., *J. Appl. Phys.*
  **46**, 4678 (1975).
Ti and V Layers Retard Interaction between Al Films and Polycrystalline Si
  K. Nakamura, S. S. Lau, M-A. Nicolet, and J. W. Mayer, *Appl. Phys. Lett.* **28**, 277 (1976).
Interaction of Metal Layers with Polycrystalline Si
  K. Nakamura, J. O. Olowolafe, S. S. Lau, M-A. Nicolet, J. W. Mayer, and R. Shima, *J.
  Appl. Phys.* **47**, 1278 (1976).
Crystallization of Ge and Si in Metal Films: I
  G. Ottaviani, D. Sigurd, V. Marrello, J. W. Mayer, and J. O. McCaldin, *J. Appl. Phys.* **45**,
  1730 (1974).
Metal–Semiconductor Interactions Studied by Ion-Backscattering Techniques
  D. Sigurd, *Inst. Phys. Conf. Ser. No. 22* 141 (1974).
Crystallization of Ge and Si in Metal Films: II
  D. Sigurd, G. Ottaviani, H. Arnal, and J. W. Mayer, *J. Appl. Phys.* **45**, 1740 (1974).
The Preparation and Analysis of Superconducting Nb–Ge Films
  L. R. Testardi, R. L. Meek, J. M. Poate, W. A. Royer, A. R. Storm, and J. H. Wernick,
  *Phys. Rev. B* **11**, 4304 (1975).
The First Phase to Nucleate in Planar Transition Metal–Germanium Interfaces
  M. Wittmer, M-A. Nicolet, and J. W. Mayer, *Thin Solid Films* **42**, 51 (1977).

## 10.6.2   Silicide Formation

An Analytical Study of Platinum Silicide Formation
  J. B. Bindell, J. W. Colby, D. R. Wonsidler, D. K. Conley, J. M. Poate, and T. C. Tisone,
  *Thin Solid Films* **37**, 441 (1976).
Effect of Oxidizing Ambients on Platinum Silicide Formation: II Auger and Backscattering
Analyses
  R. J. Blattner, C. A. Evans, S. S. Lau, J. W. Mayer, and B. M. Ullrich, *J. Electrochem. Soc.*
  **122**, 1732 (1975).
Ion-Backscattering Analysis of Tungsten Films on Heavily Doped SiGe
  J. A. Borders and J. N. Sweet, *J. Appl. Phys.* **43**, 3803 (1972).
Ion-Backscattering Study of $WSi_2$ Layer Growth in Sputtered W Contacts on Silicon
  J. A. Borders and J. N. Sweet, *in* "Applications of Ion Beams to Metals" (S. T. Picraux,
  E. P. EerNisse, and F. L. Vook, eds.), p. 179. Plenum Press, New York, 1974.
Characterization of Silicon Metallization Systems Using Ion Backscattering
  J. A. Borders and S. T. Picraux, *Proc. IEEE* **62**, 1224 (1974).
Formation Kinetics and Structure of $Pd_2Si$ Films on Si
  R. W. Bower, D. Sigurd, and R. E. Scott, *Solid-State Electron.* **16**, 1461 (1973).
Growth Kinetics Observed in the Formation of Metal Silicides on Silicon
  R. W. Bower and J. W. Mayer, *Appl. Phys. Lett.* **20**, 359 (1972).
Implanted Noble Gas Atoms as Diffusion Markers in Silicide Formation
  W. K. Chu, S. S. Lau, J. W. Mayer, H. Müller, and K. N. Tu, *Thin Solid Films* **25**, 393 (1975).
Identification of the Dominant Diffusing Species in Silicide Formation
  W. K. Chu, H. Kräutle, J. W. Mayer, H. Müller, M-A. Nicolet, and K. N. Tu, *Appl. Phys.
  Lett.* **25**, 454 (1974).

Silicide Formation in Rh–Si Schottky Barrier Diodes
  D. J. Coe, E. H. Rhoderick, P. H. Gerzon, and A. W. Tinsley, *Inst. Phys. Conf. Ser. No. 22* 74 (1974).
Reactions between the Ta–Pt–Ta–Au Metallization and PtSi Ohmic Contacts
  H. M. Day, A. Christau, W. H. Weisenberger, and J. K. Hirvonen, *J. Electrochem. Soc.* **122**, 769 (1975).
Investigation of CVD Tungsten Metallizations on Silicon by Backscattering
  P. Eichinger, H. Sauerman, and M. Wahl, *in* "Ion Beam Surface Layer Analysis" (O. Meyer, G. Linker, and F. Käppeler, eds.), Vol. I, p. 353. Plenum Press, New York, 1976.
Tungsten as a Marker in Thin Film Diffusion Studies
  G. J. van Gurp, D. Sigurd, and W. F. van der Weg, *Appl. Phys. Lett.* **29**, 159 (1976).
Studies of the Ti–W Metallization System on Si
  J. M. Harris, S. S. Lau, M-A. Nicolet, and R. S. Nowicki, *J. Electrochem. Soc.* **123**, 120 (1976).
Reactions of Thin Metal Films with Si or $SiO_2$ Substrates H. Kräutle, W. K. Chu, M-A. Nicolet, and J. W. Mayers, *in* "Applications of Ion Beams to Metals" (S. T. Picraux, E. P. Eerhisse, and F. L. Vook, *eds.*), p. 193. Plenum Press, New York, 1974.
Kinetics of Silicide Formation by Thin Films of V on Si and $SiO_2$ Substrates
  H. Kräutle, M-A. Nicolet, and J. W. Mayer, *J. Appl. Phys.* **45**, 3304 (1974).
Iron Silicide Thin Film Formation at Low Temperatures
  S. S. Lau, J. S. Y. Feng, J. O. Olowolafe, and M-A. Nicolet, *Thin Solid Films* **25**, 415 (1975).
Evaluation of Glancing Angle X-Ray Diffraction and MeV $^4$He Backscattering Analyses of Silicide Formation
  S. S. Lau, W. K. Chu, J. W. Mayer, and K. N. Tu, *Thin Solid Films* **23**, 205 (1974).
Alloying of Thin Palladium Films with Single Crystal and Amorphous Silicon
  D. H. Lee, R. R. Hart, D. A. Kiewit, and O. J. Marsh, *Phys. Status Solidi (a)* **15**, 645 (1973).
Separate Estimate of Crystallite Orientations and Scattering Centers in Polycrystals by Backscattering Techniques
  H. Ishiwara and S. Furukawa, *J. Appl. Phys.* **47**, 1686 (1976).
Analysis of Thin-Film Structures with Nuclear Backscattering and X-Ray Diffraction
  J. W. Mayer and K. N. Tu, *J. Vac. Sci. Technol.* **11**, 86 (1974).
Migration of Mo Atoms across Mo–Si Interface Induced by Ar Ion Bombardment
  H. Nishi, T. Sakuraii, T. Akamatsu, and T. Furuya, *Appl. Phys. Lett.* **25**, 337 (1974).
Influence of the Nature of the Si Substrate on Nickel Silicide Formed from Thin Ni Films
  J. O. Olowolafe, M-A. Nicolet, and J. W. Mayer, *Thin Solid Films* **38**, 143 (1976).
Formation Kinetics of $CrSi_2$ Films on Si Substrates with and without Interposed $Pd_2Si$ Layer
  J. O. Olowolafe, M-A. Nicolet, and J. W. Mayer, *J. Appl. Phys.* **47**, 5182 (1976).
Chromium Thin Films as a Barrier to the Interaction of $Pd_2Si$ with Al
  J. O. Olowolafe, M-A. Nicolet, and J. W. Mayer, *Solid State Electron.* **20**, 413 (1977).
Silicide Formation Correlated with Surface Resistivity Measurements
  S. Petersson, E. Mgbenu, and P. A. Tove, *Phys. Status Solidi (a)* **36**, 217 (1976).
Kinetics and Mechanism of Platinum Silicide Formation on Silicon
  J. M. Poate and T. C. Tisone, *Appl. Phys. Lett.* **24**, 391 (1974).
Hafnium–Silicon Surface Barriers
  A. N. Saxena, J. J. Grob, M. Hage-Ali, P. Siffert, and I. V. Mitchell, *Inst. Phys. Conf. Ser. No. 22* 160 (1974).
Studies of Formation of Silicides and Their Barrier Heights to Silicon
  K. E. Sundstrom, S. Petersson, and P. A. Tove, *Phys. Status Solidi (a)* **20**, 653 (1973).
Epitaxial Growth of Nickel Silicide $NiSi_2$ on Silicon
  K. N. Tu, E. I. Alessandrini, W. K. Chu, H. Kräutle, and J. W. Mayer, *Jpn. J. Appl. Phys. Suppl. 2* Pt. 1, 669 (1974).

Structure and Growth Kinetics of $Ni_2Si$ on Silicon
  K. N. Tu, W. K. Chu, and J. W. Mayer, *Thin Solid Films* **25**, 403 (1975).
Formation of Vanadium Silicides by the Interactions of V with Bare and Oxidized Si Wafers
  K. N. Tu, J. F. Ziegler, and C. J. Kircher, *Appl. Phys. Lett.* **23**, 493 (1973).
Kinetics of the Formation of Hafnium Silicides on Silicon
  J. F. Ziegler, J. W. Mayer, C. J. Kircher, and K. N. Tu, *J. Appl. Phys.* **44**, 3851 (1973).


## 10.6.3   Metal–GaAs

Alloying Behavior of Au–In and Au–Sn Films on Semiconductors
  L. Buene, T. Finstad, K. Timstad, O. Lonsjo, and T. Olsen, *Thin Solid Films* **34**, 149 (1976).
Depth Profiling with Ion-Induced X-Rays
  L. C. Feldman and P. J. Silverman, *in* "Ion Beam Surface Layer Analysis" (O. Meyer,
  G. Linker, and F. Käppeler, eds.), Vol. II, p. 735. Plenum, New York, 1976.
Alloying Behavior of Au and Au–Ge on GaAs
  J. Gyulai, J. W. Mayer, V. Rodriguez, A. Y. C. Yu, and H. J. Gopen, *J. Appl. Phys.* **42**,
  3578 (1971).
Reaction of Sputtered Pt Films on GaAs
  V. Kumar, *J. Phys. Chem. Solids* **36**, 535 (1975).
Effect of Alloying Behavior on the Electrical Characteristics of *n*-GaAs Schottky Diodes
Metallized with W, Au and Pt
  A. K. Sinha and J. M. Poate, *Appl. Phys. Lett.* **23**, 666 (1973).
Relative Thermal Stabilities of Thin-Film Contacts to *n*-GaAs Metallized with W, Au and Pt
  A. K. Sinha and J. M. Poate, *Jpn. J. Appl. Phys. Suppl.* **2** Pt. 1, 841 (1974).
*n*-GaAs Schottky Diodes Metallized with Ti and Pt/Ti
  A. K. Sinha, T. E. Smith, M. H. Read, and J. M. Poate, *Solid State Electron.* **19**, 489 (1976).
Thermally Induced Processes at Au–GaAs Interfaces: An Assessment by Rutherford Back-
scattering, SEM and Depth Profiling Auger Electron Spectroscopy
  C. J. Todd, G. W. B. Ashwell, J. D. Speight, and R. Heckingbottom, *Inst. Phys. Conf. Ser.
  No. 22* 171 (1974).


## 10.6.4   Metal–Dielectric

Studies on the $Al_2O_3$–Ti–Mo–Au Metallization System
  J. M. Harris, E. Lugujjo, S. U. Campisano, M-A. Nicolet, and R. Shima, *J. Vac. Sci. Technol.*
  **12**, 524 (1975).
Silicide Formation at Low-Temperature by Metal–$SiO_2$ Reaction
  H. Kräutle, M-A. Nicolet, and J. W. Mayer, *Phys. Status Solidi (a)* **20**, K33 (1973).
Migration of Gold Atoms through Thin Silicon Oxide Films
  C. J. Madams, D. V. Morgan, and M. J. Howes, *J. Appl. Phys.* **45**, 5089 (1974).
Low Temperature Migration of Gold through Thin Films of Silicon Monoxide
  D. V. Morgan, M. J. Howes, and C. J. Madams, *J. Electrochem. Soc.* **123**, 295 (1976).
Formation of Vanadium Silicides by the Interactions of V with Bare and Oxidized Si Wafers
  K. N. Tu, J. F. Ziegler, and C. J. Kircher, *Appl. Phys. Lett.* **23**, 493 (1973).


## 10.6.5   Metal–Metal

Thin Film Interdiffusion of Chromium and Copper
  J. E. E. Baglin, V. Brusic, E. Alessandrini, and J. F. Ziegler, *in* "Applications of Ion Beams to

Metals" (S. T. Picraux, E. P. EerNisse, and F. L. Vook, eds.), p. 169. Plenum Press, New York, 1974.
Thin Film Interdiffusion of Cr and Cu
  J. E. E. Baglin, V. Brusic, and E. Alessandrini, *Thin Solid Films* **25**, 449 (1975).
The Analysis of Nickel and Chromium Migration through Gold Layers
  A. Barcz, A. Turos, and L. Wielunski, *in* "Ion Beam Surface Layer Analysis" (O. Meyer, G. Linker, and F. Käppeler, eds.), Vol. I, p. 407. Plenum Press, New York, 1976.
Ion Backscattering Analysis of Interdiffusion in Cu–Au Thin Films
  J. A. Borders, *Thin Solid Films* **19**, 359 (1973).
Characteristics of Aluminum–Titanium Electrical Contacts on Silicon
  R. W. Bower, *Appl. Phys. Lett.* **23**, 99 (1973).
Low Temperature Interdiffusion in Copper–Gold Thin Films Analyzed by Helium Backscattering
  S. U. Campisano, G. Foti, F. Grasso, and E. Rimini, *Thin Solid Films* **19**, 339 (1973).
Analysis of Compound Formation in Au–Al Thin Films
  S. U. Campisano, G. Foti, F. Grasso, J. W. Mayer, and E. Rimini, *in* "Applications of Ion Beams to Metals" (S. T. Picraux, E. P. EerNisse, and F. L. Vook, eds.), p. 159. Plenum Press, New York, 1974.
Backscattering and T.E.M. Studies of Grain Boundary Diffusion in Thin Metal Films
  S. U. Campisano, E. Costanzo, G. Foti, and E. Rimini, *in* "Ion Beam Surface Layer Analysis" (O. Meyer, G. Linker, and F. Käppeler, eds.), Vol. I, p. 397. Plenum Press, New York, 1976.
Ion Backscattering Analysis of Mixing in Au–Cu and Au–Al Thin Films
  S. U. Campisano, G. Foti, F. Grasso, and E. Rimini, *Jpn. J. Appl. Phys. Suppl.* **2** Pt. 1, 637 (1974).
Kinetics of Phase Formation in Au–Al Thin Films
  S. U. Campisano, G. Foti, E. Rimini, S. S. Lau, and J. W. Mayer, *Phil. Mag.* **31**, 903 (1975).
Determination of Concentration Profiles in Thin Metallic Films: Applications and Limitations of He$^+$ Backscattering
  S. U. Campisano, G. Foti, F. Grasso, and E. Rimini, *Thin Solid Films* **25**, 431 (1975).
SEM, Auger, Spectroscopy and Ion Backscattering Techniques Applied to Analyses of Au/Refractory Metallizations
  A. Christou, L. Jarvis, W. H. Weisenberger, and J. K. Hirvonen, *J. Electron. Mater.* **4**, 329 (1975).
Low Temperature Interdiffusion in the Au–Pd and Au–Rh Thin Film Couples
  W. J. DeBonte and J. M. Poate, *Thin Solid Films* **25**, 441 (1975).
Rutherford Scattering Studies of Diffusion in Thin Multilayer Metal Films
  W. J. DeBonte, J. M. Poate, C. M. Melliar-Smith, and R. A. Levesque, *in* "Applications of Ion Beams to Metals" (S. T. Picraux, E. P. EerNisse, and F. L. Vook, eds.), p. 147. Plenum Press, New York, 1974.
Thin Film Interdiffusion II. Ti–Rh, Ti–Pt, Ti–Rh–Au and Ti–Au–Rh
  W. J. DeBonte, J. M. Poate, C. M. Melliar-Smith, and R. A. Levesque, *J. Appl. Phys.* **46**, 4284 (1975).
Diffusion Mechanisms in the Pd/Au Thin Film System and the Correlation of Resistivity Changes with Auger Electron Spectroscopy and Rutherford Backscattering Profiles
  P. M. Hall, J. M. Morabito, and J. M. Poate, *Thin Solid Films* **33**, 107 (1976).
Kinetics of Compound Formation in Thin Couples of Al and Transition Metals
  J. K. Howard, R. F. Lever, P. J. Smith, and P. S. Ho, *J. Vac. Sci. Technol.* **13**, 68 (1976).
Backscattering Investigation of Low-Temperature Migration of Chromium through Gold Films
  J. K. Hirvonen, W. H. Weisenberger, J. E. Westmoreland, and R. A. Meussner, *Appl. Phys. Lett.* **21**, 37 (1972).

Interdiffusion Mechanisms in Ag–Au Thin Film Couples
  R. G. Kirsch, J. M. Poate, and M. Eibschutz, *Appl. Phys. Lett.* **29**, 773 (1976).
Interdiffusion of Thin Cr and Au Films Deposited on Silicon
  G. Majni, G. Ottaviani, and M. Prudenziati, *Thin Solid Films* **38**, 15 (1976).
Deterioration of Contacts to Kanthal Thin Film Resistors due to Metal Diffusion Effects
  R. Mustonen, S. Petersson, and P. A. Tove, *Thin Solid Films* **24**, S47 (1974).
Mass Transport between Two Metal Layers as Studied by Ion Scattering
  S. T. Picraux, *Jpn. J. Appl. Phys. Suppl.* **2** Pt. 1, 657 (1974).
Multilayer Thin Film Analysis by Ion Backscattering
  S. T. Picraux and F. L. Vook, *Appl. Phys. Lett.* **18**, 191 (1971).
Thin-Film Interdiffusion. I. Au–Pd, Pd–Au, Ti–Pd, Ti–Au, Ti–Pd–Au and Ti–Au–Pd
  J. M. Poate, P. A. Turner, W. J. DeBonte, and J. Yahalom, *J. Appl. Phys.* **46**, 4275 (1975).
Anodic Processing for Multilevel LSI
  G. C. Schwartz and V. Platter, *J. Electrochem. Soc.* **123**, 34 (1976).
Backscattering Investigation of the Low Temperature Stability of the Aluminum–Silver System
  J. E. Westmoreland and W. H. Weisenberger, *Thin Solid Films* **19**, 349 (1973).
Thermal Stability of a Proposed Magnetic Bubble Memory
  J. F. Ziegler, J. E. E. Baglin, and A. Gangulee, *Appl. Phys. Lett.* **24**, 36 (1974).

## 10.6.6   Solid-Phase Epitaxy

Solid Phase Transport and Epitaxial Growth of Ge and Si
  C. Canali, J. W. Mayer, G. Ottaviani, D. Sigurd, and W. van der Weg, *Appl. Phys. Lett.* **25**, 3 (1974).
Solid Phase Epitaxial Growth of Si through Palladium Silicide Layers
  C. Canali, S. U. Campisano, S. S. Lau, Z. L. Liau, and J. W. Mayer, *J. Appl. Phys.* **46**, 2831 (1975).
Channeling Effect Measurements of the Recrystallization of Amorphous Si Layers on Crystal Si
  L. Csepregi, J. W. Mayer, and T. W. Sigmon, *Phys. Lett.* **54A**, 157 (1975).
Antimony Doping of Si Layers Grown by Solid-Phase Epitaxy
  S. S. Lau *et al.*, *Appl. Phys. Lett.* **28**, 148 (1976).
Kinetics of the Initial Stage of Si Transport through Pd-Silicide for Epitaxial Growth
  Z. L. Liau, S. U. Campisano, C. Canali, S. S. Lau, and J. W. Mayer, *J. Electrochem. Soc.* **122**, 1696 (1975).
Solid Phase Epitaxial Growth of Ge Layers
  V. Marrello, J. M. Caywood, M-A. Nicolet, and J. W. Mayer, *Phys. Status Solidi (a)* **13**, 531 (1972).
Some Aspects of Ge Epitaxial Growth by Solid Solution
  G. Ottaviani, C. Canali, and G. Majni, *J. Appl. Phys.* **47**, 627 (1976).

## 10.7   ION IMPLANTATION IN METALS

### 10.7.1   Lattice Site Location

The Physical State of Implanted W in Copper
  J. A. Borders, A. G. Cullis, and J. M. Poate, *Inst. Phys. Conf. Ser.* **28**, 204 (1976).
Lattice Site Location of Ion-Implanted Impurities in Cu and other fcc Metals
  J. A. Borders and J. M. Poate, *Phys. Rev. B* **13**, 696 (1976).
The Physical State of Implanted Tungsten in Cu
  A. G. Cullis, J. M. Poate, and J. A. Borders, *Appl. Phys. Lett.* **28**, 314 (1976).

The Lattice Site Location of C Implanted into Fe
L. C. Feldman, E. N. Kaufman, J. M. Poate, and W. M. Augustyniak, *in* "Ion Implantation in Semiconductors and Other Materials" (B. L. Crowder, ed.), p. 491. Plenum Press, New York, 1973.
Radiation Damage and Ion Behavior in Ion Implanted Vanadium and Nickel Single Crystals
M. Gettings, K. G. Langguth, and G. Linker, *in* "Applications of Ion Beams to Metals" (S. T. Picraux, E. P. EerNisse, and F. L. Vook, eds.), p. 241. Plenum Press, New York, 1974.
Ion Implantation and Radiation Damage in Vanadium
G. Linker, M. Gettings, and O. Meyer, *in* "Ion Implantation in Semiconductors and Other Materials" (B. L. Crowder, ed.), p. 465. Plenum Press, New York, 1973.
Formation of Substitutional Alloys by Ion Implantation in Metals
J. M. Poate, W. J. DeBonte, W. M. Augustyniak, and J. A. Borders, Appl. Phys. Lett. **25**, 698 (1974).
The Formation of Substitutional Alloys in fcc Metals by High Dose Implantations
J. M. Poate, W. J. DeBonte, W. M. Augustyniak, and J. A. Borders, *in* "Ion Implantation in Semiconductors" (S. Namba, ed), p. 361. Plenum Press, New York, 1975.
A Rutherford Backscattering and Channeling Study of Dy Implanted into Single Crystal Ni
G. A. Stephens, E. Robinson, and J. S. Williams, *in* "Ion Implantation in Semiconductors" (S. Namba, ed.), p. 375. Plenum Press, New York, 1975.
Lattice Location of Impurities Implanted into Metals
H. de Waard and L. C. Feldman, *in* "Applications of Ion Beams to Metals" (S. T. Picraux, E. P. EerNisse, and F. L. Vook, eds.), p. 317. Plenum Press, New York, 1974.

## 10.7.2 Diffusion from Implanted Layers

Application of the Backscattering Method for the Measurement of Diffusion of Zinc in Aluminum
A. Fontell, E. Arminen, and M. Turunen, *Phys. Status Solidi (a)* **15**, 113 (1973).
Implantation and Diffusion of Au in Be: Behavior During Annealing of a Low-Solubility Implant
S. M. Myers and R. A. Langley, *in* "Applications of Ion Beams to Metals" (S. T. Picraux, E. P. EerNisse, and F. L. Vook, eds.), p. 283. Plenum Press, New York, 1974.
Study of the Diffusion of Au and Ag in Be Using Ion Beams
S. M. Myers and R. A. Langley, *J. Appl. Phys.* **46**, 1034 (1975).
Enhanced Diffusion of Zn in Al under High-Flux Heavy-Ion Irradiation
S. M. Myers and S. T. Picraux, *J. Appl. Phys.* **46**, 4774 (1975).
Implantation and Diffusion of Cu in Be
S. M. Myers, W. Beezhold, and S. T. Picraux, *in* "Ion Implantation in Semiconductors and Other Materials" (B. L. Crowder, ed.), p. 455. Plenum Press, New York, 1973.
Study of Cu Diffusion in Be Using Ion Backscattering
S. M. Myers, S. T. Picraux, and T. S. Prevender, *Phys. Rev. B* **9**, 3953 (1974).
Possible Radiation Enhanced Diffusion of Nickel Ions in Titanium
J. F. Turner, W. Temple, and G. Dearnaley, *in* "Ion Implantation in Semiconductors and Other Materials" (B. L. Crowder, ed.), p. 437. Plenum Press, New York, 1973.

## 10.7.3 Disorder and Metallurgy

Ranges of Ions with $Z_1 \geq 54$ in Al and $Al_2O_3$
H. H. Andersen, J. Bøttiger, and H. W. Jorgesen, *Appl. Phys. Lett.* **26**, 678 (1975).
The Physical State of Implanted W in Copper
J. A. Borders, A. G. Cullis, and J. M. Poate, *Inst. Phys. Conf. Ser.* **28**, 204 (1976).

Dechanneling from 2-MeV He$^+$ Damage in Gold
K. L. Merkle, P. P. Pronko, D. S. Gemmell, R. C. Mikkelson, and J. R. Wrobel, *Phys. Rev.*
*B* 8, 1002 (1973).
Phase Equilibria and Diffusion in the Be–Al–Fe System Using High-Energy Ion Beams
S. M. Myers and J. E. Smugeresky, *Metall. Trans.* 7A, 795 (1976).
Implantation Metallurgy
S. T. Picraux, *Inst. Phys. Conf. Ser.* 28, 183 (1976).
The Formation of Substitutional Alloys by Ion-Implantation in Metals
J. M. Poate, W. J. DeBonte, W. M. Augustyniak, and J. A. Borders, *Appl. Phys. Lett.* 25,
698 (1974).
Dechanneling Measurements of Defect Depth Profiles and Effective Cross-Channel Distribu-
tion of Misaligned Atoms in Ion-Irradiated Gold
P. P. Pronko, *Nucl. Instrum. Methods* 132, 249 (1976).
Dechanneling From Damage Clusters in Heavy Ion Irradiated Gold
P. P. Pronko and K. L. Merkley, *in* "Applications of Ion Beams to Metals" (S. T. Picraux,
E. P. EerNisse, and F. L. Vook, eds.), p. 481. Plenum Press, New York, 1974.
Backscattering Analysis of Ion Bombardment Damage in Nb and W at Low (25°K) Temperature
P. P. Pronko, J. Bøttiger, J. A. Davies, and J. B. Mitchell, *Radiat. Effects* 21, 25 (1974).
Energy Dependence of Channeling Analysis in Implantation Damaged Al
E. Rimini, S. U. Campisano, G. Foti, P. Baeri, and S. T. Picraux, *in* "Ion Beam Surface
Layer Analysis" (O. Meyer, G. Linker, and F. Käppeler, eds.), Vol. 2, p. 597. Plenum Press,
New York, 1976.
Ion Implanted Surface Layers in Copper and Aluminum
D. K. Sood and G. Dearnaley, *Inst. Phys. Conf. Seris* 28, 196 (1976).
A Rutherford Backscattering and Channeling Study of Dy Implanted into Single Crystal Ni
G. A. Stephans, E. Robinson, and J. S. Williams, *in* "Ion Implantation in Semiconductors"
(S. Namba, ed.), p. 375. Plenum Press, New York, 1975.
Sb-Implanted Al Studied by Ion Backscattering and Electron Microscopy
G. J. Thomas and S. T. Picraux, *in* "Applications of Ion Beams to Metals" (S. T. Picraux,
E. P. EerNisse, and F. L. Vook, eds.), p. 257. Plenum Press, New York, 1974.

## 10.7.4   Corrosion, Oxidation, and Superconductivity

The Influence of Ion Implantation upon the High Temperature Oxidation of Titanium and
Stainless Steel
G. Dearnaley, P. D. Goode, W. S. Miller, and J. F. Turner, *in* "Ion Implanatation in Semi-
conductors and Other Materials" (B. L. Crowder, ed.), p. 405. Plenum Press, New York,
1973.
The Use of Ion Beams in Corrosion Science
G. Dearnaley, *in* "Applications of Ion Beams to Metals" (S. T. Picraux, E. P. EerNisse,
and F. L. Vook, eds.), p. 63. Plenum Press, New York, 1974.
Ion Implantation in Metals
G. Dearnaley, *in* "New Uses of Ion Accelerators" (J. F. Ziegler, ed.), Chapter 5, p. 283.
Plenum Press, New York, 1975.
Oxidation of Pb by Low Energy $O_2{}^+$-Bombardment
J. Geerk and O. Meyer, *Surf. Sci.* 32, 222 (1972).
Friction Changes in Ion-Implanted Steel
N. E. W. Hartley, W. E. Swindlehurst, G. Dearnaley, and J. F. Turner, *J. Mater. Sci.* 8,
900 (1973).
Frictional Changes Induced by the Ion Implantation of Steel
N. E. W. Hartley, G. Dearnaley, and J. F. Turner, *in* "Ion Implantation in Semiconductors
and Other Materials" (B. L. Crowder, ed.), p. 423. Plenum Press, New York, 1973.

Determination of Implanted Carbon Profiles in NbC Single Crystals from Random Back-scattering Spectra
  K. G. Langguth, G. Linker, and J. Geerk, *in* "Ion Beam Surface Layer Analysis" (O. Meyer, G. Linker and F. Käppeler, eds.), Vol. I, p. 273. Plenum Press, New York, 1976.
The Influence of Heavy Ion Bombardment on the Superconducting Transition Temperature of Thin Films
  G. Linker and O. Meyer, *in* "Ion Implantation in Semiconductors" (S. Namba, ed.), p. 309 Plenum Press, New York, 1975.
Superconducting Properties and Structural Transformations of Nitrogen Implanted Molybdenum Films
  G. Linker and O. Meyer, *Solid State Commun.* **20**, 695 (1976).
Ion Implantation in Superconductors
  O. Meyer, *in* "New Uses of Ion Accelerators" (J. F. Ziegler, ed.), Ch. 6, p. 323. Plenum Press, New York, 1975.
Enhancement of the Superconducting Transition Temperature by Ion Implantation in Molybdenum Thin Films
  O. Meyer, *Inst. Phys. Conf. Ser.* **28**, 168 (1976).
Ion Implantation in Superconducting Thin Films
  O. Meyer, H. Mann, and E. Phrilingos, *in* "Applications of Ion Beams to Metals" (S. T. Picraux, E. P. EerNisse, and F. L. Vook, eds.), p. 15. Plenum Press, New York, 1974.
Effect of B, C, N and Ne ion Implantation on the Oxidation of Polycrystalline Cu
  H. M. Naguib, R. J. Kriegler, J. A. Davies, and J. B. Mitchell, *J. Vac. Sci. Technol.* **13**, 396 (1976).
Impurity Effects on the Corrosion of Aluminum
  C. Towler, R. A. Collins, and G. Dearnaley, *J. Vac. Sci. Technol.* **12**, 520 (1975).


## 10.8  ION IMPLANTATION IN SEMICONDUCTORS


### 10.8.1  Silicon (1976)

Distribution across the Channel of Defects Induced by Nitrogen Bombardment in Silicon
  P. Baeri, S. U. Campisano, G. Ciavola, G. Foti, and E. Rimini, *Appl. Phys. Lett.* **28**, 9 (1976).
Radial Distribution of Ion-Induced Defects Determined by Channeling Measurements
  P. Baeri, S. U. Campisano, G. Ciavola, G. Foti, and E. Rimini, *Nucl. Instrum. Methods* **132**, 237 (1976).
Depth Distributions of Silver Ions Implanted in Si and $SiO_2$
  A. Barcz, A. Turos, L. Wielunski, W. Rosinski, and B. Wojtowicz-Natanson, *Radiat. Effects* **25**, 91 (1975).
The Influence of Sputtering, Range Shortening and Stress-Induced Outdiffusion on the Retention of Xenon Implanted in Silicon.
  P. Blank, K. Wittmaack, and F. Schulz, *Nucl. Instrum. Methods* **132**, 387 (1976).
Production of Solar Cells by Recoil Implantation
  O. Christensen and H. L. Bay, *Appl. Phys. Lett.* **28**, 491 (1976).
Dose Dependence of Residual Lattice Disorder in Ion-Implanted and Annealed Silicon
  C. E. Christodoulides, W. A. Grant, and J. S. Williams, *Appl. Phys. Lett.* **30**, 322 (1977).
Disorder Produced by High-Dose Implantation in Si
  L. Csepregi, E. F. Kennedy, S. S. Lau, J. W. Mayer, and T. W. Sigmon, *Appl. Phys. Lett.* **29**, 645 (1976).
Regrowth Behavior of Ion Implanted Amorphous Layers on ⟨111⟩ Silicon
  L. Csepregi, J. W. Mayer, and T. W. Sigmon, *Appl. Phys. Lett.* **29**, 93 (1976).

Polyatomic-Ion Implantation Damage in Si
  J. A. Davies, G. Foti, L. M. Howe, J. B. Mitchell, and K. B. Winterbon, *Phys. Rev. Lett.*
  **34**, 1441 (1975).
Range Parameters of Heavy Ions at 10 and 35 keV in Silicon
  A. Feuerstein, S. Kalbitzer, and H. Oetzmann, *Phys. Lett.* **51A**, 165 (1975).
The Application of Correlated SIMS and RBS Techniques to the Measurement of Ion Implanted
Range Profiles
  D. Fuller, J. S. Colligon, and J. S. Williams, *Surf. Sci.* **54**, 647 (1976).
Particularities of Crystalline to Amorphous State Conversion in Silicon Heavily Damaged by
140 keV Si$^{++}$ Ions
  A. Golanski, *et al.*, *Phys. Status Solidi (a)* **38**, 139 (1976).
Lattice Location and Ionization Induced Annealing of Self-Interstitials in Boron-Implanted
Silicon by Rutherford Backscattering
  G. Götz and G. Sommer, *Phys. Status Solidi (b)* **32**, K151 (1975).
Investigation of Radiation Damage in Silicon by a Backscattering Method
  G. Götz, K. Hehl, F. Schwabe, and E. Glaser, *Radiat. Effects* **25**, 27 (1975).
Measurement of Projected and Lateral Range Parameters for Low Energy Heavy Ions in
Silicon by Rutherford Backscattering
  W. A. Grant, J. S. Williams, and D. Dodds, *in* "Ion Beam Surface Layer Analysis" (O. Meyer,
  G. Linker, and F. Käppeler, eds.), Vol. I, p. 235. Plenum Press, New York, 1976.
Damage Profiles and Annealing of Si (B) Implants
  J. J. Grob, P. Siffert, R. Prisslinger, and S. Kalbitzer, *Inst. Phys. Conf. Ser.* **28**, 24 (1976).
Annealing of Damage Caused by Implantation of Group 1B Elements into Silicon
  A. Johansen, B. Svenningsen, L. T. Chadderton, and J. L. Whitton, *Inst. Phys. Conf. Ser.*
  **28**, 267 (1976).
Lattice Defects in Ion-Implanted Semiconductors
  L. C. Kimerling and J. M. Poate, *Inst. Phys. Conf. Ser. No. 23* 126 (1975).
Study of the Annealing Behavior of High Dose Implants in Silicon and Germanium Crystals
  H. Kräutle, *Radiat. Effects* **24**, 255 (1975).
Lateral Spread of Damage Formed by Ion Implantation
  H. Matsumura and S. Furukawa, *J. Appl. Phys.* **47**, 1746 (1976).
MeV He Backscattering Analysis of Ion-Implanted Si: Drive-In Diffusion and Epitaxial
Regrowth
  J. W. Mayer, L. Cespregi, J. Gyulai, T. Nagy, G. Mezey, P. Revesz, and E. Kotai, *Thin
  Solid Film* **32**, 303 (1976).
Radiation Enhanced Diffusion of Arsenic in Si
  H. Ryssel, H. Kranz, and P. Eichinger, *Inst. Phys. Conf. Ser.* **28**, 1 (1976).
Visible Interference Effects in Silicon Caused by High-Current–High-Dose Implantation
  T. E. Seidel, G. A. Pasteur, and J. C. C. Tsai, *Appl. Phys. Lett.* **29**, 648 (1976).
Recoil-Implanted Antimony-Doped Surface Layers in Silicon
  J. M. Shannon, *Inst. Phys. Conf. Ser.* **28**, 37 (1976).
Low Energy Ion Induced Damage in Silicon at 50°K
  D. A. Thompson and R. S. Walker, *Nucl. Instrum. Methods* **132**, 281 (1976).
The Measurement of Pb$^+$ Ion Collection in Si by High Resolution Rutherford Backscattering
  J. S. Williams, *Phys. Lett.* **51A**, 85 (1975).
The Application of High Resolution Rutherford Backscattering to the Measurement of Ion
Ranges in Si and Al
  J. S. Williams and W. A. Grant, *Radiat. Effects* **25**, 55 (1975).
Annealing Behavior of High Dose Rare-Gas Implantations into Si
  J. S. Williams and W. A. Grant, *Inst. Phys. Conf. Ser.* **28**, 31 (1976).

## 10.8.2   III–V and II–VI (1975–1976)

Lattice Location Studies of GaAs Implanted with Te, Tl and Zr
  J. R. Brawn and W. A. Grant, *Inst. Phys. Conf. Ser.* **28**, 59 (1976).
Nitrogen Implanted Germanium: Damage Lattice Location and Electrical Properties
  A. B. Campbell, J. B. Mitchell, J. Shewchun, D. A. Thompson, and J. A. Davies, *Can. J. Phys.* **53**, 303 (1975).
Ion Implantation of Cd and Te in GaAs Crystals
  K. Gamo, M. Takai, M. S. Lin, K. Masuda, and S. Namba, *in* "Ion Implantation in Semiconductors" (S. Namba, ed.), p. 35. Plenum Press, New York, 1975.
Behaviors of Ga and P Damages Introduced by Ion Implantation into GaP
  H. Matsumura and S. Furukawa, *in* "Ion Implantation in Semiconductors" (S. Namba, ed.), p. 125. Plenum Press, New York, 1975.
Channeling Measurements of Damage in Ion Bombarded Semiconductors at 50°K
  D. A. Thompson and R. S. Walker, *Radiat. Effects* **30**, 37 (1976).
Measurement of Damage Distributions in Ion Bombarded Si, GaP and GaAs at 50°K
  R. S. Walker and D. A. Thompson, *Nucl. Instrum. Methods* **135**, 489 (1976).


## 10.9   HYDROGEN AND HELIUM IN METALS

Location of Interstitial Deuterium in $TaD_{0.067}$ by Channeling
  M. Antonini and H. D. Carstanjen, *Phys. Status Solidi (a)* **34**, K153 (1976).
Implantation Profiles of Low Energy Helium in Niobium and the Blistering Mechanism
  R. Behrisch, J. Bøttiger, W. Eckstein, U. Littmark, J. Roth, and B. M. U. Scherzer, *Appl. Phys. Lett.* **27**, 199 (1975).
Trapping of Low-Energy Helium Ions in Niobium
  R. Behrisch, J. Bøttiger, W. Eckstein, J. Roth, and B. M. U. Scherzer, *J. Nucl. Mater*, **56**, 365 (1975).
Depth Distribution of Implanted Helium and Other Low-Z Elements in Metal Films Using Proton Backscattering
  R. S. Blewer, *Appl. Phys. Lett.* **23**, 593 (1973).
Proton Backscattering as a Technique for Light Ion Surface Interaction Studies in CTR Materials Investigations
  R. S. Blewer, *J. Nucl. Mater.* **53**, 268 (1974).
Depth Distribution and Migration of Implanted Helium in Metal Foils Using Proton Backscattering
  R. S. Blewer, *in* "Applications of Ion Beams to Metals" (S. T. Picraux, E. P. EerNisse, and F. L. Vook, eds.), p. 557. Plenum Press, New York, 1974.
Depth Distribution and Migration of Implanted Low Z Elements in Solids Using Proton Elastic Scattering
  R. S. Blewer, in "Radiation Effects on Solid Surfaces" (M. Kaminsky, ed.) p. 262, Adv. Chem. Series 158. Amer. Chem. Soc., Washington, D.C., 1976.
The Trapping of Hydrogen Ions in Zirconium for Ion Energies Between 0.3 and 6 keV
  J. Bohdansky, J. Roth, and W. P. Poschenrieder, *Inst. Phys. Conf. Ser. No.* **28**, 307 (1976).
Range Profiles of 6–16 keV Hydrogen Ions Implanted in Metal Oxides
  J. Bøttiger, J. R. Leslie, and N. Rud, *J. Appl. Phys.* **47**, 1672 (1976).
Depth Profiling of Hydrogen and Helium Isotopes in Solids by Nuclear Reaction Analysis
  J. Bøttiger, S. T. Picraux, and N. Rud, *in* "Ion Beam Surface Layer Analysis" (O. Meyer, G. Linker, and F. Käppeler, eds.), Vol. 2, p. 811. Plenum Press, New York, 1976.

Trapping of Hydrogen Isotopes in Molybdenum and Niobium Predamaged by Ion Implantation
    J. Bøttiger, S. T. Picraux, N. Rud, and T. Laursen, *J. Appl. Phys.* **48**, 920 (1977).
Location of Interstitial Deuterium Sites in Niobium by Channeling
    H. D. Carstanjen and R. Sizmann, *Phys. Lett.* **40A**, 93 (1972).
Determination of Interstitial Deuterium Locations in Nb by Channeling
    H. D. Carstanjen and R. Sizmann, *Ber. Bunsen-Gesellsch. Phys. Chem.* **76**, 1223 (1972);
    Heft 12.
Nondestructive Analysis for Trace Amounts of Hydrogen
    B. L. Cohen, C. L. Fink, and J. H. Degnan, *J. Appl. Phys.* **43**, 19 (1972).
New Precision Technique for Measuring the Concentration Versus Depth of Hydrogen in Solids
    W. A. Lanford, H. P. Trautvetter, J. F. Ziegler, and J. Keller, *Appl. Phys. Lett.* **28**, 566 (1976).
Depth Distribution Profiling of Deuterium and $^3$He
    R. A. Langley, S. T. Picraux, and F. L. Vook, *J. Nucl. Mater.* **53**, 257 (1974).
Depth Profiling of Deuterium and Helium in Metals by Elastic Proton Scattering
    R. A. Langley, *in* "Ion Beam Surface Layer Analysis" (O. Meyer, G. Linker, and F. Käppeler,
    eds.), p. 201. Plenum Press, New York, 1976.
Obsidian Hydration Profile Measurements Using a Nuclear Reaction Technique
    R. R. Lee, D. A. Leich, T. A. Tombrello, J. E. Ericson, and I. Friedman, *Nature (London)*
    **250**, 44 (1974).
A Technique for Measuring Hydrogen Concentration Versus Depth in Solid Samples
    D. A. Leich and T. A. Tombrello, *Nucl. Instrum Methods* **108**, 67 (1973).
The Depth Distribution of Hydrogen and Fluorine in Lunar Samples
    D. A. Leich, T. A. Tombrello, and D. S. Burnett, *Geochim. Cosmochim. Acta* **2**, 1597 (1973).
Ion Channeling Studies of the Lattice Location of Interstitial Impurities: Hydrogen in Metals
    S. T. Picraux, *in* "Ion Beam Surface Layer Analysis" (O. Meyer, G. Linker, and F. Käppeler,
    eds.), Vol. 2, p. 527. 1976.
Lattice Location Studies of $^2$D and $^3$He in W
    S. T. Picraux and F. L. Vook, *in* "Applications of Ion Beam in Metals" (S. T. Picraux,
    E. P. EerNisse, and F. L. Vook, eds.), p. 407. Plenum Press, New York, 1974.
Profile Studies of Hydrogen Trapping in Metals due to Ion Damage
    S. T. Picraux, J. Bøttiger, and N. Rud, *Appl. Phys. Lett.* **28**, 179 (1976).
Enhanced Hydrogen Trapping due to He Ion Damage
    S. T. Picraux, J. Bøttiger, and N. Rud, *J. Nucl. Mater.* **63**, 110 (1976).
Lattice Location of Deuterium Implanted into W and Cr
    S. T. Picraux and F. L. Vook, *in* "Ion Implantation in Semiconductors" (S. Namba, ed.),
    p. 355. Plenum Press, New York, 1975.
Deuterium Lattice Location in Cr and W
    S. T. Picraux and F. L. Vook, *Phys. Rev. Lett.* **33**, 1216 (1974).
Depth Profiling of $^3$He and $^2$H in Solids Using the $^3$He$(d, p)^4$He Resonance
    P. P. Pronko and J. G. Pronko, *Phys. Rev.* **9**, 2870 (1974).
Determination of the Depth Distribution of Implanted Helium Atoms in Niobium by Ruther-
ford Backscattering
    J. Roth, R. Behrisch, and B. M. U. Scherzer, *Appl. Phys. Lett.* **25**, 643 (1974).
Depth Profiling of Implanted $^3$He in Solids by Nuclear Reaction and Rutherford Backscattering
    J. Roth, R. Behrisch, W. Eckstein, and B. M. U. Scherzer, *in* "Ion Beam Surface Layer
    Analysis" (O. Meyer, G. Linker, and F. Käppeler, eds.), Vol. I, p. 47. Plenum Press, New
    York, 1976.
Temperature Dependence of He Trapping in Niobium
    J. Roth, S. T. Picraux, W. Eckstein, J. Bøttiger, and R. Behrisch, *J. Nucl. Mater.* **63**, 120 (1976).
Target Thickness Determinations from a Study of Contaminant Hydrogen Distributions
    Z. E. Switkowski and R. A. Dayras, *Nucl. Instrum. Methods* **128**, 9 (1975).

A Rutherford Backscattering Study of Precipitation Phenomena in Dilute Alloys of Nb–H and Nb–D
  J. L. Whitton, J. B. Mitchell, T. Schober, and H. Wenzl, *Acta Metall.* **24**, 483 (1976).

## 10.10   SPUTTERING AND BLISTERING PROCESSES

The Dose Dependence of 45 keV $V^+$ and $Bi^+$ Ion Sputtering Yield of Copper
  H. H. Andersen, *Radiat. Effects* **19**, 257 (1973).
Nonlinear Effects in Heavy-Ion Sputtering
  H. H. Andersen and H. L. Bay, *J. Appl. Phys.* **45**, 953 (1974).
The Energy Dependence of Gold Self Sputtering
  H. L. Bay, H. H. Andersen, W. O. Hofer, and O. Nielsen, *Nucl. Instrum. Methods* **132**, 301 (1976).
On the Sputtering Mechanism in the Energy Range of Rutherford Backscattering
  R. Behrisch and R. Weissmann, *Phys. Lett.* **30A**, 506 (1969).
Analysis of Surface Layers by Light Ion Backscattering and Sputtering Combined with Auger Electron Spectroscopy (AES)
  R. Behrisch, B. M. Scherzer, and P. Staib, *Thin Solid Films* **19**, 57 (1973).
Implantation Profiles of Low Energy Helium in Niobium and the Blistering Mechanism
  R. Behrisch, J. Bøttiger, W. Eckstein, U. Littmark, J. Roth, and B. M. U. Scherzer, *Appl. Phys. Lett.* **27**, 199 (1975).
Sputtering of Potassium Chloride by H, He and Ar Ions
  J. P. Biersack and E. Santner, *Nucl. Instrum. Methods* **132**, 229 (1976).
Surface Enrichment of Copper due to keV Xe Sputtering of an Al–Cu Mixture
  W. K. Chu, J. K. Howard, and R. F. Lever, *J. Appl. Phys.* **47**, 4500 (1976).
Sputtering Yields of Niobium by Deuterium in the keV Range
  W. Eckstein, B. M. U. Scherzer, and H. Verbeek, *Radiat. Effects* **18**, 135 (1973).
Light Ion Bombardment Sputtering, Stress Buildup, and Enhanced Surface Contamination
  E. P. EerNisse, *J. Nucl. Mater.* **53**, 226 (1974).
Sputtering of Au by 45 keV Ions for Different Fluences
  E. P. EerNisse, *Appl. Phys. Lett.* **29**, 14 (1976).
Sputtering of $ErD_2$: Experiment and Theory
  E. P. EerNisse and D. K. Brice, *Nucl. Instrum. Methods* **132**, 363 (1976).
Surface-Layer Composition Changes in Sputtered Alloys and Compounds
  Z. L. Liau, W. L. Brown, R. Homer, and J. M. Poate, *Appl. Phys. Lett.* **30**, 626 (1977).
The Sputtering of PtSi and NiSi
  J. M. Poate *et al. Nucl. Instrum. Methods* **132**, 345 (1976).
Blistering and Bubble Formation
  J. Roth, *Inst. Phys. Conf. Ser.* **28**, 280 (1976).
Blistering of Niobium Due to Low Energy Helium Ion Bombardment Investigated by Rutherford Backscattering
  J. Roth, R. Behrisch, and B. M. U. Scherzer, *in* "Applications of Ion Beams to Metals" (S. T. Picraux, E. P. EerNisse, and F. L. Vook, eds.), p. 573. Plenum Press, New York, 1974.
Blistering of Niobium due to 0.5 to 9 keV Helium and Hydrogen Bombardment
  J. Roth, R. Behrisch, and B. M. U. Scherzer, *J. Nucl. Mater.* **53**, 147 (1974).
A New Technique for the Measurement of Sputtering Yields
  Z. E. Switkowski, F. M. Mann, D. W. Kneff, R. W. Ollerhead, and T. A. Tombrello, *Radiat. Effects* **29**, 65 (1976).
Change of Surface Composition of $SiO_2$ Layers During Sputtering
  A. Turos, W. F. van der Weg, D. Sigurd, and J. W. Mayer, *J. Appl. Phys.* **45**, 2777 (1974).

Contributions of Backscattering Ions to Sputtering Yields Depending on Primary Ion Energy
  R. Weissmann and R. Behrisch *Radiat. Effects* **19**, 69 (1973).
Damage in the Surface Region of Silicon Produced by Sputter-Etching
  Y. Yamamoto, K. Shinada, T. Itoh, and K. Yada, *Jpn. J. Appl. Phys.* **13**, 551 (1974).

## 10.11   MICROBEAMS AND OTHER APPLICATIONS

Quantitative Measurement of Light Element Profiles in Thick Corrosion Films on Steels,
Using the Harwell Nuclear Microbeam
  C. R. Allen, G. Dearnaley, and N. E. W. Hartley, *in* "Ion Beam Surface Layer Analysis"
  (O. Meyer, G. Linker, and F. Käppeler, eds.), Vol. 2, p. 901. Plenum Press, New York, 1976.
Elemental Analysis of Biological Samples Using Deuteron Induced X-Rays and Charged
Particles
  L. Amten, L. Glantz, B. Morenius, J. Pihl, and B. Sundqvist, *in* "Ion Beam Surface Layer
  Analysis" (O. Meyer, G. Linker, and F. Käppeler, eds.), Vol. 2, p. 795. Plenum Press, New
  York 1976.
On the Conversion from an Energy Scale to a Depth Scale in Channeling Experiments
  J. Bøttiger and F. H. Eisen, *Thin Solid Films* **19**, 239 (1973).
The Use of Proton Induced X-Ray to Monitor the Near Surface Composition of Catalysts
  J. A. Cairns, A. Lurio, J. F. Ziegler, D. F. Holloway, and J. Cookson, *in* "Ion Beam Surface
  Layer Analysis" (O. Meyer, G. Linker, and F. Käppeler, eds.), Vol. 2, p. 773. Plenum Press,
  New York, 1976.
Proton Microbeams, Their Production and Use
  J. A. Cookson, A. T. G. Ferguson, and F. D. Pilling, *J. Radioanal. Chem.* **12**, 39 (1972).
Comparison of Particle and Proton Excited X-Ray Fluorescence Applied to Trace Element
Measurements on Environmental Samples
  J. A. Cooper, *Nucl. Instrum. Methods* **106**, 525 (1973).
A Rapid Method of Obsidian Characterization by Inelastic Scattering or Protons
  G. E. Coote, N. E. Whitehead, and G. J. McCallum, *J. Radioanal. Chem.* **12**, 491 (1972).
The Use of Alpha-Particle Backscattering to the Study of Printed and Coated Layers
  L. Eriksson, G. Fladda, and P. A. Johansson, *Int. Meeting Chem. Anal. Charged Particle
  Bombardment, Namur* (September 6-8, 1971).
Mesh Size Effects in Alpha-Particle Backscattering
  N. A. Eskind and H. Mark, *J. Geophys. Res.* **67**, 4867 (1962).
Investigation of an Amino Sugar-Like Compound from the Cell Walls of Bacteria Using
Backscattering of MeV Particles
  T. G. Finstad, T. Olsen, and R. Reistad, *in* "Ion Beam Surface Layer Analysis" (O. Meyer,
  G. Linker, and F. Käppeler, eds.), Vol. I, p. 437. Plenum Press, New York, 1976.
Elemental Analysis by Elastic Scattering
  R. K. Jolly and H. B. White, Jr., *Nucl. Instrum. Methods* **97**, 299 (1971).
Determination of Trace Elements in Samples by Nuclear Scattering and Reaction Techniques
  R. K. Jolly, C. R. Gruhn, and C. Maggiore, *IEEE Trans. Nucl. Sci.* **NS-18**, 91 (1971).
A Rutherford Scattering Study of Catalyst Systems for Electroless Cu Plating. I. Surface
Chemistry of Mixed Pd, Sn Colloids
  R. L. Meek, *J. Electrochem. Soc.* **122**, 1177 (1975).
A Rutherford Scattering Study of Catalyst Systems for Electroless Cu Plating. II. $SnCl_2$
Sensitization and $PdCl_2$ Activation
  R. L. Meek, *J. Electrochem. Soc.* **122**, 1478 (1975).
The Use of Si Surface Barrier Detectors for Energy Calibration of MeV Ion Accelerators
  J. B. Mitchell, S. Agami, and J. A. Davies, *Radiat. Effects* **28**, 133 (1976).

Magnesium Distributions in Aluminum Using a Deuteron Microbeam
  C. Olivier and M. Peisach, *Radiochem. Radioanal. Lett.* **19**, 227 (1974).
The Use of Alpha Particle Scattering for the Qualitative Analysis of Elements in Solution
  M. Peisach, *J. S. Afr. Chem. Inst.* **22**, 50 (1969).
The Study of the Curing of Lacquer Films on Tinplate by Alpha-Particle Backscattering
  M. Peisach, V. W. Wilson, and Z. Szecsei, *J. Radio Chem.* **29**, 343 (1976).
Comparison of Back-Scattering Parameters Using High Energy Oxygen and Helium Ions
  S. Petersson, P. A. Tove, O. Meyer, B. Sundqvist, and A. Johansson, *Thin Solid Films* **19**, 157 (1973).
The Microanalysis of Surfaces by Scanning with Charged Particle Beams
  T. B. Pierce, P. E. Peck, and D. R. A. Cuff, *Nucl. Instrum. Methods* **67**, 1 (1969).
Trace-Element Analysis of Water Samples Using 700-keV Protons
  J. Rickards and J. F. Ziegler, *Appl. Phys. Lett.* **27**, 707 (1975).
Effects of Atomic Displacements in Guinier–Preston Zone of Al–Zn Alloy on Channeling of Protons
  T. Sakurai, *J. Phys. Soc. Jpn.* **25**, 113 (1973).
Backscattering Measurements and Surface Roughness
  K. Schmid and H. Ryssel, *Nucl. Instrum. Methods* **119**, 287 (1974).
Alpha Backscattering Technique to Establish Parameters for Airing Oleoresinous Lacquers on Tinplate
  V. W. Wilson and M. Peisach, *J. Coatings Technol.* **48**, 43 (1976).

# A

# Transformation of the Rutherford Formula from Center of Mass to Laboratory Frame of Reference

The derivation of the Rutherford formula for the differential scattering cross section in the center-of-mass coordinates [Eq. (2.20)] can be found in textbooks (e.g., Goldstein, 1959; Leighton, 1959; Evans, 1955). The transformation of the formula to the laboratory frame of reference, which leads to Eq. (2.22), is usually omitted. Darwin (1914) derives the differential cross section formula for the laboratory frame of reference, but gives few details. The following is a full-length execution of this transformation, as given by Ziegler and Lever (1973).

The starting point is the relation between the scattering angle $\theta$ in the laboratory frame of reference and the scattering angle $\theta_c$ in the center-of-mass coordinate system. A commonly quoted form is

$$\cot \theta = \cot \theta_c + x \operatorname{cosec} \theta_c, \qquad (A.1)$$

where $x = M_1/M_2$. The number of particles scattered in a solid angle $2\pi \sin \theta \, d\theta$ will correspond to those observed in the solid angle $2\pi \sin \theta_c \, d\theta_c$, from which one obtains the conservation relation

$$\frac{d\sigma}{d\Omega} \sin \theta \, d\theta = \left(\frac{d\sigma}{d\Omega}\right)_c \sin \theta_c \, d\theta_c, \qquad (A.2)$$

where $d\sigma/d\Omega$ is the desired differential cross section in the laboratory co-ordinates, and $(d\sigma/d\Omega)_c$ is the differential cross section in the center-of-mass coordinates as given by Eq. (2.20). The transformation is greatly simplified if Eq. (A.1) is written in the form

$$\sin \Delta = x \sin \theta \tag{A.3}$$

where $\Delta = \theta_c - \theta$. This equation may be obtained most conveniently by applying the sine rule to the appropriate triangle of velocities. Since $\theta_c = \theta + \Delta$, we obtain, from Eq. (A.3),

$$\frac{d\theta_c}{d\theta} = 1 + \frac{x \cos \theta}{\cos \Delta} = \frac{\sin \theta \cos \Delta + x \sin \theta \cos \theta}{\sin \theta \cos \Delta}$$

$$= \frac{\sin(\theta + \Delta)}{\sin \theta \cos \Delta} = \frac{\sin \theta_c}{\sin \theta \cos \Delta}. \tag{A.4}$$

Hence, from Eq. (A.2),

$$\frac{d\sigma/d\Omega}{(d\sigma/d\Omega)_c} = \left(\frac{\sin \theta_c}{\sin \theta}\right)^2 \frac{1}{\cos \Delta}. \tag{A.5}$$

Applying this to Eq. (2.20), we obtain

$$\frac{d\sigma}{d\Omega} = \left(\frac{Z_1 Z_2 e^2}{2E}\right)^2 \left[\frac{(1 + x)\sin \theta_c}{2 \sin \theta \sin^2(\theta_c/2)}\right]^2 \Big/ \cos \Delta, \tag{A.6}$$

where $E$, the energy in the laboratory coordinates, is given by

$$E = E_c(1 + x) \tag{A.7}$$

if we assume that the scattering particle $M_2$ is initially at rest.

One may at this point simply calculate $\theta_c$ and $\Delta$ for a given $\theta$, using Eq. (A.3), and then substitute into Eq. (A.6) to obtain $d\sigma/d\Omega$. It is neater to eliminate $\theta_c$ and $\Delta$ from Eq. (A.6) as follows. From Eq. (A.3), we have $1 + x = (\sin \Delta + \sin \theta)/\sin \theta$ and also

$$\sin \theta_c/2 \sin^2 \tfrac{1}{2}\theta_c = \cot \tfrac{1}{2}\theta_c. \tag{A.8}$$

Hence, using the relation

$$\frac{\sin \theta + \sin \Delta}{\cos \theta + \cos \Delta} = \frac{2 \sin \tfrac{1}{2}(\theta + \Delta) \cos \tfrac{1}{2}(\theta - \Delta)}{2 \cos \tfrac{1}{2}(\theta + \Delta) \cos \tfrac{1}{2}(\theta - \Delta)} = \tan \tfrac{1}{2}\theta_c, \tag{A.9}$$

we obtain

$$\frac{(1 + x)\sin \theta_c}{2 \sin^2 \tfrac{1}{2}\theta_c} = \frac{\cos \theta + \cos \Delta}{\sin \theta}. \tag{A.10}$$

Hence, substituting this last equation into Eq. (A.6), we obtain

$$\frac{d\sigma}{d\Omega} = \left(\frac{Z_1 Z_2 e^2}{2E}\right)^2 \frac{(\cos\theta + \cos\Delta)^2}{\sin^4\theta\cos\Delta}. \tag{A.11}$$

Since $\cos\Delta = (1 - \sin^2\Delta)^{1/2} = (1 - x^2\sin^2\theta)^{1/2}$, we obtain

$$\frac{d\sigma}{d\Omega} = \left(\frac{Z_1 Z_2 e^2}{2E}\right)^2 \frac{[\cos\theta + (1 - x^2\sin^2\theta)^{1/2}]^2}{\sin^4\theta(1 - x^2\sin^2\theta)^{1/2}}, \tag{A.12}$$

which is the Rutherford formula given in Eq. (2.22).

### REFERENCES

Darwin, C. G., (1914). *Phil. Mag.* **28**, 499.
Evans, R. D. (1955). "The Atomic Nucleus." McGraw-Hill, New York.
Goldstein, H. (1959). "Classical Mechanics." Addison-Wesley, Reading, Massachusetts.
Leighton, R. B. (1959). "Principles of Modern Physics." McGraw-Hill, New York.
Ziegler, J. F., and Lever, R. F. (1973). *Thin Solid Films* **19**, 291.

Appendix

# *B*

# Influence of Energy Straggling on a Thin-Film Spectrum

We consider the simplest case, which is that of a thin monoelemental film. The energy dependence of the scattering and that of the stopping cross sections are neglected. Let $f(E_0, x, E)\,dE$ be the probability that an incident particle that passes through a thickness $x$ has an energy between $E$ and $E + dE$. Before the particle impinges on the target, the energy is $E_0$. After penetrating through a distance $x$, the particle energy $E$ lies somewhere between zero and $E_0$. To realistically describe energy straggling, the probability density function $f(E_0, x, E)$ must thus meet the conditions

$$\int_0^{E_0} f(E_0, x, E)\,dE = 1, \tag{B.1}$$

$$f(E_0, 0, E) = \delta(E - E_0), \tag{B.2}$$

$$f(E_0, x, E) = 0 \quad \text{if} \quad \begin{cases} E_0 < E \\ E < 0. \end{cases} \tag{B.3}$$

The number $H(E_1)$ of particles backscattered into the detector with energy $E_1$ and originating from the interval $dx$ at $x$ thus is

$$H(E_1) = \sigma(E_0)\Omega Q N \int_0^t dx \int_0^{KE_0} f(E_0, x/\cos\theta_1, E') f(KE', x/\cos\theta_2, E_1)\,dE', \tag{B.4}$$

323

where $t$ is the thickness of the film. The total number of counts in the thin-film signal of the spectrum is $A = \sigma(E_0)\Omega Q N t$ [Eq. (4.7)]; it is therefore appropriate to introduce the function

$$g(E_1) = \int_0^t dx \int_0^{KE_0} f(E_0, x/\cos\theta_1, E')f(KE', x/\cos\theta_2, E_1)\,dE' \qquad (B.5)$$

with the normalization

$$\int_0^{KE_0} g(E_1)\,dE_1 = t. \qquad (B.6)$$

When energy straggling is represented by a Gaussian energy distribution centered at $E_0 - (x/\cos\theta_1)N\varepsilon$, the function $f(E_0, x, E)$ takes the form

$$f(E_0, x, E) = (2\pi s^2 x)^{-1/2} \exp\{-[E - (E_0 - N\varepsilon x)]^2/2s^2 x\}, \qquad (B.7)$$

where $x$ in this equation is replaced by $x/\cos\theta_1$ or $x/\cos\theta_2$ along the inward or the outward track, respectively, and where

$$s^2 x \equiv \Omega_s^2. \qquad (B.8)$$

When Bohr's value of energy straggling is assumed, $s^2 = 4\pi(Z_1 e^2)^2 N Z_2$ [Eq. (2.57)], independent of energy. The choice of a Gaussian energy distribution is convenient, but a Gaussian violates the condition of Eq. (B.3), and thus leads to unphysical results outside of the range $0 < E < E_0$. That condition cannot be met exactly by any function that is symmetrical in $E$ with respect to $E_0 - N\varepsilon x$ and does not vanish somewhere in the range $0 < E < E_0$. Hence a correct description of straggling demands a nonsymmetrical energy distribution of the particles with respect to their mean energy.

For the Gaussian distribution given in Eq. (B.7), the integration over $E'$ in Eq. (B.5) can be carried out in closed form. When $K = 1$ and the limits of integration are extended to $\pm\infty$, one obtains

$$g(E_1) = \int_0^t [2\pi s^2 l(x)]^{-1/2}\,dx \exp\{-[E_1 - (E_0 - N\varepsilon l(x))]^2/2s^2 l(x)\}, \qquad (B.9)$$

where $l(x) = (x/\cos\theta_1) + (x/\cos\theta_2)$ is the total length of the inward plus outward paths of the particle through the target. In the limit $s \to 0$, i.e., in the absence of energy straggling, the integrand is a delta function of the argument $E_1 - E_0 + N\varepsilon l(x)$; $g(E_1)$ then is constant in the energy range $E_0 - N\varepsilon t(\sec\theta_1 + \sec\theta_2) < E_1 < E_0$ and zero elsewhere, as expected. In general, when $s$ and $x \neq 0$, $g(E_1)$ can be expressed in normalized units $e^*$ and $x^*$ of energy difference and distance:

$$E_0 - E_1 = 2(s^2/N\varepsilon)e^*, \qquad (B.10)$$

$$x = [2/(\sec\theta_1 + \sec\theta_2)](s/N\varepsilon)^2 x^*, \qquad (B.11)$$

and it is convenient to consider

$$g^*(e^*) \equiv N\varepsilon(\sec\theta_1 + \sec\theta_2)g(E_1) = \frac{1}{\sqrt{\pi}} \int_0^{t^*} \frac{1}{\sqrt{x^*}} \exp\left[-\frac{(e^* - x^*)^2}{x^*}\right] dx^*.$$

$$(B.12)$$

A numerical evaluation of this integral for three values of $t^* = (t/2)(\sec\theta_1 + \sec\theta_2)^{-1}(s/N\varepsilon)^2$ is shown in Fig. B.1. At $e^* = 0$ ($E_1 = E_0$), $g^*$ has the value $\text{erf}(\sqrt{t^*})$, which is very nearly unity for all values of $t^*$ substantially larger than 1. When this is so, the value of $g^*$ in the vicinity of $e^* \simeq t^*$ can be approximated by

$$g^*(e^*) \simeq (1/t^*) \int_{-\infty}^{t^*} \exp\left[-(e^* - x^*)^2/t^*\right] dx^* \qquad (B.13)$$

$$\simeq \tfrac{1}{2}\{1 - \text{erf}[(e^* - t^*)/\sqrt{t^*}]\} \qquad (B.14)$$

since $\sqrt{x^*}$ and $x^*$ vary much more slowly than $(e^* - x^*)^2$ around $e^* \simeq t^*$. This approximation is also shown in Fig. B.1 (dashed lines). In regular units, this approximation has the form

$$g(E_1) = \tfrac{1}{2} + \tfrac{1}{2}\text{erf}[(E_1 - E_{1,t})/\sqrt{2}\Omega_s]. \qquad (B.15)$$

This means that the low-energy step at $E_{1,t}$ in a thin-film spectrum is not sharp, as it is typically shown to be in simplified sketches of backscattering spectra. Rather, the step at $E_{1,t}$ is replaced by a convolution of that step with

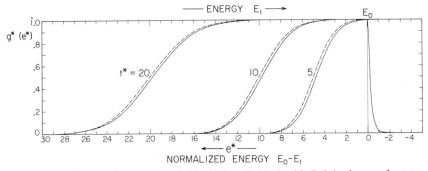

**Fig. B.1** Backscattering spectra of thin films calculated with Bohr's theory of energy straggling (solid lines) for three values of $t^*$ [Eq. (B.12)]. The parameter $t^*$ measures the film thickness in units of the distance for which the $dE/dx$ losses and the standard deviation $\Omega_s$ of energy straggling are about equal. The abscissa gives the energy difference $E_0 - E_1$ in units of $2s^2/N\varepsilon$, where $s^2$ specifies the magnitude of energy straggling, viz., $\Omega_s^2 = s^2x$. The dashed lines give the corresponding approximate solutions which assume a simple erf shape [Eq. (B.15)]. The finite value of the calculated signal for $E_1 > E_0$ is an artifact caused by unphysical assumptions in the straggling model.

a Gaussian of an energy variance $\Omega_s^2 = s^2 l$. This is the energy straggling that would be observed after transmission of the beam through a film of thickness $l$. Because of its simplicity, this result offers a convenient description of energy straggling in backscattering spectra of thin films. The result is only approximate; in reality a backscattering spectrum contains samplings of the energy profile of the penetrating beam at every depth, whereas in the transmitted beam all particles traverse the same total thickness.

To extend this result to target masses with $K < 1$, consider first the energy straggling of a beam traversing two layers in succession. Because of the assumption of Gaussian energy distributions, the resulting straggling is $(\Omega_A^2 + \Omega_B^2)^{1/2}$, where $\Omega_A$ and $\Omega_B$ are the straggling measured at the appropriate energies for layers A and B individually. In the backscattering configuration, $\Omega_B$ simply corresponds to the energy straggling in the outgoing path $\Omega_{out}$. The straggling generated in the incoming path $\Omega_{in}$ must be modified, since this path terminates with an elastic collision. It can be shown that the standard deviation of any distribution function of the particle energy is multiplied by $K$ after an elastic collision. The straggling in backscattering configuration is thus given by

$$\Omega_s^2 = K^2 \Omega_{in}^2 + \Omega_{out}^2. \tag{B.16}$$

With this relationship the energy straggling $\Omega_s$ can be calculated for particles that traverse the entire thickness of a film along an inward track, are scattered at the rear interface with a kinematic factor $K$, and then traverse the film again along an outward track to the detector. This value of $\Omega_s$ can then be inserted into Eq. (B.15) to account for energy straggling in the backscattering spectrum of any elemental film.

As is clear from inspection of Fig. B.1, substantial departures from the erf result are expected only in the range $t^* < 1$ or, in regular units, when $t(N\varepsilon)^2 (\sec\theta_1 + \sec\theta_2)/2 < s^2$. This condition demands that the rms energy variation $[\propto (s^2 t)^{1/2}]$ exceeds the energy loss ($\propto tN\varepsilon$). The former increases only as the square root of the thickness; the energy loss is proportional to the thickness. The condition is thus met only for very thin layers. A simple estimate of this thickness is found by taking Bohr's expression for energy straggling $[s^2 = 4\pi(Z_1 e^2)^2 N Z_2$, Eq. (2.57)] and the Bethe–Bloch value for the electronic energy loss $[N\varepsilon = s^2 L/m_e v_1^2$, Eq. (2.45)]. The assumption of a Gaussian profile for energy straggling then becomes questionable when

$$t < (2/L)(m_e/M_1)(E_0/N\varepsilon). \tag{B.17}$$

For typical values at 1 MeV the right-hand side is less than 100 Å for $^1$H and less than 20 Å for $^4$He. These thicknesses are well below the depth resolution achieved in standard backscattering spectrometry systems. The assumption of a Gaussian profile for energy straggling is thus quite justified.

This conclusion is well supported by experimental evidence. Measurements of energy straggling in the megaelectron volt range have been made that are based on the analysis of the low-energy edge of the backscattering spectrum obtained from thin elemental films. Gaussian energy profiles were observed in even the thinnest films for which meaningful straggling measurements could be made with a conventional Si detector and preamplifier system (Harris *et al.*, 1975).

The finite signal amplitude beyond the front edge of the normalized spectrum in Fig. B.1 is unphysical. It would correspond to particles scattered back with energies larger than the primary energy $E_0$. If the Gaussian distribution for $f(E_0, x, E)$ were replaced by a function that vanishes for $E > E_0$, the signal amplitude of $g(E_1)$ would vanish beyond the front edge at $E_0$ as well. The rear edge of $g(E_1)$ would also be modified if the Gaussian distribution were replaced by a function that vanishes below $E = 0$. How significant that modification would be depends on the position of the rear edge at $E_1 - E_t$ with respect to the origin $E_1 = 0$. For thin-film spectra, $E_t$ and $E_0$ differ little, so that the rear edge of $g(E_1)$ tends to be a better approximation to the exact solution than the leading edge.

### REFERENCE

Harris, J. M., and Nicolet, M-A., (1975). *Phys. Rev. B* **11**, 1013 *J. Vac. Technol.* **12**, 439.

Appendix

# *C*

# The True Position of the Edges of a Narrow Rectangular Signal

A very thin film generates a backscattering signal in the form of a Gaussian whose standard deviation $\Omega$ is that of the system resolution. The true position of the signal on the energy axis—that is, the position measured if $\Omega$ were very small—is given by the energy at which the Gaussian has its *maximum,* shown in Fig. C.1a. On the other hand, a film of substantial thickness generates a backscattering signal which, ideally, has the form of a pulse with sharp steps (see Fig. C.1c). These steps are smoothed out by the system resolution, and the observed signal has edges in the form of error functions whose width is again characterized by the standard deviation $\Omega$. Now, the true location of the steps is given by the *half-height point* of the signal. When the film is of an intermediate thickness, as in Fig. C.1b, the true position of the edges is somewhere between the maximum and the half-height point. There are cases when this true position must be known precisely because the positive identification of a mass or the accurate determination of an energy position may depend on the correct determination of this edge. We show here one way to accomplish this.

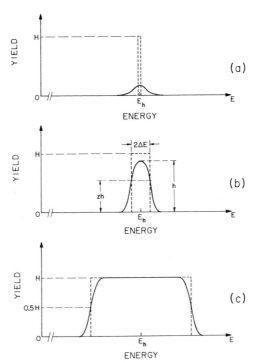

**Fig. C.1** (a) A rectangular signal whose width $2\,\Delta E$ is much less than the standard deviation $\Omega$ of the system resolution is detected as a Gaussian; (c) one that is much wider than $\Omega$ yields a flat-topped signal with error-function-like edges; (b) the proper interpretation of the case when $\Delta E \simeq \Omega$ requires some attention (see also Fig. 4.10).

Three quantities have to be measured: the number of counts $h$ at the maximum of the experimentally measured signal, the total number of counts $A$ contained in the experimentally measured signal, and the standard deviation $\Omega$ of the system resolution, expressed in numbers of channels or fractions thereof. The latter is derived most readily from the high-energy edge of a backscattering spectrum from a monoisotopic, thick target, such as Au. On this high-energy edge, the 16–84% points are $2\Omega$ apart (see Fig. 2.12). Given these three quantities, form the ratio $y = A/h\Omega$ and enter this number on the abscissa of Fig. C.2. The curve $y(x)$ shown in the figure gives a value $x$. With this value, form the number of channels $x\Omega$. This is the number of channels separating the position of the maximum $h$ on the abscissa from the true position of the edge on either side of $h$. Rather than locating the true position of the edges from the value of $x$, as just described, one can alternatively read off the value $z(x)$ given by the second curve for that same $x$. This value $z$ gives

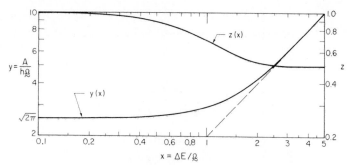

**Fig. C.2** The ratio $y$ of the total number of counts $A$ in a detected signal to the product of its maximum height $h$ and the system resolution determine a value of $x$ via $y(x)$. The width $2\,\Delta E$ of the ideal rectangular signal is then $2x\Omega$ (see Fig. C.1), and $zh$ for the value of $z$ taken at the same $x$ gives the observed signal height at the position of the edges of the ideal rectangular signal.

the ratio of the signal height *observed* at the true position of the edge divided by the maximum value $h$. For very wide signals, i.e., for large values of $x$, $z$ approaches $\frac{1}{2}$; this corresponds to the situation shown in Fig. C.1c. For very narrow signals, i.e., for small values of $x$, $z$ approaches unity; this corresponds to the situation shown in Fig. C.1a.

The functions $y(x)$ and $z(x)$ are derived as follows. Let the true pulse-like signal be centered on $E_h$ and have a height $H$ and a width $2\,\Delta E$. In practice, $H$ is expressed in numbers of counts, and $\Delta E$ in numbers of channels. For the derivation, however, we shall assume that both the idealized and the observed signals are functions of a continuous variable $E$. The observed signal then has the form

$$f(E) = [H/(2\pi\Omega^2)^{1/2}] \int_{E_h - \Delta E}^{E_h + \Delta E} \exp[-(E - E')^2/2\Omega^2]\, dE'. \qquad (C.1)$$

We can set $E_h = 0$ without loss of generality, and $E$ is then measured with respect to $E_h$ taken as the origin; then

$$f(E) = (H/\sqrt{\pi}) \int_{(E - \Delta E)/(2\Omega^2)^{1/2}}^{(E + \Delta E)/(2\Omega^2)^{1/2}} \exp(-z^2)\, dz. \qquad (C.2)$$

By symmetry, the maximum $h$ of $f$ occurs at $E_h$, that is, at $E = 0$, so that

$$h = H\,\mathrm{erf}[\Delta E/(2\Omega^2)^{1/2}]. \qquad (C.3)$$

The value of $f$ at the edges of the pulse is given by

$$f(E = \pm\Delta E) = \tfrac{1}{2}H\,\mathrm{erf}[2\Delta E/(2\Omega^2)^{1/2}]. \qquad (C.4)$$

The ratio $z$ of this value to that at the maximum hence is

$$z(x) = \tfrac{1}{2}[\mathrm{erf}(\sqrt{2}x)/\mathrm{erf}(x/\sqrt{2})], \qquad (C.5)$$

where $x \equiv \Delta E / \underset{\sim}{\Omega}$ is the distance from the position of the edge to the center point of the function, measured in units of the standard deviation $\underset{\sim}{\Omega}$.

The integral over $f(E)$ for all values of $E$ has the value $A = 2 \Delta \bar{E} H$. The ratio $y \equiv A/h \underset{\sim}{\Omega}$ thus is

$$y(x) = 2x/\mathrm{erf}(x/\sqrt{2}). \tag{C.6}$$

The two functions $z(x)$ and $y(x)$ are plotted in Fig. C.2.

Appendix

# *D*

# List of Energy-Loss Compilations

Numerous experiments measure the energy loss of protons and helium ions in solid and gaseous targets. The experimental techniques are all similar (see Chapter 9) and can be categorized in the following way:

1.   Measurements of Energy Loss by Transmission.   One can measure the energy loss incurred by the projectile passing through a thin foil of known thickness or through a gas cell with known dimension and gas pressure.

2.   Measurements of Energy Loss by Backscattering.   The technique is similar to that described under point 1, except that the detected projectile is scattered backward from a thin film on a substrate.

3.   Measurements of Backscattering Yield.   The method is discussed in Section 9.4. The spectrum height gives energy-loss information.

Here we will list the available compilations of experimental energy-loss information.

### A.   COMPILATIONS BEFORE 1970

Whaling, W. (1958). *In* "Handbuck der Physik" (S. Flugge, ed.), Vol. 34, p. 193. Springer, Berlin.

Marion, J. B. (1960). 1960 Nuclear Data Tables, Part 3, Nuclear Reaction Graphs. Nat. Acad. of Sci., Nat. Res. Council, Washington, D. C.

Bichsel, H. (1963). "American Institute of Physics Handbook," p. 8–22. McGraw-Hill, New York.

Studies in Penetration of Charged Particles in Matter, (1964). Nuclear Sci. Ser., Rep. No. 39. Nat. Acad. of Sci., Nat. Res. Council, Publ. 1133, Washington, D.C.

Janni, J. F. (1966). Air Force Weapons Lab., Rep. AFWL-TR-65–150.

Williamson, C. F., Boujot, J.–P., and Picard, J. (1966). Rep. CEA-R 3042.

## B.  COMPILATIONS SINCE 1970

Northcliffe, L. C., and Schilling, R. F. (1970). Range and stopping-power tables for heavy ions, *Nucl. Data Tables* **7**, No. 3–4, 233–463.

Range and energy loss for protons, helium, and all heavy ions in 12 different solid elements 9 gaseous elements, and the 3 compounds Mylar, $(CH_2)_n$, and water are tabulated in the energy region of 0.0125 to 12 MeV/amu. The tables are based on an investigation of the systematic relationships observed among experimental data, guided by simple theoretical expectations and extrapolated into regions where no measurements have been made.

Ziegler, J. F., and Chu, W. K. (1974) Stopping cross sections and backscattering factors for $^4$He ions in matter; $Z = 1$–92, $E(^4$He$) = 400$–4000 keV, *At. Nucl. Data Tables* **13**, 463–489.

This compilation gives elemental stopping cross sections for $^4$He ions in the energy region 0.4–4 MeV. Most of the compilation is based on the measurements made at Baylor University. Interpolations and extrapolations are made for unmeasured elements, with consideration of the $Z_2$ oscillatory structure of $\varepsilon$. The compilation is intended for backscattering applications; therefore parameters for backscattering depth calculation are also given. Tables VI–IX of this book are taken from this reference.

Andersen, H. H., and Ziegler, J. F. (1977). "Hydrogen Stopping Powers and Ranges in All Elements." Pergamon, Oxford.

This is a compilation of energy-loss data for protons as a function of energy for most elements. The data are presented in graphical form and also in analytical form for elements on which a number of energy-loss measurements have been made.

A series of compilations of energy loss data for helium ions and heavy ions and their ranges in matter will be published by Ziegler *et al.* (1978). Pergamon, Oxford.

Appendix

# E

# Rough Targets

Laterally nonuniform samples generate backscattering spectra that cannot be interpreted simply. It is therefore important to know when the sample is laterally nonuniform. Though the spectrum itself does not necessarily reveal this fact, certain features in it can serve as warning signals. Consider, for instance, the spectrum of Fig. E.1, obtained from a sample consisting of a

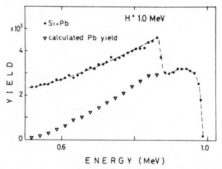

**Fig. E.1** Energy spectrum of backscattered 1.0-MeV $^1$H particles impinging on a sample as shown in Fig. E.2. Triangles show the Pb contribution to the spectrum calculated under the (wrong) assumption of a laterally uniform distribution of Pb over the entire surface. The result nicely fits a standard diffusion profile. [After Campisano *et al.* (1975).]

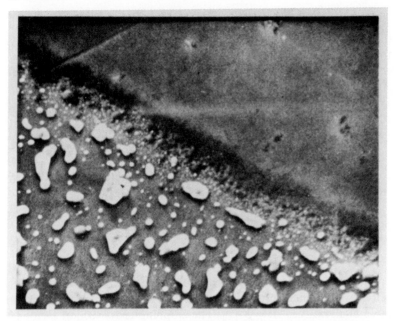

**Fig. E.2** Scanning electron micrograph of a thin film of Pb deposited on a Si substrate and subsequently annealed in vacuum for 10 min at 275°C. The average lateral dimension of the Pb islands is 2 $\mu$m. The upper right corner shows the bare Si substrate. [After Campisano *et al.* (1975).]

thin film of Pb on a Si substrate, annealed at 275°C for 10 min (Campisano *et al.*, 1975). The spectrum can be readily interpreted as indicating a laterally uniform penetration of Pb into the Si substrate (triangles in the figure). In fact, the sample is laterally highly nonuniform because during the heat treatment the Pb film broke off into many small balls of Pb resting on the Si substrate (Fig. E.2). Signals that rise sharply at an energy corresponding to the surface edge of some element and droop off toward decreasing energies are indicative of lateral nonuniformity. The following simple calculation shows why.

Consider an elemental target that consists of a sphere of radius $r$ and is irradiated by an incident particle beam whose cross section exceeds that of the sphere. For simplicity, assume that the scattered particles are detected at an angle $\theta$ of 180°, that is, along the direction of the incident beam, and that the energy loss $dE/dx$ is independent of energy and has the value $f$. The particles that contribute to the backscattering yield at some energy $E_1$ of the spectrum are all subjected to the same energy loss along their inward and outward paths. For a constant $dE/dx$ and $\theta = 180°$, this condition means that the particles are scattered from the same depth below the surface. The points that satisfy this condition within the target all lie on the section of an imaginary sphere of the same radius $r$ whose center is displaced from that of the target by a distance $l$ in the direction of the incident beam. The yield of the backscattering spectrum at the energy $E_1$ is proportional to the differential volume spanned by this locus surface of scattering points at the depth $l$ and the locus surface for the points at depth $l + dl$. This volume is equal to the projection of the locus surface for the depth $l$ on a plane perpendicular to the direction of the incident beam times $dl$. The area $S$ of this projection is

$$S = \pi[r^2 - (l/2)^2].  \tag{E.1}$$

The depth $l$ is given by $(E_0 - E)/f$ for the inward path and by $(KE - E_1)/f$ for the backward path, so that

$$E = (E_0 + E_1)/(K + 1)  \quad \text{and} \quad  l = (KE_0 - E_1)/(K + 1)f.  \tag{E.2}$$

The yield is thus [cf. Eq. (3.36)]

$$H = \sigma(E)\Omega\phi S(l)N\,\Delta l,  \tag{E.3}$$

where $\phi$ is the flux of particles (particles per area) of the incident beam and $\Delta l$ is the width that corresponds to the energy $\mathscr{E}$ of one channel of the multichannel analyzer, i.e., $\Delta l = \mathscr{E}/[\varepsilon]$. The stopping cross section factor $[\varepsilon]$ is independent of energy since, by assumption, $dE/dx = \text{const} \equiv f$ and the target is elemental, so that the volume density of atoms $N$ is the same throughout. The functional dependence of the yield is thus given by that of the scattering cross section $\sigma(E)$ and the area $S(l)$. With the previous equations and the fact that $\sigma(E) = \sigma(E_0)(E_0/E)^2$, one obtains

$$H = \sigma(E_0)\Omega\phi N\,\Delta l\,F,  \tag{E.4}$$

where

$$F = \pi r^2(E_0/E)^2[1 - (l/2r)^2]  \tag{E.5}$$

contains the energy-dependent terms. Thus $F$ can be written as

$$F = \pi r^2[(K + 1)/(x + 1)]^2[b^2 - (K - x)^2]/b^2,  \tag{E.6}$$

where $x = E_1/E_0$ is the detected energy $E_1$ of the spectrum normalized to the incident energy $E_0$. The parameter

$$b = 2r(dE/dx)(K + 1)/E_0  \tag{E.7}$$

is the energy lost by a particle traversing the sphere along the diameter of length $2r$ in the beam direction and is normalized to $E_0$ as well. This parameter thus gives the maximum width $\Delta E$ of the backscattering signal in units of $E_0$. The edge of the signal is at $x = KE_0/E_0 = K$. Beyond that point the signal is zero. Below $x = K - (\Delta E/E_0) = K - b$, the signal also vanishes. The value of $F$ at the signal edge $x = K$ is $\pi r^2$. It is convenient to normalize $F$ in terms of this value because the energy dependence of backscattering spectra for different spheres can then readily be compared.

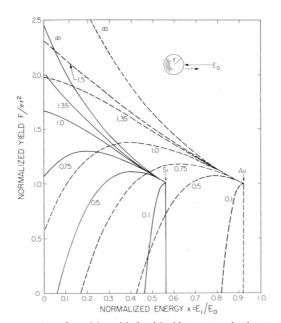

**Fig. E.3** Energy spectra of particles with fixed incident energy backscattered from an elemental sphere of varying diameter, calculated for $K = 0.9219$ (Au, dashed lines) and for $K = 0.5632$ (Si, solid lines). The energy axis is normalized with respect to the incident energy $E_0$. The yield is proportional to $F$ [Eqs. (E.4) and (E.5)] and is normalized with respect to the area $\pi r^2$ which the sphere presents to the beam. The parameter $b$ [Eq. (E.7)] measures in units of $E_0$ the radius of the sphere in terms of the energy loss $\Delta E$ of a particle that completely traverses the sphere back and forth.

Figure E.3 gives spectra for $K = 0.9219$ (Au, dashed lines) and for $K = 0.5632$ (Si, solid lines) and various $b$s, i.e., various radii for the sphere. If the sphere is small and the energy loss of the particles traversing it from end to end is relatively small ($b \ll 1$), the signal has a sharp step at the energy edge of the element and drops off rapidly to zero at lower energies. As the radius of the sphere increases, a maximum develops and the decrease is more gradual. For large spheres ($b \gg 1$) the spectrum tends to that of a planar target ($b = \infty$) where the energy dependence is due solely to the increase in the scattering cross section and the geometrical effect vanishes.

If the energy loss is a function of the energy or if the scattering angle is less than 180°, the surface forming the locus of all scattering points within the target and generating signals at a given detected energy is no longer the section of a sphere, but a surface of complex shape. The resulting spectra will be modified in the details, but the overall shape of the spectra will remain similar to those shown in the figure. In reality, laterally nonuniform targets of balled-up films such as those shown in Fig. E.2 always contain islands of many shapes and sizes. The spectrum of such a sample must be derived from a statistical description of the shapes and sizes present in the target. In principle, backscattering spectra are thus capable of furnishing such statistical information from an analysis of the shape of a backscattering spectrum. Such analyses have not been made to date, but spectra have been computed for the finite number of different shapes of islands.

**REFERENCE**

Campisano, S. U., Foti, G., Grasso, F., and Rimini, E. (1975). *Thin Solid Films* **25**, 431.

Appendix

# F

# Numerical Tables

The tables in this appendix are included for the convenience of the user of backscattering spectrometry. To date, most of the backscattering analyses have been carried out with $^4$He as a projectile. Consequently, most of our tables pertain to $^4$He. The three basic tables for that case are III (or V), VI, and X, which give the kinematic factor $K_{M_2}$ (or $\bar{K}$), the stopping cross section $\varepsilon$, and the scattering cross section $\sigma$, respectively. From the values of $K_{M_2}$ and $\varepsilon$ one obtains the stopping cross section factor $[\varepsilon_0]$ and the energy-loss factor $[S_0]$, which give the conversion of energy loss to depth in backscattering spectra. From the scattering cross section $\sigma$ and the stopping cross section factor one obtains the surface height of the backscattering yield for a beam of normal incidence. Tables III (or IV) and X are conceptually accurate to the last digit; the stopping cross sections of Table VI are derived from empirical sets of values and may be in error by as much as 10%, but are typically good to within a few percent.

### TABLE I. ELEMENTS

This table lists the elements under the element heading from hydrogen to bismuth, gives their atomic number $Z$ under the atomic number heading, and the mass of the stable isotopes under the isotopic mass heading in atomic mass units ($^{12}$C being defined as 12 amu, exactly). The relative abundances of the stable isotopes under the relative abundance heading are given in fractions of a total of unity, to four significant digits, and omitting the calcium isotope 46, whose relative abundance is less than $0.5 \times 10^{-4}$. The atomic weight column gives the product, in atomic mass units, of the isotopic mass and its abundance, summed over all the isotopes listed for that element in the preceding two columns. The atomic density column gives the number of atoms per cubic centimeter for the element. The last column gives the specific gravity of the element in grams per cubic centimeter.

The values of the isotopic masses and of the relative abundances are those given by W. H. Johnson, Jr., and A. O. Nier (1967). *in* "Handbook of Physics" (E. U. Condon and H. Odishaw, eds.), pp. 9–63. McGraw-Hill, New York. The atomic density and the specific gravity of the elements are those listed by C. Kittel (1971). "Introduction to Solid State Physics," 4th ed., p. 39. Wiley, New York. The atomic densities differ slightly from those given in Table IX. Also, in some cases the product of the atomic weight and the atomic density differs numerically from the product of the corresponding specific gravity and Avogadro's number.

### TABLE II.   $K_{M_2}$ FOR $^1$H AS PROJECTILE AND INTEGER TARGET MASS $M_2$

This table gives the kinematic factor $K_{M_2}$ defined by Eq. (2.6) for a H atom as projectile ($M_1 = 1.007825$ amu) and integral atomic masses for the target atom ($M_2 = 2$–216 amu). The parameter $\theta$ is the scattering angle measured in the laboratory frame of reference, as shown in Fig. 2.1. The content of this table is also represented graphically in Fig. 2.2.

The factor $K_{M_2}$ *is approximately described by* $1 - 4(M_1/M_2) + \delta^2(M_1/M_2)$, where $\delta$ measures the deviation of $\theta$ from 180° in arc units [see Eq. (2.12)]. A table of $\delta^2$ versus $\pi - \theta$ is given to facilitate the evaluation of $K_{M_2}$ by this expression.

### TABLE III.   $K_{M_2}$ FOR $^4$HE AS PROJECTILE AND INTEGER TARGET MASS $M_2$

This table gives the kinematic factor $K_{M_2}$ defined by Eq. (2.6) for a He atom as projectile ($M_1 = 4.002603$ amu) and integral atomic masses for the target atom ($M_2 = 6$–216 amu). The parameter $\theta$ is the scattering angle measured in the laboratory frame of reference, as shown in Fig. 2.1. The content of this table is also represented graphically in Fig. 2.2.

The factor $K_{M_2}$ is approximately described by $1 - 4(M_1/M_2) + \delta^2(M_1/M_2)$, where $\delta$ measures the deviation of $\theta$ from $180°$ in arc units [see Eq. (2.12)]. A table of $\delta^2$ versus $\pi - \theta$ is given to facilitate the evaluation of $K_{M_2}$ by this expression.

## TABLE IV.   $\bar{K}$ FOR $^1$H AS PROJECTILE

This table gives the mean kinematic factor $\bar{K}$ defined in Section 3.8 as the weighted average of the kinematic factors $K_{M_2}$ of the isotopes of an element. The projectile is a $^1$H atom ($M_1 = 1.007825$ amu). The kinematic factor $K_{M_2}$ of an individual isotope is calculated by Eq. (2.6), for the isotopic masses listed in Table I for $M_2$. For the average, the kinematic factor of each isotope listed in Table I is weighted by the relative abundance given for that isotope in the same table. The parameter $\theta$ is the scattering angle measured in the laboratory frame of reference, as shown in Fig. 2.1.

The factor $\bar{K}$ is approximately described by $1 - 4(\overline{M_1/M_2}) + \delta^2(\overline{M_1/M_2})$, where $\delta$ measures the deviation of $\theta$ from $180°$ in arc units and $\overline{M_1/M_2}$ is the average of the ratio of $M_1/M_2$ weighted by the relative abundance of each isotope given in Table I for a particular element. A table of $\delta^2$ versus $\pi - \theta$ is given to facilitate the evaluation of $\bar{K}$ by this expression.

The $\bar{K}$ given here is numerically very close to $K_{\bar{M}}$. Usually no distinction is made between the two, and the symbol $K$ is used for both (see Section 3.8).

## TABLE V.   $\bar{K}$ FOR $^4$HE AS PROJECTILE

This table gives the mean kinematic factor $\bar{K}$, defined in Section 3.8 as the weighted average of the kinematic factors $K_{M_2}$ of the isotopes of an element. The projectile is a $^4$He atom ($M_1 = 4.002603$ amu). The kinematic factor $K_{M_2}$ of an individual isotope is calculated by Eq. (2.6), for the isotopic masses listed in Table I for $M_2$. For the average, the kinematic factor of each isotope listed in Table I is weighted by the relative abundance given for that isotope in the same table. The parameter $\theta$ is the scattering angle measured in the laboratory frame of reference, as shown in Fig. 2.1.

The factor $\bar{K}$ is approximately described by $1 - 4(\overline{M_1/M_2}) + \delta^2(\overline{M_1/M_2})$, where $\delta$ measures the deviation of $\theta$ from $180°$ in arc units and where $\overline{M_1/M_2}$ is the average of the ratio of $M_1/M_2$ weighted by the relative abundance of each isotope given in Table I for a particular element. A table of $\delta^2$ versus $\pi - \theta$ is given to facilitate the evaluation of $\bar{K}$ by this expression.

$\bar{K}$ given here is numerically very close to $K_{\bar{M}}$. Usually no distinction is made between the two, and the symbol $K$ is used for both (see Section 3.8).

## TABLE VI.   $^4$HE STOPPING CROSS SECTIONS $\varepsilon$

This table gives semiempirical values of the stopping cross section $\varepsilon$ *in electron volts per* $(10^{15}$ *atoms per square centimeter)* for $^4$He at energies from 400 to 4000 keV in all elements.

The values are those given by J. F. Ziegler and W. K. Chu (1974). *At. Data Nucl. Data Tables* **13**, 481–482. The content of this table is also represented graphically in Fig. 2.8.

## TABLE VII.   POLYNOMIAL FIT TO THE $^4$HE STOPPING CROSS SECTIONS $\varepsilon$

This table gives the coefficients $A_0, A_1, A_2, A_3, A_4$, and $A_5$ *in* $10^{-15}$ *electron volt-square centimeters per atom* for the polynomial

$$\varepsilon = A_0 + A_1 E + A_2 E^2 + A_3 E^3 + A_4 E^4 + A_5 E^5,$$

with the least squares fitted to the semiempirical $^4$He stopping cross sections of Table VI. The energy $E$ in this polynomial is expressed in megaelectron volts.

The values of the coefficients $A_0$–$A_5$ are those given by J. F. Ziegler and W. K. Chu (1974). *At. Data Nucl. Data Tables* **13**, 483.

## TABLE VIII.   $^4$HE STOPPING CROSS SECTION FACTOR $[\varepsilon_0]$

This table gives the stopping cross section factor *in* $10^{-15}$ *electron volt-square centimeters per atom* in the surface energy approximation, as defined by Eq. (3.12), for $^4$He of normal incidence ($\theta_1 = 0$) and a scattering angle $\theta$ of 170° ($\theta_2 = 10°$) for incident energies $E_0$ from 1.0 to 4.0 MeV. The table is based on the semiempirical stopping cross sections given in Table VI and on values of $K_{\overline{M}}$ from J. F. Ziegler (1973). *Thin Solid Films* **19**, 289. The values of $K_{\overline{M}}$ are practically equal to those given in Table V for $\overline{K}$.

## TABLE IX.   $^4$HE ENERGY-LOSS FACTOR $[S_0]$

This table gives the energy-loss factor *in electron volts per angstrom* in the surface energy approximation, as defined by Eq. (3.11), for $^4$He of normal incidence ($\theta_1 = 0$) and a scattering angle of $\theta = 170°$ ($\theta_2 = 10°$) and for incident energies from 1.0 to 4.0 MeV. The table is based on the values given in Table VIII and the atomic densities listed in the third column. The value of $[S_0]$ depends linearly on the value of the atomic density $N$ through the relation between $dE/dx$ and $\varepsilon$; $\varepsilon \equiv (1/N) dE/dx$ [Eq. (2.36)]. Values of $K_{\overline{M}}$ were taken from J. F. Ziegler (1973). *Thin Solid Films* **19**, 289. The values of $K_M$ are practically equal to those given in Table V for $\overline{K}$.

## TABLE X.   RUTHERFORD SCATTERING CROSS SECTION OF THE ELEMENTS FOR 1 MeV $^4$HE

This table gives the differential scattering cross section $d\sigma/d\Omega$ in $10^{-24}$ *square centimeters per steradian* (*barn*) as calculated from the Rutherford scattering cross section formula (2.22) for $^4$He as a projectile, for an incident energy of 1 MeV, and for a scattering angle $\theta$ from 90 to 179.5° as measured in the laboratory system. The mass $M_1$ of $^4$He is taken as 4.0026 amu; the mass $M_2$ assumed for the scattering element is an average over the isotopic masses and is listed in the third column. The value of $d\sigma/d\Omega$ obtained in this manner differs insignificantly from the weighted average of the isotopic scattering cross sections. The content of this table is also represented graphically in Fig. 2.5.

## TABLE XI.   SURFACE HEIGHT OF BACKSCATTERING YIELD FOR $^4$HE ON ELEMENTAL TARGETS $H_0$

This table gives the backscattering yield in the surface energy approximation $H_0$ defined by Eq. (3.38), *in counts*, for $^4$He incident perpendicularly ($\theta_1 = 0$) on the target at an energy $E_0$ from 1.0 to 4.0 MeV and scattered by $\theta = 170°$. The cross section values listed in Table X for $\theta = 170°$ are used for $\sigma(E_0)$, the stopping cross section factors listed in Table VIII are used for $[\varepsilon_0]$, and it is assumed that

solid angle of detection   $\Omega = 10^{-3}$ steradian,
total incident dose      $Q = 6.24 \times 10^{12}$ $^4$He particles
   (corresponding to 1 $\mu$C of integrated $^4$He$^+$ beam),
energy per channel      $\mathscr{E} = 1.0$ keV.

For different values of $\Omega$, $Q$, and $\mathscr{E}$, the surface height $H_0$ scales in direct proportion to these parameters.

**TABLE I**
Elements

| ELE-MENT | AT. NO. (Z) | ISOTOPIC MASS (amu) | RELATIVE ABUN-DANCE | ATOMIC WEIGHT (amu) | ATOMIC DENSITY (atom/cm³) | SPECIFIC GRAVITY |
|---|---|---|---|---|---|---|
| H | 1 | 1.007825 | 0.9995 | 1.008 | | 0.088 |
| HE | 2 | 4.002603 | 1.0000 | 4.003 | | 0.205 |
| LI | 3 | 6.015125 | 0.0756 | 6.940 | 4.700E 22 | 0.542 |
| | | 7.016004 | 0.9244 | | | |
| BE | 4 | 9.012186 | 1.0000 | 9.012 | 1.21 E 23 | 1.82 |
| B | 5 | 10.012939 | 0.1961 | 10.814 | 1.30 E 23 | 2.47 |
| | | 11.009305 | 0.8039 | | | |
| C | 6 | 12.000000 | 0.9889 | 12.011 | 1.76 E 23 | 3.516 |
| | | 13.003354 | 0.0111 | | | |
| N | 7 | 14.003074 | 0.9963 | 14.007 | | 1.03 |
| | | 15.000108 | 0.0037 | | | |
| C | 8 | 15.994915 | 0.9976 | 15.999 | | |
| | | 16.999133 | 0.0004 | | | |
| | | 17.999160 | 0.0020 | | | |
| F | 9 | 18.998405 | 1.0000 | 18.998 | | |
| NE | 10 | 19.992441 | 0.9092 | 20.171 | 4.36 E 22 | 1.51 |
| | | 20.993849 | 0.0026 | | | |
| | | 21.991385 | 0.0882 | | | |
| NA | 11 | 22.989771 | 1.0000 | 22.990 | 2.652E 22 | 1.013 |
| MG | 12 | 23.985042 | 0.7870 | 24.310 | 4.30 E 22 | 1.74 |
| | | 24.985839 | 0.1013 | | | |
| | | 25.982593 | 0.1117 | | | |
| AL | 13 | 26.981539 | 1.0000 | 26.982 | 6.02 E 22 | 2.70 |
| SI | 14 | 27.976929 | 0.9221 | 28.086 | 5.00 E 22 | 2.33 |
| | | 28.976496 | 0.0470 | | | |
| | | 29.973763 | 0.0309 | | | |
| P | 15 | 30.973765 | 1.0000 | 30.974 | | |
| S | 16 | 31.972074 | 0.9500 | 32.061 | | |
| | | 32.971462 | 0.0076 | | | |
| | | 33.967865 | 0.0422 | | | |
| | | 35.967090 | 0.0001 | | | |
| CL | 17 | 34.968851 | 0.7577 | 35.453 | | 2.03 |
| | | 36.965899 | 0.2423 | | | |
| AR | 18 | 35.967545 | 0.0034 | 39.948 | 2.66 E 22 | 1.77 |
| | | 37.962728 | 0.0006 | | | |
| | | 39.962384 | 0.9960 | | | |
| K | 19 | 38.963710 | 0.9310 | 39.097 | 1.402E 22 | 0.910 |
| | | 39.964000 | 0.0001 | | | |
| | | 40.961832 | 0.0688 | | | |
| CA | 20 | 39.962589 | 0.9697 | 40.081 | 2.30 E 22 | 1.53 |
| | | 41.958625 | 0.0064 | | | |
| | | 42.958780 | 0.0015 | | | |
| | | 43.955491 | 0.0206 | | | |
| | | 47.952531 | 0.0019 | | | |
| SC | 21 | 44.955919 | 1.0000 | 44.956 | 4.27 E 22 | 2.99 |
| TI | 22 | 45.952632 | 0.0793 | 47.879 | 5.66 E 22 | 4.51 |
| | | 46.951769 | 0.0728 | | | |
| | | 47.947950 | 0.7394 | | | |
| | | 48.947870 | 0.0551 | | | |
| | | 49.944786 | 0.0534 | | | |
| V | 23 | 49.947164 | 0.0024 | 50.942 | 7.22 E 22 | 6.09 |
| | | 50.943961 | 0.9976 | | | |
| CR | 24 | 49.946055 | 0.0435 | 51.996 | 8.33 E 22 | 7.19 |
| | | 51.940513 | 0.8376 | | | |
| | | 52.940653 | 0.0951 | | | |
| | | 53.938882 | 0.0238 | | | |
| MN | 25 | 54.938050 | 1.0000 | 54.938 | 8.18 E 22 | 7.47 |
| FE | 26 | 53.939617 | 0.0582 | 55.847 | 8.50 E 22 | 7.87 |
| | | 55.934936 | 0.9166 | | | |

**TABLE I (Continued)**

| ELE-MENT | AT. NO. (Z) | ISOTOPIC MASS (amu) | RELATIVE ABUN-DANCE | ATOMIC WEIGHT (amu) | ATOMIC DENSITY (atom/cm³) | SPECIFIC GRAVITY |
|---|---|---|---|---|---|---|
|  |  | 56.935398 | 0.0219 |  |  |  |
|  |  | 57.933282 | 0.0033 |  |  |  |
| CO | 27 | 58.933189 | 1.0000 | 58.933 | 8.97 E 22 | 8.9 |
| NI | 28 | 57.935342 | 0.6788 | 58.728 | 9.14 E 22 | 8.91 |
|  |  | 59.930787 | 0.2623 |  |  |  |
|  |  | 60.931056 | 0.0119 |  |  |  |
|  |  | 61.928342 | 0.0366 |  |  |  |
|  |  | 63.927958 | 0.0108 |  |  |  |
| CU | 29 | 62.929592 | 0.6917 | 63.546 | 8.45 E 22 | 8.93 |
|  |  | 64.927786 | 0.3083 |  |  |  |
| ZN | 30 | 63.929145 | 0.4889 | 65.387 | 6.55 E 22 | 7.13 |
|  |  | 65.926052 | 0.2781 |  |  |  |
|  |  | 66.927145 | 0.0411 |  |  |  |
|  |  | 67.924857 | 0.1857 |  |  |  |
|  |  | 69.925334 | 0.0062 |  |  |  |
| GA | 31 | 68.925574 | 0.6040 | 69.717 | 5.10 E 22 | 5.91 |
|  |  | 70.924706 | 0.3960 |  |  |  |
| GE | 32 | 69.924252 | 0.2052 | 72.638 | 4.42 E 22 | 5.32 |
|  |  | 71.922082 | 0.2743 |  |  |  |
|  |  | 72.923463 | 0.0776 |  |  |  |
|  |  | 73.921181 | 0.3654 |  |  |  |
|  |  | 75.921405 | 0.0776 |  |  |  |
| AS | 33 | 74.921596 | 1.0000 | 74.922 | 4.65 E 22 | 5.77 |
| SE | 34 | 73.922476 | 0.0087 | 78.990 | 3.67 E 22 | 4.81 |
|  |  | 75.919207 | 0.0902 |  |  |  |
|  |  | 76.919911 | 0.0758 |  |  |  |
|  |  | 77.917314 | 0.2352 |  |  |  |
|  |  | 79.916527 | 0.4982 |  |  |  |
|  |  | 81.916707 | 0.0919 |  |  |  |
| BR | 35 | 78.918329 | 0.5069 | 79.904 | 2.36 E 22 | 4.05 |
|  |  | 80.916292 | 0.4931 |  |  |  |
| KR | 36 | 77.920403 | 0.0035 | 83.801 | 2.17 E 22 | 3.09 |
|  |  | 79.916380 | 0.0227 |  |  |  |
|  |  | 81.913482 | 0.1156 |  |  |  |
|  |  | 82.914131 | 0.1155 |  |  |  |
|  |  | 83.911503 | 0.5690 |  |  |  |
|  |  | 85.910616 | 0.1737 |  |  |  |
| RB | 37 | 84.911800 | 0.7215 | 85.468 | 1.148E 22 | 1.629 |
|  |  | 86.909187 | 0.2785 |  |  |  |
| SR | 38 | 83.913430 | 0.0056 | 87.616 | 1.78 E 22 | 2.58 |
|  |  | 85.909285 | 0.0986 |  |  |  |
|  |  | 86.908892 | 0.0702 |  |  |  |
|  |  | 87.905641 | 0.8256 |  |  |  |
| Y | 39 | 88.905872 | 1.0000 | 88.906 | 3.02 E 22 | 4.48 |
| ZR | 40 | 89.904700 | 0.5146 | 91.224 | 4.29 E 22 | 6.51 |
|  |  | 90.905642 | 0.1123 |  |  |  |
|  |  | 91.905031 | 0.1711 |  |  |  |
|  |  | 93.906313 | 0.1740 |  |  |  |
|  |  | 95.908286 | 0.0280 |  |  |  |
| NB | 41 | 92.906382 | 1.0000 | 92.906 | 5.56 E 22 | 8.58 |
| MO | 42 | 91.906810 | 0.1584 | 95.890 | 6.42 E 22 | 10.22 |
|  |  | 93.905090 | 0.0904 |  |  |  |
|  |  | 94.905839 | 0.1572 |  |  |  |
|  |  | 95.904674 | 0.1653 |  |  |  |
|  |  | 96.906022 | 0.0946 |  |  |  |
|  |  | 97.905409 | 0.2378 |  |  |  |
|  |  | 99.907475 | 0.0963 |  |  |  |
| TC | 43 | 0.0 | 0.0 |  | 7.04 E 22 | 11.50 |
| RU | 44 | 95.907598 | 0.0551 | 101.046 | 7.36 E 22 | 12.36 |
|  |  | 97.905289 | 0.0187 |  |  |  |
|  |  | 98.905936 | 0.1272 |  |  |  |
|  |  | 99.904218 | 0.1262 |  |  |  |

(Continued)

## TABLE I (Continued)

| ELE-MENT | AT. NO. (Z) | ISOTOPIC MASS (amu) | RELATIVE ABUN-DANCE | ATOMIC WEIGHT (amu) | ATOMIC DENSITY (atom/cm$^3$) | SPECIFIC GRAVITY |
|---|---|---|---|---|---|---|
| | | 100.905577 | 0.1707 | | | |
| | | 101.904348 | 0.3161 | | | |
| | | 103.905430 | 0.1858 | | | |
| RH | 45 | 102.905511 | 1.0000 | 102.906 | 7.26 E 22 | 12.42 |
| PD | 46 | 101.905609 | 0.0096 | 106.441 | 6.80 E 22 | 12.00 |
| | | 103.904011 | 0.1097 | | | |
| | | 104.905064 | 0.2223 | | | |
| | | 105.903479 | 0.2733 | | | |
| | | 107.903891 | 0.2671 | | | |
| | | 109.905164 | 0.1181 | | | |
| AG | 47 | 106.905094 | 0.5183 | 107.868 | 5.85 E 22 | 10.50 |
| | | 108.904756 | 0.4817 | | | |
| CD | 48 | 105.906463 | 0.0122 | 112.434 | 4.64 E 22 | 8.65 |
| | | 107.904187 | 0.0088 | | | |
| | | 109.903012 | 0.1239 | | | |
| | | 110.904188 | 0.1275 | | | |
| | | 111.902763 | 0.2407 | | | |
| | | 112.904409 | 0.1226 | | | |
| | | 113.903360 | 0.2886 | | | |
| | | 115.904762 | 0.0758 | | | |
| IN | 49 | 112.904089 | 0.0428 | 114.818 | 3.83 E 22 | 7.29 |
| | | 114.903871 | 0.9572 | | | |
| SN | 50 | 111.904835 | 0.0096 | 118.734 | 3.62 E 22 | 5.76 |
| | | 113.902773 | 0.0066 | | | |
| | | 114.903346 | 0.0035 | | | |
| | | 115.901745 | 0.1430 | | | |
| | | 116.902958 | 0.0761 | | | |
| | | 117.901606 | 0.2403 | | | |
| | | 118.903313 | 0.0858 | | | |
| | | 119.902198 | 0.3285 | | | |
| | | 121.903441 | 0.0472 | | | |
| | | 123.905272 | 0.0594 | | | |
| SB | 51 | 120.903816 | 0.5725 | 121.755 | 3.31 E 22 | 6.69 |
| | | 122.904213 | 0.4275 | | | |
| TE | 52 | 119.904023 | 0.0009 | 127.628 | 2.94 E 22 | 6.25 |
| | | 121.903066 | 0.0246 | | | |
| | | 122.904277 | 0.0087 | | | |
| | | 123.902842 | 0.0461 | | | |
| | | 124.904418 | 0.0699 | | | |
| | | 125.903322 | 0.1871 | | | |
| | | 127.904476 | 0.3179 | | | |
| | | 129.906238 | 0.3448 | | | |
| I | 53 | 126.904470 | 1.0000 | 126.904 | 2.36 E 22 | 4.95 |
| XE | 54 | 123.906120 | 0.0010 | 131.305 | 1.64 E 22 | 3.78 |
| | | 125.904288 | 0.0009 | | | |
| | | 127.903540 | 0.0192 | | | |
| | | 128.904784 | 0.2644 | | | |
| | | 129.903509 | 0.0408 | | | |
| | | 130.905085 | 0.2118 | | | |
| | | 131.904161 | 0.2689 | | | |
| | | 133.905397 | 0.1044 | | | |
| | | 135.907221 | 0.0887 | | | |
| CS | 55 | 132.905355 | 1.0000 | 132.905 | 9.05 E 21 | 1.997 |
| BA | 56 | 129.906245 | 0.0010 | 137.327 | 1.60 E 22 | 3.59 |
| | | 131.905120 | 0.0010 | | | |
| | | 133.904612 | 0.0242 | | | |
| | | 134.905550 | 0.0659 | | | |
| | | 135.904300 | 0.0781 | | | |
| | | 136.905500 | 0.1132 | | | |
| | | 137.905000 | 0.7166 | | | |
| LA | 57 | 137.906910 | 0.0009 | 138.905 | 2.70 E 22 | 6.17 |
| | | 138.906140 | 0.9991 | | | |

**TABLE I (Continued)**

| ELE-MENT | AT. NO. (Z) | ISOTOPIC MASS (amu) | RELATIVE ABUN-DANCE | ATOMIC WEIGHT (amu) | ATOMIC DENSITY (atom/cm$^3$) | SPECIFIC GRAVITY |
|---|---|---|---|---|---|---|
| CE | 58 | 135.907100 | 0.0019 | 140.101 | 2.91 E 22 | 6.77 |
|    |    | 137.905830 | 0.0025 |         |           |      |
|    |    | 139.905392 | 0.8848 |         |           |      |
|    |    | 141.909140 | 0.1107 |         |           |      |
| PR | 59 | 140.907596 | 1.0000 | 140.908 | 2.92 E 22 | 6.78 |
| ND | 60 | 141.907663 | 0.2711 | 144.241 | 2.93 E 22 | 7.00 |
|    |    | 142.909779 | 0.1217 |         |           |      |
|    |    | 143.910039 | 0.2385 |         |           |      |
|    |    | 144.912538 | 0.0830 |         |           |      |
|    |    | 145.913086 | 0.1722 |         |           |      |
|    |    | 147.916869 | 0.0573 |         |           |      |
|    |    | 149.920915 | 0.0562 |         |           |      |
| PM | 61 | 0.0        | 0.0    |         |           |      |
| SM | 62 | 143.911989 | 0.0309 | 150.363 | 3.030E 22 | 7.54 |
|    |    | 146.914867 | 0.1497 |         |           |      |
|    |    | 147.914791 | 0.1124 |         |           |      |
|    |    | 148.917180 | 0.1383 |         |           |      |
|    |    | 149.917276 | 0.0744 |         |           |      |
|    |    | 151.919756 | 0.2672 |         |           |      |
|    |    | 153.922282 | 0.2271 |         |           |      |
| EU | 63 | 150.919838 | 0.4782 | 151.964 | 2.04 E 22 | 5.25 |
|    |    | 152.921242 | 0.5218 |         |           |      |
| GD | 64 | 151.919794 | 0.0020 | 157.256 | 3.02 E 22 | 7.89 |
|    |    | 153.920929 | 0.0215 |         |           |      |
|    |    | 154.922664 | 0.1473 |         |           |      |
|    |    | 155.922175 | 0.2047 |         |           |      |
|    |    | 156.924025 | 0.1568 |         |           |      |
|    |    | 157.924178 | 0.2487 |         |           |      |
|    |    | 159.927115 | 0.2190 |         |           |      |
| TB | 65 | 158.925351 | 1.0000 | 158.925 | 3.22 E 22 | 8.27 |
| DY | 66 | 155.923930 | 0.0005 | 162.484 | 3.17 E 22 | 8.53 |
|    |    | 157.924449 | 0.0009 |         |           |      |
|    |    | 159.925202 | 0.0229 |         |           |      |
|    |    | 160.926945 | 0.1888 |         |           |      |
|    |    | 161.926803 | 0.2553 |         |           |      |
|    |    | 162.928755 | 0.2497 |         |           |      |
|    |    | 163.929200 | 0.2818 |         |           |      |
| HO | 67 | 164.930421 | 1.0000 | 164.930 | 3.22 E 22 | 8.80 |
| ER | 68 | 161.928740 | 0.0014 | 167.261 | 3.26 E 22 | 9.04 |
|    |    | 163.929287 | 0.0156 |         |           |      |
|    |    | 165.930307 | 0.3341 |         |           |      |
|    |    | 166.932060 | 0.2294 |         |           |      |
|    |    | 167.932383 | 0.2707 |         |           |      |
|    |    | 169.935560 | 0.1488 |         |           |      |
| TM | 69 | 168.934245 | 1.0000 | 168.934 | 3.32 E 22 | 9.32 |
| YB | 70 | 167.934160 | 0.0014 | 173.036 | 3.02 E 22 | 6.97 |
|    |    | 169.935020 | 0.0303 |         |           |      |
|    |    | 170.936430 | 0.1431 |         |           |      |
|    |    | 171.936360 | 0.2182 |         |           |      |
|    |    | 172.938060 | 0.1613 |         |           |      |
|    |    | 173.938740 | 0.3184 |         |           |      |
|    |    | 175.942680 | 0.1273 |         |           |      |
| LU | 71 | 174.940640 | 0.9741 | 174.967 | 3.39 E 22 | 9.84 |
|    |    | 175.942660 | 0.0259 |         |           |      |
| HF | 72 | 173.940360 | 0.0018 | 178.509 | 4.52 E 22 | 13.20 |
|    |    | 175.941570 | 0.0520 |         |           |      |
|    |    | 176.943400 | 0.1850 |         |           |      |
|    |    | 177.943880 | 0.2714 |         |           |      |
|    |    | 178.946030 | 0.1375 |         |           |      |
|    |    | 179.946820 | 0.3524 |         |           |      |
| TA | 73 | 179.947544 | 0.0001 | 180.948 | 5.55 E 22 | 16.66 |
|    |    | 180.948007 | 0.9999 |         |           |      |

(Continued)

**TABLE I (Continued)**

| ELE-<br>MENT | AT.<br>NO.<br>(Z) | ISOTOPIC<br>MASS<br>(amu) | RELATIVE<br>ABUN-<br>DANCE | ATOMIC<br>WEIGHT<br>(amu) | ATOMIC<br>DENSITY<br>(atom/cm³) | SPECIFIC<br>GRAVITY |
|---|---|---|---|---|---|---|
| W | 74 | 179.947000<br>181.948301<br>182.950324<br>183.951025<br>185.954440 | 0.0014<br>0.2641<br>0.1440<br>0.3064<br>0.2841 | 183.842 | 6.30 E 22 | 19.25 |
| RE | 75 | 184.953059<br>186.955833 | 0.3707<br>0.6293 | 186.213 | 6.80 E 22 | 21.03 |
| OS | 76 | 183.952750<br>185.953870<br>186.955832<br>187.956081<br>188.958300<br>189.958630<br>191.961450 | 0.0002<br>0.0159<br>0.0164<br>0.1330<br>0.1610<br>0.2640<br>0.4100 | 190.333 | 7.14 E 22 | 22.58 |
| IR | 77 | 190.960640<br>192.963012 | 0.3730<br>0.6270 | 192.216 | 7.06 E 22 | 22.55 |
| PT | 78 | 189.959950<br>191.961150<br>193.962725<br>194.964813<br>195.964967<br>197.967895 | 0.0001<br>0.0078<br>0.3290<br>0.3380<br>0.2530<br>0.0721 | 195.081 | 6.62 E 22 | 21.47 |
| AU | 79 | 196.966541 | 1.0000 | 196.967 | 5.90 E 22 | 19.28 |
| HG | 80 | 195.965820<br>197.966756<br>198.968279<br>199.968327<br>200.970308<br>201.970642<br>203.973495 | 0.0015<br>0.1002<br>0.1684<br>0.2313<br>0.1322<br>0.2980<br>0.0685 | 200.617 | 4.26 E 22 | 14.26 |
| TL | 81 | 202.972353<br>204.974442 | 0.2950<br>0.7050 | 204.384 | 3.50 E 22 | 11.87 |
| PB | 82 | 203.973044<br>205.974468<br>206.975903<br>207.976650 | 0.0148<br>0.2360<br>0.2260<br>0.5230 | 207.177 | 3.30 E 22 | 11.34 |
| BI | 83 | 208.980394 | 1.0000 | 208.980 | 2.82 E 22 | 9.80 |

TABLE II

$K_{M_2}$ for $^1H$ as Projectile and Integer Target Mass $M_2$

| ATOMIC MASS $M_2$ (amu) | SCATTERING ANGLE $\theta$ | | | | | | | | | |
|---|---|---|---|---|---|---|---|---|---|---|
| | 180° | 170° | 160° | 150° | 140° | 130° | 120° | 110° | 100° | 90 |
| 2 | 0.1088 | 0.1105 | 0.1157 | 0.1248 | 0.1387 | 0.1584 | 0.1855 | 0.2219 | 0.2695 | 0.329 |
| 3 | 0.2471 | 0.2496 | 0.2573 | 0.2705 | 0.2897 | 0.3155 | 0.3486 | 0.3897 | 0.4392 | 0.497 |
| 4 | 0.3570 | 0.3598 | 0.3680 | 0.3820 | 0.4020 | 0.4282 | 0.4609 | 0.5001 | 0.5459 | 0.597 |
| 5 | 0.4416 | 0.4443 | 0.4524 | 0.4661 | 0.4854 | 0.5104 | 0.5411 | 0.5773 | 0.6187 | 0.664 |
| 6 | 0.5075 | 0.5101 | 0.5179 | 0.5309 | 0.5491 | 0.5725 | 0.6009 | 0.6340 | 0.6714 | 0.712 |
| 7 | 0.5599 | 0.5624 | 0.5698 | 0.5820 | 0.5990 | 0.6208 | 0.6471 | 0.6774 | 0.7114 | 0.748 |
| 8 | 0.6025 | 0.6048 | 0.6118 | 0.6232 | 0.6392 | 0.6594 | 0.6837 | 0.7117 | 0.7427 | 0.776 |
| 9 | 0.6378 | 0.6399 | 0.6464 | 0.6572 | 0.6721 | 0.6910 | 0.7135 | 0.7394 | 0.7679 | 0.798 |
| 10 | 0.6673 | 0.6694 | 0.6755 | 0.6856 | 0.6996 | 0.7172 | 0.7382 | 0.7622 | 0.7887 | 0.816 |
| 11 | 0.6925 | 0.6944 | 0.7002 | 0.7097 | 0.7228 | 0.7394 | 0.7590 | 0.7814 | 0.8060 | 0.832 |
| 12 | 0.7141 | 0.7159 | 0.7214 | 0.7304 | 0.7427 | 0.7583 | 0.7768 | 0.7977 | 0.8207 | 0.845 |
| 13 | 0.7329 | 0.7346 | 0.7398 | 0.7483 | 0.7600 | 0.7747 | 0.7921 | 0.8118 | 0.8333 | 0.856 |
| 14 | 0.7494 | 0.7511 | 0.7560 | 0.7640 | 0.7751 | 0.7890 | 0.8054 | 0.8240 | 0.8443 | 0.865 |
| 15 | 0.7640 | 0.7656 | 0.7702 | 0.7779 | 0.7884 | 0.8016 | 0.8172 | 0.8347 | 0.8539 | 0.874 |
| 16 | 0.7770 | 0.7785 | 0.7829 | 0.7902 | 0.8003 | 0.8128 | 0.8276 | 0.8442 | 0.8624 | 0.881 |
| 17 | 0.7887 | 0.7901 | 0.7943 | 0.8013 | 0.8109 | 0.8228 | 0.8369 | 0.8527 | 0.8699 | 0.888 |
| 18 | 0.7992 | 0.8005 | 0.8046 | 0.8112 | 0.8204 | 0.8318 | 0.8452 | 0.8603 | 0.8767 | 0.894 |
| 19 | 0.8087 | 0.8100 | 0.8139 | 0.8202 | 0.8290 | 0.8399 | 0.8527 | 0.8672 | 0.8828 | 0.899 |
| 20 | 0.8173 | 0.8186 | 0.8223 | 0.8284 | 0.8368 | 0.8473 | 0.8596 | 0.8734 | 0.8883 | 0.904 |
| 21 | 0.8252 | 0.8264 | 0.8300 | 0.8359 | 0.8440 | 0.8540 | 0.8658 | 0.8790 | 0.8934 | 0.908 |
| 22 | 0.8325 | 0.8336 | 0.8371 | 0.8427 | 0.8505 | 0.8602 | 0.8715 | 0.8842 | 0.8980 | 0.912 |
| 23 | 0.8391 | 0.8403 | 0.8436 | 0.8490 | 0.8565 | 0.8658 | 0.8767 | 0.8890 | 0.9022 | 0.916 |
| 24 | 0.8453 | 0.8464 | 0.8496 | 0.8549 | 0.8621 | 0.8710 | 0.8816 | 0.8933 | 0.9061 | 0.919 |
| 25 | 0.8510 | 0.8520 | 0.8552 | 0.8602 | 0.8672 | 0.8759 | 0.8860 | 0.8974 | 0.9097 | 0.922 |
| 26 | 0.8563 | 0.8573 | 0.8603 | 0.8652 | 0.8720 | 0.8804 | 0.8902 | 0.9011 | 0.9130 | 0.925 |
| 27 | 0.8612 | 0.8622 | 0.8651 | 0.8699 | 0.8764 | 0.8845 | 0.8940 | 0.9046 | 0.9161 | 0.928 |
| 28 | 0.8659 | 0.8668 | 0.8696 | 0.8742 | 0.8806 | 0.8884 | 0.8976 | 0.9079 | 0.9189 | 0.930 |
| 29 | 0.8702 | 0.8711 | 0.8738 | 0.8783 | 0.8844 | 0.8920 | 0.9010 | 0.9109 | 0.9216 | 0.932 |
| 30 | 0.8742 | 0.8751 | 0.8778 | 0.8821 | 0.8881 | 0.8955 | 0.9041 | 0.9137 | 0.9241 | 0.935 |
| 31 | 0.8780 | 0.8789 | 0.8815 | 0.8857 | 0.8915 | 0.8987 | 0.9070 | 0.9164 | 0.9265 | 0.937 |
| 32 | 0.8816 | 0.8824 | 0.8850 | 0.8891 | 0.8947 | 0.9017 | 0.9098 | 0.9189 | 0.9287 | 0.938 |
| 33 | 0.8850 | 0.8858 | 0.8882 | 0.8922 | 0.8977 | 0.9045 | 0.9124 | 0.9213 | 0.9308 | 0.940 |
| 34 | 0.8882 | 0.8890 | 0.8913 | 0.8952 | 0.9006 | 0.9072 | 0.9149 | 0.9235 | 0.9328 | 0.942 |
| 35 | 0.8912 | 0.8920 | 0.8943 | 0.8981 | 0.9033 | 0.9097 | 0.9172 | 0.9256 | 0.9346 | 0.944 |
| 36 | 0.8940 | 0.8948 | 0.8971 | 0.9008 | 0.9058 | 0.9121 | 0.9194 | 0.9276 | 0.9364 | 0.945 |
| 37 | 0.8967 | 0.8975 | 0.8997 | 0.9033 | 0.9083 | 0.9144 | 0.9215 | 0.9295 | 0.9380 | 0.947 |
| 38 | 0.8993 | 0.9000 | 0.9022 | 0.9057 | 0.9106 | 0.9165 | 0.9235 | 0.9313 | 0.9396 | 0.948 |
| 39 | 0.9018 | 0.9025 | 0.9046 | 0.9080 | 0.9127 | 0.9186 | 0.9254 | 0.9330 | 0.9411 | 0.949 |
| 40 | 0.9041 | 0.9048 | 0.9069 | 0.9102 | 0.9148 | 0.9205 | 0.9272 | 0.9346 | 0.9426 | 0.950 |
| 41 | 0.9063 | 0.9070 | 0.9090 | 0.9123 | 0.9168 | 0.9224 | 0.9289 | 0.9361 | 0.9439 | 0.952 |
| 42 | 0.9085 | 0.9091 | 0.9111 | 0.9143 | 0.9187 | 0.9242 | 0.9305 | 0.9376 | 0.9452 | 0.953 |
| 43 | 0.9105 | 0.9111 | 0.9131 | 0.9162 | 0.9205 | 0.9259 | 0.9321 | 0.9390 | 0.9465 | 0.954 |
| 44 | 0.9124 | 0.9131 | 0.9150 | 0.9181 | 0.9223 | 0.9275 | 0.9336 | 0.9404 | 0.9476 | 0.955 |
| 45 | 0.9143 | 0.9149 | 0.9168 | 0.9198 | 0.9239 | 0.9290 | 0.9350 | 0.9416 | 0.9488 | 0.956 |
| 46 | 0.9161 | 0.9167 | 0.9185 | 0.9215 | 0.9255 | 0.9305 | 0.9364 | 0.9429 | 0.9499 | 0.957 |
| 47 | 0.9178 | 0.9184 | 0.9202 | 0.9231 | 0.9270 | 0.9320 | 0.9377 | 0.9441 | 0.9509 | 0.958 |
| 48 | 0.9194 | 0.9200 | 0.9218 | 0.9246 | 0.9285 | 0.9333 | 0.9389 | 0.9452 | 0.9519 | 0.958 |
| 49 | 0.9210 | 0.9216 | 0.9233 | 0.9261 | 0.9299 | 0.9346 | 0.9402 | 0.9463 | 0.9529 | 0.959 |
| 50 | 0.9225 | 0.9231 | 0.9248 | 0.9275 | 0.9313 | 0.9359 | 0.9413 | 0.9473 | 0.9538 | 0.960 |
| 51 | 0.9240 | 0.9245 | 0.9262 | 0.9289 | 0.9326 | 0.9371 | 0.9424 | 0.9483 | 0.9547 | 0.961 |
| 52 | 0.9254 | 0.9259 | 0.9276 | 0.9302 | 0.9338 | 0.9383 | 0.9435 | 0.9493 | 0.9555 | 0.962 |
| 53 | 0.9267 | 0.9273 | 0.9289 | 0.9315 | 0.9350 | 0.9394 | 0.9445 | 0.9502 | 0.9563 | 0.962 |
| 54 | 0.9281 | 0.9286 | 0.9301 | 0.9327 | 0.9362 | 0.9405 | 0.9455 | 0.9511 | 0.9571 | 0.96 |
| 55 | 0.9293 | 0.9298 | 0.9314 | 0.9339 | 0.9373 | 0.9416 | 0.9465 | 0.9520 | 0.9579 | 0.966 |
| 56 | 0.9305 | 0.9310 | 0.9326 | 0.9350 | 0.9384 | 0.9426 | 0.9474 | 0.9528 | 0.9586 | 0.96 |
| 57 | 0.9317 | 0.9322 | 0.9337 | 0.9361 | 0.9395 | 0.9436 | 0.9483 | 0.9536 | 0.9593 | 0.965 |
| 58 | 0.9328 | 0.9333 | 0.9348 | 0.9372 | 0.9405 | 0.9445 | 0.9492 | 0.9544 | 0.9600 | 0.965 |

| $\theta$ | 180° | 178° | 176° | 174° | 172° | 170° | 168° | 166° | 164° | 162° | 160° |
|---|---|---|---|---|---|---|---|---|---|---|---|
| $\delta^2$ | 0 | 1.218 $\times10^{-3}$ | 4.874 $\times10^{-3}$ | 1.097 $\times10^{-2}$ | 1.950 $\times10^{-2}$ | 3.046 $\times10^{-2}$ | 4.386 $\times10^{-2}$ | 5.971 $\times10^{-2}$ | 7.798´ $\times10^{-2}$ | 9.870 $\times10^{-2}$ | 1.218 $\times10^{-1}$ |

**TABLE II (Continued)**

| TOMIC ASS $M_2$ (amu) | SCATTERING ANGLE $\theta$ | | | | | | | | | |
|---|---|---|---|---|---|---|---|---|---|---|
| | 180° | 170° | 160° | 150° | 140° | 130° | 120° | 110° | 100° | 90° |
| 59 | 0.9339 | 0.9344 | 0.9359 | 0.9382 | 0.9414 | 0.9454 | 0.9500 | 0.9552 | 0.9607 | 0.9664 |
| 60 | 0.9350 | 0.9355 | 0.9369 | 0.9392 | 0.9424 | 0.9463 | 0.9509 | 0.9559 | 0.9613 | 0.9670 |
| 61 | 0.9360 | 0.9365 | 0.9379 | 0.9402 | 0.9433 | 0.9472 | 0.9516 | 0.9566 | 0.9620 | 0.9675 |
| 62 | 0.9370 | 0.9375 | 0.9389 | 0.9411 | 0.9442 | 0.9480 | 0.9524 | 0.9573 | 0.9626 | 0.9680 |
| 63 | 0.9380 | 0.9385 | 0.9398 | 0.9420 | 0.9451 | 0.9488 | 0.9531 | 0.9580 | 0.9631 | 0.9685 |
| 64 | 0.9389 | 0.9394 | 0.9407 | 0.9429 | 0.9459 | 0.9496 | 0.9539 | 0.9586 | 0.9637 | 0.9690 |
| 65 | 0.9399 | 0.9403 | 0.9416 | 0.9438 | 0.9467 | 0.9503 | 0.9545 | 0.9592 | 0.9643 | 0.9695 |
| 66 | 0.9407 | 0.9412 | 0.9425 | 0.9446 | 0.9475 | 0.9511 | 0.9552 | 0.9598 | 0.9648 | 0.9699 |
| 67 | 0.9416 | 0.9420 | 0.9433 | 0.9454 | 0.9483 | 0.9518 | 0.9559 | 0.9604 | 0.9653 | 0.9704 |
| 68 | 0.9424 | 0.9429 | 0.9441 | 0.9462 | 0.9490 | 0.9525 | 0.9565 | 0.9610 | 0.9658 | 0.9708 |
| 69 | 0.9432 | 0.9437 | 0.9449 | 0.9469 | 0.9497 | 0.9531 | 0.9571 | 0.9616 | 0.9663 | 0.9712 |
| 70 | 0.9440 | 0.9444 | 0.9457 | 0.9477 | 0.9504 | 0.9538 | 0.9577 | 0.9621 | 0.9668 | 0.9716 |
| 71 | 0.9448 | 0.9452 | 0.9464 | 0.9484 | 0.9511 | 0.9544 | 0.9583 | 0.9626 | 0.9672 | 0.9720 |
| 72 | 0.9455 | 0.9459 | 0.9471 | 0.9491 | 0.9518 | 0.9550 | 0.9589 | 0.9631 | 0.9677 | 0.9724 |
| 73 | 0.9463 | 0.9467 | 0.9478 | 0.9498 | 0.9524 | 0.9557 | 0.9594 | 0.9636 | 0.9681 | 0.9728 |
| 74 | 0.9470 | 0.9474 | 0.9485 | 0.9504 | 0.9530 | 0.9562 | 0.9600 | 0.9641 | 0.9685 | 0.9731 |
| 75 | 0.9477 | 0.9481 | 0.9492 | 0.9511 | 0.9536 | 0.9568 | 0.9605 | 0.9646 | 0.9689 | 0.9735 |
| 76 | 0.9483 | 0.9487 | 0.9499 | 0.9517 | 0.9542 | 0.9574 | 0.9610 | 0.9650 | 0.9694 | 0.9738 |
| 77 | 0.9490 | 0.9494 | 0.9505 | 0.9523 | 0.9548 | 0.9579 | 0.9615 | 0.9655 | 0.9697 | 0.9742 |
| 78 | 0.9496 | 0.9500 | 0.9511 | 0.9529 | 0.9554 | 0.9584 | 0.9620 | 0.9659 | 0.9701 | 0.9745 |
| 79 | 0.9502 | 0.9506 | 0.9517 | 0.9535 | 0.9559 | 0.9589 | 0.9624 | 0.9663 | 0.9705 | 0.9748 |
| 80 | 0.9509 | 0.9512 | 0.9523 | 0.9541 | 0.9565 | 0.9595 | 0.9629 | 0.9668 | 0.9709 | 0.9751 |
| 81 | 0.9514 | 0.9518 | 0.9529 | 0.9546 | 0.9570 | 0.9599 | 0.9634 | 0.9672 | 0.9712 | 0.9754 |
| 82 | 0.9520 | 0.9524 | 0.9534 | 0.9552 | 0.9575 | 0.9604 | 0.9638 | 0.9675 | 0.9716 | 0.9757 |
| 83 | 0.9526 | 0.9529 | 0.9540 | 0.9557 | 0.9580 | 0.9609 | 0.9642 | 0.9679 | 0.9719 | 0.9760 |
| 84 | 0.9531 | 0.9535 | 0.9545 | 0.9562 | 0.9585 | 0.9613 | 0.9646 | 0.9683 | 0.9722 | 0.9763 |
| 85 | 0.9537 | 0.9540 | 0.9550 | 0.9567 | 0.9590 | 0.9618 | 0.9651 | 0.9687 | 0.9726 | 0.9766 |
| 86 | 0.9542 | 0.9545 | 0.9556 | 0.9572 | 0.9595 | 0.9622 | 0.9655 | 0.9690 | 0.9729 | 0.9768 |
| 87 | 0.9547 | 0.9551 | 0.9561 | 0.9577 | 0.9599 | 0.9627 | 0.9658 | 0.9694 | 0.9732 | 0.9771 |
| 88 | 0.9552 | 0.9556 | 0.9565 | 0.9582 | 0.9604 | 0.9631 | 0.9662 | 0.9697 | 0.9735 | 0.9774 |
| 89 | 0.9557 | 0.9560 | 0.9570 | 0.9586 | 0.9608 | 0.9635 | 0.9666 | 0.9701 | 0.9738 | 0.9776 |
| 90 | 0.9562 | 0.9565 | 0.9575 | 0.9591 | 0.9612 | 0.9639 | 0.9670 | 0.9704 | 0.9741 | 0.9779 |
| 91 | 0.9567 | 0.9570 | 0.9579 | 0.9595 | 0.9616 | 0.9643 | 0.9673 | 0.9707 | 0.9743 | 0.9781 |
| 92 | 0.9571 | 0.9574 | 0.9584 | 0.9599 | 0.9620 | 0.9646 | 0.9677 | 0.9710 | 0.9746 | 0.9783 |
| 93 | 0.9576 | 0.9579 | 0.9588 | 0.9604 | 0.9624 | 0.9650 | 0.9680 | 0.9713 | 0.9749 | 0.9786 |
| 94 | 0.9580 | 0.9583 | 0.9593 | 0.9608 | 0.9628 | 0.9654 | 0.9683 | 0.9716 | 0.9751 | 0.9788 |
| 95 | 0.9585 | 0.9588 | 0.9597 | 0.9612 | 0.9632 | 0.9657 | 0.9687 | 0.9719 | 0.9754 | 0.9790 |
| 96 | 0.9589 | 0.9592 | 0.9601 | 0.9616 | 0.9636 | 0.9661 | 0.9690 | 0.9722 | 0.9757 | 0.9792 |
| 97 | 0.9593 | 0.9596 | 0.9605 | 0.9620 | 0.9640 | 0.9664 | 0.9693 | 0.9725 | 0.9759 | 0.9794 |
| 98 | 0.9597 | 0.9600 | 0.9609 | 0.9623 | 0.9643 | 0.9668 | 0.9696 | 0.9728 | 0.9761 | 0.9796 |
| 99 | 0.9601 | 0.9604 | 0.9613 | 0.9627 | 0.9647 | 0.9671 | 0.9699 | 0.9730 | 0.9764 | 0.9798 |
| 100 | 0.9605 | 0.9608 | 0.9617 | 0.9631 | 0.9650 | 0.9674 | 0.9702 | 0.9733 | 0.9766 | 0.9800 |
| 101 | 0.9609 | 0.9612 | 0.9620 | 0.9634 | 0.9654 | 0.9677 | 0.9705 | 0.9736 | 0.9768 | 0.9802 |
| 102 | 0.9612 | 0.9615 | 0.9624 | 0.9638 | 0.9657 | 0.9681 | 0.9708 | 0.9738 | 0.9771 | 0.9804 |
| 103 | 0.9616 | 0.9619 | 0.9628 | 0.9641 | 0.9660 | 0.9684 | 0.9711 | 0.9741 | 0.9773 | 0.9806 |
| 104 | 0.9620 | 0.9623 | 0.9631 | 0.9645 | 0.9663 | 0.9687 | 0.9713 | 0.9743 | 0.9775 | 0.9808 |
| 105 | 0.9623 | 0.9626 | 0.9634 | 0.9648 | 0.9667 | 0.9690 | 0.9716 | 0.9746 | 0.9777 | 0.9810 |
| 106 | 0.9627 | 0.9630 | 0.9638 | 0.9651 | 0.9670 | 0.9692 | 0.9719 | 0.9748 | 0.9779 | 0.9812 |
| 107 | 0.9630 | 0.9633 | 0.9641 | 0.9655 | 0.9673 | 0.9695 | 0.9721 | 0.9750 | 0.9781 | 0.9813 |
| 108 | 0.9634 | 0.9636 | 0.9644 | 0.9658 | 0.9676 | 0.9698 | 0.9724 | 0.9753 | 0.9783 | 0.9815 |
| 109 | 0.9637 | 0.9640 | 0.9648 | 0.9661 | 0.9679 | 0.9701 | 0.9726 | 0.9755 | 0.9785 | 0.9817 |
| 110 | 0.9640 | 0.9643 | 0.9651 | 0.9664 | 0.9682 | 0.9703 | 0.9729 | 0.9757 | 0.9787 | 0.9818 |
| 111 | 0.9643 | 0.9646 | 0.9654 | 0.9667 | 0.9684 | 0.9706 | 0.9731 | 0.9759 | 0.9789 | 0.9820 |
| 112 | 0.9646 | 0.9649 | 0.9657 | 0.9670 | 0.9687 | 0.9709 | 0.9734 | 0.9761 | 0.9791 | 0.9822 |
| 113 | 0.9650 | 0.9652 | 0.9660 | 0.9673 | 0.9690 | 0.9711 | 0.9736 | 0.9763 | 0.9793 | 0.9823 |
| 114 | 0.9653 | 0.9655 | 0.9663 | 0.9675 | 0.9693 | 0.9714 | 0.9738 | 0.9766 | 0.9795 | 0.9825 |

(Continued)

| $\theta$ | 180° | 178° | 176° | 174° | 172° | 170° | 168° | 166° | 164° | 162° | 160° |
|---|---|---|---|---|---|---|---|---|---|---|---|
| $\delta^2$ | 0 | $1.218 \times 10^{-3}$ | $4.874 \times 10^{-3}$ | $1.097 \times 10^{-2}$ | $1.950 \times 10^{-2}$ | $3.046 \times 10^{-2}$ | $4.386 \times 10^{-2}$ | $5.971 \times 10^{-2}$ | $7.798 \times 10^{-2}$ | $9.870 \times 10^{-2}$ | $1.218 \times 10^{-1}$ |

**TABLE II (Continued)**

| ATOMIC MASS $M_2$ (amu) | SCATTERING ANGLE $\theta$ | | | | | | | | | |
|---|---|---|---|---|---|---|---|---|---|---|
| | 180° | 170° | 160° | 150° | 140° | 130° | 120° | 110° | 100° | 90 |
| 115 | 0.9656 | 0.9658 | 0.9666 | 0.9678 | 0.9695 | 0.9716 | 0.9741 | 0.9768 | 0.9796 | 0.982 |
| 116 | 0.9658 | 0.9661 | 0.9669 | 0.9681 | 0.9698 | 0.9719 | 0.9743 | 0.9769 | 0.9798 | 0.982 |
| 117 | 0.9661 | 0.9664 | 0.9671 | 0.9684 | 0.9700 | 0.9721 | 0.9745 | 0.9771 | 0.9800 | 0.982 |
| 118 | 0.9664 | 0.9667 | 0.9674 | 0.9686 | 0.9703 | 0.9723 | 0.9747 | 0.9773 | 0.9802 | 0.983 |
| 119 | 0.9667 | 0.9669 | 0.9677 | 0.9689 | 0.9705 | 0.9726 | 0.9749 | 0.9775 | 0.9803 | 0.983 |
| 120 | 0.9670 | 0.9672 | 0.9679 | 0.9691 | 0.9708 | 0.9728 | 0.9751 | 0.9777 | 0.9805 | 0.983 |
| 121 | 0.9672 | 0.9675 | 0.9682 | 0.9694 | 0.9710 | 0.9730 | 0.9753 | 0.9779 | 0.9806 | 0.98 |
| 122 | 0.9675 | 0.9677 | 0.9685 | 0.9696 | 0.9712 | 0.9732 | 0.9755 | 0.9781 | 0.9808 | 0.98 |
| 123 | 0.9678 | 0.9680 | 0.9687 | 0.9699 | 0.9715 | 0.9734 | 0.9757 | 0.9782 | 0.9810 | 0.98 |
| 124 | 0.9680 | 0.9683 | 0.9690 | 0.9701 | 0.9717 | 0.9736 | 0.9759 | 0.9784 | 0.9811 | 0.98 |
| 125 | 0.9683 | 0.9685 | 0.9692 | 0.9704 | 0.9719 | 0.9739 | 0.9761 | 0.9786 | 0.9813 | 0.98 |
| 126 | 0.9685 | 0.9687 | 0.9694 | 0.9706 | 0.9721 | 0.9741 | 0.9763 | 0.9788 | 0.9814 | 0.98 |
| 127 | 0.9688 | 0.9690 | 0.9697 | 0.9708 | 0.9724 | 0.9743 | 0.9765 | 0.9789 | 0.9815 | 0.98 |
| 128 | 0.9690 | 0.9692 | 0.9699 | 0.9710 | 0.9726 | 0.9745 | 0.9767 | 0.9791 | 0.9817 | 0.98 |
| 129 | 0.9692 | 0.9695 | 0.9701 | 0.9713 | 0.9728 | 0.9747 | 0.9768 | 0.9792 | 0.9818 | 0.98 |
| 130 | 0.9695 | 0.9697 | 0.9704 | 0.9715 | 0.9730 | 0.9748 | 0.9770 | 0.9794 | 0.9820 | 0.98 |
| 131 | 0.9697 | 0.9699 | 0.9706 | 0.9717 | 0.9732 | 0.9750 | 0.9772 | 0.9796 | 0.9821 | 0.98 |
| 132 | 0.9699 | 0.9701 | 0.9708 | 0.9719 | 0.9734 | 0.9752 | 0.9774 | 0.9797 | 0.9822 | 0.98 |
| 133 | 0.9701 | 0.9704 | 0.9710 | 0.9721 | 0.9736 | 0.9754 | 0.9775 | 0.9799 | 0.9824 | 0.98 |
| 134 | 0.9704 | 0.9706 | 0.9712 | 0.9723 | 0.9738 | 0.9756 | 0.9777 | 0.9800 | 0.9825 | 0.98 |
| 135 | 0.9706 | 0.9708 | 0.9715 | 0.9725 | 0.9740 | 0.9758 | 0.9779 | 0.9802 | 0.9826 | 0.98 |
| 136 | 0.9708 | 0.9710 | 0.9717 | 0.9727 | 0.9742 | 0.9759 | 0.9780 | 0.9803 | 0.9828 | 0.98 |
| 137 | 0.9710 | 0.9712 | 0.9719 | 0.9729 | 0.9744 | 0.9761 | 0.9782 | 0.9804 | 0.9829 | 0.98 |
| 138 | 0.9712 | 0.9714 | 0.9721 | 0.9731 | 0.9745 | 0.9763 | 0.9783 | 0.9806 | 0.9830 | 0.98 |
| 139 | 0.9714 | 0.9716 | 0.9723 | 0.9733 | 0.9747 | 0.9765 | 0.9785 | 0.9807 | 0.9831 | 0.98 |
| 140 | 0.9716 | 0.9718 | 0.9725 | 0.9735 | 0.9749 | 0.9766 | 0.9786 | 0.9809 | 0.9832 | 0.98 |
| 141 | 0.9718 | 0.9720 | 0.9727 | 0.9737 | 0.9751 | 0.9768 | 0.9788 | 0.9810 | 0.9834 | 0.98 |
| 142 | 0.9720 | 0.9722 | 0.9728 | 0.9739 | 0.9752 | 0.9770 | 0.9789 | 0.9811 | 0.9835 | 0.98 |
| 143 | 0.9722 | 0.9724 | 0.9730 | 0.9740 | 0.9754 | 0.9771 | 0.9791 | 0.9813 | 0.9836 | 0.98 |
| 144 | 0.9724 | 0.9726 | 0.9732 | 0.9742 | 0.9756 | 0.9773 | 0.9792 | 0.9814 | 0.9837 | 0.98 |
| 145 | 0.9726 | 0.9728 | 0.9734 | 0.9744 | 0.9757 | 0.9774 | 0.9794 | 0.9815 | 0.9838 | 0.98 |
| 146 | 0.9728 | 0.9730 | 0.9736 | 0.9746 | 0.9759 | 0.9776 | 0.9795 | 0.9816 | 0.9839 | 0.98 |
| 147 | 0.9729 | 0.9732 | 0.9738 | 0.9747 | 0.9761 | 0.9777 | 0.9796 | 0.9818 | 0.9840 | 0.98 |
| 148 | 0.9731 | 0.9733 | 0.9739 | 0.9749 | 0.9762 | 0.9779 | 0.9798 | 0.9819 | 0.9841 | 0.98 |
| 149 | 0.9733 | 0.9735 | 0.9741 | 0.9751 | 0.9764 | 0.9780 | 0.9799 | 0.9820 | 0.9842 | 0.98 |
| 150 | 0.9735 | 0.9737 | 0.9743 | 0.9752 | 0.9765 | 0.9782 | 0.9800 | 0.9821 | 0.9844 | 0.98 |
| 151 | 0.9737 | 0.9739 | 0.9744 | 0.9754 | 0.9767 | 0.9783 | 0.9802 | 0.9822 | 0.9845 | 0.98 |
| 152 | 0.9738 | 0.9740 | 0.9746 | 0.9756 | 0.9769 | 0.9785 | 0.9803 | 0.9824 | 0.9846 | 0.98 |
| 153 | 0.9740 | 0.9742 | 0.9748 | 0.9757 | 0.9770 | 0.9786 | 0.9804 | 0.9825 | 0.9847 | 0.98 |
| 154 | 0.9742 | 0.9744 | 0.9749 | 0.9759 | 0.9771 | 0.9787 | 0.9806 | 0.9826 | 0.9848 | 0.98 |
| 155 | 0.9743 | 0.9745 | 0.9751 | 0.9760 | 0.9773 | 0.9789 | 0.9807 | 0.9827 | 0.9849 | 0.98 |
| 156 | 0.9745 | 0.9747 | 0.9752 | 0.9762 | 0.9774 | 0.9790 | 0.9808 | 0.9828 | 0.9849 | 0.98 |
| 157 | 0.9746 | 0.9748 | 0.9754 | 0.9763 | 0.9776 | 0.9791 | 0.9809 | 0.9829 | 0.9850 | 0.98 |
| 158 | 0.9748 | 0.9750 | 0.9756 | 0.9765 | 0.9777 | 0.9793 | 0.9810 | 0.9830 | 0.9851 | 0.98 |
| 159 | 0.9750 | 0.9752 | 0.9757 | 0.9766 | 0.9779 | 0.9794 | 0.9812 | 0.9831 | 0.9852 | 0.98 |
| 160 | 0.9751 | 0.9753 | 0.9759 | 0.9768 | 0.9780 | 0.9795 | 0.9813 | 0.9832 | 0.9853 | 0.98 |
| 161 | 0.9753 | 0.9755 | 0.9760 | 0.9769 | 0.9781 | 0.9796 | 0.9814 | 0.9833 | 0.9854 | 0.98 |
| 162 | 0.9754 | 0.9756 | 0.9762 | 0.9770 | 0.9783 | 0.9798 | 0.9815 | 0.9834 | 0.9855 | 0.98 |
| 163 | 0.9756 | 0.9758 | 0.9763 | 0.9772 | 0.9784 | 0.9799 | 0.9816 | 0.9835 | 0.9856 | 0.98 |
| 164 | 0.9757 | 0.9759 | 0.9764 | 0.9773 | 0.9785 | 0.9800 | 0.9817 | 0.9836 | 0.9857 | 0.99 |
| 165 | 0.9759 | 0.9760 | 0.9766 | 0.9775 | 0.9787 | 0.9801 | 0.9818 | 0.9837 | 0.9858 | 0.98 |
| 166 | 0.9760 | 0.9762 | 0.9767 | 0.9776 | 0.9788 | 0.9802 | 0.9820 | 0.9838 | 0.9858 | 0.98 |
| 167 | 0.9761 | 0.9763 | 0.9769 | 0.9777 | 0.9789 | 0.9804 | 0.9821 | 0.9839 | 0.9859 | 0.98 |
| 168 | 0.9763 | 0.9765 | 0.9770 | 0.9779 | 0.9790 | 0.9805 | 0.9822 | 0.9840 | 0.9860 | 0.98 |
| 169 | 0.9764 | 0.9766 | 0.9771 | 0.9780 | 0.9792 | 0.9806 | 0.9823 | 0.9841 | 0.9861 | 0.98 |
| 170 | 0.9766 | 0.9767 | 0.9773 | 0.9781 | 0.9793 | 0.9807 | 0.9824 | 0.9842 | 0.9862 | 0.98 |
| 171 | 0.9767 | 0.9769 | 0.9774 | 0.9782 | 0.9794 | 0.9808 | 0.9825 | 0.9843 | 0.9863 | 0.98 |

| $\theta$ | 180° | 178° | 176° | 174° | 172° | 170° | 168° | 166° | 164° | 162° | 160° |
|---|---|---|---|---|---|---|---|---|---|---|---|
| $\delta^2$ | 0 | 1.218 $\times 10^{-3}$ | 4.874 $\times 10^{-3}$ | 1.097 $\times 10^{-2}$ | 1.950 $\times 10^{-2}$ | 3.046 $\times 10^{-2}$ | 4.386 $\times 10^{-2}$ | 5.971 $\times 10^{-2}$ | 7.798 $\times 10^{-2}$ | 9.870 $\times 10^{-2}$ | 1.218 $\times 10^{-1}$ |

**TABLE II (Continued)**

| ATOMIC MASS $M_2$ (amu) | SCATTERING ANGLE $\theta$ | | | | | | | | | |
|---|---|---|---|---|---|---|---|---|---|---|
| | 180° | 170° | 160° | 150° | 140° | 130° | 120° | 110° | 100° | 90° |
| 172 | 0.9768 | 0.9770 | 0.9775 | 0.9784 | 0.9795 | 0.9809 | 0.9826 | 0.9844 | 0.9863 | 0.9883 |
| 173 | 0.9770 | 0.9771 | 0.9777 | 0.9785 | 0.9796 | 0.9810 | 0.9827 | 0.9845 | 0.9864 | 0.9884 |
| 174 | 0.9771 | 0.9773 | 0.9778 | 0.9786 | 0.9797 | 0.9811 | 0.9828 | 0.9846 | 0.9865 | 0.9885 |
| 175 | 0.9772 | 0.9774 | 0.9779 | 0.9787 | 0.9799 | 0.9813 | 0.9829 | 0.9847 | 0.9866 | 0.9885 |
| 176 | 0.9774 | 0.9775 | 0.9780 | 0.9789 | 0.9800 | 0.9814 | 0.9830 | 0.9847 | 0.9866 | 0.9886 |
| 177 | 0.9775 | 0.9777 | 0.9782 | 0.9790 | 0.9801 | 0.9815 | 0.9831 | 0.9848 | 0.9867 | 0.9887 |
| 178 | 0.9776 | 0.9778 | 0.9783 | 0.9791 | 0.9802 | 0.9816 | 0.9832 | 0.9849 | 0.9868 | 0.9887 |
| 179 | 0.9777 | 0.9779 | 0.9784 | 0.9792 | 0.9803 | 0.9817 | 0.9833 | 0.9850 | 0.9869 | 0.9888 |
| 180 | 0.9779 | 0.9780 | 0.9785 | 0.9793 | 0.9804 | 0.9818 | 0.9833 | 0.9851 | 0.9869 | 0.9889 |
| 181 | 0.9780 | 0.9781 | 0.9786 | 0.9794 | 0.9805 | 0.9819 | 0.9834 | 0.9852 | 0.9870 | 0.9889 |
| 182 | 0.9781 | 0.9783 | 0.9787 | 0.9795 | 0.9806 | 0.9820 | 0.9835 | 0.9852 | 0.9871 | 0.9890 |
| 183 | 0.9782 | 0.9784 | 0.9789 | 0.9797 | 0.9807 | 0.9821 | 0.9836 | 0.9853 | 0.9872 | 0.9890 |
| 184 | 0.9783 | 0.9785 | 0.9790 | 0.9798 | 0.9808 | 0.9822 | 0.9837 | 0.9854 | 0.9872 | 0.9891 |
| 185 | 0.9784 | 0.9786 | 0.9791 | 0.9799 | 0.9809 | 0.9823 | 0.9838 | 0.9855 | 0.9873 | 0.9892 |
| 186 | 0.9786 | 0.9787 | 0.9792 | 0.9800 | 0.9810 | 0.9824 | 0.9839 | 0.9856 | 0.9874 | 0.9892 |
| 187 | 0.9787 | 0.9788 | 0.9793 | 0.9801 | 0.9811 | 0.9824 | 0.9840 | 0.9856 | 0.9874 | 0.9893 |
| 188 | 0.9788 | 0.9789 | 0.9794 | 0.9802 | 0.9812 | 0.9825 | 0.9840 | 0.9857 | 0.9875 | 0.9893 |
| 189 | 0.9789 | 0.9791 | 0.9795 | 0.9803 | 0.9813 | 0.9826 | 0.9841 | 0.9858 | 0.9876 | 0.9894 |
| 190 | 0.9790 | 0.9792 | 0.9796 | 0.9804 | 0.9814 | 0.9827 | 0.9842 | 0.9859 | 0.9876 | 0.9894 |
| 191 | 0.9791 | 0.9793 | 0.9797 | 0.9805 | 0.9815 | 0.9828 | 0.9843 | 0.9859 | 0.9877 | 0.9895 |
| 192 | 0.9792 | 0.9794 | 0.9798 | 0.9806 | 0.9816 | 0.9829 | 0.9844 | 0.9860 | 0.9878 | 0.9896 |
| 193 | 0.9793 | 0.9795 | 0.9799 | 0.9807 | 0.9817 | 0.9830 | 0.9845 | 0.9861 | 0.9878 | 0.9896 |
| 194 | 0.9794 | 0.9796 | 0.9800 | 0.9808 | 0.9818 | 0.9831 | 0.9845 | 0.9862 | 0.9879 | 0.9897 |
| 195 | 0.9795 | 0.9797 | 0.9801 | 0.9809 | 0.9819 | 0.9832 | 0.9846 | 0.9862 | 0.9879 | 0.9897 |
| 196 | 0.9796 | 0.9798 | 0.9802 | 0.9810 | 0.9820 | 0.9832 | 0.9847 | 0.9863 | 0.9880 | 0.9898 |
| 197 | 0.9797 | 0.9799 | 0.9803 | 0.9811 | 0.9821 | 0.9833 | 0.9848 | 0.9864 | 0.9881 | 0.9898 |
| 198 | 0.9798 | 0.9800 | 0.9804 | 0.9812 | 0.9822 | 0.9834 | 0.9848 | 0.9864 | 0.9881 | 0.9899 |
| 199 | 0.9799 | 0.9801 | 0.9805 | 0.9813 | 0.9823 | 0.9835 | 0.9849 | 0.9865 | 0.9882 | 0.9899 |
| 200 | 0.9800 | 0.9802 | 0.9806 | 0.9814 | 0.9824 | 0.9836 | 0.9850 | 0.9866 | 0.9882 | 0.9900 |
| 201 | 0.9801 | 0.9803 | 0.9807 | 0.9815 | 0.9824 | 0.9837 | 0.9851 | 0.9866 | 0.9883 | 0.9900 |
| 202 | 0.9802 | 0.9804 | 0.9808 | 0.9816 | 0.9825 | 0.9837 | 0.9851 | 0.9867 | 0.9884 | 0.9901 |
| 203 | 0.9803 | 0.9805 | 0.9809 | 0.9816 | 0.9826 | 0.9838 | 0.9852 | 0.9868 | 0.9884 | 0.9901 |
| 204 | 0.9804 | 0.9806 | 0.9810 | 0.9817 | 0.9827 | 0.9839 | 0.9853 | 0.9868 | 0.9885 | 0.9902 |
| 205 | 0.9805 | 0.9807 | 0.9811 | 0.9818 | 0.9828 | 0.9840 | 0.9854 | 0.9869 | 0.9885 | 0.9902 |
| 206 | 0.9806 | 0.9808 | 0.9812 | 0.9819 | 0.9829 | 0.9841 | 0.9854 | 0.9870 | 0.9886 | 0.9903 |
| 207 | 0.9807 | 0.9809 | 0.9813 | 0.9820 | 0.9830 | 0.9841 | 0.9855 | 0.9870 | 0.9886 | 0.9903 |
| 208 | 0.9808 | 0.9809 | 0.9814 | 0.9821 | 0.9830 | 0.9842 | 0.9856 | 0.9871 | 0.9887 | 0.9904 |
| 209 | 0.9809 | 0.9810 | 0.9815 | 0.9822 | 0.9831 | 0.9843 | 0.9856 | 0.9871 | 0.9887 | 0.9904 |
| 210 | 0.9810 | 0.9811 | 0.9816 | 0.9822 | 0.9832 | 0.9844 | 0.9857 | 0.9872 | 0.9888 | 0.9904 |
| 211 | 0.9811 | 0.9812 | 0.9816 | 0.9823 | 0.9833 | 0.9844 | 0.9858 | 0.9873 | 0.9889 | 0.9905 |
| 212 | 0.9812 | 0.9813 | 0.9817 | 0.9824 | 0.9833 | 0.9845 | 0.9858 | 0.9873 | 0.9889 | 0.9905 |
| 213 | 0.9813 | 0.9814 | 0.9818 | 0.9825 | 0.9834 | 0.9846 | 0.9859 | 0.9874 | 0.9890 | 0.9906 |
| 214 | 0.9813 | 0.9815 | 0.9819 | 0.9826 | 0.9835 | 0.9846 | 0.9860 | 0.9874 | 0.9890 | 0.9906 |
| 215 | 0.9814 | 0.9816 | 0.9820 | 0.9827 | 0.9836 | 0.9847 | 0.9860 | 0.9875 | 0.9891 | 0.9907 |
| 216 | 0.9815 | 0.9816 | 0.9821 | 0.9827 | 0.9837 | 0.9848 | 0.9861 | 0.9876 | 0.9891 | 0.9907 |

| $\theta$ | 180° | 178° | 176° | 174° | 172° | 170° | 168° | 166° | 164° | 162° | 160° |
|---|---|---|---|---|---|---|---|---|---|---|---|
| $\delta^2$ | 0 | $1.218 \times 10^{-3}$ | $4.874 \times 10^{-3}$ | $1.097 \times 10^{-2}$ | $1.950 \times 10^{-2}$ | $3.046 \times 10^{-2}$ | $4.386 \times 10^{-2}$ | $5.971 \times 10^{-2}$ | $7.798 \times 10^{-2}$ | $9.870 \times 10^{-2}$ | $1.218 \times 10^{-1}$ |

**TABLE III**

$K_{M_2}$ for $^4$He as Projectile and Integer Target Mass $M_2$

| ATOMIC MASS $M_2$ (amu) | SCATTERING ANGLE $\theta$ | | | | | | | | | |
|---|---|---|---|---|---|---|---|---|---|---|
| | 180° | 170° | 160° | 150° | 140° | 130° | 120° | 110° | 100° | 90 |
| 6 | 0.0399 | 0.0407 | 0.0433 | 0.0479 | 0.0554 | 0.0668 | 0.0838 | 0.1092 | 0.1465 | 0.199 |
| 7 | 0.0742 | 0.0755 | 0.0796 | 0.0868 | 0.0980 | 0.1143 | 0.1375 | 0.1699 | 0.2140 | 0.272 |
| 8 | 0.1109 | 0.1126 | 0.1179 | 0.1271 | 0.1411 | 0.1610 | 0.1883 | 0.2249 | 0.2726 | 0.333 |
| 9 | 0.1477 | 0.1497 | 0.1559 | 0.1667 | 0.1827 | 0.2051 | 0.2351 | 0.2741 | 0.3235 | 0.384 |
| 10 | 0.1834 | 0.1857 | 0.1926 | 0.2044 | 0.2220 | 0.2460 | 0.2777 | 0.3180 | 0.3681 | 0.428 |
| 11 | 0.2175 | 0.2200 | 0.2273 | 0.2400 | 0.2586 | 0.2837 | 0.3164 | 0.3573 | 0.4073 | 0.466 |
| 12 | 0.2498 | 0.2523 | 0.2600 | 0.2733 | 0.2925 | 0.3183 | 0.3515 | 0.3926 | 0.4420 | 0.499 |
| 13 | 0.2800 | 0.2827 | 0.2907 | 0.3043 | 0.3239 | 0.3501 | 0.3834 | 0.4243 | 0.4730 | 0.529 |
| 14 | 0.3084 | 0.3111 | 0.3192 | 0.3331 | 0.3530 | 0.3793 | 0.4125 | 0.4530 | 0.5007 | 0.555 |
| 15 | 0.3349 | 0.3377 | 0.3459 | 0.3599 | 0.3798 | 0.4062 | 0.4391 | 0.4790 | 0.5257 | 0.578 |
| 16 | 0.3598 | 0.3625 | 0.3708 | 0.3848 | 0.4047 | 0.4309 | 0.4636 | 0.5027 | 0.5483 | 0.599 |
| 17 | 0.3830 | 0.3857 | 0.3940 | 0.4080 | 0.4279 | 0.4538 | 0.4860 | 0.5244 | 0.5689 | 0.618 |
| 18 | 0.4047 | 0.4075 | 0.4157 | 0.4296 | 0.4493 | 0.4750 | 0.5067 | 0.5444 | 0.5877 | 0.636 |
| 19 | 0.4251 | 0.4278 | 0.4360 | 0.4498 | 0.4693 | 0.4947 | 0.5258 | 0.5627 | 0.6050 | 0.652 |
| 20 | 0.4442 | 0.4469 | 0.4551 | 0.4687 | 0.4880 | 0.5129 | 0.5435 | 0.5796 | 0.6209 | 0.666 |
| 21 | 0.4622 | 0.4648 | 0.4729 | 0.4864 | 0.5055 | 0.5300 | 0.5600 | 0.5953 | 0.6355 | 0.679 |
| 22 | 0.4791 | 0.4817 | 0.4897 | 0.5030 | 0.5218 | 0.5459 | 0.5754 | 0.6099 | 0.6491 | 0.692 |
| 23 | 0.4950 | 0.4976 | 0.5055 | 0.5186 | 0.5371 | 0.5608 | 0.5897 | 0.6235 | 0.6617 | 0.703 |
| 24 | 0.5100 | 0.5126 | 0.5203 | 0.5333 | 0.5515 | 0.5748 | 0.6031 | 0.6361 | 0.6734 | 0.714 |
| 25 | 0.5242 | 0.5267 | 0.5344 | 0.5472 | 0.5650 | 0.5879 | 0.6157 | 0.6480 | 0.6843 | 0.724 |
| 26 | 0.5376 | 0.5401 | 0.5476 | 0.5602 | 0.5778 | 0.6003 | 0.6275 | 0.6591 | 0.6946 | 0.733 |
| 27 | 0.5503 | 0.5527 | 0.5602 | 0.5726 | 0.5899 | 0.6120 | 0.6386 | 0.6695 | 0.7042 | 0.741 |
| 28 | 0.5623 | 0.5647 | 0.5721 | 0.5843 | 0.6013 | 0.6230 | 0.6491 | 0.6793 | 0.7132 | 0.749 |
| 29 | 0.5737 | 0.5761 | 0.5833 | 0.5954 | 0.6121 | 0.6334 | 0.6590 | 0.6886 | 0.7217 | 0.757 |
| 30 | 0.5846 | 0.5869 | 0.5941 | 0.6059 | 0.6223 | 0.6432 | 0.6683 | 0.6973 | 0.7296 | 0.764 |
| 31 | 0.5949 | 0.5972 | 0.6042 | 0.6159 | 0.6320 | 0.6525 | 0.6772 | 0.7056 | 0.7372 | 0.771 |
| 32 | 0.6047 | 0.6070 | 0.6139 | 0.6254 | 0.6413 | 0.6614 | 0.6856 | 0.7134 | 0.7443 | 0.777 |
| 33 | 0.6141 | 0.6164 | 0.6232 | 0.6344 | 0.6500 | 0.6698 | 0.6936 | 0.7208 | 0.7511 | 0.783 |
| 34 | 0.6231 | 0.6253 | 0.6320 | 0.6431 | 0.6584 | 0.6779 | 0.7012 | 0.7279 | 0.7575 | 0.789 |
| 35 | 0.6316 | 0.6338 | 0.6404 | 0.6513 | 0.6664 | 0.6855 | 0.7084 | 0.7346 | 0.7636 | 0.794 |
| 36 | 0.6398 | 0.6420 | 0.6485 | 0.6592 | 0.6740 | 0.6928 | 0.7153 | 0.7410 | 0.7694 | 0.799 |
| 37 | 0.6476 | 0.6498 | 0.6562 | 0.6667 | 0.6813 | 0.6998 | 0.7218 | 0.7470 | 0.7749 | 0.804 |
| 38 | 0.6551 | 0.6572 | 0.6635 | 0.6739 | 0.6883 | 0.7064 | 0.7281 | 0.7529 | 0.7802 | 0.805 |
| 39 | 0.6623 | 0.6644 | 0.6706 | 0.6808 | 0.6950 | 0.7128 | 0.7341 | 0.7584 | 0.7852 | 0.813 |
| 40 | 0.6692 | 0.6713 | 0.6774 | 0.6874 | 0.7014 | 0.7189 | 0.7398 | 0.7637 | 0.7900 | 0.818 |
| 41 | 0.6759 | 0.6779 | 0.6839 | 0.6938 | 0.7075 | 0.7248 | 0.7453 | 0.7688 | 0.7946 | 0.822 |
| 42 | 0.6822 | 0.6842 | 0.6901 | 0.6999 | 0.7134 | 0.7304 | 0.7506 | 0.7736 | 0.7990 | 0.826 |
| 43 | 0.6884 | 0.6903 | 0.6962 | 0.7058 | 0.7191 | 0.7358 | 0.7557 | 0.7783 | 0.8033 | 0.829 |
| 44 | 0.6943 | 0.6962 | 0.7019 | 0.7114 | 0.7245 | 0.7410 | 0.7605 | 0.7828 | 0.8072 | 0.833 |
| 45 | 0.7000 | 0.7019 | 0.7075 | 0.7169 | 0.7297 | 0.7459 | 0.7652 | 0.7871 | 0.8111 | 0.836 |
| 46 | 0.7054 | 0.7073 | 0.7129 | 0.7221 | 0.7348 | 0.7507 | 0.7697 | 0.7912 | 0.8148 | 0.839 |
| 47 | 0.7107 | 0.7126 | 0.7181 | 0.7271 | 0.7396 | 0.7554 | 0.7740 | 0.7952 | 0.8184 | 0.843 |
| 48 | 0.7158 | 0.7176 | 0.7231 | 0.7320 | 0.7443 | 0.7598 | 0.7782 | 0.7990 | 0.8218 | 0.846 |
| 49 | 0.7207 | 0.7225 | 0.7279 | 0.7367 | 0.7488 | 0.7641 | 0.7822 | 0.8027 | 0.8251 | 0.849 |
| 50 | 0.7255 | 0.7273 | 0.7325 | 0.7412 | 0.7532 | 0.7682 | 0.7860 | 0.8062 | 0.8283 | 0.851 |
| 51 | 0.7301 | 0.7318 | 0.7370 | 0.7456 | 0.7574 | 0.7722 | 0.7898 | 0.8097 | 0.8314 | 0.854 |
| 52 | 0.7345 | 0.7363 | 0.7414 | 0.7499 | 0.7615 | 0.7761 | 0.7934 | 0.8130 | 0.8344 | 0.857 |
| 53 | 0.7389 | 0.7405 | 0.7456 | 0.7540 | 0.7654 | 0.7798 | 0.7969 | 0.8162 | 0.8373 | 0.859 |
| 54 | 0.7430 | 0.7447 | 0.7497 | 0.7579 | 0.7693 | 0.7835 | 0.8003 | 0.8193 | 0.8400 | 0.862 |
| 55 | 0.7471 | 0.7487 | 0.7536 | 0.7618 | 0.7729 | 0.7870 | 0.8035 | 0.8222 | 0.8427 | 0.864 |
| 56 | 0.7510 | 0.7526 | 0.7575 | 0.7655 | 0.7765 | 0.7903 | 0.8067 | 0.8251 | 0.8453 | 0.866 |
| 57 | 0.7548 | 0.7564 | 0.7612 | 0.7691 | 0.7800 | 0.7936 | 0.8097 | 0.8279 | 0.8478 | 0.868 |
| 58 | 0.7584 | 0.7600 | 0.7648 | 0.7726 | 0.7834 | 0.7968 | 0.8127 | 0.8306 | 0.8502 | 0.870 |
| 59 | 0.7620 | 0.7636 | 0.7683 | 0.7760 | 0.7866 | 0.7999 | 0.8156 | 0.8333 | 0.8526 | 0.872 |
| 60 | 0.7655 | 0.7670 | 0.7717 | 0.7793 | 0.7898 | 0.8029 | 0.8184 | 0.8358 | 0.8548 | 0.874 |
| 61 | 0.7689 | 0.7704 | 0.7750 | 0.7825 | 0.7928 | 0.8058 | 0.8211 | 0.8383 | 0.8571 | 0.876 |
| 62 | 0.7721 | 0.7737 | 0.7782 | 0.7856 | 0.7958 | 0.8086 | 0.8237 | 0.8407 | 0.8592 | 0.878 |

| $\theta$ | 180° | 178° | 176° | 174° | 172° | 170° | 168° | 166° | 164° | 162° | 160° |
|---|---|---|---|---|---|---|---|---|---|---|---|
| $\delta^2$ | 0 | $1.218 \times 10^{-3}$ | $4.874 \times 10^{-3}$ | $1.097 \times 10^{-2}$ | $1.950 \times 10^{-2}$ | $3.046 \times 10^{-2}$ | $4.386 \times 10^{-2}$ | $5.971 \times 10^{-2}$ | $7.798 \times 10^{-2}$ | $9.870 \times 10^{-2}$ | $1.218 \times 10^{-1}$ |

**TABLE III (Continued)**

| ATOMIC MASS $M_2$ (amu) | SCATTERING ANGLE $\theta$ | | | | | | | | | |
|---|---|---|---|---|---|---|---|---|---|---|
| | 180° | 170° | 160° | 150° | 140° | 130° | 120° | 110° | 100° | 90° |
| 63 | 0.7753 | 0.7768 | 0.7813 | 0.7886 | 0.7987 | 0.8113 | 0.8262 | 0.8430 | 0.8613 | 0.8805 |
| 64 | 0.7784 | 0.7799 | 0.7843 | 0.7916 | 0.8015 | 0.8140 | 0.8287 | 0.8453 | 0.8633 | 0.8823 |
| 65 | 0.7814 | 0.7829 | 0.7873 | 0.7944 | 0.8043 | 0.8166 | 0.8311 | 0.8475 | 0.8652 | 0.8840 |
| 66 | 0.7844 | 0.7858 | 0.7901 | 0.7972 | 0.8070 | 0.8191 | 0.8334 | 0.8496 | 0.8672 | 0.8856 |
| 67 | 0.7872 | 0.7887 | 0.7929 | 0.7999 | 0.8095 | 0.8216 | 0.8357 | 0.8517 | 0.8690 | 0.8873 |
| 68 | 0.7900 | 0.7914 | 0.7956 | 0.8026 | 0.8121 | 0.8240 | 0.8379 | 0.8537 | 0.8708 | 0.8888 |
| 69 | 0.7927 | 0.7941 | 0.7983 | 0.8051 | 0.8145 | 0.8263 | 0.8401 | 0.8557 | 0.8726 | 0.8903 |
| 70 | 0.7954 | 0.7967 | 0.8009 | 0.8076 | 0.8169 | 0.8285 | 0.8422 | 0.8576 | 0.8743 | 0.8918 |
| 71 | 0.7979 | 0.7993 | 0.8034 | 0.8101 | 0.8193 | 0.8307 | 0.8442 | 0.8594 | 0.8759 | 0.8933 |
| 72 | 0.8004 | 0.8018 | 0.8058 | 0.8125 | 0.8215 | 0.8329 | 0.8462 | 0.8612 | 0.8775 | 0.8947 |
| 73 | 0.8029 | 0.8042 | 0.8082 | 0.8148 | 0.8238 | 0.8350 | 0.8482 | 0.8630 | 0.8791 | 0.8960 |
| 74 | 0.8053 | 0.8066 | 0.8105 | 0.8170 | 0.8259 | 0.8370 | 0.8501 | 0.8647 | 0.8806 | 0.8974 |
| 75 | 0.8076 | 0.8089 | 0.8128 | 0.8192 | 0.8280 | 0.8390 | 0.8519 | 0.8664 | 0.8821 | 0.8987 |
| 76 | 0.8099 | 0.8112 | 0.8151 | 0.8214 | 0.8301 | 0.8410 | 0.8537 | 0.8681 | 0.8836 | 0.8999 |
| 77 | 0.8121 | 0.8134 | 0.8172 | 0.8235 | 0.8321 | 0.8429 | 0.8555 | 0.8697 | 0.8850 | 0.9012 |
| 78 | 0.8143 | 0.8156 | 0.8193 | 0.8256 | 0.8341 | 0.8447 | 0.8572 | 0.8712 | 0.8864 | 0.9024 |
| 79 | 0.8164 | 0.8177 | 0.8214 | 0.8276 | 0.8360 | 0.8465 | 0.8589 | 0.8727 | 0.8878 | 0.9036 |
| 80 | 0.8185 | 0.8197 | 0.8234 | 0.8295 | 0.8379 | 0.8483 | 0.8605 | 0.8742 | 0.8891 | 0.9047 |
| 81 | 0.8205 | 0.8217 | 0.8254 | 0.8315 | 0.8397 | 0.8500 | 0.8621 | 0.8757 | 0.8904 | 0.9058 |
| 82 | 0.8225 | 0.8237 | 0.8274 | 0.8333 | 0.8415 | 0.8517 | 0.8637 | 0.8771 | 0.8917 | 0.9069 |
| 83 | 0.8244 | 0.8257 | 0.8293 | 0.8352 | 0.8433 | 0.8534 | 0.8652 | 0.8785 | 0.8929 | 0.9080 |
| 84 | 0.8263 | 0.8275 | 0.8311 | 0.8370 | 0.8450 | 0.8550 | 0.8667 | 0.8799 | 0.8941 | 0.9090 |
| 85 | 0.8282 | 0.8294 | 0.8329 | 0.8387 | 0.8467 | 0.8565 | 0.8682 | 0.8812 | 0.8953 | 0.9101 |
| 86 | 0.8300 | 0.8312 | 0.8347 | 0.8404 | 0.8483 | 0.8581 | 0.8696 | 0.8825 | 0.8964 | 0.9111 |
| 87 | 0.8318 | 0.8330 | 0.8364 | 0.8421 | 0.8499 | 0.8596 | 0.8710 | 0.8837 | 0.8976 | 0.9120 |
| 88 | 0.8335 | 0.8347 | 0.8381 | 0.8438 | 0.8515 | 0.8611 | 0.8724 | 0.8850 | 0.8987 | 0.9130 |
| 89 | 0.8353 | 0.8364 | 0.8398 | 0.8454 | 0.8530 | 0.8625 | 0.8737 | 0.8862 | 0.8997 | 0.9139 |
| 90 | 0.8369 | 0.8381 | 0.8414 | 0.8470 | 0.8545 | 0.8640 | 0.8750 | 0.8874 | 0.9008 | 0.9148 |
| 91 | 0.8386 | 0.8397 | 0.8430 | 0.8485 | 0.8560 | 0.8654 | 0.8763 | 0.8886 | 0.9018 | 0.9157 |
| 92 | 0.8402 | 0.8413 | 0.8446 | 0.8500 | 0.8575 | 0.8667 | 0.8776 | 0.8897 | 0.9029 | 0.9166 |
| 93 | 0.8418 | 0.8429 | 0.8461 | 0.8515 | 0.8589 | 0.8681 | 0.8788 | 0.8908 | 0.9039 | 0.9175 |
| 94 | 0.8433 | 0.8444 | 0.8476 | 0.8530 | 0.8603 | 0.8694 | 0.8800 | 0.8919 | 0.9048 | 0.9183 |
| 95 | 0.8448 | 0.8459 | 0.8491 | 0.8544 | 0.8616 | 0.8706 | 0.8812 | 0.8930 | 0.9058 | 0.9191 |
| 96 | 0.8463 | 0.8474 | 0.8506 | 0.8558 | 0.8630 | 0.8719 | 0.8824 | 0.8941 | 0.9067 | 0.9200 |
| 97 | 0.8478 | 0.8488 | 0.8520 | 0.8572 | 0.8643 | 0.8731 | 0.8835 | 0.8951 | 0.9076 | 0.9207 |
| 98 | 0.8492 | 0.8503 | 0.8534 | 0.8585 | 0.8656 | 0.8743 | 0.8846 | 0.8961 | 0.9085 | 0.9215 |
| 99 | 0.8506 | 0.8516 | 0.8548 | 0.8599 | 0.8669 | 0.8755 | 0.8857 | 0.8971 | 0.9094 | 0.9223 |
| 100 | 0.8520 | 0.8530 | 0.8561 | 0.8612 | 0.8681 | 0.8767 | 0.8868 | 0.8981 | 0.9103 | 0.9230 |
| 101 | 0.8533 | 0.8544 | 0.8574 | 0.8624 | 0.8693 | 0.8778 | 0.8878 | 0.8990 | 0.9111 | 0.9238 |
| 102 | 0.8547 | 0.8557 | 0.8587 | 0.8637 | 0.8705 | 0.8790 | 0.8889 | 0.9000 | 0.9120 | 0.9245 |
| 103 | 0.8560 | 0.8570 | 0.8600 | 0.8649 | 0.8717 | 0.8801 | 0.8899 | 0.9009 | 0.9128 | 0.9252 |
| 104 | 0.8573 | 0.8583 | 0.8612 | 0.8661 | 0.8728 | 0.8812 | 0.8909 | 0.9018 | 0.9136 | 0.9259 |
| 105 | 0.8585 | 0.8595 | 0.8625 | 0.8673 | 0.8740 | 0.8822 | 0.8919 | 0.9027 | 0.9144 | 0.9266 |
| 106 | 0.8598 | 0.8607 | 0.8637 | 0.8685 | 0.8751 | 0.8833 | 0.8928 | 0.9036 | 0.9151 | 0.9272 |
| 107 | 0.8610 | 0.8619 | 0.8649 | 0.8696 | 0.8762 | 0.8843 | 0.8938 | 0.9044 | 0.9159 | 0.9279 |
| 108 | 0.8622 | 0.8631 | 0.8660 | 0.8708 | 0.8772 | 0.8853 | 0.8947 | 0.9053 | 0.9166 | 0.9285 |
| 109 | 0.8633 | 0.8643 | 0.8672 | 0.8719 | 0.8783 | 0.8863 | 0.8956 | 0.9061 | 0.9174 | 0.9292 |
| 110 | 0.8645 | 0.8654 | 0.8683 | 0.8730 | 0.8793 | 0.8873 | 0.8965 | 0.9069 | 0.9181 | 0.9298 |
| 111 | 0.8656 | 0.8666 | 0.8694 | 0.8740 | 0.8804 | 0.8882 | 0.8974 | 0.9077 | 0.9188 | 0.9304 |
| 112 | 0.8667 | 0.8677 | 0.8705 | 0.8751 | 0.8814 | 0.8892 | 0.8983 | 0.9085 | 0.9195 | 0.9310 |
| 113 | 0.8678 | 0.8688 | 0.8716 | 0.8761 | 0.8823 | 0.8901 | 0.8991 | 0.9093 | 0.9202 | 0.9316 |
| 114 | 0.8689 | 0.8699 | 0.8726 | 0.8771 | 0.8833 | 0.8910 | 0.9000 | 0.9100 | 0.9209 | 0.9322 |
| 115 | 0.8700 | 0.8709 | 0.8736 | 0.8781 | 0.8843 | 0.8919 | 0.9008 | 0.9108 | 0.9215 | 0.9327 |
| 116 | 0.8710 | 0.8719 | 0.8747 | 0.8791 | 0.8852 | 0.8928 | 0.9016 | 0.9115 | 0.9222 | 0.9333 |
| 117 | 0.8721 | 0.8730 | 0.8757 | 0.8801 | 0.8861 | 0.8936 | 0.9024 | 0.9122 | 0.9228 | 0.9338 |
| 118 | 0.8731 | 0.8740 | 0.8767 | 0.8810 | 0.8870 | 0.8945 | 0.9032 | 0.9129 | 0.9234 | 0.9344 |
| 119 | 0.8741 | 0.8750 | 0.8776 | 0.8820 | 0.8879 | 0.8953 | 0.9040 | 0.9136 | 0.9241 | 0.9349 |

(Continued)

| $\theta$ | 180° | 178° | 176° | 174° | 172° | 170° | 168° | 166° | 164° | 162° | 160° |
|---|---|---|---|---|---|---|---|---|---|---|---|
| $\delta^2$ | 0 | $1.218 \times 10^{-3}$ | $4.874 \times 10^{-3}$ | $1.097 \times 10^{-2}$ | $1.950 \times 10^{-2}$ | $3.046 \times 10^{-2}$ | $4.386 \times 10^{-2}$ | $5.971 \times 10^{-2}$ | $7.798 \times 10^{-2}$ | $9.870 \times 10^{-2}$ | $1.218 \times 10^{-1}$ |

**TABLE III (Continued)**

| ATOMIC MASS $M_2$ (amu) | 180° | 170° | 160° | 150° | 140° | 130° | 120° | 110° | 100° | 90° |
|---|---|---|---|---|---|---|---|---|---|---|
| 120 | 0.8751 | 0.8759 | 0.8786 | 0.8829 | 0.8888 | 0.8962 | 0.9047 | 0.9143 | 0.9247 | 0.935 |
| 121 | 0.8760 | 0.8769 | 0.8795 | 0.8838 | 0.8897 | 0.8970 | 0.9055 | 0.9150 | 0.9253 | 0.936 |
| 122 | 0.8770 | 0.8778 | 0.8804 | 0.8847 | 0.8905 | 0.8978 | 0.9062 | 0.9157 | 0.9259 | 0.936 |
| 123 | 0.8779 | 0.8788 | 0.8814 | 0.8856 | 0.8914 | 0.8986 | 0.9070 | 0.9163 | 0.9264 | 0.937 |
| 124 | 0.8788 | 0.8797 | 0.8823 | 0.8865 | 0.8922 | 0.8993 | 0.9077 | 0.9170 | 0.9270 | 0.937 |
| 125 | 0.8797 | 0.8806 | 0.8831 | 0.8873 | 0.8930 | 0.9001 | 0.9084 | 0.9176 | 0.9276 | 0.937 |
| 126 | 0.8806 | 0.8815 | 0.8840 | 0.8882 | 0.8938 | 0.9009 | 0.9091 | 0.9182 | 0.9281 | 0.938 |
| 127 | 0.8815 | 0.8824 | 0.8849 | 0.8890 | 0.8946 | 0.9016 | 0.9098 | 0.9189 | 0.9287 | 0.938 |
| 128 | 0.8824 | 0.8832 | 0.8857 | 0.8858 | 0.8954 | 0.9023 | 0.9104 | 0.9195 | 0.9292 | 0.939 |
| 129 | 0.8832 | 0.8841 | 0.8866 | 0.8906 | 0.8962 | 0.9030 | 0.9111 | 0.9201 | 0.9297 | 0.939 |
| 130 | 0.8841 | 0.8849 | 0.8874 | 0.8914 | 0.8969 | 0.9038 | 0.9117 | 0.9207 | 0.9303 | 0.940 |
| 131 | 0.8849 | 0.8857 | 0.8882 | 0.8922 | 0.8977 | 0.9045 | 0.9124 | 0.9212 | 0.9308 | 0.940 |
| 132 | 0.8857 | 0.8866 | 0.8890 | 0.8930 | 0.8984 | 0.9051 | 0.9130 | 0.9218 | 0.9313 | 0.941 |
| 133 | 0.8866 | 0.8874 | 0.8898 | 0.8937 | 0.8991 | 0.9058 | 0.9136 | 0.9224 | 0.9318 | 0.941 |
| 134 | 0.8873 | 0.8882 | 0.8906 | 0.8945 | 0.8998 | 0.9065 | 0.9143 | 0.9229 | 0.9323 | 0.942 |
| 135 | 0.8881 | 0.8889 | 0.8913 | 0.8952 | 0.9005 | 0.9072 | 0.9149 | 0.9235 | 0.9328 | 0.942 |
| 136 | 0.8889 | 0.8897 | 0.8921 | 0.8959 | 0.9012 | 0.9078 | 0.9155 | 0.9240 | 0.9332 | 0.942 |
| 137 | 0.8897 | 0.8905 | 0.8928 | 0.8967 | 0.9019 | 0.9084 | 0.9161 | 0.9246 | 0.9337 | 0.943 |
| 138 | 0.8904 | 0.8912 | 0.8936 | 0.8974 | 0.9026 | 0.9091 | 0.9166 | 0.9251 | 0.9342 | 0.943 |
| 139 | 0.8912 | 0.8920 | 0.8943 | 0.8981 | 0.9033 | 0.9097 | 0.9172 | 0.9256 | 0.9346 | 0.944 |
| 140 | 0.8919 | 0.8927 | 0.8950 | 0.8988 | 0.9039 | 0.9103 | 0.9178 | 0.9261 | 0.9351 | 0.944 |
| 141 | 0.8926 | 0.8934 | 0.8957 | 0.8994 | 0.9046 | 0.9109 | 0.9183 | 0.9266 | 0.9355 | 0.944 |
| 142 | 0.8933 | 0.8941 | 0.8964 | 0.9001 | 0.9052 | 0.9115 | 0.9189 | 0.9271 | 0.9360 | 0.945 |
| 143 | 0.8941 | 0.8948 | 0.8971 | 0.9008 | 0.9058 | 0.9121 | 0.9194 | 0.9276 | 0.9364 | 0.945 |
| 144 | 0.8947 | 0.8955 | 0.8978 | 0.9014 | 0.9065 | 0.9127 | 0.9200 | 0.9281 | 0.9368 | 0.945 |
| 145 | 0.8954 | 0.8962 | 0.8984 | 0.9021 | 0.9071 | 0.9133 | 0.9205 | 0.9286 | 0.9372 | 0.946 |
| 146 | 0.8961 | 0.8969 | 0.8991 | 0.9027 | 0.9077 | 0.9138 | 0.9210 | 0.9290 | 0.9377 | 0.946 |
| 147 | 0.8968 | 0.8975 | 0.8997 | 0.9034 | 0.9083 | 0.9144 | 0.9215 | 0.9295 | 0.9381 | 0.947 |
| 148 | 0.8974 | 0.8982 | 0.9004 | 0.9040 | 0.9089 | 0.9150 | 0.9220 | 0.9300 | 0.9385 | 0.947 |
| 149 | 0.8981 | 0.8988 | 0.9010 | 0.9046 | 0.9095 | 0.9155 | 0.9226 | 0.9304 | 0.9389 | 0.947 |
| 150 | 0.8987 | 0.8995 | 0.9016 | 0.9052 | 0.9100 | 0.9160 | 0.9230 | 0.9309 | 0.9393 | 0.948 |
| 151 | 0.8994 | 0.9001 | 0.9023 | 0.9058 | 0.9106 | 0.9166 | 0.9235 | 0.9313 | 0.9397 | 0.948 |
| 152 | 0.9000 | 0.9007 | 0.9029 | 0.9064 | 0.9112 | 0.9171 | 0.9240 | 0.9317 | 0.9400 | 0.948 |
| 153 | 0.9006 | 0.9013 | 0.9035 | 0.9070 | 0.9117 | 0.9176 | 0.9245 | 0.9322 | 0.9404 | 0.949 |
| 154 | 0.9012 | 0.9019 | 0.9041 | 0.9075 | 0.9123 | 0.9181 | 0.9250 | 0.9326 | 0.9408 | 0.949 |
| 155 | 0.9018 | 0.9025 | 0.9047 | 0.9081 | 0.9128 | 0.9186 | 0.9254 | 0.9330 | 0.9412 | 0.949 |
| 156 | 0.9024 | 0.9031 | 0.9052 | 0.9087 | 0.9133 | 0.9191 | 0.9259 | 0.9334 | 0.9415 | 0.950 |
| 157 | 0.9030 | 0.9037 | 0.9058 | 0.9092 | 0.9139 | 0.9196 | 0.9264 | 0.9338 | 0.9419 | 0.950 |
| 158 | 0.9036 | 0.9043 | 0.9064 | 0.9098 | 0.9144 | 0.9201 | 0.9268 | 0.9343 | 0.9423 | 0.950 |
| 159 | 0.9042 | 0.9049 | 0.9069 | 0.9103 | 0.9149 | 0.9206 | 0.9272 | 0.9347 | 0.9426 | 0.950 |
| 160 | 0.9048 | 0.9054 | 0.9075 | 0.9108 | 0.9154 | 0.9211 | 0.9277 | 0.9350 | 0.9430 | 0.951 |
| 161 | 0.9053 | 0.9060 | 0.9080 | 0.9114 | 0.9159 | 0.9215 | 0.9281 | 0.9354 | 0.9433 | 0.951 |
| 162 | 0.9059 | 0.9066 | 0.9086 | 0.9119 | 0.9164 | 0.9220 | 0.9285 | 0.9358 | 0.9436 | 0.951 |
| 163 | 0.9064 | 0.9071 | 0.9091 | 0.9124 | 0.9169 | 0.9225 | 0.9290 | 0.9362 | 0.9440 | 0.952 |
| 164 | 0.9070 | 0.9076 | 0.9096 | 0.9129 | 0.9174 | 0.9229 | 0.9294 | 0.9366 | 0.9443 | 0.952 |
| 165 | 0.9075 | 0.9082 | 0.9102 | 0.9134 | 0.9179 | 0.9234 | 0.9298 | 0.9370 | 0.9446 | 0.952 |
| 166 | 0.9080 | 0.9087 | 0.9107 | 0.9139 | 0.9183 | 0.9238 | 0.9302 | 0.9373 | 0.9450 | 0.952 |
| 167 | 0.9086 | 0.9092 | 0.9112 | 0.9144 | 0.9188 | 0.9243 | 0.9306 | 0.9377 | 0.9453 | 0.953 |
| 168 | 0.9091 | 0.9097 | 0.9117 | 0.9149 | 0.9193 | 0.9247 | 0.9310 | 0.9380 | 0.9456 | 0.953 |
| 169 | 0.9096 | 0.9103 | 0.9122 | 0.9154 | 0.9197 | 0.9251 | 0.9314 | 0.9384 | 0.9459 | 0.953 |
| 170 | 0.9101 | 0.9108 | 0.9127 | 0.9159 | 0.9202 | 0.9255 | 0.9318 | 0.9387 | 0.9462 | 0.954 |
| 171 | 0.9106 | 0.9113 | 0.9132 | 0.9163 | 0.9206 | 0.9260 | 0.9322 | 0.9391 | 0.9465 | 0.954 |
| 172 | 0.9111 | 0.9117 | 0.9137 | 0.9168 | 0.9211 | 0.9264 | 0.9326 | 0.9394 | 0.9468 | 0.954 |
| 173 | 0.9116 | 0.9122 | 0.9141 | 0.9173 | 0.9215 | 0.9268 | 0.9329 | 0.9398 | 0.9471 | 0.954 |
| 174 | 0.9121 | 0.9127 | 0.9146 | 0.9177 | 0.9219 | 0.9272 | 0.9333 | 0.9401 | 0.9474 | 0.955 |
| 175 | 0.9126 | 0.9132 | 0.9151 | 0.9182 | 0.9224 | 0.9276 | 0.9337 | 0.9404 | 0.9477 | 0.955 |
| 176 | 0.9130 | 0.9137 | 0.9155 | 0.9186 | 0.9228 | 0.9280 | 0.9340 | 0.9408 | 0.9480 | 0.955 |

| $\theta$ | 180° | 178° | 176° | 174° | 172° | 170° | 168° | 166° | 164° | 162° | 160° |
|---|---|---|---|---|---|---|---|---|---|---|---|
| $\delta^2$ | 0 | 1.218 $\times 10^{-3}$ | 4.874 $\times 10^{-3}$ | 1.097 $\times 10^{-2}$ | 1.950 $\times 10^{-2}$ | 3.046 $\times 10^{-2}$ | 4.386 $\times 10^{-2}$ | 5.971 $\times 10^{-2}$ | 7.798 $\times 10^{-2}$ | 9.870 $\times 10^{-2}$ | 1.218 $\times 10^{-1}$ |

**TABLE III (Continued)**

| ATOMIC MASS $M_2$ (amu) | SCATTERING ANGLE $\theta$ | | | | | | | | | |
|---|---|---|---|---|---|---|---|---|---|---|
| | 180° | 170° | 160° | 150° | 140° | 130° | 120° | 110° | 100° | 90° |
| 177 | 0.9135 | 0.9141 | 0.9160 | 0.9191 | 0.9232 | 0.9284 | 0.9344 | 0.9411 | 0.9483 | 0.9558 |
| 178 | 0.9140 | 0.9146 | 0.9164 | 0.9195 | 0.9236 | 0.9288 | 0.9348 | 0.9414 | 0.9486 | 0.9560 |
| 179 | 0.9144 | 0.9150 | 0.9169 | 0.9199 | 0.9240 | 0.9292 | 0.9351 | 0.9417 | 0.9489 | 0.9563 |
| 180 | 0.9149 | 0.9155 | 0.9173 | 0.9203 | 0.9245 | 0.9295 | 0.9355 | 0.9421 | 0.9491 | 0.9565 |
| 181 | 0.9153 | 0.9159 | 0.9178 | 0.9208 | 0.9249 | 0.9299 | 0.9358 | 0.9424 | 0.9494 | 0.9567 |
| 182 | 0.9158 | 0.9164 | 0.9182 | 0.9212 | 0.9252 | 0.9303 | 0.9361 | 0.9427 | 0.9497 | 0.9570 |
| 183 | 0.9162 | 0.9168 | 0.9186 | 0.9216 | 0.9256 | 0.9306 | 0.9365 | 0.9430 | 0.9499 | 0.9572 |
| 184 | 0.9167 | 0.9173 | 0.9191 | 0.9220 | 0.9260 | 0.9310 | 0.9368 | 0.9433 | 0.9502 | 0.9574 |
| 185 | 0.9171 | 0.9177 | 0.9195 | 0.9224 | 0.9264 | 0.9314 | 0.9371 | 0.9436 | 0.9505 | 0.9576 |
| 186 | 0.9175 | 0.9181 | 0.9199 | 0.9228 | 0.9268 | 0.9317 | 0.9375 | 0.9439 | 0.9507 | 0.9579 |
| 187 | 0.9179 | 0.9185 | 0.9203 | 0.9232 | 0.9272 | 0.9321 | 0.9378 | 0.9442 | 0.9510 | 0.9581 |
| 188 | 0.9184 | 0.9189 | 0.9207 | 0.9236 | 0.9275 | 0.9324 | 0.9381 | 0.9444 | 0.9512 | 0.9583 |
| 189 | 0.9188 | 0.9194 | 0.9211 | 0.9240 | 0.9279 | 0.9328 | 0.9384 | 0.9447 | 0.9515 | 0.9585 |
| 190 | 0.9192 | 0.9198 | 0.9215 | 0.9244 | 0.9283 | 0.9331 | 0.9387 | 0.9450 | 0.9517 | 0.9587 |
| 191 | 0.9196 | 0.9202 | 0.9219 | 0.9248 | 0.9286 | 0.9335 | 0.9391 | 0.9453 | 0.9520 | 0.9589 |
| 192 | 0.9200 | 0.9206 | 0.9223 | 0.9251 | 0.9290 | 0.9338 | 0.9394 | 0.9456 | 0.9522 | 0.9592 |
| 193 | 0.9204 | 0.9210 | 0.9227 | 0.9255 | 0.9294 | 0.9341 | 0.9397 | 0.9458 | 0.9525 | 0.9594 |
| 194 | 0.9208 | 0.9214 | 0.9231 | 0.9259 | 0.9297 | 0.9344 | 0.9400 | 0.9461 | 0.9527 | 0.9596 |
| 195 | 0.9212 | 0.9217 | 0.9234 | 0.9262 | 0.9301 | 0.9348 | 0.9403 | 0.9464 | 0.9530 | 0.9598 |
| 196 | 0.9216 | 0.9221 | 0.9238 | 0.9266 | 0.9304 | 0.9351 | 0.9406 | 0.9467 | 0.9532 | 0.9600 |
| 197 | 0.9219 | 0.9225 | 0.9242 | 0.9270 | 0.9307 | 0.9354 | 0.9409 | 0.9469 | 0.9534 | 0.9602 |
| 198 | 0.9223 | 0.9229 | 0.9246 | 0.9273 | 0.9311 | 0.9357 | 0.9411 | 0.9472 | 0.9537 | 0.9604 |
| 199 | 0.9227 | 0.9233 | 0.9249 | 0.9277 | 0.9314 | 0.9360 | 0.9414 | 0.9474 | 0.9539 | 0.9606 |
| 200 | 0.9231 | 0.9236 | 0.9253 | 0.9280 | 0.9317 | 0.9364 | 0.9417 | 0.9477 | 0.9541 | 0.9608 |
| 201 | 0.9234 | 0.9240 | 0.9256 | 0.9284 | 0.9321 | 0.9367 | 0.9420 | 0.9479 | 0.9543 | 0.9610 |
| 202 | 0.9238 | 0.9243 | 0.9260 | 0.9287 | 0.9324 | 0.9370 | 0.9423 | 0.9482 | 0.9545 | 0.9611 |
| 203 | 0.9242 | 0.9247 | 0.9264 | 0.9290 | 0.9327 | 0.9373 | 0.9426 | 0.9484 | 0.9548 | 0.9613 |
| 204 | 0.9245 | 0.9251 | 0.9267 | 0.9294 | 0.9330 | 0.9376 | 0.9428 | 0.9487 | 0.9550 | 0.9615 |
| 205 | 0.9249 | 0.9254 | 0.9270 | 0.9297 | 0.9334 | 0.9379 | 0.9431 | 0.9489 | 0.9552 | 0.9617 |
| 206 | 0.9252 | 0.9258 | 0.9274 | 0.9300 | 0.9337 | 0.9381 | 0.9434 | 0.9492 | 0.9554 | 0.9619 |
| 207 | 0.9256 | 0.9261 | 0.9277 | 0.9304 | 0.9340 | 0.9384 | 0.9436 | 0.9494 | 0.9556 | 0.9621 |
| 208 | 0.9259 | 0.9264 | 0.9281 | 0.9307 | 0.9343 | 0.9387 | 0.9439 | 0.9497 | 0.9558 | 0.9622 |
| 209 | 0.9262 | 0.9268 | 0.9284 | 0.9310 | 0.9346 | 0.9390 | 0.9442 | 0.9499 | 0.9560 | 0.9624 |
| 210 | 0.9266 | 0.9271 | 0.9287 | 0.9313 | 0.9349 | 0.9393 | 0.9444 | 0.9501 | 0.9562 | 0.9626 |
| 211 | 0.9269 | 0.9275 | 0.9290 | 0.9316 | 0.9352 | 0.9396 | 0.9447 | 0.9504 | 0.9564 | 0.9628 |
| 212 | 0.9273 | 0.9278 | 0.9294 | 0.9320 | 0.9355 | 0.9398 | 0.9449 | 0.9506 | 0.9566 | 0.9629 |
| 213 | 0.9276 | 0.9281 | 0.9297 | 0.9323 | 0.9358 | 0.9401 | 0.9452 | 0.9508 | 0.9568 | 0.9631 |
| 214 | 0.9279 | 0.9284 | 0.9300 | 0.9326 | 0.9361 | 0.9404 | 0.9454 | 0.9510 | 0.9570 | 0.9633 |
| 215 | 0.9282 | 0.9288 | 0.9303 | 0.9329 | 0.9364 | 0.9407 | 0.9457 | 0.9513 | 0.9572 | 0.9634 |
| 216 | 0.9286 | 0.9291 | 0.9306 | 0.9332 | 0.9366 | 0.9409 | 0.9459 | 0.9515 | 0.9574 | 0.9636 |

| $\theta$ | 180° | 178° | 176° | 174° | 172° | 170° | 168° | 166° | 164° | 162° | 160° |
|---|---|---|---|---|---|---|---|---|---|---|---|
| $\delta^2$ | 0 | 1.218 $\times 10^{-3}$ | 4.874 $\times 10^{-3}$ | 1.097 $\times 10^{-2}$ | 1.950 $\times 10^{-2}$ | 3.046 $\times 10^{-2}$ | 4.386 $\times 10^{-2}$ | 5.971 $\times 10^{-2}$ | 7.798 $\times 10^{-2}$ | 9.870 $\times 10^{-2}$ | 1.218 $\times 10^{-1}$ |

**TABLE IV**

$\bar{K}$ for $^1$H as Projectile

| ELEMENT | AT. NO. ($Z_2$) | SCATTERING ANGLE $\theta$ | | | | | | | | | |
|---|---|---|---|---|---|---|---|---|---|---|---|
| | | 90° | 100° | 110° | 120° | 130° | 140° | 150° | 160° | 170° | 180° |
| HE | 2 | 0.5977 | 0.5461 | 0.5004 | 0.4611 | 0.4284 | 0.4023 | 0.3823 | 0.3683 | 0.3600 | 0.3573 |
| LI | 3 | 0.7461 | 0.7090 | 0.6748 | 0.6442 | 0.6179 | 0.5960 | 0.5789 | 0.5666 | 0.5592 | 0.5567 |
| BE | 4 | 0.7988 | 0.7682 | 0.7397 | 0.7139 | 0.6913 | 0.6725 | 0.6576 | 0.6468 | 0.6403 | 0.6381 |
| B | 5 | 0.8293 | 0.8027 | 0.7778 | 0.7551 | 0.7352 | 0.7185 | 0.7052 | 0.6955 | 0.6897 | 0.6878 |
| C | 6 | 0.8452 | 0.8208 | 0.7979 | 0.7769 | 0.7585 | 0.7429 | 0.7306 | 0.7216 | 0.7161 | 0.7143 |
| N | 7 | 0.8658 | 0.8443 | 0.8241 | 0.8055 | 0.7891 | 0.7752 | 0.7641 | 0.7561 | 0.7512 | 0.7495 |
| O | 8 | 0.8815 | 0.8624 | 0.8442 | 0.8276 | 0.8128 | 0.8003 | 0.7902 | 0.7829 | 0.7785 | 0.7770 |
| F | 9 | 0.8992 | 0.8828 | 0.8672 | 0.8527 | 0.8399 | 0.8290 | 0.8202 | 0.8138 | 0.8100 | 0.8086 |
| NE | 10 | 0.9048 | 0.8892 | 0.8743 | 0.8606 | 0.8484 | 0.8380 | 0.8296 | 0.8236 | 0.8198 | 0.8186 |
| NA | 11 | 0.9160 | 0.9022 | 0.8889 | 0.8767 | 0.8658 | 0.8565 | 0.8490 | 0.8435 | 0.8402 | 0.8391 |
| MG | 12 | 0.9203 | 0.9072 | 0.8946 | 0.8829 | 0.8725 | 0.8636 | 0.8565 | 0.8513 | 0.8481 | 0.8470 |
| AL | 13 | 0.9280 | 0.9160 | 0.9046 | 0.8939 | 0.8845 | 0.8763 | 0.8698 | 0.8650 | 0.8621 | 0.8612 |
| SI | 14 | 0.9307 | 0.9192 | 0.9081 | 0.8979 | 0.8887 | 0.8809 | 0.8746 | 0.8700 | 0.8672 | 0.8662 |
| P | 15 | 0.9370 | 0.9264 | 0.9163 | 0.9070 | 0.8986 | 0.8914 | 0.8856 | 0.8814 | 0.8788 | 0.8779 |
| S | 16 | 0.9390 | 0.9287 | 0.9190 | 0.9099 | 0.9017 | 0.8948 | 0.8892 | 0.8851 | 0.8826 | 0.8817 |
| CL | 17 | 0.9447 | 0.9354 | 0.9265 | 0.9182 | 0.9108 | 0.9044 | 0.8993 | 0.8955 | 0.8932 | 0.8924 |
| AR | 18 | 0.9508 | 0.9425 | 0.9345 | 0.9271 | 0.9204 | 0.9147 | 0.9101 | 0.9067 | 0.9047 | 0.9040 |
| K | 19 | 0.9496 | 0.9412 | 0.9330 | 0.9255 | 0.9187 | 0.9129 | 0.9082 | 0.9047 | 0.9026 | 0.9019 |
| CA | 20 | 0.9510 | 0.9427 | 0.9348 | 0.9274 | 0.9208 | 0.9151 | 0.9105 | 0.9071 | 0.9050 | 0.9044 |
| SC | 21 | 0.9561 | 0.9487 | 0.9416 | 0.9349 | 0.9290 | 0.9239 | 0.9197 | 0.9167 | 0.9148 | 0.9142 |
| TI | 22 | 0.9588 | 0.9518 | 0.9450 | 0.9388 | 0.9331 | 0.9283 | 0.9244 | 0.9216 | 0.9198 | 0.9192 |
| V | 23 | 0.9612 | 0.9546 | 0.9483 | 0.9424 | 0.9371 | 0.9325 | 0.9288 | 0.9261 | 0.9245 | 0.9239 |
| CR | 24 | 0.9620 | 0.9555 | 0.9493 | 0.9435 | 0.9383 | 0.9338 | 0.9302 | 0.9275 | 0.9259 | 0.9254 |
| MN | 25 | 0.9640 | 0.9578 | 0.9519 | 0.9464 | 0.9415 | 0.9373 | 0.9338 | 0.9313 | 0.9298 | 0.9292 |
| FE | 26 | 0.9645 | 0.9585 | 0.9527 | 0.9473 | 0.9424 | 0.9382 | 0.9349 | 0.9324 | 0.9309 | 0.9303 |
| CO | 27 | 0.9664 | 0.9606 | 0.9551 | 0.9500 | 0.9454 | 0.9414 | 0.9382 | 0.9358 | 0.9344 | 0.9339 |
| NI | 28 | 0.9666 | 0.9609 | 0.9553 | 0.9502 | 0.9455 | 0.9415 | 0.9383 | 0.9359 | 0.9345 | 0.9340 |
| CU | 29 | 0.9688 | 0.9634 | 0.9583 | 0.9535 | 0.9492 | 0.9455 | 0.9425 | 0.9403 | 0.9390 | 0.9385 |
| ZN | 30 | 0.9696 | 0.9644 | 0.9594 | 0.9548 | 0.9506 | 0.9470 | 0.9441 | 0.9419 | 0.9406 | 0.9402 |
| GA | 31 | 0.9715 | 0.9666 | 0.9619 | 0.9575 | 0.9536 | 0.9502 | 0.9475 | 0.9454 | 0.9442 | 0.9438 |
| GE | 32 | 0.9727 | 0.9680 | 0.9635 | 0.9593 | 0.9555 | 0.9522 | 0.9496 | 0.9477 | 0.9465 | 0.9461 |
| AS | 33 | 0.9735 | 0.9689 | 0.9645 | 0.9604 | 0.9568 | 0.9536 | 0.9510 | 0.9492 | 0.9480 | 0.9476 |
| SE | 34 | 0.9748 | 0.9705 | 0.9663 | 0.9624 | 0.9589 | 0.9559 | 0.9535 | 0.9517 | 0.9506 | 0.9502 |
| BR | 35 | 0.9751 | 0.9708 | 0.9667 | 0.9629 | 0.9594 | 0.9564 | 0.9540 | 0.9522 | 0.9512 | 0.9508 |
| KR | 36 | 0.9762 | 0.9722 | 0.9682 | 0.9646 | 0.9612 | 0.9584 | 0.9561 | 0.9544 | 0.9534 | 0.9530 |
| RB | 37 | 0.9767 | 0.9727 | 0.9688 | 0.9652 | 0.9620 | 0.9592 | 0.9569 | 0.9553 | 0.9543 | 0.9539 |
| SR | 38 | 0.9773 | 0.9734 | 0.9696 | 0.9661 | 0.9629 | 0.9602 | 0.9580 | 0.9564 | 0.9554 | 0.9550 |
| Y | 39 | 0.9776 | 0.9737 | 0.9700 | 0.9666 | 0.9634 | 0.9607 | 0.9586 | 0.9570 | 0.9560 | 0.9557 |
| ZR | 40 | 0.9781 | 0.9744 | 0.9708 | 0.9674 | 0.9643 | 0.9617 | 0.9596 | 0.9580 | 0.9571 | 0.9568 |
| NB | 41 | 0.9785 | 0.9749 | 0.9713 | 0.9680 | 0.9650 | 0.9624 | 0.9603 | 0.9588 | 0.9579 | 0.9575 |

| | 180° | 178° | 176° | 174° | 172° | 170° | 168° | 166° | 164° | 162° | 160° |
|---|---|---|---|---|---|---|---|---|---|---|---|
| MO 42 | 0.9586 | 0.9588 | 0.9591 | 0.9600 | 0.9615 | 0.9635 | 0.9660 | 0.9689 | 0.9722 | 0.9756 | 0.9792 |
| TC 43 | 0.0 | 0.0 | 0.0 | 0.0 | 0.0 | 0.0 | 0.0 | 0.0 | 0.0 | 0.0 | 0.0 |
| RU 44 | 0.9605 | 0.9607 | 0.9610 | 0.9618 | 0.9633 | 0.9652 | 0.9676 | 0.9703 | 0.9734 | 0.9767 | 0.9800 |
| RH 45 | 0.9614 | 0.9616 | 0.9619 | 0.9627 | 0.9641 | 0.9660 | 0.9683 | 0.9710 | 0.9741 | 0.9773 | 0.9806 |
| PD 46 | 0.9627 | 0.9629 | 0.9632 | 0.9640 | 0.9654 | 0.9672 | 0.9695 | 0.9721 | 0.9750 | 0.9781 | 0.9813 |
| AG 47 | 0.9631 | 0.9633 | 0.9636 | 0.9644 | 0.9657 | 0.9675 | 0.9698 | 0.9724 | 0.9752 | 0.9783 | 0.9815 |
| CD 48 | 0.9647 | 0.9649 | 0.9651 | 0.9659 | 0.9672 | 0.9689 | 0.9711 | 0.9736 | 0.9763 | 0.9793 | 0.9823 |
| IN 49 | 0.9653 | 0.9655 | 0.9658 | 0.9665 | 0.9678 | 0.9695 | 0.9716 | 0.9740 | 0.9767 | 0.9796 | 0.9826 |
| SN 50 | 0.9664 | 0.9666 | 0.9669 | 0.9676 | 0.9688 | 0.9705 | 0.9725 | 0.9748 | 0.9775 | 0.9803 | 0.9832 |
| SB 51 | 0.9672 | 0.9674 | 0.9677 | 0.9684 | 0.9696 | 0.9712 | 0.9732 | 0.9755 | 0.9780 | 0.9808 | 0.9836 |
| TE 52 | 0.9687 | 0.9689 | 0.9691 | 0.9698 | 0.9710 | 0.9725 | 0.9744 | 0.9766 | 0.9790 | 0.9816 | 0.9843 |
| I 53 | 0.9685 | 0.9687 | 0.9690 | 0.9696 | 0.9708 | 0.9723 | 0.9742 | 0.9765 | 0.9789 | 0.9815 | 0.9842 |
| XE 54 | 0.9697 | 0.9699 | 0.9701 | 0.9707 | 0.9718 | 0.9733 | 0.9752 | 0.9773 | 0.9797 | 0.9822 | 0.9849 |
| CS 55 | 0.9699 | 0.9701 | 0.9703 | 0.9710 | 0.9721 | 0.9736 | 0.9754 | 0.9775 | 0.9799 | 0.9824 | 0.9849 |
| BA 56 | 0.9708 | 0.9710 | 0.9713 | 0.9719 | 0.9730 | 0.9744 | 0.9762 | 0.9782 | 0.9805 | 0.9829 | 0.9854 |
| LA 57 | 0.9711 | 0.9713 | 0.9716 | 0.9722 | 0.9733 | 0.9747 | 0.9764 | 0.9785 | 0.9807 | 0.9831 | 0.9856 |
| CE 58 | 0.9713 | 0.9715 | 0.9718 | 0.9724 | 0.9734 | 0.9748 | 0.9765 | 0.9786 | 0.9808 | 0.9832 | 0.9856 |
| PR 59 | 0.9716 | 0.9718 | 0.9720 | 0.9726 | 0.9737 | 0.9751 | 0.9768 | 0.9788 | 0.9810 | 0.9834 | 0.9858 |
| ND 60 | 0.9722 | 0.9724 | 0.9726 | 0.9733 | 0.9743 | 0.9756 | 0.9773 | 0.9793 | 0.9814 | 0.9837 | 0.9861 |
| PM 61 | 0.0 | 0.0 | 0.0 | 0.0 | 0.0 | 0.0 | 0.0 | 0.0 | 0.0 | 0.0 | 0.0 |
| SM 62 | 0.9733 | 0.9735 | 0.9737 | 0.9743 | 0.9753 | 0.9766 | 0.9782 | 0.9801 | 0.9822 | 0.9844 | 0.9867 |
| EU 63 | 0.9736 | 0.9738 | 0.9740 | 0.9746 | 0.9756 | 0.9768 | 0.9784 | 0.9803 | 0.9824 | 0.9846 | 0.9868 |
| GD 64 | 0.9745 | 0.9747 | 0.9749 | 0.9754 | 0.9764 | 0.9776 | 0.9792 | 0.9810 | 0.9829 | 0.9851 | 0.9873 |
| TB 65 | 0.9748 | 0.9750 | 0.9751 | 0.9757 | 0.9766 | 0.9778 | 0.9794 | 0.9812 | 0.9831 | 0.9852 | 0.9874 |
| DY 66 | 0.9752 | 0.9754 | 0.9756 | 0.9761 | 0.9770 | 0.9782 | 0.9797 | 0.9815 | 0.9834 | 0.9854 | 0.9876 |
| HO 67 | 0.9757 | 0.9759 | 0.9761 | 0.9766 | 0.9775 | 0.9786 | 0.9801 | 0.9818 | 0.9837 | 0.9858 | 0.9879 |
| ER 68 | 0.9760 | 0.9762 | 0.9764 | 0.9769 | 0.9778 | 0.9789 | 0.9804 | 0.9821 | 0.9840 | 0.9860 | 0.9880 |
| TM 69 | 0.9762 | 0.9764 | 0.9766 | 0.9771 | 0.9780 | 0.9791 | 0.9806 | 0.9823 | 0.9841 | 0.9861 | 0.9881 |
| YB 70 | 0.9768 | 0.9770 | 0.9771 | 0.9777 | 0.9785 | 0.9796 | 0.9810 | 0.9827 | 0.9845 | 0.9864 | 0.9884 |
| LU 71 | 0.9770 | 0.9772 | 0.9774 | 0.9779 | 0.9787 | 0.9799 | 0.9813 | 0.9829 | 0.9847 | 0.9866 | 0.9885 |
| HF 72 | 0.9776 | 0.9778 | 0.9779 | 0.9784 | 0.9792 | 0.9804 | 0.9817 | 0.9833 | 0.9851 | 0.9869 | 0.9889 |
| TA 73 | 0.9778 | 0.9780 | 0.9781 | 0.9786 | 0.9794 | 0.9805 | 0.9819 | 0.9834 | 0.9852 | 0.9870 | 0.9889 |
| W 74 | 0.9781 | 0.9783 | 0.9785 | 0.9790 | 0.9797 | 0.9808 | 0.9821 | 0.9837 | 0.9854 | 0.9872 | 0.9891 |
| RE 75 | 0.9784 | 0.9786 | 0.9787 | 0.9793 | 0.9800 | 0.9811 | 0.9824 | 0.9839 | 0.9856 | 0.9874 | 0.9892 |
| OS 76 | 0.9790 | 0.9792 | 0.9792 | 0.9799 | 0.9806 | 0.9817 | 0.9829 | 0.9844 | 0.9860 | 0.9878 | 0.9896 |
| IR 77 | 0.9793 | 0.9795 | 0.9794 | 0.9801 | 0.9809 | 0.9819 | 0.9832 | 0.9846 | 0.9862 | 0.9878 | 0.9896 |
| PT 78 | 0.9795 | 0.9797 | 0.9799 | 0.9803 | 0.9811 | 0.9821 | 0.9833 | 0.9848 | 0.9864 | 0.9879 | 0.9897 |
| AU 79 | 0.9800 | 0.9802 | 0.9803 | 0.9808 | 0.9815 | 0.9825 | 0.9837 | 0.9851 | 0.9864 | 0.9881 | 0.9898 |
| HG 80 | 0.9803 | 0.9805 | 0.9804 | 0.9811 | 0.9818 | 0.9827 | 0.9839 | 0.9853 | 0.9867 | 0.9884 | 0.9901 |
| TL 81 | 0.9804 | 0.9806 | 0.9806 | 0.9811 | 0.9818 | 0.9828 | 0.9840 | 0.9853 | 0.9869 | 0.9885 | 0.9902 |
| PB 82 | 0.9805 | 0.9807 | 0.9807 | 0.9811 | 0.9818 | 0.9828 | 0.9840 | 0.9853 | 0.9868 | 0.9885 | 0.9901 |
| BI 83 | 0.9808 | 0.9810 | 0.9810 | 0.9815 | 0.9822 | 0.9831 | 0.9843 | 0.9856 | 0.9871 | 0.9887 | 0.9904 |

| $\theta$ | 180° | 178° | 176° | 174° | 172° | 170° | 168° | 166° | 164° | 162° | 160° |
|---|---|---|---|---|---|---|---|---|---|---|---|
| $\delta^2$ | 0 | $1.218 \times 10^{-3}$ | $4.874 \times 10^{-3}$ | $1.097 \times 10^{-2}$ | $1.950 \times 10^{-2}$ | $3.046 \times 10^{-2}$ | $4.386 \times 10^{-2}$ | $5.971 \times 10^{-2}$ | $7.798 \times 10^{-2}$ | $9.870 \times 10^{-2}$ | $1.218 \times 10^{-1}$ |

**TABLE V**

$K$ for ⁴He as Projectile

SCATTERING ANGLE $\theta$

| ELEMENT | AT. NO. ($Z_2$) | 90° | 100° | 110° | 120° | 130° | 140° | 150° | 160° | 170° | 180° |
|---|---|---|---|---|---|---|---|---|---|---|---|
| B | 5 | 0.4593 | 0.4000 | 0.3500 | 0.3091 | 0.2767 | 0.2517 | 0.2334 | 0.2209 | 0.2136 | 0.2112 |
| C | 6 | 0.5001 | 0.4423 | 0.3929 | 0.3518 | 0.3187 | 0.2929 | 0.2736 | 0.2604 | 0.2526 | 0.2501 |
| N | 7 | 0.5555 | 0.5009 | 0.4532 | 0.4127 | 0.3795 | 0.3531 | 0.3333 | 0.3194 | 0.3113 | 0.3086 |
| O | 8 | 0.5998 | 0.5483 | 0.5027 | 0.4635 | 0.4309 | 0.4047 | 0.3848 | 0.3708 | 0.3625 | 0.3597 |
| F | 9 | 0.6520 | 0.6050 | 0.5627 | 0.5258 | 0.4946 | 0.4693 | 0.4498 | 0.4360 | 0.4278 | 0.4251 |
| NE | 10 | 0.6687 | 0.6233 | 0.5822 | 0.5463 | 0.5158 | 0.4909 | 0.4717 | 0.4580 | 0.4499 | 0.4472 |
| NA | 11 | 0.7034 | 0.6615 | 0.6233 | 0.5896 | 0.5607 | 0.5370 | 0.5185 | 0.5053 | 0.4974 | 0.4948 |
| MG | 12 | 0.7171 | 0.6767 | 0.6397 | 0.6069 | 0.5788 | 0.5556 | 0.5375 | 0.5246 | 0.5169 | 0.5143 |
| AL | 13 | 0.7416 | 0.7040 | 0.6693 | 0.6384 | 0.6117 | 0.5897 | 0.5724 | 0.5600 | 0.5525 | 0.5500 |
| SI | 14 | 0.7505 | 0.7139 | 0.6801 | 0.6499 | 0.6238 | 0.6022 | 0.5852 | 0.5730 | 0.5657 | 0.5632 |
| P | 15 | 0.7711 | 0.7370 | 0.7054 | 0.6770 | 0.6523 | 0.6318 | 0.6156 | 0.6040 | 0.5970 | 0.5946 |
| S | 16 | 0.7779 | 0.7447 | 0.7138 | 0.6860 | 0.6619 | 0.6417 | 0.6259 | 0.6144 | 0.6076 | 0.6053 |
| CL | 17 | 0.7970 | 0.7662 | 0.7374 | 0.7114 | 0.6887 | 0.6658 | 0.6548 | 0.6440 | 0.6374 | 0.6352 |
| AR | 18 | 0.8179 | 0.7897 | 0.7634 | 0.7395 | 0.7186 | 0.7010 | 0.6871 | 0.6770 | 0.6709 | 0.6689 |
| K | 19 | 0.8142 | 0.7856 | 0.7588 | 0.7346 | 0.7133 | 0.6955 | 0.6814 | 0.6712 | 0.6650 | 0.6630 |
| CA | 20 | 0.8184 | 0.7904 | 0.7641 | 0.7403 | 0.7194 | 0.7019 | 0.6880 | 0.6779 | 0.6718 | 0.6698 |
| SC | 21 | 0.8365 | 0.8109 | 0.7869 | 0.7650 | 0.7457 | 0.7295 | 0.7166 | 0.7073 | 0.7016 | 0.6997 |
| TI | 22 | 0.8457 | 0.8214 | 0.7985 | 0.7776 | 0.7592 | 0.7437 | 0.7314 | 0.7224 | 0.7170 | 0.7152 |
| V | 23 | 0.8543 | 0.8312 | 0.8095 | 0.7896 | 0.7720 | 0.7572 | 0.7454 | 0.7368 | 0.7316 | 0.7298 |
| CR | 24 | 0.8570 | 0.8344 | 0.8129 | 0.7934 | 0.7761 | 0.7615 | 0.7498 | 0.7414 | 0.7362 | 0.7345 |
| MN | 25 | 0.8642 | 0.8425 | 0.8221 | 0.8033 | 0.7867 | 0.7727 | 0.7615 | 0.7534 | 0.7485 | 0.7468 |
| FE | 26 | 0.8662 | 0.8449 | 0.8247 | 0.8062 | 0.7898 | 0.7760 | 0.7649 | 0.7569 | 0.7520 | 0.7504 |
| CO | 27 | 0.8728 | 0.8524 | 0.8331 | 0.8154 | 0.7997 | 0.7864 | 0.7758 | 0.7681 | 0.7634 | 0.7618 |
| NI | 28 | 0.8726 | 0.8522 | 0.8328 | 0.8150 | 0.7992 | 0.7859 | 0.7752 | 0.7675 | 0.7628 | 0.7612 |
| CU | 29 | 0.8815 | 0.8624 | 0.8442 | 0.8276 | 0.8128 | 0.8002 | 0.7902 | 0.7829 | 0.7785 | 0.7770 |
| ZN | 30 | 0.8846 | 0.8659 | 0.8482 | 0.8319 | 0.8175 | 0.8052 | 0.7954 | 0.7883 | 0.7839 | 0.7825 |
| GA | 31 | 0.8914 | 0.8738 | 0.8570 | 0.8416 | 0.8279 | 0.8162 | 0.8069 | 0.8001 | 0.7960 | 0.7946 |
| GE | 32 | 0.8956 | 0.8786 | 0.8624 | 0.8475 | 0.8342 | 0.8229 | 0.8139 | 0.8073 | 0.8033 | 0.8020 |
| AS | 33 | 0.8986 | 0.8820 | 0.8663 | 0.8518 | 0.8389 | 0.8279 | 0.8191 | 0.8126 | 0.8087 | 0.8074 |
| SE | 34 | 0.9035 | 0.8877 | 0.8727 | 0.8588 | 0.8464 | 0.8359 | 0.8275 | 0.8213 | 0.8176 | 0.8163 |
| BR | 35 | 0.9046 | 0.8890 | 0.8741 | 0.8603 | 0.8481 | 0.8377 | 0.8293 | 0.8232 | 0.8195 | 0.8183 |
| KR | 36 | 0.9088 | 0.8938 | 0.8796 | 0.8664 | 0.8546 | 0.8446 | 0.8366 | 0.8307 | 0.8271 | 0.8259 |
| RB | 37 | 0.9105 | 0.8958 | 0.8818 | 0.8688 | 0.8573 | 0.8474 | 0.8395 | 0.8337 | 0.8302 | 0.8290 |
| SR | 38 | 0.9126 | 0.8982 | 0.8845 | 0.8718 | 0.8605 | 0.8509 | 0.8431 | 0.8375 | 0.8340 | 0.8329 |
| Y | 39 | 0.9138 | 0.8996 | 0.8861 | 0.8736 | 0.8624 | 0.8529 | 0.8452 | 0.8396 | 0.8362 | 0.8351 |
| ZR | 40 | 0.9159 | 0.9020 | 0.8888 | 0.8765 | 0.8656 | 0.8563 | 0.8488 | 0.8433 | 0.8400 | 0.8389 |
| NB | 41 | 0.9174 | 0.9038 | 0.8907 | 0.8787 | 0.8679 | 0.8588 | 0.8514 | 0.8460 | 0.8427 | 0.8416 |

| | 180° | 178° | 176° | 174° | 172° | 170° | 168° | 166° | 164° | 162° | 160° |
|---|---|---|---|---|---|---|---|---|---|---|---|
| MO 42 | 0.8461 | 0.8471 | 0.8503 | 0.8556 | 0.8628 | 0.8717 | 0.8822 | 0.8939 | 0.9066 | 0.9134 | 0.9198 |
| TC 43 | 0.0 | 0.0 | 0.0 | 0.0 | 0.0 | 0.0 | 0.0 | 0.0 | 0.0 | 0.0 | 0.0 |
| RU 44 | 0.8532 | 0.8542 | 0.8573 | 0.8623 | 0.8692 | 0.8777 | 0.8877 | 0.8989 | 0.9110 | 0.9175 | 0.9236 |
| RH 45 | 0.8558 | 0.8569 | 0.8599 | 0.8648 | 0.8716 | 0.8800 | 0.8898 | 0.9008 | 0.9127 | 0.9191 | 0.9251 |
| PD 46 | 0.8603 | 0.8613 | 0.8642 | 0.8690 | 0.8756 | 0.8838 | 0.8933 | 0.9040 | 0.9155 | 0.9217 | 0.9276 |
| AG 47 | 0.8620 | 0.8630 | 0.8659 | 0.8706 | 0.8771 | 0.8852 | 0.8946 | 0.9052 | 0.9165 | 0.9226 | 0.9284 |
| CD 48 | 0.8673 | 0.8682 | 0.8710 | 0.8756 | 0.8818 | 0.8896 | 0.8987 | 0.9089 | 0.9199 | 0.9258 | 0.9313 |
| IN 49 | 0.8698 | 0.8707 | 0.8735 | 0.8780 | 0.8841 | 0.8917 | 0.9007 | 0.9106 | 0.9214 | 0.9272 | 0.9326 |
| SN 50 | 0.8738 | 0.8747 | 0.8773 | 0.8817 | 0.8877 | 0.8951 | 0.9037 | 0.9134 | 0.9239 | 0.9296 | 0.9348 |
| SB 51 | 0.8767 | 0.8776 | 0.8802 | 0.8845 | 0.8903 | 0.8976 | 0.9060 | 0.9155 | 0.9257 | 0.9312 | 0.9363 |
| TE 52 | 0.8820 | 0.8829 | 0.8854 | 0.8895 | 0.8951 | 0.9020 | 0.9102 | 0.9192 | 0.9290 | 0.9343 | 0.9392 |
| I 53 | 0.8814 | 0.8823 | 0.8848 | 0.8889 | 0.8945 | 0.9015 | 0.9097 | 0.9188 | 0.9286 | 0.9339 | 0.9388 |
| XE 54 | 0.8852 | 0.8860 | 0.8885 | 0.8925 | 0.8979 | 0.9047 | 0.9126 | 0.9215 | 0.9310 | 0.9362 | 0.9409 |
| CS 55 | 0.8865 | 0.8873 | 0.8897 | 0.8937 | 0.8991 | 0.9058 | 0.9136 | 0.9223 | 0.9317 | 0.9368 | 0.9415 |
| BA 56 | 0.8899 | 0.8907 | 0.8931 | 0.8969 | 0.9021 | 0.9086 | 0.9162 | 0.9247 | 0.9338 | 0.9388 | 0.9434 |
| LA 57 | 0.8911 | 0.8919 | 0.8942 | 0.8980 | 0.9032 | 0.9096 | 0.9172 | 0.9256 | 0.9346 | 0.9395 | 0.9440 |
| CE 58 | 0.8919 | 0.8927 | 0.8950 | 0.8988 | 0.9039 | 0.9103 | 0.9178 | 0.9261 | 0.9350 | 0.9399 | 0.9444 |
| PR 59 | 0.8926 | 0.8933 | 0.8956 | 0.8994 | 0.9045 | 0.9109 | 0.9183 | 0.9266 | 0.9355 | 0.9404 | 0.9448 |
| ND 60 | 0.8949 | 0.8956 | 0.8979 | 0.9016 | 0.9066 | 0.9128 | 0.9201 | 0.9282 | 0.9369 | 0.9417 | 0.9460 |
| PM 61 | 0.0 | 0.0 | 0.0 | 0.0 | 0.0 | 0.0 | 0.0 | 0.0 | 0.0 | 0.0 | 0.0 |
| SM 62 | 0.8989 | 0.8997 | 0.9018 | 0.9054 | 0.9102 | 0.9162 | 0.9232 | 0.9310 | 0.9394 | 0.9439 | 0.9481 |
| EU 63 | 0.9000 | 0.9007 | 0.9028 | 0.9064 | 0.9111 | 0.9171 | 0.9240 | 0.9317 | 0.9400 | 0.9445 | 0.9487 |
| GD 64 | 0.9032 | 0.9039 | 0.9059 | 0.9094 | 0.9140 | 0.9197 | 0.9265 | 0.9339 | 0.9420 | 0.9464 | 0.9504 |
| TB 65 | 0.9041 | 0.9048 | 0.9069 | 0.9103 | 0.9149 | 0.9206 | 0.9272 | 0.9346 | 0.9426 | 0.9470 | 0.9509 |
| DY 66 | 0.9061 | 0.9067 | 0.9088 | 0.9121 | 0.9166 | 0.9221 | 0.9287 | 0.9359 | 0.9437 | 0.9480 | 0.9518 |
| HO 67 | 0.9075 | 0.9081 | 0.9101 | 0.9134 | 0.9178 | 0.9233 | 0.9298 | 0.9369 | 0.9446 | 0.9488 | 0.9526 |
| ER 68 | 0.9087 | 0.9094 | 0.9113 | 0.9145 | 0.9189 | 0.9244 | 0.9307 | 0.9378 | 0.9454 | 0.9496 | 0.9533 |
| TM 69 | 0.9096 | 0.9102 | 0.9122 | 0.9154 | 0.9197 | 0.9251 | 0.9314 | 0.9384 | 0.9459 | 0.9500 | 0.9537 |
| YB 70 | 0.9116 | 0.9122 | 0.9142 | 0.9173 | 0.9215 | 0.9268 | 0.9329 | 0.9398 | 0.9471 | 0.9512 | 0.9548 |
| LU 71 | 0.9125 | 0.9132 | 0.9151 | 0.9182 | 0.9224 | 0.9276 | 0.9337 | 0.9404 | 0.9477 | 0.9517 | 0.9553 |
| HF 72 | 0.9143 | 0.9149 | 0.9168 | 0.9198 | 0.9239 | 0.9290 | 0.9350 | 0.9417 | 0.9488 | 0.9527 | 0.9562 |
| TA 73 | 0.9153 | 0.9159 | 0.9178 | 0.9207 | 0.9248 | 0.9299 | 0.9358 | 0.9423 | 0.9494 | 0.9533 | 0.9567 |
| W 74 | 0.9166 | 0.9172 | 0.9190 | 0.9219 | 0.9260 | 0.9309 | 0.9368 | 0.9432 | 0.9502 | 0.9540 | 0.9574 |
| RE 75 | 0.9176 | 0.9182 | 0.9200 | 0.9229 | 0.9269 | 0.9318 | 0.9375 | 0.9439 | 0.9508 | 0.9546 | 0.9579 |
| OS 76 | 0.9197 | 0.9203 | 0.9221 | 0.9249 | 0.9288 | 0.9337 | 0.9394 | 0.9456 | 0.9523 | 0.9560 | 0.9593 |
| IR 77 | 0.9201 | 0.9207 | 0.9224 | 0.9252 | 0.9291 | 0.9339 | 0.9394 | 0.9456 | 0.9523 | 0.9560 | 0.9592 |
| PT 78 | 0.9212 | 0.9218 | 0.9235 | 0.9263 | 0.9301 | 0.9348 | 0.9403 | 0.9464 | 0.9530 | 0.9566 | 0.9598 |
| AU 79 | 0.9219 | 0.9225 | 0.9242 | 0.9270 | 0.9307 | 0.9354 | 0.9408 | 0.9469 | 0.9534 | 0.9570 | 0.9602 |
| HG 80 | 0.9234 | 0.9239 | 0.9256 | 0.9283 | 0.9320 | 0.9366 | 0.9420 | 0.9479 | 0.9543 | 0.9579 | 0.9610 |
| TL 81 | 0.9246 | 0.9252 | 0.9268 | 0.9295 | 0.9332 | 0.9377 | 0.9429 | 0.9488 | 0.9551 | 0.9586 | 0.9616 |
| PB 82 | 0.9255 | 0.9260 | 0.9276 | 0.9303 | 0.9339 | 0.9383 | 0.9435 | 0.9493 | 0.9555 | 0.9589 | 0.9619 |
| BI 83 | 0.9262 | 0.9268 | 0.9284 | 0.9310 | 0.9346 | 0.9390 | 0.9442 | 0.9499 | 0.9560 | 0.9593 | 0.9624 |

| $\theta$ | 180° | 178° | 176° | 174° | 172° | 170° | 168° | 166° | 164° | 162° | 160° |
|---|---|---|---|---|---|---|---|---|---|---|---|
| $\delta^2$ | 0 | $1.218 \times 10^{-3}$ | $4.874 \times 10^{-3}$ | $1.097 \times 10^{-2}$ | $1.950 \times 10^{-2}$ | $3.046 \times 10^{-2}$ | $4.386 \times 10^{-2}$ | $5.971 \times 10^{-2}$ | $7.798 \times 10^{-2}$ | $9.870 \times 10^{-2}$ | $1.218 \times 10^{-1}$ |

**TABLE VI.** $^4$He Stopping Cross Section $\varepsilon$ (Values in $10^{-15}$ eV cm$^2$)

| ELE-MENT | AT. NO. (Z$_2$) | 400 | 600 | 800 | 1000 | 1200 | 1400 | 1600 |
|---|---|---|---|---|---|---|---|---|
| | | | | ENERGY (in keV) of $^4$He | | | | |
| H | 1 | 14.02 | 14.11 | 13.5 | 12.49 | 11.34 | 10.19 | 9.154 |
| HE | 2 | 16.72 | 17.88 | 18.03 | 17.52 | 16.63 | 15.56 | 14.46 |
| LI | 3 | 22.28 | 21.99 | 21.46 | 20.64 | 19.6 | 18.42 | 17.28 |
| BE | 4 | 27.09 | 26.76 | 25.89 | 24.71 | 23.4 | 22.06 | 20.8 |
| B | 5 | 32.6 | 33.49 | 32.67 | 31.27 | 29.48 | 27.59 | 25.74 |
| C | 6 | 33.32 | 36.58 | 37.21 | 36.19 | 34.27 | 31.99 | 29.72 |
| N | 7 | 46.23 | 48.45 | 48.12 | 46.24 | 43.54 | 40.58 | 37.71 |
| O | 8 | 44.34 | 47.72 | 48.39 | 47.34 | 45.29 | 42.81 | 40.27 |
| F | 9 | 40.07 | 43.99 | 45.66 | 45.73 | 44.76 | 43.15 | 41.24 |
| NE | 10 | 39.32 | 43.59 | 45.54 | 45.86 | 45.1 | 43.68 | 41.92 |
| NA | 11 | 42.02 | 44.08 | 44.95 | 44.88 | 44.24 | 43.14 | 41.96 |
| MG | 12 | 56.04 | 57.26 | 56.78 | 55.26 | 53.21 | 50.99 | 48.82 |
| AL | 13 | 55.39 | 54.86 | 53.81 | 52.43 | 50.85 | 49.18 | 47.5 |
| SI | 14 | 70.15 | 71.09 | 69.44 | 66.3 | 62.5 | 58.62 | 55.02 |
| P | 15 | 64.66 | 68.45 | 67.57 | 65.13 | 62.06 | 58.88 | 55.87 |
| S | 16 | 62.12 | 68.61 | 69.72 | 67.75 | 64.72 | 61.48 | 58.37 |
| CL | 17 | 83.26 | 86.5 | 84.96 | 80.68 | 75.18 | 69.52 | 64.36 |
| AR | 18 | 83.61 | 88.7 | 87.82 | 83.47 | 77.52 | 71.27 | 65.56 |
| K | 19 | 83.16 | 89.08 | 90.61 | 88.88 | 85.56 | 80.68 | 75.61 |
| CA | 20 | 93.78 | 97.3 | 97.14 | 94.47 | 90.5 | 85.61 | 80.58 |
| SC | 21 | 92.58 | 96.27 | 96.3 | 93.86 | 90.42 | 85.73 | 81.12 |
| TI | 22 | 91.07 | 95.41 | 95.76 | 93.54 | 89.87 | 85.55 | 81.14 |
| V | 23 | 86.19 | 90.13 | 90.55 | 88.7 | 85.58 | 81.89 | 78.13 |
| CR | 24 | 79.42 | 84.81 | 86.62 | 85.97 | 83.76 | 80.68 | 77.24 |
| MN | 25 | 77.08 | 82.69 | 84.4 | 83.6 | 81.35 | 78.41 | 75.3 |
| FE | 26 | 80.15 | 86.9 | 89.26 | 88.64 | 86.13 | 82.59 | 78.65 |
| CO | 27 | 72.11 | 79.07 | 82.04 | 82.29 | 80.82 | 78.38 | 75.5 |
| NI | 28 | 68.29 | 74.6 | 77.74 | 78.66 | 78.07 | 76.56 | 74.54 |
| CU | 29 | 62.41 | 68.2 | 71.77 | 73.58 | 74.05 | 73.5 | 72.24 |
| ZN | 30 | 65.53 | 70.47 | 72.85 | 73.47 | 72.98 | 71.71 | 69.97 |
| GA | 31 | 74.23 | 78.12 | 79.41 | 79.12 | 77.79 | 75.86 | 73.6 |
| GE | 32 | 77.76 | 81.9 | 82.76 | 82.1 | 80.26 | 77.76 | 75.18 |
| AS | 33 | 81.41 | 87.03 | 87.98 | 87.02 | 84.84 | 82.01 | 79.14 |
| SE | 34 | 83.2 | 89.4 | 89.8 | 87.8 | 84.9 | 81.6 | 78.4 |
| BR | 35 | 95.55 | 101.1 | 101.1 | 97.91 | 93.04 | 87.7 | 82.65 |
| KR | 36 | 102.2 | 108.2 | 108 | 104.2 | 98.67 | 92.74 | 87.26 |
| RB | 37 | 98.18 | 108.3 | 110.1 | 107.4 | 102.6 | 97.34 | 92.4 |
| SR | 38 | 109 | 117 | 117.4 | 114.2 | 109 | 103.1 | 97.75 |
| Y | 39 | 110 | 120.4 | 121.1 | 117.3 | 111.6 | 105.5 | 99.6 |
| ZR | 40 | 115.4 | 126 | 126.8 | 123.2 | 117.9 | 112 | 106.2 |
| NB | 41 | 118.1 | 128.2 | 128.7 | 125.1 | 119.8 | 114 | 108.4 |
| MO | 42 | 109.8 | 120.5 | 122.2 | 119.6 | 115.1 | 110 | 104.8 |
| TC | 43 | 116 | 126.8 | 128.9 | 126.3 | 121.2 | 115.4 | 109.2 |
| RU | 44 | 104.1 | 116.8 | 120.5 | 119.5 | 116 | 111.3 | 105.8 |
| RH | 45 | 100.9 | 113.6 | 117.7 | 117.2 | 113.9 | 109.5 | 104.6 |
| PD | 46 | 89.09 | 104.9 | 111.9 | 112.9 | 110.3 | 105.8 | 100.5 |
| AG | 47 | 88.63 | 101.9 | 108.4 | 110.2 | 108.8 | 105.4 | 100.9 |
| CD | 48 | 96.33 | 107 | 112 | 113 | 111.4 | 108.1 | 103.8 |
| IN | 49 | 104.3 | 110.1 | 113.7 | 115.2 | 114.8 | 112.7 | 109.3 |
| SN | 50 | 108.2 | 115.8 | 118.6 | 118.3 | 115.8 | 112.1 | 107.9 |
| SB | 51 | 116.2 | 122.2 | 122.2 | 119.9 | 116.8 | 113.3 | 110 |
| TE | 52 | 121.3 | 127.2 | 126.5 | 123.4 | 119.4 | 115.3 | 111.2 |
| I | 53 | 135 | 141.7 | 141 | 135.8 | 128.5 | 120.6 | 113.2 |
| XE | 54 | 144.7 | 149.7 | 148.2 | 143 | 136 | 128.7 | 122 |
| CS | 55 | 129.7 | 141.5 | 143.1 | 139.7 | 134.4 | 128.7 | 123.2 |
| BA | 56 | 141.2 | 150.7 | 151.4 | 147.4 | 141.3 | 134.9 | 128.7 |
| LA | 57 | 144.7 | 156.5 | 156.9 | 152.3 | 145.7 | 138.6 | 131.7 |
| CE | 58 | 136.4 | 146.1 | 147.7 | 144.5 | 139.1 | 133.2 | 127.4 |
| PR | 59 | 134.1 | 143.8 | 145.7 | 142.9 | 137.8 | 132.2 | 126.5 |
| ND | 60 | 131.9 | 141.6 | 143.5 | 141 | 136.5 | 131 | 125.5 |
| PM | 61 | 129.7 | 139.4 | 141.4 | 139.2 | 135.1 | 129.8 | 124.4 |
| SM | 62 | 127.7 | 137.7 | 139.4 | 137.4 | 133.2 | 128.4 | 123.3 |
| EU | 63 | 125.8 | 135.6 | 137.4 | 135.6 | 131.7 | 127.1 | 122.1 |
| GD | 64 | 130.1 | 139.2 | 141.7 | 139.9 | 135.8 | 131.1 | 125.8 |
| TB | 65 | 122.2 | 131.7 | 133.6 | 132.2 | 128.7 | 124.5 | 119.7 |
| DY | 66 | 111.5 | 123.9 | 128.1 | 127.9 | 125.4 | 121.8 | 117.7 |
| HO | 67 | 107.5 | 118.4 | 122.4 | 122.3 | 120.1 | 117 | 113.2 |
| ER | 68 | 106.1 | 116.8 | 120.8 | 120.7 | 118.6 | 115.7 | 112 |
| TM | 69 | 104.7 | 115.2 | 119.2 | 119.2 | 117.2 | 114.1 | 110.8 |
| YB | 70 | 103.5 | 113.8 | 117.7 | 117.8 | 115.9 | 113 | 109.7 |
| LU | 71 | 106.3 | 116.9 | 120.4 | 120.2 | 118.1 | 115 | 111.6 |
| HF | 72 | 109.7 | 120.8 | 124.5 | 124.3 | 122.2 | 118.9 | 115.4 |
| TA | 73 | 105.8 | 117.5 | 121.7 | 121.8 | 119.8 | 116.7 | 113.1 |
| W | 74 | 103.4 | 114.2 | 118 | 118.2 | 116.5 | 113.9 | 110.9 |
| RE | 75 | 114.4 | 125.8 | 129.8 | 129.8 | 127.3 | 124.1 | 120.4 |
| OS | 76 | 112.5 | 124.5 | 129 | 129.5 | 127.3 | 124.3 | 120.7 |
| IR | 77 | 110.7 | 123.2 | 128.2 | 129.3 | 127.3 | 124.5 | 121.1 |
| PT | 78 | 103.1 | 117.6 | 124.2 | 126.2 | 125.6 | 123.7 | 121.2 |
| AU | 79 | 109.9 | 122.7 | 128 | 129.1 | 127.9 | 125.3 | 122.3 |
| HG | 80 | 103.5 | 116.9 | 122.7 | 124.2 | 123.5 | 121.6 | 119.2 |
| TL | 81 | 113.4 | 125 | 129.5 | 130 | 128.4 | 125.8 | 122.9 |
| PB | 82 | 126.4 | 138.1 | 141.9 | 141.6 | 139.1 | 135.9 | 132.4 |
| BI | 83 | 124.6 | 136 | 139.2 | 138.3 | 135.4 | 131.9 | 128.3 |
| PO | 84 | 127.3 | 140 | 143 | 141.8 | 138.4 | 134.6 | 130.6 |
| AT | 85 | 128.2 | 142.7 | 146.1 | 145.1 | 141.1 | 137 | 132.7 |
| RN | 86 | 127.7 | 144.4 | 148.7 | 147.7 | 143.5 | 139.2 | 134.7 |
| FR | 87 | 143.7 | 158.2 | 160.9 | 158.5 | 153 | 147.6 | 142.2 |
| RA | 88 | 155.2 | 167.8 | 169.8 | 166.1 | 160.1 | 154 | 147.9 |
| AC | 89 | 158.1 | 171.3 | 173.8 | 170 | 164.1 | 157.6 | 151 |
| TH | 90 | 159.4 | 173.4 | 176.6 | 173.2 | 167.2 | 160.5 | 153.7 |
| PA | 91 | 153.1 | 166.8 | 170.4 | 168 | 162.8 | 156.8 | 150.8 |
| U | 92 | 150.7 | 164.4 | 168.4 | 166.6 | 161.9 | 156.1 | 150.3 |

**TABLE VI (Continued)**

| ELE-MENT | AT. NO. (Z$_2$) | ENERGY (in keV) of $^4$He | | | | | |
|---|---|---|---|---|---|---|---|
| | | 1800 | 2000 | 2400 | 2800 | 3200 | 3600 | 4000 |
| H | 1 | 8.289 | 7.606 | 6.75 | 6.081 | 5.534 | 5.108 | 4.683 |
| HE | 2 | 13.44 | 12.52 | 11.12 | 10.02 | 9.117 | 8.416 | 7.714 |
| LI | 3 | 16.24 | 15.35 | 13.63 | 12.3 | 11.22 | 10.33 | 9.587 |
| BE | 4 | 19.65 | 18.64 | 16.55 | 14.93 | 13.61 | 12.61 | 11.68 |
| B | 5 | 24.1 | 22.7 | 20.09 | 18.09 | 16.51 | 15.22 | 14.14 |
| C | 6 | 27.68 | 25.97 | 23.1 | 20.8 | 18.99 | 17.59 | 16.36 |
| N | 7 | 35.15 | 32.98 | 29.39 | 26.68 | 24.45 | 22.6 | 21.05 |
| O | 8 | 37.91 | 35.84 | 32.39 | 29.54 | 27.21 | 25.18 | 23.55 |
| F | 9 | 39.24 | 37.31 | 34.44 | 31.79 | 29.46 | 27.45 | 25.75 |
| NE | 10 | 40.06 | 38.24 | 35.79 | 33.56 | 31.55 | 29.66 | 27.99 |
| NA | 11 | 40.73 | 39.56 | 36.57 | 34.04 | 31.86 | 29.95 | 28.24 |
| MG | 12 | 46.85 | 45.11 | 41.57 | 38.67 | 36.2 | 34.16 | 32.33 |
| AL | 13 | 45.85 | 44.25 | 40.38 | 37.38 | 34.96 | 32.92 | 31.08 |
| SI | 14 | 51.88 | 49.26 | 44.71 | 41.27 | 38.44 | 36.11 | 34.09 |
| P | 15 | 53.08 | 50.67 | 45.88 | 42.17 | 39.14 | 36.66 | 34.55 |
| S | 16 | 55.41 | 52.89 | 47.81 | 43.88 | 40.68 | 38.06 | 35.83 |
| CL | 17 | 60.04 | 56.65 | 51.15 | 46.98 | 43.53 | 40.68 | 38.34 |
| AR | 18 | 60.81 | 57.13 | 51.75 | 47.56 | 44.07 | 41.18 | 38.79 |
| K | 19 | 70.84 | 66.64 | 60.13 | 54.97 | 50.8 | 47.36 | 44.45 |
| CA | 20 | 75.64 | 71.19 | 64.19 | 58.63 | 54.1 | 50.37 | 47.21 |
| SC | 21 | 76.65 | 72.53 | 65.71 | 60.17 | 55.6 | 51.8 | 48.58 |
| TI | 22 | 77 | 73.31 | 67.61 | 62.56 | 58.04 | 54.17 | 50.84 |
| V | 23 | 74.6 | 71.45 | 66.28 | 61.75 | 57.64 | 53.95 | 50.79 |
| CR | 24 | 73.78 | 70.51 | 66.16 | 62.13 | 58.54 | 55.17 | 52.12 |
| MN | 25 | 72.36 | 69.72 | 65.29 | 61.39 | 58.01 | 54.85 | 52 |
| FE | 26 | 74.71 | 71.05 | 66.57 | 62.86 | 59.59 | 56.64 | 54.02 |
| CO | 27 | 72.55 | 69.77 | 65.5 | 62.01 | 58.96 | 56.23 | 53.73 |
| NI | 28 | 72.3 | 70.04 | 66 | 61 | 56.55 | 52.64 | 49.94 |
| CU | 29 | 70.51 | 68.48 | 64.9 | 62 | 59.34 | 56.91 | 54.72 |
| ZN | 30 | 68.12 | 66.04 | 62.25 | 59.06 | 56.35 | 53.91 | 51.77 |
| GA | 31 | 71.25 | 68.82 | 64.46 | 60.89 | 57.91 | 55.29 | 53 |
| GE | 32 | 72.46 | 69.8 | 65.08 | 61.25 | 58.1 | 55.38 | 53.02 |
| AS | 33 | 75.97 | 73.05 | 67.85 | 63.66 | 60.24 | 57.32 | 54.8 |
| SE | 34 | 75.3 | 72.4 | 67.06 | 62.88 | 59.43 | 56.5 | 53.99 |
| BR | 35 | 78.29 | 74.75 | 69.06 | 64.64 | 60.96 | 57.9 | 55.17 |
| KR | 36 | 82.66 | 79.04 | 73.04 | 68.14 | 64.21 | 60.94 | 58 |
| RB | 37 | 87.67 | 83.47 | 76.51 | 71.01 | 66.59 | 62.85 | 59.7 |
| SR | 38 | 92.56 | 87.93 | 80.3 | 74.29 | 69.51 | 65.46 | 62.04 |
| Y | 39 | 93.9 | 88.8 | 84.35 | 77.66 | 72.11 | 67.92 | 64.37 |
| ZR | 40 | 100.7 | 95.7 | 87.64 | 81.02 | 75.72 | 71.2 | 67.44 |
| NB | 41 | 103 | 98.1 | 90.26 | 83.66 | 78.21 | 73.66 | 69.68 |
| MO | 42 | 99.88 | 95.21 | 88.09 | 81.86 | 76.49 | 72.01 | 68.18 |
| TC | 43 | 103.3 | 97.85 | 90.14 | 83.58 | 78.09 | 73.43 | 69.48 |
| RU | 44 | 100.7 | 95.86 | 89.05 | 82.95 | 77.71 | 73.21 | 69.35 |
| RH | 45 | 99.85 | 95.42 | 89.21 | 83.44 | 78.33 | 73.87 | 70.03 |
| PD | 46 | 95.32 | 90.65 | 86.26 | 81.88 | 77.72 | 73.77 | 70.04 |
| AG | 47 | 96.02 | 91.22 | 86.66 | 82.54 | 78.53 | 73.98 | 68.91 |
| CD | 48 | 99.24 | 94.71 | 89.47 | 84.81 | 80.43 | 76.31 | 72.53 |
| IN | 49 | 105 | 100 | 94.46 | 89.7 | 85.69 | 81.81 | 78.17 |
| SN | 50 | 103.6 | 99.49 | 93.82 | 89.04 | 84.81 | 81.14 | 77.7 |
| SB | 51 | 106.8 | 103.7 | 97.58 | 92.45 | 88.01 | 84.26 | 80.61 |
| TE | 52 | 107.4 | 103.9 | 97.5 | 92.23 | 87.74 | 83.82 | 80.34 |
| I | 53 | 106.7 | 101.5 | 95.02 | 89.84 | 85.31 | 81.53 | 78.07 |
| XE | 54 | 116.2 | 111.5 | 104.1 | 98.28 | 93.26 | 89.05 | 85.31 |
| CS | 55 | 117.9 | 113.1 | 105 | 98.36 | 92.91 | 88.36 | 84.29 |
| BA | 56 | 122.9 | 117.6 | 108.7 | 101.6 | 95.71 | 90.86 | 86.53 |
| LA | 57 | 125.1 | 119 | 109.9 | 102.6 | 96.71 | 91.72 | 87.5 |
| CE | 58 | 121.8 | 116.7 | 108 | 101 | 95.22 | 90.56 | 86.27 |
| PR | 59 | 121 | 116 | 107.5 | 100.5 | 94.84 | 90.27 | 86.02 |
| ND | 60 | 120.2 | 115.3 | 106.9 | 99.99 | 94.4 | 89.8 | 85.71 |
| PM | 61 | 119.2 | 114.5 | 106.2 | 99.41 | 93.91 | 89.35 | 85.35 |
| SM | 62 | 118.3 | 113.6 | 105.5 | 98.8 | 93.37 | 88.86 | 84.96 |
| EU | 63 | 117.3 | 112.7 | 104.7 | 98.15 | 92.81 | 88.34 | 84.53 |
| GD | 64 | 120.8 | 116 | 107.7 | 100.8 | 95.23 | 90.57 | 86.59 |
| TB | 65 | 115.2 | 110.8 | 103.1 | 96.8 | 91.62 | 87.24 | 83.63 |
| DY | 66 | 113.5 | 109.5 | 102.3 | 96.4 | 91.36 | 87.11 | 83.41 |
| HO | 67 | 109.5 | 105.7 | 98.56 | 92.71 | 87.83 | 83.66 | 80.21 |
| ER | 68 | 108.4 | 104.8 | 97.7 | 91.98 | 87.17 | 83.07 | 79.65 |
| TM | 69 | 107.2 | 103.7 | 96.82 | 91.22 | 86.46 | 82.43 | 79.06 |
| YB | 70 | 106.2 | 102.9 | 96.06 | 90.57 | 85.87 | 81.89 | 78.54 |
| LU | 71 | 108 | 104.5 | 97.42 | 91.75 | 86.89 | 82.79 | 79.34 |
| HF | 72 | 111.6 | 108 | 100.6 | 94.7 | 89.62 | 85.32 | 81.72 |
| TA | 73 | 109.4 | 105.6 | 98.96 | 93.14 | 88.4 | 84.19 | 80.53 |
| W | 74 | 107.6 | 104.4 | 97.87 | 92.17 | 87.42 | 83.2 | 79.61 |
| RE | 75 | 116.5 | 112.7 | 105.2 | 98.97 | 93.58 | 88.98 | 85.12 |
| OS | 76 | 117 | 113.3 | 106 | 99.75 | 94.32 | 89.69 | 85.79 |
| IR | 77 | 117.7 | 114 | 106.8 | 100.7 | 95.29 | 90.62 | 86.67 |
| PT | 78 | 118.2 | 115 | 108.4 | 102.5 | 97.21 | 92.6 | 88.64 |
| AU | 79 | 118.9 | 115.5 | 110 | 104.9 | 99.92 | 95.42 | 91.39 |
| HG | 80 | 116.3 | 113.3 | 107.1 | 101.5 | 96.45 | 91.9 | 87.95 |
| TL | 81 | 119.6 | 116.4 | 109.8 | 104 | 98.82 | 94.11 | 89.99 |
| PB | 82 | 128.7 | 125 | 117.7 | 111.5 | 106 | 100.9 | 96.47 |
| BI | 83 | 124.5 | 120.8 | 113.6 | 107.6 | 102.2 | 97.4 | 93.07 |
| PO | 84 | 126.5 | 122.6 | 115.2 | 109 | 103.6 | 98.76 | 94.38 |
| AT | 85 | 128.5 | 124.4 | 116.7 | 110.4 | 104.9 | 100 | 95.6 |
| RN | 86 | 130.3 | 126 | 118.1 | 111.7 | 106 | 101.2 | 96.72 |
| FR | 87 | 137 | 132.1 | 123.2 | 116 | 109.9 | 104.7 | 99.92 |
| RA | 88 | 142.1 | 136.8 | 127.1 | 119.4 | 112.9 | 107.4 | 102.5 |
| AC | 89 | 144.9 | 139.4 | 129.2 | 121.3 | 114.5 | 108.9 | 103.9 |
| TH | 90 | 147.4 | 141.7 | 131.2 | 123 | 116 | 110.2 | 105.2 |
| PA | 91 | 145 | 139.5 | 129.6 | 121.8 | 115.1 | 109.5 | 104.7 |
| U | 92 | 144.6 | 139.3 | 129.5 | 121.7 | 115.2 | 109.7 | 105 |

**TABLE VII**

Polynomial Fit to $^4$He Stopping Cross Sections $\varepsilon$

| ELE-MENT | AT. NO. $(Z_2)$ | COEFFICIENT (in $10^{-15}$ eV cm$^2$/atom) | | | | | |
|---|---|---|---|---|---|---|---|
| | | $A_0$ | $A_1$ | $A_2$ | $A_3$ | $A_4$ | $A_5$ |
| H | 1 | 10.6 | 16.62 | −24.33 | 11.96 | −2.575 | 0.2054 |
| HE | 2 | 9.852 | 27.54 | −30.72 | 13.26 | −2.618 | 0.1962 |
| LI | 3 | 20.59 | 8.638 | −13.35 | 5.697 | −1.077 | 0.07722 |
| BE | 4 | 26.09 | 7.082 | −13.27 | 5.857 | −1.137 | 0.08396 |
| B | 5 | 26.08 | 29.34 | −38.08 | 17.19 | −3.507 | 0.2709 |
| C | 6 | 15.95 | 69.35 | −76.59 | 33.79 | −6.859 | 0.5291 |
| N | 7 | 31.01 | 64 | −76.97 | 34.76 | −7.127 | 0.5523 |
| O | 8 | 25.9 | 73.3 | −80.5 | 35.17 | −7.1 | 0.5462 |
| F | 9 | 21.33 | 70.24 | −69.08 | 28.29 | −5.455 | 0.4052 |
| NE | 10 | 19.32 | 74.35 | −72.1 | 29.56 | −5.694 | 0.4207 |
| NA | 11 | 33.29 | 32.27 | −29.79 | 10.88 | −1.918 | 0.1339 |
| MG | 12 | 47.43 | 37.15 | −45.69 | 20.13 | −4.078 | 0.3154 |
| AL | 13 | 55.94 | 0.6773 | −4.752 | 0.3401 | 0.2662 | −0.04046 |
| SI | 14 | 57.97 | 56.59 | −77.66 | 36.41 | −7.624 | 0.5995 |
| P | 15 | 47.29 | 74.15 | −88.48 | 40.09 | −8.329 | 0.657 |
| S | 16 | 33.29 | 115.6 | −126.5 | 56.27 | −11.61 | 0.912 |
| CL | 17 | 57.34 | 114 | −146.7 | 69.63 | −14.83 | 1.183 |
| AR | 18 | 48.33 | 150.7 | −186.6 | 88.28 | −18.81 | 1.502 |
| K | 19 | 48.69 | 135.1 | −145.4 | 61.75 | −12.08 | 0.9 |
| CA | 20 | 69.25 | 99.92 | −114.5 | 48.6 | −9.422 | 0.6956 |
| SC | 21 | 69.01 | 95.47 | −108 | 45.7 | −8.889 | 0.6607 |
| TI | 22 | 64.01 | 109.8 | −125.5 | 55.67 | −11.37 | 0.8827 |
| V | 23 | 62.45 | 95.97 | −108.8 | 48.18 | −9.853 | 0.7664 |
| CR | 24 | 50.64 | 110.8 | −115.6 | 49.1 | −9.69 | 0.729 |
| MN | 25 | 49.19 | 108.4 | −114.2 | 49.4 | −9.962 | 0.7663 |
| FE | 26 | 44.36 | 137.5 | −143.1 | 61.02 | −12.03 | 0.9011 |
| CO | 27 | 40.04 | 120.1 | −117.6 | 48.42 | −9.34 | 0.6905 |
| NI | 28 | 41.59 | 97.79 | −91.19 | 37.32 | −7.482 | 0.5893 |
| CU | 29 | 40.72 | 73.99 | −56.66 | 18.06 | −2.656 | 0.1452 |
| ZN | 30 | 45.6 | 71.9 | −63.75 | 23.68 | −4.188 | 0.2886 |
| GA | 31 | 56.87 | 65.24 | −63.23 | 24.41 | −4.429 | 0.3112 |
| GE | 32 | 58.71 | 73.52 | −75.13 | 30.36 | −5.739 | 0.4182 |
| AS | 33 | 57.16 | 94.44 | −97.78 | 40.73 | −7.931 | 0.5938 |
| SE | 34 | 56.66 | 107.1 | −117.8 | 51.99 | −10.64 | 0.8308 |
| BR | 35 | 61.73 | 140.2 | −165.5 | 76 | −15.88 | 1.25 |
| KR | 36 | 65.45 | 153.7 | −184 | 85.61 | −18.08 | 1.437 |
| RB | 37 | 50.69 | 189.8 | −208.3 | 93.13 | −19.22 | 1.506 |
| SR | 38 | 68.59 | 164.5 | −186 | 82.93 | −16.98 | 1.32 |
| Y | 39 | 56.26 | 220.1 | −254.3 | 119.1 | −25.56 | 2.068 |
| ZR | 40 | 66.91 | 196.3 | −218.8 | 97.94 | −20.22 | 1.585 |
| NB | 41 | 71.85 | 188.1 | −211.2 | 94.92 | −19.66 | 1.546 |
| MO | 42 | 61.36 | 193 | −210.2 | 93.62 | −19.33 | 1.518 |
| TC | 43 | 62.92 | 210.4 | −228.6 | 100.7 | −20.49 | 1.586 |
| RU | 44 | 46.61 | 222 | −230 | 99.79 | −20.18 | 1.557 |
| RH | 45 | 43.18 | 221.8 | −228.5 | 99.53 | −20.25 | 1.572 |
| PD | 46 | 13.57 | 285.2 | −287.4 | 124.9 | −25.24 | 1.939 |

**TABLE VII (Continued)**

| ELE-MENT | AT. NO. | COEFFICIENT (in $10^{-15}$ eV cm$^2$/atom) | | | | | |
|---|---|---|---|---|---|---|---|
| | $(Z_2)$ | $A_0$ | $A_1$ | $A_2$ | $A_3$ | $A_4$ | $A_5$ |
| AG | 47 | 26.6 | 228.9 | −219.8 | 90.5 | −17.35 | 1.261 |
| CD | 48 | 46 | 186.6 | −180.3 | 73.86 | −14.16 | 1.037 |
| IN | 49 | 76.23 | 96.67 | −78.42 | 23.5 | −2.83 | 0.08811 |
| SN | 50 | 70.2 | 144.2 | −145.8 | 60.37 | −11.63 | 0.8555 |
| SB | 51 | 91.18 | 102.1 | −113.7 | 50.29 | −10.36 | 0.8144 |
| TE | 52 | 94.56 | 111.9 | −130.3 | 59.01 | −12.31 | 0.9746 |
| I | 53 | 87.62 | 199.8 | −243.6 | 114.2 | −24.12 | 1.909 |
| XE | 54 | 108.1 | 158.1 | −198.2 | 93.19 | −19.75 | 1.571 |
| CS | 55 | 77.48 | 210 | −231 | 103.5 | −21.42 | 1.685 |
| BA | 56 | 94.41 | 190.9 | −216.1 | 96.76 | −19.94 | 1.559 |
| LA | 57 | 89.09 | 227 | −256.7 | 115.6 | −23.91 | 1.874 |
| CE | 58 | 89.21 | 188.9 | −207.6 | 91.31 | −18.57 | 1.437 |
| PR | 59 | 87.3 | 185.9 | −201.7 | 87.98 | −17.77 | 1.368 |
| ND | 60 | 85.98 | 181.4 | −195 | 84.51 | −17 | 1.307 |
| PM | 61 | 84.65 | 177.4 | −188.8 | 81.33 | −16.29 | 1.248 |
| SM | 62 | 83.62 | 173.6 | −184.3 | 79.37 | −15.93 | 1.223 |
| EU | 63 | 82.63 | 169.3 | −178.4 | 76.46 | −15.29 | 1.171 |
| GD | 64 | 86.29 | 171.6 | −180.7 | 77.22 | −15.39 | 1.176 |
| TB | 65 | 80.92 | 160.9 | −167.4 | 71.16 | −14.15 | 1.08 |
| DY | 66 | 60.25 | 194.3 | −192.6 | 81.11 | −16.1 | 1.229 |
| HO | 67 | 62.97 | 167.8 | −163.6 | 67.57 | −13.22 | 1 |
| ER | 68 | 62.51 | 163.8 | −159 | 65.48 | −12.78 | 0.9652 |
| TM | 69 | 61.99 | 160.6 | −155.9 | 64.31 | −12.59 | 0.9533 |
| YB | 70 | 61.54 | 157.3 | −152.2 | 62.67 | −12.26 | 0.9276 |
| LU | 71 | 64.13 | 159.7 | −156.5 | 65 | −12.8 | 0.9748 |
| HF | 72 | 65.33 | 167.6 | −164 | 68.08 | −13.41 | 1.021 |
| TA | 73 | 58.79 | 177.1 | −172.4 | 71.69 | −14.11 | 1.071 |
| W | 74 | 61.69 | 156.6 | −150.9 | 62.45 | −12.33 | 0.9421 |
| RE | 75 | 67.95 | 175.3 | −171.8 | 71.58 | −14.16 | 1.082 |
| OS | 76 | 64.24 | 181 | −174.9 | 72.58 | −14.34 | 1.096 |
| IR | 77 | 60.99 | 185.1 | −176.4 | 72.77 | −14.33 | 1.094 |
| PT | 78 | 48.78 | 198.6 | −181.1 | 73.5 | −14.38 | 1.095 |
| AU | 79 | 57.99 | 193.2 | −185.4 | 77.98 | −15.59 | 1.199 |
| HG | 80 | 53.2 | 184.9 | −170.8 | 70.02 | −13.82 | 1.058 |
| TL | 81 | 67.92 | 170.2 | −163.3 | 68.01 | −13.53 | 1.041 |
| PB | 82 | 79.64 | 177.7 | −176.3 | 74.69 | −15.01 | 1.16 |
| BI | 83 | 78.46 | 177.6 | −180.6 | 77.75 | −15.78 | 1.229 |
| PO | 84 | 76.75 | 196.3 | −202.4 | 88.29 | −18.09 | 1.418 |
| AT | 85 | 70.31 | 225.1 | −232.9 | 102.3 | −21.08 | 1.659 |
| RN | 86 | 61.48 | 256.9 | −264.8 | 116.8 | −24.12 | 1.902 |
| FR | 87 | 82.69 | 241.2 | −257.8 | 114.6 | −23.71 | 1.868 |
| RA | 88 | 98.73 | 226.4 | −248.1 | 110.7 | −22.88 | 1.8 |
| AC | 89 | 98.14 | 239.4 | −260.8 | 115.8 | −23.83 | 1.868 |
| TH | 90 | 95.29 | 253.9 | −273.5 | 120.7 | −24.73 | 1.931 |
| PA | 91 | 91.87 | 239.7 | −252.7 | 110.1 | −22.36 | 1.736 |
| U | 92 | 89.72 | 236.7 | −245.9 | 106.1 | −21.39 | 1.651 |

**TABLE VIII.** $^4$He Stopping Cross Section Factor [$\varepsilon_0$] (Values in $10^{-15}$ eV cm$^2$)

| ELE-MENT | AT. NO. $(Z_2)$ | ENERGY (in MeV) OF INCIDENT $^4$He | | | | | | | | | | |
|---|---|---|---|---|---|---|---|---|---|---|---|---|
| | | 1.0 | 1.2 | 1.4 | 1.6 | 1.8 | 2.0 | 2.4 | 2.8 | 3.2 | 3.6 | 4.0 |
| BE | 4 | 31.0 | 30.9 | 30.8 | 30.7 | 30.5 | 30.4 | 30.1 | 29.8 | 29.5 | 29.2 | 28.9 |
| B | 5 | 37.9 | 38.2 | 38.3 | 38.3 | 38.3 | 38.2 | 38.0 | 37.7 | 37.2 | 36.5 | 35.9 |
| C | 6 | 38.7 | 40.0 | 40.9 | 41.6 | 42.1 | 42.4 | 42.8 | 42.9 | 42.5 | 41.7 | 40.8 |
| N | 7 | 59.5 | 60.1 | 60.3 | 60.2 | 59.8 | 59.3 | 58.1 | 56.6 | 54.6 | 52.4 | 50.2 |
| O | 8 | 61.3 | 62.3 | 62.8 | 62.8 | 62.5 | 62.0 | 60.5 | 58.7 | 56.4 | 53.9 | 51.5 |
| F | 9 | 61.0 | 62.4 | 63.1 | 63.3 | 63.1 | 62.6 | 61.0 | 59.0 | 56.7 | 54.2 | 51.8 |
| NE | 10 | 61.9 | 63.6 | 64.4 | 64.4 | 64.1 | 63.9 | 62.4 | 60.4 | 58.2 | 55.7 | 53.3 |
| NA | 11 | 66.2 | 66.7 | 66.8 | 66.5 | 65.9 | 65.1 | 63.2 | 61.0 | 58.6 | 56.3 | 54.1 |
| MG | 12 | 86.4 | 85.6 | 84.3 | 82.7 | 80.9 | 79.0 | 75.2 | 71.5 | 67.9 | 64.5 | 61.7 |
| AL | 13 | 84.8 | 83.4 | 81.9 | 80.3 | 78.6 | 76.8 | 73.1 | 69.5 | 66.1 | 62.9 | 59.9 |
| SI | 14 | 109.6 | 107.0 | 103.7 | 100.1 | 96.3 | 92.6 | 85.8 | 79.7 | 74.3 | 69.6 | 65.9 |
| P | 15 | 107.6 | 106.0 | 103.5 | 100.3 | 96.9 | 93.5 | 86.9 | 81.0 | 75.6 | 70.8 | 67.3 |
| S | 16 | 110.3 | 109.8 | 107.8 | 104.8 | 101.3 | 97.8 | 90.9 | 84.5 | 78.7 | 73.6 | 70.0 |
| CL | 17 | 138.9 | 134.5 | 128.8 | 122.6 | 116.3 | 110.3 | 99.8 | 91.4 | 84.4 | 78.5 | 74.6 |
| AR | 18 | 145.8 | 140.9 | 134.2 | 126.9 | 119.5 | 112.6 | 101.0 | 92.1 | 85.1 | 79.2 | 75.6 |
| K | 19 | 150.6 | 148.7 | 144.8 | 139.6 | 133.3 | 128.0 | 116.8 | 107.2 | 99.2 | 92.3 | 87.0 |
| CA | 20 | 162.6 | 159.3 | 154.5 | 148.7 | 142.5 | 136.2 | 124.5 | 114.4 | 105.8 | 98.5 | 92.6 |
| SC | 21 | 164.0 | 160.7 | 155.9 | 150.2 | 144.1 | 138.0 | 126.5 | 116.6 | 108.2 | 101.0 | 95.2 |
| TI | 22 | 164.5 | 161.2 | 156.3 | 150.6 | 144.6 | 138.7 | 128.1 | 119.2 | 110.6 | 104.8 | 99.3 |
| V | 23 | 156.9 | 154.0 | 149.7 | 144.7 | 139.4 | 134.3 | 124.9 | 117.0 | 110.1 | 103.8 | 98.7 |
| CR | 24 | 151.1 | 149.6 | 146.3 | 142.1 | 137.4 | 132.7 | 124.1 | 116.8 | 110.0 | 105.0 | 100.2 |
| MN | 25 | 148.0 | 146.4 | 143.3 | 139.2 | 134.7 | 130.3 | 122.2 | 115.5 | 109.7 | 104.3 | 99.8 |
| FE | 26 | 157.2 | 155.4 | 151.1 | 146.3 | 140.8 | 135.4 | 125.7 | 118.2 | 112.2 | 107.0 | 102.8 |
| CO | 27 | 145.8 | 145.4 | 143.0 | 139.5 | 135.3 | 131.1 | 123.1 | 116.5 | 111.4 | 106.2 | 102.1 |
| NI | 28 | 138.5 | 139.2 | 138.2 | 136.1 | 133.2 | 130.0 | 123.0 | 116.3 | 111.4 | 105.0 | 102.1 |
| CU | 29 | 130.0 | 131.9 | 132.2 | 131.2 | 129.4 | 127.0 | 121.6 | 116.3 | 110.2 | 105.3 | 97.4 |
| ZN | 30 | 131.5 | 131.8 | 130.8 | 128.7 | 126.1 | 123.1 | 117.1 | 111.4 | 106.5 | 102.8 | 103.3 |
| GA | 31 | 143.6 | 142.5 | 140.0 | 137.0 | 133.4 | 129.6 | 122.0 | 115.7 | 110.2 | 105.3 | 98.1 |
| GE | 32 | 150.0 | 148.2 | 145.0 | 141.0 | 136.6 | 132.2 | 124.0 | 117.0 | 111.1 | 105.9 | 101.1 |
| AS | 33 | 159.8 | 157.6 | 153.7 | 149.0 | 143.3 | 138.9 | 129.7 | 122.1 | 115.7 | 110.0 | 101.6 |
| SE | 34 | 162.9 | 159.6 | 154.6 | 149.0 | 143.3 | 137.8 | 128.4 | 120.2 | 114.6 | 108.8 | 105.3 |
| BR | 35 | 182.4 | 176.2 | 168.2 | 159.7 | 151.4 | 144.1 | 132.4 | 124.2 | 117.8 | 111.7 | 104.1 |
| KR | 36 | 195.1 | 187.8 | 178.8 | 169.4 | 160.4 | 152.4 | 140.0 | 131.4 | 124.6 | 117.9 | 106.8 |
| RB | 37 | 200.5 | 195.2 | 187.2 | 178.3 | 169.3 | 161.1 | 147.6 | 137.8 | 129.8 | 124.6 | 112.6 |
| SR | 38 | 214.0 | 207.4 | 198.6 | 188.8 | 179.2 | 170.3 | 155.5 | 144.6 | 135.7 | 127.6 | 116.3 |
| Y | 39 | 220.7 | 213.2 | 203.2 | 192.5 | 182.4 | 173.5 | 159.8 | 157.9 | 148.3 | 132.9 | 121.1 |
| ZR | 40 | 231.9 | 225.2 | 215.7 | 205.3 | 194.9 | 185.3 | 169.6 | 157.7 | 148.0 | 139.1 | 126.0 |
| NB | 41 | 235.8 | 228.9 | 219.6 | 209.3 | 199.1 | 189.7 | 174.3 | 162.7 | 153.2 | 143.8 | 131.8 |
| MO | 42 | 225.2 | 219.7 | 211.5 | 202.2 | 192.9 | 184.2 | 170.0 | 159.2 | 149.9 | 140.8 | 136.3 |
| TC | 43 | 238.2 | 231.7 | 222.2 | 211.5 | 200.1 | 190.9 | 174.7 | 161.7 | 153.0 | 143.7 | 133.4 |
| RU | 44 | 224.8 | 220.7 | 213.0 | 204.4 | 195.1 | 186.1 | 171.4 | 161.1 | 152.1 | 143.2 | 136.1 |
| RH | 45 | 220.3 | 216.7 | 210.0 | 201.7 | 193.2 | 185.1 | 171.8 | 161.8 | 153.2 | 144.4 | 135.0 |
| PD | 46 | 211.8 | 209.2 | 202.7 | 194.2 | 185.5 | 177.5 | 165.1 | 157.2 | 150.8 | 143.7 | 137.0 |
| AG | 47 | 206.5 | 205.4 | 200.5 | 193.7 | 186.2 | 178.9 | 166.9 | 158.3 | 151.7 | 144.7 | 136.9 |
| CD | 48 | 212.7 | 211.0 | 206.1 | 199.4 | 192.2 | 185.1 | 172.9 | 163.5 | 156.0 | 148.9 | 141.9 |

| | Z | 160° | 162° | 164° | 166° | 168° | 170° | 172° | 174° | 176° | 178° | 180° |
|---|---|---|---|---|---|---|---|---|---|---|---|---|
| IN | 49 | 152.5 | 159.0 | 165.1 | 173.1 | 183.7 | 196.2 | 202.6 | 208.5 | 213.3 | 216.4 | 217.0 |
| SN | 50 | 151.3 | 157.7 | 164.7 | 172.2 | 181.5 | 196.7 | 200.7 | 208.0 | 215.0 | 220.8 | 224.1 |
| SB | 51 | 156.9 | 163.7 | 171.7 | 179.8 | 188.6 | 199.3 | 205.5 | 212.0 | 218.5 | 224.5 | 228.9 |
| TE | 52 | 156.5 | 163.0 | 167.0 | 179.7 | 188.9 | 196.3 | 207.7 | 215.2 | 223.8 | 230.7 | 236.7 |
| I | 53 | 152.1 | 159.0 | 167.0 | 174.3 | 183.7 | 199.1 | 209.7 | 222.2 | 235.8 | 249.4 | 261.0 |
| XE | 54 | 166.3 | 173.8 | 182.8 | 191.5 | 201.9 | 217.0 | 227.2 | 238.8 | 251.3 | 263.9 | 274.9 |
| CS | 55 | 164.7 | 172.7 | 182.6 | 192.7 | 204.1 | 219.6 | 229.4 | 240.1 | 251.1 | 261.2 | 268.5 |
| BA | 56 | 169.4 | 177.9 | 188.6 | 199.3 | 212.0 | 229.1 | 239.8 | 251.6 | 263.5 | 274.8 | 283.6 |
| LA | 57 | 171.2 | 179.8 | 190.6 | 201.4 | 214.4 | 232.8 | 244.5 | 257.6 | 271.2 | 284.0 | 294.0 |
| CE | 58 | 168.9 | 177.3 | 187.5 | 198.1 | 210.7 | 227.5 | 237.6 | 248.7 | 260.0 | 270.3 | 277.9 |
| PR | 59 | 168.4 | 176.7 | 186.7 | 197.2 | 209.7 | 226.2 | 236.2 | 246.9 | 257.8 | 267.5 | 271.3 |
| ND | 60 | 167.8 | 175.9 | 185.8 | 196.3 | 208.6 | 224.9 | 234.6 | 245.0 | 255.4 | 264.1 | 268.1 |
| PM | 61 | 167.1 | 175.1 | 184.9 | 195.2 | 207.4 | 224.4 | 232.6 | 242.9 | 253.0 | 261.9 | 261.4 |
| SM | 62 | 166.3 | 174.2 | 183.9 | 194.1 | 206.1 | 221.7 | 230.8 | 240.6 | 250.3 | 258.9 | 255.5 |
| EU | 63 | 165.5 | 173.7 | 183.0 | 192.8 | 204.7 | 219.7 | 228.8 | 238.3 | 247.7 | 255.5 | 261.8 |
| GD | 64 | 169.7 | 177.7 | 187.7 | 198.3 | 210.8 | 226.7 | 236.0 | 245.9 | 255.6 | 264.2 | 270.0 |
| TB | 65 | 163.8 | 171.2 | 180.3 | 189.3 | 201.8 | 216.4 | 224.9 | 233.8 | 242.6 | 250.2 | 255.2 |
| DY | 66 | 163.5 | 170.9 | 180.0 | 189.4 | 200.1 | 213.6 | 221.5 | 229.6 | 237.9 | 243.6 | 246.7 |
| HO | 67 | 157.2 | 164.2 | 173.0 | 182.4 | 193.1 | 206.1 | 213.5 | 221.0 | 227.9 | 233.3 | 235.9 |
| ER | 68 | 156.1 | 163.1 | 171.8 | 180.9 | 191.4 | 204.2 | 211.4 | 218.7 | 225.4 | 230.6 | 233.0 |
| TM | 69 | 155.0 | 161.8 | 170.4 | 179.5 | 189.7 | 202.2 | 209.1 | 216.2 | 222.7 | 227.7 | 230.0 |
| YB | 70 | 154.1 | 160.8 | 169.3 | 178.3 | 188.3 | 200.5 | 207.3 | 214.2 | 220.5 | 225.4 | 227.6 |
| LU | 71 | 155.7 | 162.7 | 171.5 | 180.7 | 191.1 | 203.7 | 210.7 | 217.9 | 224.6 | 230.0 | 232.6 |
| HF | 72 | 160.5 | 167.8 | 174.5 | 186.7 | 197.6 | 210.6 | 218.0 | 225.6 | 232.6 | 238.1 | 238.0 |
| TA | 73 | 158.3 | 165.6 | 174.5 | 183.6 | 193.9 | 207.2 | 213.8 | 221.2 | 228.1 | 233.5 | 233.5 |
| W | 74 | 156.4 | 163.7 | 172.4 | 181.5 | 191.1 | 206.6 | 213.8 | 221.2 | 228.6 | 233.5 | 235.9 |
| RE | 75 | 167.4 | 176.6 | 185.1 | 196.3 | 208.1 | 221.4 | 227.8 | 236.4 | 242.9 | 247.1 | 251.5 |
| OS | 76 | 168.7 | 178.5 | 188.6 | 198.8 | 209.8 | 221.4 | 228.0 | 236.7 | 243.3 | 248.6 | 250.0 |
| IR | 77 | 170.5 | 182.3 | 192.3 | 202.3 | 212.6 | 224.4 | 230.0 | 237.2 | 241.8 | 244.9 | 244.1 |
| PT | 78 | 174.4 | 187.0 | 197.8 | 206.1 | 215.2 | 226.2 | 232.5 | 236.7 | 244.4 | 249.3 | 250.3 |
| AU | 79 | 179.8 | 187.0 | 197.3 | 206.1 | 215.2 | 226.6 | 232.0 | 239.0 | 244.9 | 249.3 | 250.7 |
| HG | 80 | 173.1 | 181.0 | 190.8 | 200.0 | 210.1 | 221.2 | 227.0 | 232.0 | 237.2 | 241.1 | 241.1 |
| TL | 81 | 177.3 | 185.5 | 195.5 | 205.3 | 215.6 | 227.4 | 227.0 | 234.0 | 240.6 | 251.1 | 252.7 |
| PB | 82 | 190.1 | 190.8 | 199.5 | 220.3 | 231.3 | 244.4 | 234.0 | 259.5 | 266.8 | 272.5 | 275.7 |
| BI | 83 | 183.4 | 183.9 | 209.8 | 209.5 | 223.2 | 236.1 | 243.7 | 251.6 | 259.4 | 265.9 | 269.7 |
| PO | 84 | 185.9 | 191.9 | 202.5 | 212.6 | 226.3 | 239.7 | 247.7 | 256.3 | 264.9 | 270.1 | 276.8 |
| AT | 85 | 188.2 | 196.4 | 205.4 | 215.6 | 229.2 | 243.0 | 251.5 | 260.1 | 270.1 | 278.1 | 283.0 |
| RN | 86 | 190.4 | 196.9 | 208.1 | 218.6 | 232.2 | 246.4 | 255.3 | 265.2 | 275.1 | 283.7 | 288.7 |
| FR | 87 | 196.9 | 206.3 | 218.6 | 230.0 | 242.4 | 258.7 | 269.0 | 280.3 | 292.0 | 302.0 | 310.1 |
| RA | 88 | 202.0 | 211.8 | 224.6 | 237.0 | 250.5 | 268.0 | 279.5 | 291.4 | 304.7 | 316.6 | 325.6 |
| AC | 89 | 204.9 | 214.8 | 228.0 | 240.7 | 254.8 | 273.6 | 285.4 | 298.4 | 311.8 | 324.2 | 333.5 |
| TH | 90 | 207.5 | 217.7 | 231.1 | 244.2 | 258.8 | 278.2Z | 290.4 | 303.9 | 317.7 | 330.3 | 339.6 |
| PA | 91 | 206.6 | 216.2 | 228.2 | 241.1 | 255.6 | 274.1 | 285.4 | 297.7 | 310.1 | 321.2 | 328.9 |
| U | 92 | 207.1 | 216.6 | 229.1 | 241.5 | 255.6 | 273.9 | 284.9 | 296.8 | 308.7 | 319.2 | 326.2 |

| θ | 180° | 178° | 176° | 174° | 172° | 170° | 168° | 166° | 164° | 162° | 160° |
|---|---|---|---|---|---|---|---|---|---|---|---|
| δ | 0 | 1.218 ×10⁻³ | 4.874 ×10⁻³ | 1.097 ×10⁻² | 1.950 ×10⁻² | 3.046 ×10⁻² | 4.386 ×10⁻² | 5.971 ×10⁻² | 7.798 ×10⁻² | 9.870 ×10⁻² | 1.218 ×10⁻¹ |

**TABLE IX.** ⁴He Energy-Loss Factor [$S_0$] (Values in eV/Å for $\theta_1 = 0°$ and $\theta_2 = 10°$)

| ELEMENT | AT. NO. (Z₂) | ENERGY (in MeV) OF INCIDENT ⁴He | | | | | | | | | | |
|---|---|---|---|---|---|---|---|---|---|---|---|---|
| | | 1.0 | 1.2 | 1.4 | 1.6 | 1.8 | 2.0 | 2.4 | 2.8 | 3.2 | 3.6 | 4.0 |
| BE | 4 | 12.046 | 37.3 | 37.2 | 37.1 | 36.9 | 36.8 | 36.6 | 36.2 | 35.9 | 35.5 | 34.8 |
| B | 5 | 13.093 | 49.7 | 50.0 | 50.1 | 50.2 | 50.1 | 50.0 | 49.7 | 48.7 | 47.8 | 46.9 |
| C | 6 | 11.364 | 44.0 | 45.4 | 46.5 | 47.3 | 47.8 | 48.2 | 48.7 | 48.7 | 48.3 | 46.4 |
| N | 7 | 3.482 | 20.7 | 20.8 | 21.0 | 20.9 | 20.8 | 20.7 | 20.2 | 19.7 | 19.0 | 17.5 |
| O | 8 | 4.302 | 26.4 | 26.8 | 27.0 | 27.0 | 26.9 | 26.7 | 26.0 | 25.2 | 24.3 | 22.1 |
| F | 9 | 3.522 | 21.5 | 22.0 | 22.2 | 22.3 | 22.2 | 22.0 | 21.5 | 20.8 | 20.0 | 18.2 |
| NE | 10 | 3.585 | 22.2 | 22.8 | 23.1 | 23.2 | 23.1 | 22.9 | 22.4 | 21.7 | 20.8 | 19.1 |
| NA | 11 | 2.541 | 16.8 | 17.0 | 17.0 | 16.9 | 16.7 | 16.5 | 16.1 | 15.5 | 14.9 | 13.7 |
| MG | 12 | 4.302 | 37.2 | 36.8 | 36.3 | 35.5 | 34.8 | 34.0 | 32.3 | 30.7 | 29.2 | 26.5 |
| AL | 13 | 6.023 | 51.1 | 50.3 | 49.4 | 48.4 | 47.3 | 46.2 | 44.0 | 41.8 | 39.8 | 36.1 |
| SI | 14 | 4.978 | 54.5 | 53.3 | 51.6 | 49.8 | 47.9 | 46.1 | 42.7 | 39.7 | 37.0 | 32.8 |
| P | 15 | 3.543 | 38.1 | 37.6 | 36.7 | 35.8 | 34.3 | 33.1 | 30.8 | 28.7 | 26.8 | 25.1 |
| S | 16 | 3.886 | 42.9 | 42.7 | 41.9 | 40.7 | 39.4 | 38.0 | 35.3 | 32.8 | 30.6 | 28.6 |
| CL | 17 | 3.221 | 44.7 | 43.3 | 41.5 | 39.5 | 37.5 | 35.5 | 32.2 | 29.4 | 27.2 | 24.0 |
| AR | 18 | 2.489 | 36.3 | 35.1 | 33.4 | 31.6 | 29.7 | 28.0 | 25.1 | 22.9 | 21.2 | 18.8 |
| K | 19 | 1.330 | 20.0 | 19.8 | 19.3 | 18.6 | 17.8 | 17.0 | 15.5 | 14.3 | 13.2 | 11.6 |
| CA | 20 | 2.014 | 32.8 | 32.1 | 31.1 | 29.3 | 28.7 | 27.4 | 25.1 | 23.0 | 21.3 | 18.7 |
| SC | 21 | 4.015 | 65.9 | 64.5 | 62.6 | 60.3 | 57.2 | 55.4 | 50.8 | 46.8 | 43.4 | 38.2 |
| TI | 22 | 5.682 | 93.4 | 91.6 | 88.8 | 85.6 | 82.2 | 78.8 | 72.8 | 67.7 | 63.4 | 56.4 |
| V | 23 | 7.213 | 113.2 | 111.1 | 108.0 | 104.4 | 100.6 | 96.8 | 90.1 | 84.4 | 79.4 | 71.2 |
| CR | 24 | 8.331 | 125.8 | 124.6 | 121.9 | 118.1 | 114.5 | 110.6 | 103.3 | 97.3 | 92.1 | 83.5 |
| MN | 25 | 8.150 | 120.6 | 119.4 | 116.8 | 113.4 | 109.8 | 106.2 | 99.6 | 94.1 | 89.4 | 81.4 |
| FE | 26 | 8.483 | 133.3 | 131.8 | 128.5 | 124.1 | 119.5 | 114.9 | 106.7 | 100.3 | 95.2 | 87.2 |
| CO | 27 | 8.990 | 131.1 | 130.7 | 128.1 | 125.4 | 121.6 | 117.6 | 110.7 | 104.8 | 99.8 | 91.8 |
| NI | 28 | 9.126 | 126.3 | 127.0 | 126.1 | 124.2 | 121.6 | 118.6 | 112.4 | 106.1 | 99.8 | 88.9 |
| CU | 29 | 8.483 | 110.3 | 111.9 | 112.1 | 111.3 | 109.8 | 107.8 | 103.1 | 98.6 | 94.5 | 87.6 |
| ZN | 30 | 6.547 | 86.1 | 86.3 | 85.6 | 84.4 | 82.5 | 80.6 | 76.6 | 73.0 | 69.7 | 64.3 |
| GA | 31 | 5.104 | 73.3 | 72.8 | 71.5 | 69.9 | 68.1 | 66.1 | 62.4 | 59.1 | 56.2 | 51.6 |
| GE | 32 | 4.429 | 66.4 | 65.6 | 64.2 | 62.4 | 60.5 | 58.6 | 54.9 | 51.8 | 49.2 | 45.0 |
| AS | 33 | 4.598 | 73.5 | 72.5 | 70.7 | 68.5 | 66.2 | 63.9 | 59.6 | 56.1 | 53.2 | 48.4 |
| SE | 34 | 3.650 | 59.5 | 58.2 | 56.4 | 54.4 | 52.3 | 50.3 | 46.9 | 44.1 | 41.8 | 38.0 |
| BR | 35 | 2.563 | 46.8 | 45.1 | 43.1 | 40.9 | 38.8 | 36.9 | 33.9 | 31.8 | 30.2 | 27.4 |
| KR | 36 | 1.870 | 36.5 | 35.1 | 33.4 | 31.7 | 30.0 | 28.5 | 26.2 | 24.6 | 23.3 | 21.1 |
| RB | 37 | 1.077 | 21.6 | 21.0 | 20.2 | 19.2 | 18.2 | 17.4 | 16.2 | 14.8 | 14.0 | 12.5 |
| SR | 38 | 1.787 | 38.2 | 37.1 | 35.5 | 33.8 | 32.0 | 30.4 | 27.8 | 25.8 | 24.3 | 21.6 |
| Y | 39 | 3.042 | 67.1 | 64.8 | 61.8 | 58.7 | 55.5 | 52.8 | 48.6 | 45.7 | 43.2 | 38.3 |
| ZR | 40 | 4.272 | 99.1 | 96.2 | 92.2 | 87.7 | 83.3 | 79.2 | 72.4 | 67.5 | 63.3 | 56.3 |
| NB | 41 | 5.577 | 131.5 | 127.7 | 122.4 | 116.7 | 111.0 | 105.8 | 97.2 | 90.8 | 85.4 | 76.0 |
| MO | 42 | 6.407 | 144.3 | 140.8 | 135.5 | 129.6 | 123.0 | 118.0 | 108.9 | 102.0 | 96.0 | 85.5 |
| TC | 43 | 0.000 | 0.0 | 0.0 | 0.0 | 0.0 | 0.0 | 0.0 | 0.0 | 0.0 | 0.0 | 0.0 |
| RU | 44 | 7.257 | 163.2 | 160.2 | 154.8 | 148.3 | 141.6 | 135.3 | 124.8 | 116.9 | 110.3 | 98.5 |
| RH | 45 | 7.257 | 159.8 | 157.2 | 152.1 | 146.4 | 140.2 | 134.3 | 124.7 | 117.4 | 111.2 | 99.4 |
| PD | 46 | 6.767 | 143.3 | 141.6 | 137.1 | 131.4 | 125.6 | 120.1 | 111.8 | 106.4 | 102.1 | 92.7 |
| AG | 47 | 5.848 | 120.7 | 120.1 | 117.2 | 113.3 | 108.9 | 104.6 | 97.6 | 92.6 | 88.7 | 79.8 |

The columns below give, for each element (symbol and atomic number $Z$), a leading value column followed by the quantities tabulated at scattering angles $\theta = 180^\circ$ through $160^\circ$.

| Elem | $Z$ | value | 180° | 178° | 176° | 174° | 172° | 170° | 168° | 166° | 164° | 162° | 160° |
|---|---|---|---|---|---|---|---|---|---|---|---|---|---|
| CD | 48 | 4.598 | 97.8 | 97.0 | 94.7 | 91.7 | 88.4 | 85.1 | 79.5 | 75.2 | 71.7 | 68.4 | 65.2 |
| IN | 49 | 3.836 | 83.3 | 83.3 | 81.8 | 80.0 | 77.7 | 75.3 | 70.5 | 66.4 | 63.6 | 61.0 | 58.5 |
| SN | 50 | 3.695 | 82.8 | 82.8 | 81.6 | 79.5 | 76.9 | 74.2 | 71.6 | 67.1 | 63.6 | 60.8 | 55.9 |
| SB | 51 | 3.273 | 74.9 | 73.5 | 71.5 | 69.4 | 67.3 | 65.2 | 61.7 | 58.8 | 56.2 | 53.6 | 51.4 |
| TE | 52 | 2.938 | 61.2 | 58.4 | 55.3 | 53.3 | 51.1 | 49.0 | 46.7 | 43.1 | 40.9 | 39.1 | 46.0 |
| I | 53 | 2.344 | 38.6 | 37.0 | 35.3 | 33.5 | 31.9 | 30.5 | 28.3 | 26.9 | 25.7 | 24.4 | 35.7 |
| XE | 54 | 1.404 | 23.1 | 22.5 | 21.6 | 21.6 | 20.7 | 19.7 | 18.9 | 17.6 | 16.6 | 15.7 | 23.3 |
| CS | 55 | 0.860 | 43.8 | 43.8 | 42.4 | 40.7 | 38.8 | 37.0 | 35.4 | 32.7 | 30.8 | 29.1 | 14.2 |
| BA | 56 | 1.544 | 78.7 | 76.0 | 76.0 | 72.6 | 69.0 | 65.5 | 62.3 | 57.4 | 53.9 | 51.0 | 26.2 |
| LA | 57 | 2.677 | 79.7 | 78.7 | 77.5 | 74.6 | 71.3 | 68.2 | 65.2 | 60.4 | 56.8 | 54.1 | 45.8 |
| CE | 58 | 2.868 | 79.5 | 79.7 | 77.5 | 74.6 | 71.5 | 68.4 | 65.5 | 60.7 | 57.1 | 54.1 | 48.5 |
| PR | 59 | 2.896 | 79.3 | 79.5 | 77.4 | 74.7 | 71.6 | 68.6 | 65.7 | 61.0 | 57.4 | 54.3 | 48.8 |
| ND | 60 | 2.924 | 79.1 | 77.6 | 77.4 | 74.7 | 71.6 | 68.6 | 65.7 | 61.0 | 57.4 | 54.3 | 49.1 |
| PM | 61 | 0.000 | 0.0 | 0.0 | 0.0 | 0.0 | 0.0 | 0.0 | 0.0 | 0.0 | 0.0 | 0.0 | 0.0 |
| SM | 62 | 3.027 | 80.1 | 78.4 | 75.8 | 72.8 | 69.9 | 67.1 | 64.2 | 58.8 | 55.7 | 52.7 | 50.3 |
| EU | 63 | 2.084 | 54.1 | 53.3 | 51.6 | 49.7 | 47.7 | 45.8 | 42.7 | 40.2 | 38.1 | 36.1 | 34.4 |
| GD | 64 | 3.027 | 81.7 | 80.0 | 77.7 | 74.4 | 71.4 | 68.6 | 63.8 | 60.0 | 56.8 | 53.8 | 51.4 |
| TB | 65 | 3.137 | 80.1 | 78.5 | 76.1 | 73.3 | 70.2 | 67.9 | 63.3 | 59.7 | 56.6 | 53.7 | 51.8 |
| DY | 66 | 3.170 | 78.2 | 77.2 | 75.4 | 72.8 | 70.2 | 67.7 | 63.4 | 60.0 | 56.6 | 54.2 | 51.8 |
| HO | 67 | 3.221 | 76.0 | 75.5 | 73.4 | 71.6 | 68.8 | 66.4 | 62.2 | 58.7 | 55.7 | 55.1 | 50.6 |
| ER | 68 | 3.273 | 76.3 | 75.5 | 73.8 | 71.6 | 69.2 | 66.9 | 62.7 | 59.2 | 56.2 | 56.2 | 51.1 |
| TM | 69 | 3.328 | 76.0 | 75.7 | 74.1 | 71.0 | 69.6 | 67.3 | 63.1 | 59.7 | 56.7 | 56.7 | 51.6 |
| YB | 70 | 2.429 | 55.3 | 54.7 | 53.6 | 52.0 | 50.3 | 48.7 | 45.7 | 43.3 | 41.1 | 41.1 | 37.4 |
| LU | 71 | 3.384 | 78.7 | 77.8 | 76.0 | 73.7 | 71.3 | 68.9 | 64.7 | 61.2 | 58.0 | 58.0 | 52.7 |
| HF | 72 | 4.429 | 106.6 | 105.5 | 103.0 | 99.9 | 96.6 | 93.3 | 87.5 | 82.7 | 78.4 | 74.3 | 71.1 |
| TA | 73 | 5.526 | 130.4 | 129.0 | 126.0 | 122.2 | 118.1 | 114.1 | 107.1 | 101.5 | 96.4 | 91.5 | 87.5 |
| W | 74 | 6.320 | 144.7 | 143.6 | 140.7 | 136.9 | 132.8 | 128.7 | 121.3 | 115.0 | 109.1 | 103.4 | 98.9 |
| RE | 75 | 6.806 | 171.2 | 169.2 | 165.0 | 160.0 | 155.0 | 149.9 | 140.6 | 132.0 | 126.0 | 119.2 | 113.9 |
| OS | 76 | 7.145 | 179.3 | 177.6 | 173.8 | 168.9 | 158.2 | 148.7 | 140.0 | 133.4 | 126.2 | 119.2 | 120.6 |
| IR | 77 | 7.053 | 161.8 | 175.3 | 171.9 | 167.3 | 162.2 | 157.1 | 148.0 | 140.2 | 133.0 | 126.2 | 120.3 |
| PT | 78 | 6.619 | 147.9 | 162.1 | 160.0 | 156.7 | 152.7 | 148.5 | 141.0 | 133.9 | 127.3 | 120.7 | 115.4 |
| AU | 79 | 5.905 | 98.1 | 147.2 | 144.6 | 141.1 | 137.3 | 133.6 | 127.1 | 121.7 | 116.5 | 111.0 | 106.2 |
| HG | 80 | 4.070 | 88.5 | 98.1 | 96.8 | 94.8 | 92.4 | 90.0 | 85.5 | 81.9 | 77.0 | 73.7 | 62.1 |
| TL | 81 | 3.502 | 90.7 | 88.5 | 87.1 | 84.2 | 81.9 | 79.6 | 75.5 | 71.9 | 68.5 | 68.5 | 62.6 |
| PB | 82 | 3.291 | 76.3 | 89.7 | 87.8 | 85.4 | 82.9 | 80.4 | 76.1 | 72.5 | 69.1 | 65.5 | 51.9 |
| BI | 83 | 2.828 | 73.4 | 75.2 | 73.3 | 71.2 | 68.9 | 66.8 | 63.1 | 60.0 | 57.5 | 54.4 | 49.3 |
| PO | 84 | 2.653 | 72.2 | 72.2 | 70.3 | 68.0 | 65.7 | 63.6 | 60.1 | 57.2 | 54.5 | 51.9 | 49.3 |
| AT | 85 | 0.000 | 0.0 | 0.0 | 0.0 | 0.0 | 0.0 | 0.0 | 0.0 | 0.0 | 0.0 | 0.0 | 0.0 |
| RN | 86 | 0.000 | 0.0 | 0.0 | 0.0 | 0.0 | 0.0 | 0.0 | 0.0 | 0.0 | 0.0 | 0.0 | 0.0 |
| FR | 87 | 0.000 | 0.0 | 0.0 | 0.0 | 0.0 | 0.0 | 0.0 | 0.0 | 0.0 | 0.0 | 0.0 | 27.0 |
| RA | 88 | 1.338 | 43.6 | 42.4 | 40.8 | 39.1 | 37.4 | 35.9 | 33.5 | 31.7 | 30.1 | 28.3 | 0.0 |
| AC | 89 | 0.000 | 0.0 | 0.0 | 0.0 | 0.0 | 0.0 | 0.0 | 0.0 | 0.0 | 0.0 | 0.0 | 0.0 |
| TH | 90 | 3.027 | 102.8 | 100.0 | 96.2 | 92.0 | 87.9 | 84.2 | 78.3 | 73.9 | 69.9 | 65.9 | 62.8 |
| PA | 91 | 4.015 | 132.1 | 129.0 | 124.5 | 119.5 | 114.6 | 110.1 | 102.6 | 97.0 | 91.9 | 86.8 | 82.9 |
| U | 92 | 4.818 | 157.2 | 153.8 | 148.8 | 143.0 | 137.3 | 132.0 | 123.1 | 116.4 | 110.4 | 104.4 | 99.8 |

| $\theta$ | 180° | 178° | 176° | 174° | 172° | 170° | 168° | 166° | 164° | 162° | 160° |
|---|---|---|---|---|---|---|---|---|---|---|---|
| $\delta^2$ | 0 | $1.218 \times 10^{-3}$ | $4.874 \times 10^{-3}$ | $1.097 \times 10^{-2}$ | $1.950 \times 10^{-2}$ | $3.046 \times 10^{-2}$ | $4.386 \times 10^{-2}$ | $5.971 \times 10^{-2}$ | $7.798 \times 10^{-2}$ | $9.870 \times 10^{-2}$ | $1.218 \times 10^{-1}$ |

**TABLE X.** Rutherford Scattering Cross Section of the Elements for 1 MeV $^4$He

$d\sigma/d\Omega$ (in $10^{-24}$ cm$^2$/steradian)

| ELEMENT | AT. NO. ($Z_2$) | AVRG MASS (amu) | 90° | 100° | 110° | 120° | 130° | 140° | 150° | 160° | 170° | 179.5° |
|---|---|---|---|---|---|---|---|---|---|---|---|---|
| BE | 4 | 9.012 | 0.2972 | 0.2065 | 0.1503 | 0.1143 | 0.09069 | 0.075 | 0.06467 | 0.05817 | 0.05458 | 0.05344 |
| B | 5 | 10.81 | 0.4814 | 0.3395 | 0.2514 | 0.1945 | 0.1569 | 0.1316 | 0.1149 | 0.1043 | 0.09837 | 0.09648 |
| C | 6 | 12.01 | 0.7036 | 0.4992 | 0.3722 | 0.29 | 0.2354 | 0.1988 | 0.1744 | 0.1588 | 0.1502 | 0.1474 |
| N | 7 | 14.01 | 0.9734 | 0.6952 | 0.522 | 0.4098 | 0.3351 | 0.2849 | 0.2513 | 0.2299 | 0.218 | 0.2142 |
| O | 8 | 16 | 1.285 | 0.921 | 0.6947 | 0.548 | 0.4503 | 0.3844 | 0.3403 | 0.3122 | 0.2965 | 0.2915 |
| F | 9 | 19 | 1.641 | 1.181 | 0.8948 | 0.7089 | 0.585 | 0.5014 | 0.4455 | 0.4097 | 0.3898 | 0.3834 |
| NE | 10 | 20.18 | 2.032 | 1.464 | 1.11 | 0.8805 | 0.7275 | 0.6242 | 0.5551 | 0.5109 | 0.4862 | 0.4783 |
| NA | 11 | 22.99 | 2.47 | 1.783 | 1.355 | 1.077 | 0.8915 | 0.7664 | 0.6827 | 0.6291 | 0.5993 | 0.5897 |
| MG | 12 | 24.31 | 2.944 | 2.126 | 1.617 | 1.286 | 1.066 | 0.9169 | 0.8172 | 0.7534 | 0.7178 | 0.7064 |
| AL | 13 | 26.98 | 3.465 | 2.505 | 1.907 | 1.518 | 1.26 | 1.085 | 0.9678 | 0.893 | 0.8512 | 0.8378 |
| SI | 14 | 28.09 | 4.022 | 2.908 | 2.215 | 1.765 | 1.464 | 1.262 | 1.126 | 1.039 | 0.9905 | 0.975 |
| P | 15 | 30.97 | 4.625 | 3.347 | 2.551 | 2.034 | 1.689 | 1.457 | 1.301 | 1.201 | 1.145 | 1.127 |
| S | 16 | 32.06 | 5.265 | 3.811 | 2.905 | 2.317 | 1.925 | 1.66 | 1.483 | 1.369 | 1.306 | 1.286 |
| CL | 17 | 35.45 | 5.953 | 4.311 | 3.288 | 2.624 | 2.182 | 1.883 | 1.683 | 1.554 | 1.483 | 1.46 |
| AR | 18 | 39.95 | 6.683 | 4.842 | 3.696 | 2.951 | 2.455 | 2.12 | 1.895 | 1.752 | 1.671 | 1.646 |
| K | 19 | 39.1 | 7.444 | 5.394 | 4.116 | 3.287 | 2.734 | 2.36 | 2.11 | 1.95 | 1.861 | 1.832 |
| CA | 20 | 40.08 | 8.251 | 5.978 | 4.563 | 3.644 | 3.031 | 2.617 | 2.34 | 2.163 | 2.064 | 2.032 |
| SC | 21 | 44.96 | 9.106 | 6.601 | 5.04 | 4.027 | 3.351 | 2.895 | 2.589 | 2.394 | 2.285 | 2.249 |
| TI | 22 | 47.9 | 9.998 | 7.249 | 5.536 | 4.424 | 3.683 | 3.182 | 2.846 | 2.632 | 2.512 | 2.474 |
| V | 23 | 50.94 | 10.93 | 7.927 | 6.055 | 4.84 | 4.03 | 3.482 | 3.116 | 2.881 | 2.75 | 2.708 |
| CR | 24 | 52 | 11.91 | 8.633 | 6.594 | 5.271 | 4.389 | 3.793 | 3.394 | 3.138 | 2.996 | 2.95 |
| MN | 25 | 54.94 | 12.92 | 9.372 | 7.159 | 5.724 | 4.766 | 4.12 | 3.687 | 3.409 | 3.255 | 3.205 |
| FE | 26 | 55.85 | 13.98 | 10.14 | 7.745 | 6.192 | 5.157 | 4.457 | 3.989 | 3.689 | 3.521 | 3.468 |
| CO | 27 | 58.93 | 15.08 | 10.94 | 8.356 | 6.682 | 5.565 | 4.811 | 4.305 | 3.982 | 3.801 | 3.743 |
| NI | 28 | 58.71 | 16.21 | 11.76 | 8.986 | 7.185 | 5.984 | 5.173 | 4.63 | 4.282 | 4.088 | 4.026 |
| CU | 29 | 63.54 | 17.4 | 12.62 | 9.645 | 7.714 | 6.425 | 5.555 | 4.972 | 4.599 | 4.391 | 4.324 |
| ZN | 30 | 65.37 | 18.62 | 13.51 | 10.32 | 8.257 | 6.878 | 5.947 | 5.323 | 4.924 | 4.701 | 4.63 |
| GA | 31 | 69.72 | 19.89 | 14.43 | 11.03 | 8.821 | 7.349 | 6.355 | 5.688 | 5.262 | 5.024 | 4.948 |
| GE | 32 | 72.59 | 21.2 | 15.38 | 11.75 | 9.402 | 7.833 | 6.774 | 6.064 | 5.61 | 5.356 | 5.275 |
| AS | 33 | 74.92 | 22.54 | 16.36 | 12.5 | 10 | 8.333 | 7.206 | 6.451 | 5.968 | 5.698 | 5.612 |
| SE | 34 | 78.96 | 23.93 | 17.37 | 13.27 | 10.62 | 8.849 | 7.653 | 6.851 | 6.339 | 6.052 | 5.96 |
| BR | 35 | 79.91 | 25.36 | 18.4 | 14.07 | 11.25 | 9.378 | 8.11 | 7.261 | 6.718 | 6.414 | 6.317 |
| KR | 36 | 83.8 | 26.84 | 19.47 | 14.89 | 11.91 | 9.924 | 8.583 | 7.685 | 7.11 | 6.789 | 6.686 |
| RB | 37 | 85.47 | 28.35 | 20.57 | 15.73 | 12.58 | 10.48 | 9.068 | 8.119 | 7.512 | 7.173 | 7.064 |
| SR | 38 | 87.62 | 29.9 | 21.7 | 16.59 | 13.27 | 11.06 | 9.566 | 8.565 | 7.925 | 7.567 | 7.453 |
| Y | 39 | 88.91 | 31.5 | 22.86 | 17.47 | 13.98 | 11.65 | 10.08 | 9.023 | 8.348 | 7.972 | 7.851 |
| ZR | 40 | 91.22 | 33.14 | 24.05 | 18.38 | 14.71 | 12.26 | 10.6 | 9.494 | 8.784 | 8.388 | 8.26 |
| NB | 41 | 92.91 | 34.81 | 25.27 | 19.32 | 15.46 | 12.88 | 11.14 | 9.975 | 9.23 | 8.813 | 8.68 |
| MO | 42 | 95.94 | 36.54 | 26.52 | 20.27 | 16.22 | 13.52 | 11.69 | 10.47 | 9.687 | 9.251 | 9.11 |
| TC | 43 | 99 | 38.3 | 27.8 | 21.25 | 17 | 14.17 | 12.26 | 10.98 | 10.16 | 9.698 | 9.552 |
| RU | 44 | 101.1 | 40.1 | 29.1 | 22.25 | 17.81 | 14.84 | 12.84 | 11.49 | 10.64 | 10.16 | 10.46 |
| RH | 45 | 102.9 | 41.95 | 30.44 | 23.28 | 18.63 | 15.52 | 13.43 | 12.02 | 11.13 | 10.62 | 10.46 |
| PD | 46 | 106.4 | 43.83 | 31.81 | 24.32 | 19.46 | 16.22 | 14.03 | 12.57 | 11.63 | 11.1 | 10.94 |
| AG | 47 | 107.9 | 45.76 | 33.21 | 25.39 | 20.32 | 16.94 | 14.65 | 13.12 | 12.14 | 11.59 | 11.42 |

| | Z | A | 160° | 162° | 164° | 166° | 168° | 170° | 172° | 174° | 176° | 178° |
|---|---|---|---|---|---|---|---|---|---|---|---|---|
| CD | 48 | 112.4 | 47.73 | 34.64 | 26.49 | 21.2 | 17.67 | 15.28 | 13.69 | 12.66 | 12.09 | 11.91 |
| IN | 49 | 114.8 | 49.74 | 36.1 | 27.61 | 22.09 | 18.41 | 15.93 | 14.26 | 13.2 | 12.6 | 12.41 |
| SN | 50 | 118.7 | 51.8 | 37.59 | 28.75 | 23 | 19.17 | 16.59 | 14.85 | 13.74 | 13.13 | 12.93 |
| SB | 51 | 121.8 | 53.89 | 39.11 | 29.91 | 23.93 | 19.95 | 17.26 | 15.46 | 14.3 | 13.66 | 13.45 |
| TE | 52 | 127.6 | 56.03 | 40.67 | 31.1 | 24.89 | 20.74 | 17.94 | 16.07 | 14.87 | 14.2 | 13.99 |
| I | 53 | 126.9 | 58.2 | 42.25 | 32.3 | 25.85 | 21.55 | 18.64 | 16.69 | 15.45 | 14.75 | 14.53 |
| XE | 54 | 131.3 | 60.42 | 43.86 | 33.54 | 26.84 | 22.37 | 19.35 | 17.33 | 16.04 | 15.32 | 15.08 |
| CS | 55 | 132.9 | 62.68 | 45.5 | 34.79 | 27.84 | 23.21 | 20.08 | 17.98 | 16.64 | 15.89 | 15.65 |
| BA | 56 | 137.3 | 64.98 | 47.17 | 36.07 | 28.87 | 24.06 | 20.82 | 18.64 | 17.25 | 16.47 | 16.23 |
| LA | 57 | 138.9 | 67.32 | 48.87 | 37.37 | 29.91 | 24.93 | 21.57 | 19.31 | 17.87 | 17.07 | 16.81 |
| CE | 58 | 140.1 | 69.71 | 50.6 | 38.69 | 30.97 | 25.81 | 22.33 | 20 | 18.51 | 17.67 | 17.41 |
| PR | 59 | 140.9 | 72.13 | 52.36 | 40.04 | 32.04 | 26.71 | 23.11 | 20.69 | 19.15 | 18.29 | 18.01 |
| ND | 60 | 144.2 | 74.6 | 54.15 | 41.41 | 33.14 | 27.62 | 23.9 | 21.4 | 19.81 | 18.91 | 18.63 |
| PM | 61 | 147 | 77.11 | 55.97 | 42.8 | 34.25 | 28.55 | 24.7 | 22.12 | 20.47 | 19.55 | 19.26 |
| SM | 62 | 150.4 | 79.66 | 57.82 | 44.22 | 35.39 | 29.5 | 25.52 | 22.86 | 21.15 | 20.2 | 19.89 |
| EU | 63 | 152 | 82.25 | 59.7 | 45.66 | 36.54 | 30.46 | 26.35 | 23.6 | 21.84 | 20.86 | 20.54 |
| GD | 64 | 157.3 | 84.88 | 61.62 | 47.12 | 37.71 | 31.44 | 27.2 | 24.36 | 22.54 | 21.53 | 21.2 |
| TB | 65 | 158.9 | 87.56 | 63.56 | 48.6 | 38.9 | 32.43 | 28.05 | 25.13 | 23.25 | 22.2 | 21.87 |
| DY | 66 | 162.5 | 90.27 | 65.53 | 50.11 | 40.11 | 33.43 | 28.92 | 25.91 | 23.97 | 22.89 | 22.55 |
| HO | 67 | 164.9 | 93.03 | 67.53 | 51.64 | 41.33 | 34.45 | 29.81 | 26.7 | 24.71 | 23.59 | 23.24 |
| ER | 68 | 167.3 | 95.83 | 69.56 | 53.2 | 42.58 | 35.49 | 30.71 | 27.5 | 25.45 | 24.3 | 23.94 |
| TM | 69 | 168.9 | 98.67 | 71.62 | 54.77 | 43.84 | 36.54 | 31.62 | 28.32 | 26.2 | 25.03 | 24.65 |
| YB | 70 | 173 | 101.5 | 73.72 | 56.37 | 45.12 | 37.61 | 32.54 | 29.14 | 26.97 | 25.76 | 25.37 |
| LU | 71 | 175 | 104.5 | 75.84 | 58 | 46.42 | 38.69 | 33.48 | 29.98 | 27.75 | 26.5 | 26.1 |
| HF | 72 | 178.5 | 107.4 | 77.99 | 59.64 | 47.73 | 39.79 | 34.43 | 30.84 | 28.54 | 27.25 | 26.84 |
| TA | 73 | 181 | 110.4 | 80.17 | 61.31 | 49.07 | 40.91 | 35.39 | 31.7 | 29.33 | 28.01 | 27.59 |
| W | 74 | 183.9 | 113.5 | 82.38 | 63 | 50.43 | 42.04 | 36.37 | 32.57 | 30.14 | 28.79 | 28.35 |
| RE | 75 | 186.2 | 116.6 | 84.63 | 64.72 | 51.8 | 43.18 | 37.36 | 33.46 | 30.97 | 29.57 | 29.13 |
| OS | 76 | 190.2 | 119.7 | 86.9 | 66.46 | 53.19 | 44.34 | 38.36 | 34.36 | 31.8 | 30.37 | 29.91 |
| IR | 77 | 192.2 | 122.9 | 89.2 | 68.22 | 54.6 | 45.52 | 39.38 | 35.27 | 32.64 | 31.17 | 30.7 |
| PT | 78 | 195.1 | 126.1 | 91.53 | 70 | 56.03 | 46.71 | 40.41 | 36.19 | 33.49 | 31.99 | 31.5 |
| AU | 79 | 197 | 129.3 | 93.9 | 71.81 | 57.47 | 47.91 | 41.45 | 37.13 | 34.36 | 32.81 | 32.32 |
| HG | 80 | 200.6 | 132.6 | 96.29 | 73.64 | 58.94 | 49.13 | 42.51 | 38.08 | 35.24 | 33.65 | 33.14 |
| TL | 81 | 204.4 | 136 | 98.71 | 75.49 | 60.42 | 50.37 | 43.58 | 39.03 | 36.12 | 34.5 | 33.98 |
| PB | 82 | 207.2 | 139.4 | 101.2 | 77.37 | 61.92 | 51.62 | 44.67 | 40 | 37.02 | 35.36 | 34.82 |
| BI | 83 | 209 | 142.8 | 103.7 | 79.27 | 63.44 | 52.89 | 45.76 | 40.99 | 37.93 | 36.22 | 35.68 |
| PO | 84 | 210 | 146.2 | 106.2 | 81.19 | 64.98 | 54.17 | 46.87 | 41.98 | 38.85 | 37.1 | 36.54 |
| AT | 85 | 210 | 149.7 | 108.7 | 83.13 | 66.54 | 55.47 | 47.99 | 42.99 | 39.78 | 37.99 | 37.42 |
| RN | 86 | 222 | 153.3 | 111.3 | 85.1 | 68.12 | 56.79 | 49.13 | 44.01 | 40.73 | 38.89 | 38.31 |
| FR | 87 | 223 | 156.9 | 113.9 | 87.1 | 69.71 | 58.11 | 50.28 | 45.04 | 41.68 | 39.8 | 39.2 |
| RA | 88 | 226 | 160.5 | 116.5 | 89.11 | 71.32 | 59.46 | 51.45 | 46.08 | 42.64 | 40.72 | 40.11 |
| AC | 89 | 227 | 164.2 | 119.2 | 91.15 | 72.95 | 60.82 | 52.62 | 47.13 | 43.62 | 41.66 | 41.03 |
| TH | 90 | 232 | 167.9 | 121.9 | 93.21 | 74.6 | 62.19 | 53.81 | 48.2 | 44.6 | 42.6 | 41.95 |
| PA | 91 | 231 | 171.6 | 124.6 | 95.29 | 76.27 | 63.58 | 55.01 | 49.27 | 45.6 | 43.55 | 42.89 |
| U | 92 | 238 | 175.4 | 127.4 | 97.4 | 77.96 | 64.99 | 56.23 | 50.36 | 46.61 | 44.51 | 43.84 |

| $\theta$ | 180° | 178° | 176° | 174° | 172° | 170° | 168° | 166° | 164° | 162° | 160° |
|---|---|---|---|---|---|---|---|---|---|---|---|
| $\delta^2$ | 0 | $1.218 \times 10^{-3}$ | $4.874 \times 10^{-3}$ | $1.097 \times 10^{-2}$ | $1.950 \times 10^{-2}$ | $3.046 \times 10^{-2}$ | $4.386 \times 10^{-2}$ | $5.971 \times 10^{-2}$ | $7.798 \times 10^{-2}$ | $9.870 \times 10^{-2}$ | $1.218 \times 10^{-1}$ |

**TABLE XI.** Surface Height $H_0$ in Counts for $^4$He at $\theta_1 = 0°$ and $\theta_2 = 10$ (Conditions Given in Table Heading)

| ELEMENT | AT. NO. ($Z_2$) | AVRG MASS (amu) | INCIDENT ENERGY (in MeV) OF $^4$He | | | | | | | | | | |
|---|---|---|---|---|---|---|---|---|---|---|---|---|---|
| | | | 1.0 | 1.2 | 1.4 | 1.6 | 1.8 | 2.0 | 2.4 | 2.8 | 3.2 | 3.6 | 4.0 |
| BE | 4 | 9.012 | 11 | 7.664 | 5.653 | 4.347 | 3.45 | 2.808 | 1.969 | 1.461 | 1.13 | 0.9023 | 0.7389 |
| B | 5 | 10.81 | 16.21 | 11.18 | 8.193 | 6.268 | 4.957 | 4.022 | 2.809 | 2.083 | 1.616 | 1.299 | 1.072 |
| C | 6 | 12.01 | 24.26 | 16.31 | 11.71 | 8.819 | 6.887 | 5.531 | 3.804 | 2.793 | 2.158 | 1.737 | 1.438 |
| N | 7 | 14.01 | 22.89 | 15.73 | 11.53 | 8.846 | 7.029 | 5.739 | 4.07 | 3.073 | 2.435 | 2.006 | 1.696 |
| O | 8 | 16 | 30.24 | 20.64 | 15.06 | 11.53 | 9.153 | 7.475 | 5.314 | 4.029 | 3.208 | 2.654 | 2.25 |
| F | 9 | 19 | 39.96 | 27.1 | 19.60 | 15.03 | 11.92 | 9.735 | 6.932 | 5.265 | 4.197 | 3.47 | 2.94 |
| NE | 10 | 20.18 | 49.07 | 33.19 | 24.08 | 18.37 | 14.56 | 11.89 | 8.456 | 6.415 | 5.103 | 4.211 | 3.562 |
| NA | 11 | 22.99 | 56.6 | 38.99 | 26.62 | 22.01 | 17.54 | 14.38 | 10.29 | 7.836 | 6.239 | 5.136 | 4.331 |
| MG | 12 | 24.31 | 51.94 | 36.41 | 27.16 | 21.2 | 17.12 | 14.2 | 10.36 | 8.009 | 6.456 | 5.367 | 4.548 |
| AL | 13 | 26.98 | 62.76 | 40.28 | 33.12 | 25.88 | 20.9 | 17.32 | 12.64 | 9.767 | 7.862 | 6.523 | 5.546 |
| SI | 14 | 28.09 | 56.5 | 40.18 | 30.46 | 24.17 | 19.84 | 16.71 | 12.53 | 9.905 | 8.133 | 6.865 | 5.868 |
| P | 15 | 30.97 | 66.51 | 46.88 | 35.3 | 27.87 | 22.79 | 19.14 | 14.3 | 11.27 | 9.245 | 7.798 | 6.647 |
| S | 16 | 32.06 | 74 | 51.62 | 38.64 | 30.41 | 24.84 | 20.86 | 15.6 | 12.32 | 10.13 | 8.562 | 7.293 |
| CL | 17 | 35.45 | 66.74 | 47.85 | 36.7 | 29.53 | 24.59 | 21.19 | 16.12 | 12.93 | 10.72 | 9.112 | 7.762 |
| AR | 18 | 39.95 | 71.64 | 51.49 | 39.7 | 32.16 | 26.97 | 23.19 | 17.96 | 13.83 | 11.99 | 10.18 | 8.635 |
| K | 19 | 39.1 | 79.2 | 54.29 | 40.98 | 32.53 | 26.81 | 22.72 | 17.28 | 14.46 | 11.45 | 9.722 | 8.357 |
| CA | 20 | 40.08 | 79.32 | 56.21 | 42.6 | 33.89 | 27.94 | 23.67 | 17.99 | 15.62 | 11.9 | 10.1 | 8.701 |
| SC | 21 | 44.96 | 87.06 | 61.66 | 46.72 | 37.14 | 30.59 | 25.88 | 19.6 | 16.8 | 12.89 | 10.91 | 9.378 |
| TI | 22 | 47.9 | 87.46 | 67.65 | 51.25 | 40.73 | 33.51 | 28.29 | 21.27 | 18.0 | 13.74 | 11.56 | 9.878 |
| V | 23 | 50.94 | 95.46 | 77.5 | 56.27 | 46.73 | 38.05 | 31.5 | 23.89 | 20.45 | 15.24 | 12.77 | 10.88 |
| CR | 24 | 52 | 109.9 | 86.94 | 65.29 | 51.47 | 42.05 | 35.27 | 26.2 | 22.46 | 16.53 | 13.76 | 11.68 |
| MN | 25 | 54.94 | 123.9 | 96.36 | 72.44 | 57.09 | 46.6 | 39.03 | 28.89 | 23.75 | 18.11 | 15.05 | 12.74 |
| FE | 26 | 55.85 | 137.5 | 98.36 | 74.15 | 58.74 | 48.23 | 40.64 | 30.39 | 26 | 19.16 | 15.87 | 13.38 |
| CO | 27 | 58.93 | 140 | 105.9 | 84.75 | 66.55 | 54.18 | 45.32 | 33.5 | 28.03 | 20.89 | 17.26 | 14.54 |
| NI | 28 | 58.71 | 162.9 | 113.5 | 94.33 | 73.35 | 59.19 | 49.13 | 36.01 | 30.13 | 22.81 | 19.18 | 16.39 |
| CU | 29 | 63.54 | 182.5 | 127.5 | 105.9 | 81.7 | 65.46 | 54.01 | 39.19 | 33.63 | 24.06 | 19.76 | 16.61 |
| ZN | 30 | 65.37 | 211.2 | 144.8 | 114.6 | 89.16 | 71.92 | 59.65 | 43.57 | 36.5 | 26.95 | 22.21 | 18.69 |
| GA | 31 | 69.72 | 216.6 | 154.8 | 114.8 | 89.57 | 71.62 | 60.58 | 46.87 | 37.22 | 27.83 | 23.01 | 19.41 |
| GE | 32 | 72.59 | 218.1 | 153 | 117.8 | 92.77 | 75.63 | 63.3 | 47.67 | 39.9 | 29.44 | 24.39 | 20.6 |
| AS | 33 | 74.92 | 223.1 | 156.9 | 118.2 | 93.36 | 76.37 | 64.1 | 51.16 | 41.19 | 30.07 | 24.98 | 21.13 |
| SE | 34 | 78.96 | 232.8 | 164.6 | 124.8 | 99.17 | 81.65 | 69.53 | 52.56 | 43.15 | 32.23 | 26.83 | 22.7 |
| BR | 35 | 79.91 | 232.2 | 158 | 121.1 | 98.05 | 81.64 | 69.58 | 52.63 | 43.6 | 33.24 | 27.69 | 23.45 |
| KR | 36 | 83.8 | 219.8 | 156.9 | 121.4 | 97.86 | 81.71 | 69.43 | 52.79 | 43.75 | 33.26 | 27.77 | 23.55 |
| RB | 37 | 85.47 | 217.5 | 159.5 | 122.1 | 98.23 | 81.45 | 71.8 | 54.12 | 45.28 | 33.73 | 28.31 | 24.1 |
| SR | 38 | 87.62 | 223.6 | 158.4 | 122.5 | 97.43 | 83.01 | 70.72 | 53.67 | 44.73 | 34.03 | 28.93 | 24.72 |
| Y | 39 | 88.91 | 221 | 162.3 | 124.5 | 101.1 | 85.39 | 72.58 | 54.12 | 45.28 | 34.28 | 29.08 | 24.85 |
| ZR | 40 | 91.22 | 225.7 | 167.1 | 128 | 102.8 | 85.3 | 78.45 | 59.06 | 46.34 | 34.53 | 29.55 | 25.26 |
| NB | 41 | 92.91 | 226 | 181.7 | 139.2 | 111.9 | 92.51 | 79.36 | 60.25 | 47.51 | 35.12 | 31.69 | 27.1 |
| MO | 42 | 95.94 | 233.6 | 181.4 | 151.8 | 121.3 | 93.14 | 85.11 | 64.09 | 50.25 | 37.67 | 32.51 | 27.84 |
| TC | 43 | 99 | 256.7 | 199.7 | 161.3 | 126.6 | 100.1 | 89.67 | 67.11 | 52.35 | 38.69 | 34.21 | 29.23 |
| RU | 44 | 101.1 | 254.5 | 212.7 | 174.7 | 139.6 | 106.1 | 93.14 | 70.72 | 54.35 | 40.77 | 35.48 | 30.29 |
| RH | 45 | 102.9 | 301.3 | 230.3 | 177.7 | 143.6 | 106.1 | 97.74 | 72.95 | 56.32 | 42.33 | 35.48 | 31.68 |
| PD | 46 | 106.4 | 327.7 | 245 | 184.4 | 139.6 | 115.5 | 97.74 | 72.95 | 56.32 | 44.93 | 37.26 | 31.68 |
| AG | 47 | 107.9 | 350.9 | 245 | 184.4 | 146.1 | 120.1 | 101.2 | 75.39 | 58.36 | 46.93 | 38.64 | 33.17 |

| | Z | | | | | | | | | | | | | |
|---|---|---|---|---|---|---|---|---|---|---|---|---|---|---|
| CD | 48 | 33.3 | 39.18 | 47.3 | 58.95 | 75.92 | 102.1 | 121.4 | 145 | 167.2 | 248.6 | 355.3 | 112.4 |
| IX | 49 | 32.29 | 38.23 | 46.58 | 58.05 | 74.47 | 100.9 | 120 | 147.6 | 168.7 | 252.8 | 363 | 114.8 |
| SN | 50 | 33.89 | 40.14 | 48.65 | 60.78 | 78.48 | 105.9 | 126.1 | 154.7 | 191.3 | 258 | 366 | 118.7 |
| SB | 51 | 33.99 | 40.24 | 48.55 | 60.56 | 78.57 | 107.1 | 128.2 | 157.3 | 199.3 | 264.7 | 372.9 | 121.8 |
| TE | 52 | 35.45 | 41.75 | 50.55 | 62.99 | 81.55 | 110.6 | 135.7 | 161 | 202.2 | 267.2 | 375 | 127.6 |
| I | 53 | 37.88 | 44.75 | 53.92 | 67.45 | 87.12 | 115.8 | 135.7 | 162.1 | 195.5 | 256.9 | 353.2 | 126.9 |
| XE | 54 | 35.97 | 42.51 | 51.14 | 63.77 | 82.33 | 110.2 | 130.6 | 156.6 | 194.3 | 251.9 | 348.9 | 131.3 |
| CS | 55 | 37.99 | 44.36 | 53.04 | 65.73 | 84.33 | 113.3 | 130.5 | 160.6 | 201.8 | 264.2 | 369.9 | 132.3 |
| BA | 56 | 37.69 | 44.65 | 53.33 | 65.89 | 86.03 | 112.3 | 134.6 | 161.8 | 199.7 | 260.2 | 363.1 | 137.3 |
| LA | 57 | 38.96 | 46.05 | 55.66 | 67.56 | 88.09 | 114.4 | 134.6 | 166.8 | 210.7 | 283.8 | 397.5 | 138.9 |
| CE | 58 | 40.87 | 48.07 | 57.52 | 71.11 | 91.03 | 121.3 | 143.4 | 173.3 | 216.6 | 296.7 | 416.2 | 140.1 |
| PP | 59 | 42.41 | 49.92 | 59.78 | 73.93 | 94.65 | 126.3 | 149.2 | 188.5 | 226.2 | 310.1 | 435.7 | 144.2 |
| ND | 60 | 42.03 | 51.86 | 62.12 | 76.83 | 98.38 | 131.9 | 155.5 | 188.5 | 236.6 | 324.1 | 455.9 | 147 |
| PM | 61 | 45.7 | 53.85 | 64.55 | 79.85 | 102.3 | 136.8 | 162 | 196.5 | 257.3 | 338.7 | 476.8 | 150.4 |
| SM | 62 | 47.23 | 55.93 | 67.04 | 82.95 | 106.3 | 142.1 | 168.8 | 205 | 268.6 | 353.6 | 498.2 | 152.7 |
| EU | 63 | 49.23 | 58.05 | 69.66 | 86.22 | 110.8 | 148.2 | 175.8 | 213.7 | 268.5 | 353.8 | 498.8 | 157.3 |
| GD | 64 | 49.5 | 58.1 | 69.99 | 86.53 | 110.8 | 148.2 | 190.5 | 213.7 | 291.5 | 385.2 | 543.8 | 158.9 |
| TB | 65 | 52.96 | 62.56 | 75.12 | 93.01 | 119.2 | 160.3 | 199.9 | 241.8 | 307.5 | 407.9 | 580 | 162.5 |
| DY | 66 | 54.71 | 64.71 | 77.61 | 96.37 | 124.2 | 167.5 | 213.2 | 243.9 | 330.1 | 438.9 | 625.1 | 164.9 |
| HO | 67 | 58.63 | 69.29 | 83.23 | 103.1 | 132.6 | 178.8 | 221.8 | 260.7 | 343.9 | 457.9 | 652 | 167.3 |
| ER | 68 | 60.08 | 71.87 | 86.37 | 107.2 | 137.8 | 185.5 | 230.9 | 271.4 | 358.4 | 476.9 | 680 | 168.9 |
| TM | 69 | 63.08 | 74.58 | 89.63 | 111.2 | 143.4 | 193 | 239.7 | 282.6 | 372.5 | 496 | 707.5 | 173 |
| YB | 70 | 65.31 | 77.24 | 92.85 | 115.2 | 148.6 | 200.7 | 241.2 | 293.6 | 376.2 | 500.1 | 712.2 | 175 |
| LU | 71 | 66.49 | 78.57 | 94.31 | 116.9 | 149.5 | 203.3 | 242.1 | 295 | 373.6 | 496.6 | 707.3 | 178.5 |
| HF | 72 | 66.35 | 78.1 | 93.96 | 116.6 | 149.7 | 202.1 | 252.8 | 309.2 | 391.7 | 520.8 | 742.2 | 181 |
| TA | 73 | 69.14 | 81.6 | 97.98 | 121.6 | 156.8 | 211.9 | 250.4 | 324.5 | 412.4 | 496.7 | 785.8 | 183.9 |
| W | 74 | 71.87 | 84.83 | 101.8 | 126.2 | 162.8 | 220.9 | 256 | 306.4 | 388.2 | 516.1 | 734.6 | 186.9 |
| RE | 75 | 69.03 | 81.1 | 97.52 | 120.7 | 155.3 | 209.9 | 267.5 | 313.7 | 398 | 530.1 | 756.6 | 190.2 |
| OS | 76 | 70.3 | 82.91 | 99.3 | 122.9 | 158.3 | 214.3 | 272.2 | 320.9 | 407.9 | 544.9 | 778.8 | 192.2 |
| IR | 77 | 71.41 | 84.61 | 100.5 | 125 | 163.2 | 218.6 | 285.8 | 329.9 | 421.8 | 566.9 | 818.1 | 195.1 |
| PT | 78 | 71.66 | 84.61 | 101.5 | 126.1 | 163.4 | 222.7 | 285.8 | 335.2 | 427.2 | 571.4 | 818.8 | 197 |
| AU | 79 | 71.28 | 84.22 | 107.7 | 126.9 | 165.4 | 226.6 | 284 | 352.7 | 451 | 605.5 | 872.4 | 200.6 |
| HG | 80 | 76.92 | 89.65 | 107.7 | 129.6 | 173.8 | 237.7 | 270.9 | 350.1 | 446.1 | 596.3 | −853.1 | 204.4 |
| TL | 81 | 76.01 | 89.68 | 109.2 | 133.9 | 173.6 | 237 | 286.9 | 337.2 | 422.6 | 562.8 | 801.6 | 207.2 |
| PB | 82 | 72.64 | 85.68 | 110.3 | 133.8 | 165.9 | 226.1 | 288.9 | 351.5 | 446.7 | 591.3 | 839.4 | 209 |
| BI | 83 | 72.15 | 91.01 | 111.1 | 135.8 | 176.1 | 239.7 | 291.4 | 353 | 448.6 | 592.9 | 837.9 | 210 |
| PO | 84 | 77.97 | 91.99 | 110.7 | 137.2 | 177.9 | 244.8 | 293.8 | 355.7 | 450.8 | 595.3 | 839 | 210 |
| AT | 85 | 78.87 | 93.07 | 111.4 | 138.7 | 179.9 | 246.6 | 285.1 | 358.1 | 434.6 | 592.9 | 841.9 | 223 |
| RN | 86 | 79.79 | 94.18 | 112.6 | 140.1 | 181.6 | 240.4 | 281.1 | 346.6 | 450.8 | 595.3 | 802.2 | 223 |
| FR | 87 | 78.96 | 93.06 | 111.1 | 137.9 | 178.2 | 237.1 | 270.9 | 340.6 | 434.6 | 570.8 | 781.6 | 226 |
| RA | 88 | 78.74 | 92.74 | 110.7 | 137 | 176.4 | 239.1 | 286.9 | 346.2 | 426.2 | 557.7 | 781.8 | 227 |
| AC | 89 | 79.42 | 94.39 | 111.5 | 139.1 | 177.9 | 248.2 | 288.9 | 340.3 | 426.1 | 559.7 | 780.8 | 232 |
| TH | 90 | 83.51 | 97.13 | 112.5 | 143.7 | 178.6 | 239.1 | 291.4 | 342.2 | 427.5 | 588.4 | 784.1 | 231 |
| PA | 91 | 80.18 | 94.39 | 116.1 | 143.7 | 184.9 | 248.2 | 294.3 | 357.1 | 447.8 | 588.4 | 827.6 | 231 |
| U | 92 | 82.37 | 99.1 | 118.6 | 146.9 | 189 | 254 | 301.1 | 366.1 | 459.8 | 605.3 | 853 | 238 |

| $\theta$ | 160° | 162° | 164° | 166° | 168° | 170° | 172° | 174° | 176° | 178° | 180° |
|---|---|---|---|---|---|---|---|---|---|---|---|
| $\delta^2$ | $1.218$ $\times10^{-1}$ | $9.870$ $\times10^{-2}$ | $7.798$ $\times10^{-2}$ | $5.971$ $\times10^{-2}$ | $4.386$ $\times10^{-2}$ | $3.046$ $\times10^{-2}$ | $1.950$ $\times10^{-2}$ | $1.097$ $\times10^{-2}$ | $4.874$ $\times10^{-3}$ | $1.218$ $\times10^{-3}$ | $0$ |

# Index of Definitions
# and Notation

$\mathscr{E}$    Energy interval corresponding to one channel, 57

$\mathscr{E}'$    Energy interval, inside the target, which is equivalent to $\mathscr{E}$ at the detector, 72–74

$e$    Charge on the electron, 29

$H, H(E_1)$    Number of counts in a channel of a backscattering spectrum corresponding to the energy $E_1$, 58, 73, 74, 76, 83

$H_A$    Number of counts in a channel of a backscattering spectrum corresponding to the surface signal of element A, 81–84, 102, 103

$H_i$    Number of counts in channel $i$ of a backscattering spectrum, 55, 69, 70, 76

$H_0$    Number of counts in a channel of a backscattering spectrum corresponding to backscattering events in the surface region of the target, 70

$I$    Ionization–excitation potential, 41

$K, K_{M_2}$    Kinematic factor, 23, 85

$L$    Stopping number, 41

$m$    Reduced film thickness, 245

$M$    Molecular weight, 37

$\overline{M}$    Average mass, 85

$M_1$    Mass of projectile, 22

$M_2$    Mass of target atom, 22

$\Delta M_2$    Difference in target masses; mass resolution, 25, 186

$m_e$    Rest mass of electron, 40

$N$    Number of atoms per unit volume, 26

$N^{AB}$    Number of molecules $A_m B_n$ per unit volume, 76

$N_i$    Volume density of atoms in a target at depth $x_i$

$N_0$    Avogadro's number, 37

$Nt$    Number of atoms per unit area, 26, 92

$P_\perp$    Transverse momentum transfer, 40

$Q$    Total number of particles incident on the target, 26

$dQ$    Number of particles registered by a detector, 26

$q$    Charge on the electron, 29

$R_p$    Projected range, 139

$\Delta R_p$    Projected range straggling, 140

$S$    Area, 27

$[S]$    Energy loss factor, 62

$[S(E)]$    Energy loss factor evaluated at $E$ and $KE$, 74

$[\overline{S}]$    Energy loss factor evaluated at mean energies along the incoming and the outgoing paths, 64

$[S_0]$    Energy loss factor evaluated at $E_0$ and $KE_0$, 63

$t$    Thickness of a thin target, 26

$u_1$    Thermal vibration amplitude, 236

$v$    Velocity of particle, 22

$v_0$    Bohr velocity, 42

$\mathbf{v}_0$    Incident velocity of a particle, 22

$\mathbf{v}_1$    Velocity of a scattered particle, 22

$\mathbf{v}_2$    Velocity of a recoiled target atom, 22

$x$    Mass ratio $M_1/M_2$ (only in Chapter 2), 23

$x$    Depth within a target where a backscattering event occurs, 59, 60, 74

$x_i$    Depth within a target where particles are scattered that will be collected in channel $i$, 69

$\Delta x$    Depth increment, 35

$Z_1$    Atomic number of the projectile atom, 29

$Z_2$    Atomic number of the target atom, 29

$\alpha$    Ratio of energy loss along outward to inward path, 65

$\beta$    Ratio of path length in the outgoing to that in the incoming direction, 65

$\delta$    $(\pi - \theta)$ in radians, 24

$\epsilon, \epsilon(E)$    Stopping cross section (evaluated at an energy $E$), 36

$\epsilon^A, \epsilon^B$    Stopping cross section of element A or element B, respectively, 44

$\epsilon^{A_m B_n}$    Stopping cross section of compound $A_m B_n$, 44

$[\epsilon]$    Stopping cross section factor, 62

$[\epsilon(E)]$    Stopping cross section factor evaluated at energy $E$ for the incoming path and at energy $KE$ for the outgoing path, 74

$[\bar{\epsilon}]$    Stopping cross section factor evaluated at mean energies along both the incoming and outgoing paths, 64, 80

$[\epsilon_0]$    Stopping cross section factor evaluated at the surface, 63, 79

$[\epsilon]_A^A=[\epsilon]$    Stopping cross section factor for scattering and stopping in element A, 79

$[\epsilon]_A^{A_mB_n}$ $=[\epsilon]_A^{AB}$    Stopping cross section factor for scattering from element A and stopping in medium $A_mB_n$, 79, 106

$\phi$    Recoil angle in laboratory system of reference, 22

$\psi_{\frac{1}{2}}$    Half-angle in channeling, 233

$\psi_1$    Characteristic angle in channeling, 233

$\theta$    Scattering angle in laboratory system of reference, 22, 59, 203

$\theta_c$    Scattering angle in the center-of-mass system of reference, 23

$\theta_1$    Angle between the direction of the incident beam and the target normal, 59

$\theta_2$    Angle between the direction of the scattered particle and the target normal, 59

$\theta_D$    Debye temperature, 236

$\rho$    Mass density, 37

$\Sigma$    Integral scattering cross section, 27

$\sigma$    Average differential scattering cross section in laboratory system of reference, 28

$\sigma(E)$    Value of $\sigma$ evaluated at energy $E$, 73

$\sigma_D(\psi_{\frac{1}{2}})$    Cross section to deflect particles through angles greater than $\psi_{\frac{1}{2}}$, 240

$\dfrac{d\sigma}{d\Omega}$    Differential scattering cross section in laboratory system of reference, 26

$\left(\dfrac{d\sigma}{d\Omega}\right)_c$    Differential scattering cross section in center-of-mass system of reference, 29

$\tau$    Thickness of a target slab located at depth $x$, 73

$\tau_i$    Thickness of a target slab, located at depth $x_i$, whose thickness corresponds to one channel, 69

$\tau_0$    Value of $\tau$ for the surface slab, 70

$\Omega$    Finite solid angle; often that of a detector, 28

$d\Omega$    Differential solid angle, 26

$\tilde{\Omega}$    Energy straggling, 116–118

$\tilde{\Omega}_B$    Energy straggling according to Bohr's theory, 46, 118

$\chi$    Reduced energy variable, 49

$\chi_{min}$    Minimum yield, 232

$\xi$    Normalized distance of closest approach, 235

# Index

A
B
C 8
D 9
E 0
F 1
G 2
H 3
I 4
J 5